INTRODUCTION TO
SAFETY ENGINEERING

This book is to be returned on or before
the last date stamped below.

INTRODUCTION TO SAFETY ENGINEERING

DAVID S. GLOSS

MDHR Associates, Wellesley, Massachusetts

MIRIAM GAYLE WARDLE

Boston College

A Wiley-Interscience Publication

JOHN WILEY & SONS

New York   Chichester   Brisbane   Toronto / Singapore

Copyright © 1984 by John Wiley & Sons, Inc.

All rights reserved. Published simultaneously in Canada.

Library of Congress Cataloging in Publication Data:

Gloss, David S.
    Introduction to safety engineering.

    Includes index.
    1. Industrial safety. 2. System safety. I. Wardle,
Miriam Gayle. II. Title.

T55.G584 1983   363.1'1'0973   83-16751
ISBN 0-471-87667-4

Printed in the United States of America

10 9 8 7 6 5 4 3 2 1

PREFACE

Ramazzini, the father of occupational medicine, had a dream three hundred years ago of a safe and healthful workplace for all workers. The dream has not been fulfilled. We have come a long way, but we have even a longer way to go. Health and safety are recognized as having merit. However, prevention of disability related to workplace accidents or injuries is not always considered a priority. How will this be accomplished? The answer lies in placing more well-trained safety engineers on the job and permitting them to effectively function.

This brings us to the purpose of this book. When one of the authors was chairperson of an occupational safety and health program, there were no textbooks to use in the introductory courses in occupational safety and health and accident conditions and controls. We used the National Safety Council's *Accident Prevention Manual*, the Code of Federal Regulations 29 CFR 1910, and various OSHA pamphlets for our main sources. This book is the starting place for teaching and learning about occupational safety and health.

The purpose of this book is to provide the basic information the safety engineer needs in order to function effectively in the workplace and to highlight the safety engineer's functions.

We have included an overview of safety engineering and the related professions; a description of typical workplace accidents and injuries, including psychological hazards; preventive measures, including appropriate personnel protective equipment; some specific content on initiating, developing, and evaluating a safety training program; and detailed discussions of several industries (construction, shipbuilding, transportation, and petrochemical) to illustrate how this safety information can be utilized after identifying target components for each industry.

This book is not only for safety engineers but for anyone whose vocation or avocation is concerned with safety and health. This includes industrial hygienists, industrial nurses, occupational physicians, human engineers, logisticians, top management, and students in any health and safety or related field/area.

Hopefully, it will provide the momentum for more insights into this most important area of workplace health and safety.

We would like to acknowledge the assistance provided by V. R. Lundy Jr., J. Conway, F. Lohimer, and D. Breen in developing the outline of the book. We would like also to acknowledge the editorial assistance provided by J. Jack for her helpful comments on the first draft of this book.

Ontario, Canada's Workmen's Compensation Board provided permission to use any of their forms in this book as did the State of Oregon's Workers' Compensation Department. Colorado's Department of Labor likewise gave permission to cite their work on illness and injury rates in this manuscript.

DAVID S. GLOSS
MIRIAM GAYLE WARDLE

Wellesley, Massachusetts
Chestnut Hill, Massachusetts
November 1983

CONTENTS

PART 2. HAZARD RECOGNITION AND CONTROL

PART 3.   SPECIFIC HAZARDS

11 Fire Conditions   269

12 Fire Protection and Control   293

13 Radiation   323

14 Noise   343

PART 4.   COMPONENTS OF SAFETY ADMINISTRATION

INTRODUCTION TO
SAFETY ENGINEERING

PART 1
FOUNDATIONS OF INDUSTRIAL SAFETY

1 INTRODUCTION

This chapter introduces safety engineering as a profession. It defines the role of a safety engineer and the rewards a safety engineer may receive for his or her effort. The chapter briefly explains the safety engineer's duties, such as accident investigation and safety training; the people to be dealt with, such as management and safety committees; and the legislative forces which shape the profession, like the Occupational Safety and Health Act (OSHA) and the Toxic Substances Control Act. This chapter also defines the safety profession, factors which may influence this profession, and how it has gained credibility.

WHAT IS SAFETY?

Safety is the measure of the relative freedom from risks or dangers. Safety is the degree of freedom from risks and hazards in any environment—home, office, factory, mine, schools, or their environs. Some risks are more easily reduced or eliminated than others. Safety engineering is the discipline that attempts to reduce the risks by eliminating or controlling the hazards. "How safe is safe?" Safety is relative—nothing is 100% safe under all conditions. There is always some case in which a relatively safe material or piece of equipment becomes hazardous. Drinking water is usually considered safe but can become hazardous if one drinks too much, since this can cause kidney failure. There is no such thing as "absolute safety," nor can it ever be achieved. Safety is a relative quantity that is a function of the situation in which it is measured.

GOALS OF SAFETY ENGINEERING

The goal of safety engineering is to reduce accidents and control or eliminate hazards in the workplace. Ideally, if accidents could be entirely eliminated, there would be no need for fire or casualty insurance, workers' compensation insurance claims would be a thing of the past, and safety engineers would be standing in the unemployment line. For the present, however, it is more realistic to aim for annual reductions in the number of accidents and in the number of workers' compensation insurance claims. These goals of safety engineering can be accomplished only with the support of top management. A safe working environment, with the cooperation of workers who have been well trained in safety, will show through cost-benefit analysis that safety pays.

Top Management Support

It is axiomatic that no corporate safety program can succeed without visible, unqualified support from top management. If the chief operating officer is not safety-conscious, then even the best-conceived safety plan will probably fail. Moreover, if safety engineering is to have credibility within the organization, the safety manager must report directly to top management, rather than to a middle-level staff position (see Figure 1.1).

A Safe Working Environment

Historically, safety referred only to machinery in plants and was addressed only as it served to keep the production line operating. Factories had to be protected from fire, and equipment, from damage or destruction. Workers were rarely considered. A major objective of safety engineering is to keep the workplace free of physical and health hazards. This is theoretically possible since most hazards may be identified and removed. However, some hazards may only be controlled. For example, the hazard posed by electric current cannot be removed in a modern operation. It may be controlled by barriers, circuit breakers, interlocks, ground fault interrupters, and warning signs. These will be more fully discussed in a later chapter.

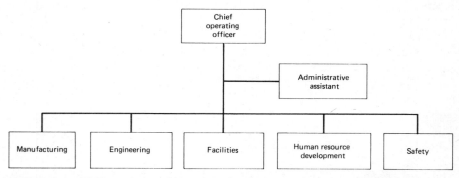

FIGURE 1.1.   The organizational chart for a corporation with an effective safety department.

Some hazards can be easily reduced or eliminated. Housekeeping tasks like picking up trash or keeping aisles clear are relatively simple, straightforward activities which both workers and supervisors understand. However, the supervisor must constantly monitor the worker's behavior in this area. Putting a mechanical safety device on a machine to keep worker's hands or bodies from injury is another straightforward safety measure. This is called "machine guarding." Ironically, the greatest hazard in the workplace is the worker; human error accounts for more accidents than any other single factor. However, we can do our best to minimize the opportunities for error by intensive training and by good workplace design.[6]

A Safety-Conscious Worker

A worker trained to be safety-conscious will be more productive and have fewer sick days and fewer accidents than a non–safety-trained worker. Workers trained in safety procedures will be more responsible in terms of continuing safe work practices and will tend to continue to keep abreast of developments in the field. Management concern for safety appears to be a motivating factor.

Costs and Benefits

From the corporate as well as the human viewpoint it pays to have workers adequately trained and safety-oriented. The cost of accidents is high in terms of medical insurance costs, lost time on the job, legal fees for court cases, workers' compensation, disability insurance, and unemployment compensation. The cost of preventing occupational accidents or illnesses is much less than the cost of treatment or workers' compensation.

THE SAFETY ENGINEER'S JOB

The safety engineer is a professional committed to making the work environment as safe as possible by focusing on any or all of the following jobs: accident prevention; human factors; the interface between the workplace and the environment; design and layout of equipment; management and supervision of safety training; and by being safety consultant to foremen, supervisors, and management.

The duties of the safety engineer depend partly on the type of industry and partly on what the company conceives the job to be, that is, the job description. The safety engineer's duties often overlap with other responsibilities, such as security or ergonomics. "Security" is the activity concerned with making an organization free from the dangers of crime and sabotage. "Ergonomics" is concerned with designing for mankind. It examines the person–machine interface in the attempt to design a healthy, stress-free place, for good safety implies good ergonomics.[6] Although at times it may be tempting to feel discouraged and frustrated, safety engineers know the importance of their work and the constructive results of effective safety measures.

One of the safety engineer's most important responsibilities is control of hazards. Without knowledge of workplace hazards acquired through training and experience, it is difficult, if not impossible, to identify them. And it is crucial to recognize them before an accident occurs, not after.

Accident investigation is another important component of the safety engineer's job. It is necessary to obtain a detailed account of the accident, to identify events that seem to have a bearing on it, and to determine whether accidents of this kind have occurred before and, if so, how often. It should also be determined how the injured person was handled and whether anything could have been done differently before, during, or after the accident. Forms must be completed and careful records kept.

WHO BECOMES A SAFETY ENGINEER?

In the past people were assigned to (or volunteered for) the role of safety engineer in response to vaguely stated needs.

Whether one can be a safety engineer with no safety training is debatable. Is on-the-job training sufficient? Should it be preceded by, or supplemented with, formal academic training? How much training is necessary depends on one's goals, how one wants to develop professionally, the goals of the organization, and one's motivation. Probably the wisest course is to allow some flexibility while setting minimum standards for licensing.

The difference between incorporating some specific applied techniques of the physical sciences in technical courses and requiring a basic knowledge of the physical sciences as a prerequisite for technical safety courses is the difference between education and training. In general terms, "education is usually thought of as the development of intellectual capabilities, while training is intended to impart specific and limited skills" (Orn,[10] p. 19).

Education should provide us with the necessary tools to meet the changing demands of the physical sciences, and mathematics components help provide this education. Today, the safety engineer is more likely to be someone who has a degree in safety engineering or a closely related field like human factors, ergonomics, biomedical engineering, or industrial nursing. The impetus for this shift is attributable to several factors, the most important one being the Occupational Safety and Health Act, which mandates that the person functioning as a safety professional possess substantial knowledge of safety and health in the workplace. Other factors include the consumer movement, the union, increased workers' compensation benefits, and the increased number of suits brought by employees.

One of the eligibility requirements for licensing as a certified safety professional is practical experience in the safety field. This can only be achieved by taking an entry-level position as a "safety professional"; thus a safety engineer must work in the field for a number of years before being certified as a safety professional.

CREDIBILITY OF THE FIELD

Safety engineering is slowly maturing as a recognized profession. In the early twentieth century industry realized that lack of safety could affect its profits. Du Pont exemplifies what can be accomplished when a corporation voluntarily assumes responsibility for the safety and health of its employees. Long before OSHA, Du Pont provided one of the safest workplaces in the United States. This is shown by its annual citation by the National Safety Council for an excellent safety record.

Safety engineering is a relatively new profession and reflects our mounting concerns for the environment, the consumer, and the rights of workers. Because this professional group is growing faster than controls on the profession can be developed or organizations can apply them, to achieve credibility safety professionals will have to exercise tact and diplomacy while sticking to what they believe. They must also accommodate the system, that is, work within it, so their ideas can be applied. They should do research and publish findings as to what works and what doesn't, join professional organizations, and attend conferences to keep up in their field, both locally and nationally.

Safety's credibility has come from a number of sources within and outside the profession which have made it the career it is today and a career which will not become obsolete in the future. Some of the forces which have given credibility to the safety profession include:

1. Major accidents and disasters
2. Legislation
3. Safety education
4. Certified safety professionals
5. Safety organizations
6. Professional journals
7. Environmental concern
8. Gross national product

Major Accidents and Disasters

More managers are becoming convinced of the importance of safety engineering, not only because of proliferating government regulations that require a specialist to interpret but also because of incidents like those at

Love Canal and Three Mile Island. The litigation costs alone arising from these disasters probably exceed by several orders of magnitude what it would have cost to install adequate safety precautions in the first place. Data are available in increasing quantities to show the relationship between improving safety and increasing productivity. This factor more than any other will convince management that safety pays and that it must now include a safety professional in the corporate hierarchy. "Knowing what to do is the task of safety engineers. Knowing how to get it done is the task of safety managers" (Orn,[10] p.4). Accidents and disasters will be discussed further in Chapter 21, "Planning for Emergencies."

Legislation

In response to pressure from labor unions and consumers, Congress passed the Occupational Safety and Health Act (OSHA) in 1970. It defined industrial safety requirements for everything from ladders to ventilation. Subsequent bureaucratic excesses led to proliferating regulations and enforcement abuses and prompted a thorough investigation of the Occupational Safety and Health Administration's practices. This resulted in an easing of some of the more blatantly irrelevant regulations (like specifying the shape of toilet seats) and an overhaul of OSHA enforcement policies. It has been suggested that this agency be dismantled or that its charter be transferred to another government department. However, even if this were to happen, many of the gains effected by OSHA would remain; programs set in motion under its impetus would probably be continued because of investments in plant and personnel; and safety engineers will still be needed to clarify these issues and to consult with management on safety matters.

The Resource Conservation and Recovery Act (RCRA) was another legislative decision that has had far-reaching effects on the worker and the environment. It requires that all hazardous materials be controlled from their point of origin, through all stages of processing and use to their final disposal. This means hazardous material will be tracked from "cradle to grave."

Safety Education

Professional education provides the minimum standard for credibility. The fact that the person has an earned degree does not mean that person is smarter than the one who lacks a degree. It does mean that the degree holder has accomplished a series of standardized tasks by which he or she may be measured. Undergraduate and graduate programs in safety are rather new. The National Safety Council publishes every other year a list of colleges and universities which offer training in safety.[9]

The Board of Certified Safety Professionals has suggested the following baccalaureate curriculum for the safety professional. Fourteen of the suggested courses are directly related to the functions and responsibilities of the certified safety professional. These are as follows:

1. Introduction to industrial processes
2. Introduction to safety and health
3. General concepts in safety and health
4. Air quality for the safety profession
5. Hazardous materials
6. Physical exposures
7. Flammable materials
8. Mechanical and electrical systems
9. System safety engineering
10. Education and training methods
11. Facilities design
12. Evaluation and control methods in occupational safety and health
13. Management techniques in safety and health
14. Materials-handling system

The 14 courses and mathematics, chemistry, physics, and communication skills are combined into a curriculum in Figure 1.2. The Board of Certified Safety Professionals does not infer that the suggested curriculum is the only academic program that can successfully prepare individuals who plan to enter the safety profession. The content included in this proposed curriculum is minimal for meeting the educational requirements for being certified as a safety professional.[3]

Certified Safety Professional

The Board of Certified Safety Professionals is the organization that sets the standards for registering safety engineers as certified safety professionals. There are two levels of certification: certified safety professional (CSP) and associate safety professional (ASP).

1. *Certified Safety Professional (CSP).* A certified safety professional is a person who has completed the following:
   a. Graduation from an accredited college or university with a bachelor's degree in engineering or some other field such as occupational safety and health
   b. Completion of at least five years of professional safety experience

SAMPLE BACCALAUREATE DEGREE PROGRAM IN SAFETY

Freshman Year

1st Semester	Lecture	Lab	Credit Hours
Calculus I	3	0	3
Chemistry I	3	3	4
Biology I	3	3	4
Composition	3	0	3
Graphics	1	3	2
			16

2nd Semester	Lecture	Lab	Credit Hours
Calculus II	3	0	3
Chemistry II	3	3	4
Biology II	3	3	4
Technical Writing	3	0	3
Introduction to Safety & Health I	3	0	3
			17

Sophomore Year

1st Semester	Lecture	Lab	Credit Hours
Physics I	3	3	4
Introductory Psychology	3	0	3
Statistics	3	0	3
Economics	3	0	3
*Electives	3		3
			16

2nd Semester	Lecture	Lab	Credit Hours
Physics II	3	3	4
Public Speaking	3	0	3
Industrial Psych.	3	0	3
Introduction to Safety & Health II	3	0	3
*Electives	3		3
			16

Junior Year

Applied Mechanics	3	0	3
Materials & Processes	3	0	3
Intro. to Mgmt.	3	0	3
Computer Science	2	3	3
*Electives			3
			15
System Safety	3	0	3
Human Factors/ Ergonomics	2	3	3
Industrial Hygiene and Toxicology	3	3	4
*Electives			6
			16

(Possible Junior-Senior Elective—Internship between Junior & Senior Years)

Senior Year

Fire Prevention & Protection	3	0	3
Safety Engineering	4	0	4
Industrial Hygiene Engineering	3	0	3
*Electives			6
			16
Managing the Safety Function	3	0	3
*Electives			12
			15

TOTAL SEMESTER HOURS: 127   TOTAL ELECTIVES: 33 sem. hrs.

*Where electives are shown, institutional requirements may be substituted.

FIGURE 1.2. Sample baccalaureate degree program in safety (*courtesy of Board of Certified Safety Professionals*).

    c.   Achievement of a passing score on both the core examination and the specialty examination.

2. ***Associate Safety Professional (ASP).***   Associate safety professional is the designation given to a person who has completed the following:

    a.   Graduation from an accredited college or university with a bachelor's degree in engineering or some other acceptable field such as occupational safety and health

    b.   Completion of at least one year of professional safety experience

    c.   Achievement of a passing score on the core examination.

A major revision to the core examination has been made by the Board of Certified Safety Professionals which will be used for the first time in 1983.[2]

The new core examination will consist of the following topics together with their weightings:

1. Basic and applied sciences (36%)
2. Program management and evaluation (13%)
3. Fire prevention and protection (13%)
4. Equipment and facilities (12%)
5. Environmental aspects (14%)
6. System safety and product safety (12%)

There are many variations to the above requirements, and changes may be made by the board from time to time. The person planning to become a safety professional should obtain a copy of "Certification Procedures and Requirements" from the board.[1]

Safety Organizations

Professional safety organizations provide additional credibility to safety engineering. They provide a forum for safety engineers to exchange ideas and information, a focal point for outsiders to seek guidance, and an organizational structure to give the public safety information and to lobby for passage of safety legislation.

The major national safety organizations include the following:

1. National Safety Council
2. American Society of Safety Engineers
3. System Safety Society

Professional Journals

Professional information, new techniques, and ideas must be communicated for a vital profession, and professional journals are a means of performing these communications. They are also a means of finding out who is doing what and where and of announcing professional meetings and continuing education programs.

Safety journals have proliferated in the past decade giving further credence to this rapidly growing field. Some of these are *Professional Safety, Concern, Journal of Safety Research, National Safety News, Journal of Occupational Accidents,* and *Occupational Health and Safety.*

Environmental Concern

There has always been concern for health issues in the United States. However, it has taken some sort of a crisis to produce a major effect. In 1906 Upton Sinclair wrote a book which exposed the horrendously filthy conditions in the meat-packing industry,[12] and this ultimately led to the Meat Inspection Act of 1906 and the Pure Food and Drug Act. Rachel Carson wrote a book which awakened the public consciousness to the insidious effects of the pesticide DDT on the environment.[4] One thing DDT did was to cause eggshells of certain species of birds to be so thin that they could not survive, which led to several species of birds becoming endangered species, including the bald eagle, our national symbol. Carson pointed out the dangers of other chemicals to the environment. Almost alone she triggered the ecology movement of the 1960s and 1970s that culminated in the passage of the Environmental Protection Act of 1969. This important piece of legislation affected people, business, industry, and the environment. Earth Day, celebrated annually in the late 1960s and early 1970s, was an outgrowth of the ecology movement. It was a day set aside to educate people about the effects of humankind's depredations to the earth's biosphere. In the early 1980s, citizens collectively and individually are still speaking out on environmental issues. They do not always win, as in the case of the Seabrook nuclear plant, but these groups serve to maintain public awareness of the issues.

Gross National Product and Safety

Accidents cost the American public and industry a staggering amount every year. In 1980 over 105,000 persons were killed accidentally, and over 10 million suffered a disabling injury. The dollar value of these accidents was $83.2 billion.[8] This is over 8% of our gross national product. If safety engineering could start a new total safety program to reduce the toll by 50%, what a major boost this would be to the gross national product!

HALLMARKS OF A PROFESSION

If saftey engineering is to be considered a profession, then it must meet the criteria for professionalization. Greenwood proposed that professions have specific characteristics.

1. A well-defined theoretical base
2. Recognition as a profession by the clientele
3. Community sanction for professionalization
4. A code of ethics, which regulates the professional's relationships with peers, clients, and the world at large
5. A professional organization[14]

Theoretical Base

Many professions have a strong theoretical basis against which to function. Psychology and physics have systematic theories that explain many of the observations. In these professions theory construction is important.

Psychological theories have played a major role in integrating the vast amount of data that have accumulated on behavior. Psychology has several theories to explain learning, perception, and child development.[7] In astronomy the theories of Kepler and Galileo concerning the earth moving around the sun have withstood several centuries of scrutiny. However, the "big bang" theory of the creation of the universe requires extensive support before it will become accepted.

Other professions, such as nursing, have only a limited theoretical infrastructure, and workers in these fields are striving feverishly to fill the void and be recognized as members of a profession by their clients. There is little question that both law and medicine are recognized as professions; however, an occupation like plumbing is usually not considered a profession.

Community Sanction

Community sanction usually takes the form of educational requirements and licensure. Orn stated that "a professional is educated for his profession, while a technician is trained for his occupation"(Orn,[10] p. 19). This statement indicated that there is a difference between education and training. The same difference is used in describing persons who are mentally retarded. Education is the higher form of learning.

Vollmer and Mills stated that every profession attempts to persuade the community to grant the profession authority within certain powers and privileges.[14] The profession's control over its education process is one of its power centers. This is achieved through an accrediting process. By granting or withholding accreditation, a profession can regulate its schools as to their number, location, curriculum content, and caliber of instruction. The American Medical Association regulates medical education very closely. This is achieved via two routes. First, the profession convinces the community that no one should be allowed to have a professional title unless it has been conferred by an accredited professional school. No one can be a medical doctor (M.D.) without graduating from an accredited school. Another prerequisite may be an examination before a professional board whose personnel have been drawn from those who have already achieved this professional status.[14]

Some industries seek engineering graduates who can immediately fit in with the production needs of the organization without costly, time-consuming, on-the-job training. Such industries support the maintenance of engineering specializations that are closely tied to the needs of industry. Those who tend to support continued specialization also tend to deemphasize the importance of advanced degrees for engineers.

Support for the continuation of engineering specializations against the proposal to make a graduate degree the first professional degree is substantial in both industry and the professional associations. The views of top management from a national sample of organizations employing engineers indicate that approximately one-half of them wish to maintain programs of specialization in the traditional areas of engineering rather than move to a common program with little specialization. In addition, almost one-half of the organizations indicate that a bachelor's degree is all that is needed for engineers to work in their firms.[11]

Ethics Governing All Relationships

Freidson states that when an occupation has become fully professionalized, even if its work characteristically goes on in an organization, management can control the information sources related to the work but cannot control what the engineers do and how they do it. The ethics governing colleague relationships demand behavior that is cooperative, egalitarian, and supportive. Members of a profession share technical knowledge with each other. Any advance in theory and practice made by one professional is quickly disseminated to colleagues through the professional association.[5]

Engineers now serve dominant industrial and governmental interests, and they usually do not attempt to determine whether or not their work contributes to pollution, to international tension, to urban decay, or to the creation of a totalitarian social system.[11]

The existence of an engineering profession with a strong sense of social responsibility and with a commitment to human welfare will not guarantee that our technological society will be directed toward more humane objectives. It might, however, cast the engineering profession in the much needed role of advocate for the needs and rights of persons in a technological age.[11]

THE FUTURE SAFETY ENGINEER

Many demands will be made on safety engineers. They will be able to fulfill them basing their decisions and functioning on their educational background, their training, and the expanding society in this technological age. We can assume that the demand for safety engineers will increase and that they will gain greater authority and be given a larger voice in decisions. The safety engineer will become an indispensable member of management and of design teams.

To some extent this may create another kind of problem. Safety engineers will have to be aware of their own limitations and be careful not to attempt to solve problems which they are not qualified to address. It is one thing to know one's limits and be willing to expand one's knowledge; it is another to assume responsibility in areas outside one's knowledge. Continuing education will help keep safety engineers up-to-date and enable them to add to their capabilities.

GROWTH FACTORS IN SAFETY ENGINEERING

Factors that have promoted the growth of safety engineering include consumer movements resulting in heightened awareness of industrial safety issues (for example, toxic chemicals) and increased pressure on legislators to force businesses to adopt safety measures and provide a hazard-free environment.

JOB DESCRIPTION OF A TYPICAL SAFETY ENGINEER

The safety professional performs many of the following activities in the course of carrying out the job: safety inspections, safety training, collaboration with safety committees, working with management, accident investigation, and processing workers' compensation claims.

Safety Inspection

Safety inspection is one of the primary tasks of safety engineers who are expected to assess plants either by making safety inspection tours themselves or supervising others who make these inspections. Safety engineers also serve as consultants and as guides if an outside agency makes a safety inspection tour. They are responsible to management for inspection findings, regardless of who does them.

Safety Training

In collaboration with members of the health-care team, safety engineers are in charge of safety training. This training covers not only hazardous materials but also day-to-day safety hazards like wet floors, cluttered aisles, ventilation, illumination, noise protection, first aid, and safe operation of forklift trucks.

Collaboration with Safety Committees

Safety engineers should work closely with all safety committees. Since committees may be set up to handle a variety of tasks, safety engineers need to know the purpose of each as it affects safety issues so there is no unnecessary duplication of responsibilities, no contradictory information being disseminated, and no decisions being made or carried out that are not congruent with the safety philosophy of the organization.

Working with Management

As has already been emphasized in this chapter, it is imperative that safety engineers be directly responsible to top management rather than middle management if they are to be effective.

The contribution of the safety professional is in the management decision-making process. Problems in this area deal with the optimum utilization of people, materials, equipment, and energy to achieve the objectives of an organization. The organization may be a single industrial plant, a hospital, a governmental agency, a restaurant chain, an entire multiplant corporation, or any other group organized to produce a product, perform a service, or carry out a function. The managers of these organizations need factual information along with well-defined alternatives and predicted consequences to help them recognize and solve existing or potential accident problems.

> The safety professional collects, analyzes, and arranges this information in such a way as to fulfill management's decision-making needs. He assists various levels of management by originating and developing policy recommendations, operating plans, programs, measures and controls which will permit the effective accident-free use of human and material resources.[13]

Accident Investigation

Accidents must be investigated thoroughly both in terms of reconstructing what happened, so that preventive measures can be taken, and providing information to insurance carriers, medical personnel, and others. This is most often undertaken by the safety engineer.

Accident data include time of injury, place where injury occurred, the name and identification number of the injured person, nature of the injury (bruise, laceration, abrasion, burn), location of the injury, the severity, names of witnesses, and details of witnesses' observations.

Workplace accidents require not only that the injured be given prompt medical attention but also that witnesses to accidents be given an opportunity to discuss their feelings and reactions to what happened. We cannot expect a worker to go back to work immediately after witnessing a serious accident. In fact, one component of a safety program is to evaluate witnesses to make sure they are psychologically able to return to work. If a worker returns prematurely, when still emotionally upset, it may very well cause another accident because of inattentiveness and distraction.

Industry is universally concerned about the number of workers' compensation claims in terms of money, time lost, motivation, and publicity. Safety engineers are responsible for the paperwork for these claims, forms for which may vary depending on the agency. They must see that the injured employee receives medical attention and prepares a detailed history to reconstruct the events that led to the accident. Other aspects include noting the accident conditions at the time, discovering whether something happened that was unexpected or unavoidable, and describing the effects of the accident. For example, was there any factor in the work environment that can be altered so that this type of accident will have a lower probability of recurrence? Workers' compensation will be thoroughly discussed in Chapter 4.

CAREER PATH OF THE SAFETY ENGINEER

Safety engineering like any other profession has a career path. Its direction depends on the entry position, the individual's long-term goals, and the decision on how best to achieve these goals. The means may be by further education and experience or perhaps relocating to a different organization with more career possibilities that fit the specific safety engineer's work objectives. A typical career path for a safety engineer might include the following positions: safety engineer, safety manager, director of safety and health, director of human resources, and, finally, chief executive officer. One of the most critical aspects in career planning is knowing where you want to go, proceeding accordingly, and then being satisfied with gradual attainment of goals.

SUGGESTED LEARNING EXPERIENCES

1. Develop a list of goals and objectives that you would like to attain by the time you have completed this book.

2. Begin keeping a safety engineering notebook. Include articles from newspapers, magazines, safety documents, and any other relevant sources. You will add to this as you go through each chapter. At the conclusion of this book, you will have a useful compilation of safety literature.

3. Write a paper on how you made the decision to become involved in systems safety engineering.

4. Prepare an open industry forum where top management is presenting a new safety policy to its employees.

5. Make a poster depicting human error as a workplace hazard.

6. Illustrate how you perceive each employee in a work system is weighted in the implementation of safety.

7. Do a critique of professions, and based upon that critique, take a hypothetical time frame when you will call yourself a *professional*.

8. If you were an editor of a safety journal, what criteria would you use to accept or reject submitted manuscripts?

9. Evaluate the criteria to become a certified safety professional. Are they fair?

10. Is safety truly a profession?

REFERENCES

1. Board of Certified Safety Professionals. *Certification Procedures and Requirements*, n.d.
2. Board of Certified Safety Professionals. *BCSP Newsletter*, 1982, **8.** (Dec.)
3. Board of Certified Safety Professionals Ad Hoc Committee on Academic Guidelines. *Curricula Guidelines for Baccalaureate Degree Programs in Safety* (Tech. Rep. 2). BCSP, 1981.
4. Carson, R. *Silent Spring*. Boston: Houghton Mifflin, 1962.
5. Freidson, E. (Ed.). *The Professions and Their Prospects*. Beverly Hills, Calif.: Sage Publications, 1973.
6. International Organization for Standardization. *Ergonomic Principles in the Design of Work Systems*. Geneva, 1981.
7. Marx, M. H., and Hillix, J. *Systems and Theories in Psychology* (3rd ed.). New York: McGraw-Hill, 1979.
8. National Safety Council, *Accident Facts*. Chicago, 1981.
9. National Safety Council. *1981–82 College and University Safety Courses*. Chicago, 1981.
10. Orn, M. K. The education of the safety professional. *Professional Safety*, 1980, **25,** 18–22. (Dec.)
11. Perrucci, R., Lebold, W. K., and Howland, W. E. The engineer in industry and government. *Journal of Engineering Education*, 1965, **56,** 237–273.
12. Sinclair, U. *The Jungle*. New York: New American Library, 1980.
13. Tarrants, W. Preparation growth development in safety. *Professional Safety*, 1981, **26,** 35–39.
14. Vollmer, H., and Mills, D. *Professionalization*. Englewood Cliffs, N.J.: Prentice-Hall, 1966.

2 HISTORY OF HEALTH HAZARDS, ACCIDENTS, AND SAFETY ISSUES IN THE WORKPLACE

The connection between certain occupations and specific diseases has long been known. In the fourth century B.C., Hippocrates identified lead poisoning. Nearly 2,000 years ago, Pliny the Elder (A.D. 23–79) wrote an encyclopedia of the natural sciences, *Historia Naturalis,* in which he referred to the dust from mercury ore grinding and the fumes from lead. He recommended what is probably the first piece of personnel protective equipment when he proposed that workers wear a protective mask made from a bladder to prevent inhalation of toxic substances.

SAFETY AND HEALTH IN THE PREINDUSTRIAL WORLD

In preindustrial society hazards were everywhere, and accidents occurred so frequently that death and disability were facts of everyday life. Accidents that seldom occur today were commonplace then, including such accidents as being kicked by a horse, gored by a bull, scalded from carrying hot water, and burned while making soap with hot lard.

Studies of population growth (demography) showed minimal increases in population during the Middle Ages. The time for the world's population to double was measured in hundreds of years. Epidemics of bubonic plague, smallpox, cholera, and diphtheria ravaged Europe, wiping out cities more effectively than any army. Considering the enormous tolls taken by these scourges, however, occupational diseases and injuries were not significant enough to be a cause of concern. Life just went on as well as it could; misery was rampant, and infant mortality claimed many lives. Work just continued, and if a worker survived, fine; if not, that was how it was.

Mining has always been considered one of the most hazardous of occupations, and the mine was the first place where safety and health were considered. Accidents in the mine (cave-ins, floodings, slips and falls, fires, explosions, and asphyxiation) are common. Black lung disease was pandemic. In the sixteenth century, mining and its associated hazards were detailed by George Bauer (1492–1555), better known as Agricola, the "father of mineralogy." He wrote a 12-volume series on mining and metallurgy and described the diseases now know as silicosis, tuberculosis, asthma, and lung cancer. He was very concerned with mine ventilation and described a number of methods for improving it, for example, huge bellows.

Among the most beautiful artifacts of the Middle Ages are the illuminated manuscripts drawn by medieval scribes. The paints were usually either lead- or mercury-based, and the artists were poisoned from pointing their paintbrushes with their tongues. The substances in the paint may have changed, but the practice of pointing paintbrushes has not. In the twentieth century, workers painted watch and clock dials with radium so they could be seen at night and developed osteosarcoma (bone cancer). Artists using benzene as a paint solvent developed leukemia (a type of cancer, often fatal, which involves an uncontrollable multiplication of white cells).

In preindustrial society, swords were sharpened on sandstone wheels. This caused a form of pneumoconiosis, called Grinder's Disease, which is the accumulation of dust in the lungs. Grinder's Disease was also associated with marble cutting, pyrite roasting, dyeing, mercury smelting, silvering of mirrors, and glass blowing.[4] Workers afflicted today by diseases similar to Grinder's Disease are talc and asbestos workers and coal miners.

The Mad Hatter is familiar to us from *Alice in Wonderland*; however, being "mad as a hatter" was an occupational hazard associated with making felt hats. The syndrome, which apparently resulted from constant exposure to the mercury used in making felt from fur, caused personality changes, as well as such physical symptoms as tremors (an involuntary trembling motion of the body), weight loss, gingivitis (inflammation of the gums), and insomnia (inability to sleep). Today, occupations such as dentistry, caustic soda production, jewelry crafting, papermaking, photography, and taxidermy may include the use of mercury and may cause similar symptoms from exposure.

BEGINNINGS OF OCCUPATIONAL MEDICINE

Bernadino Ramazzini (1633–1714) was the first to systematically describe occupational diseases. He believed workers should be able to earn a living without being at risk for injury or disease. Ramazzini might be considered the "father of occupational health and safety" because he observed a variety of workplaces, described the medical problems associated with them, and proposed sound preventive measures. Three hundred years after Ramazzini the Occupational Safety and Health Act's General Duty Clause almost paraphrased him.

The principal method of home heating prior to the invention of oil and gas furnaces was the fireplace. Burning wood or coal creates an accumulation of soot in the chimney which must be cleaned periodically. Little boys were hired to climb down chimneys and scrub off the soot. This work caused many a young man to develop cancer of the scrotum. The treatment was castration. (The chimney sweep has been glorified in song in the movie *Mary Poppins*.) The link between soot and cancer was first identified in 1775 by Percival Pott. Today, coke oven workers are at increased risk of developing lung or kidney cancer.

INDUSTRIAL REVOLUTION

The industrial revolution was the rise of modern technology. Identified by such factors as power-driven machines (steam, water), it introduced more hazards, such as developments in chemistry that presented disease risk as well as physical injury from explosions and poisoning. The industrial revolution was much more than a few inventions and the introduction of the factory system. It brought about the growth of big business; created work for millions; and was responsible for the shifting of people from the country to the city, creating new social classes and new social problems. Workers in the factories were considered expendable. They were often treated worse than slaves. Slaves cost a great deal of money, and owners wanted to protect their investment, but workers cost nothing to hire or replace.

Historians list 1750 as the beginning of the industrial revolution in England. In more general terms, it began with the invention of the printing press by Gutenberg in Germany about 1450 and was an evolving process throughout the Renaissance and early modern period. Within 20 years, printing presses were in operation throughout Europe, and probably many printers, journeymen, and apprentices had been injured. Many of the injuries were hands crushed by the presses. This problem has been solved by making the paper feed at intervals and putting guards on the presses. From England, the industrial revolution spread to Europe and then westward to North America. By the early twentieth century, it penetrated the

Union of Soviet Socialist Republics, China, and Japan. In the latter half of the twentieth century, the industrial revolution had spread to what are presently called Third World countries: those in Asia, Africa, and South and Central America.

The industrial revolution took hold in England because that country had many advantages over other European countries. England had raw materials available and markets for its products in its vast colonies throughout the world. The invention of the flying shuttle by John Kay in 1773 increased the speed of weaving, reduced the number of workers required, and made it possible to weave wider cloth. The flying shuttle was a very unsafe machine. It took five or six spinners to make enough thread to keep one weaver busy at the loom. A weaver who did not have that many spinners at home went from house to house in search of spinners who could be hired to spin some of the wool into thread. James Hargreave invented the spinning jenny. With the new jenny one worker could spin six or seven threads at the same time, thus accomplishing the work normally done by several workers using spinning wheels. James Watt improved the steam engine so that it was practical enough to provide steam power for factories. As a result of his invention, many large power-driven factories were established, some of them employing hundreds or thousands of workers. The factories were totally unsafe. Some of the hazards included unguarded machines, open belt drives, flying shuttles, and open holes. There were no fire escapes, and the illumination was inadequate. These hazards may not even have been the most serious. During the industrial revolution there were hundreds if not thousands of hazards in any factory. Every basic concept of safety and health was violated, and hardly a day went by without some worker being maimed or killed.

By the late Middle Ages surface mining of coal was exhausted. This necessitated underground mining, which resulted in a whole new set of problems, including the depth of the mine, raising the coal, water in the mine shaft, and methane gas. Methane (also known as "fire damp," "marsh damp," and "marsh gas") is a colorless, odorless, and tasteless gas which itself is an asphyxiant but when exposed to heat or flame is extremely flammable. The open-flame illumination used in the mines ignited the methane, and the resultant explosions caused many deaths. The miner's safety lamp, invented by Humphrey Davy in response to a request by the Society for the Prevention of Accidents in Coal Mines, enclosed the flame within a cylinder of wire gauze through which insufficient heat passed to ignite the methane. The safety lamp was the subject of a poem by Thomas Moore:

> O for that lamp's metallic gauze.
> That curtain of protecting wire which Davy
> delicately draws around illicit, dangerous fire:
> The wall he sets twixt flame and air

(Lie that which banned young, this be's bliss),
Through whose small holes, this dangerous pair
May see each other, but not kiss.[7]

Electric illumination, of course, did away with the flame problem altogether, but mine explosions still take many lives worldwide.

THE INDUSTRIAL REVOLUTION COMES TO THE UNITED STATES

Between 1776 and 1850, the United States grew from a small agrarian country to a large industrial and agricultural one. As the period drew to a close, there were signs of impending social upheaval. Increasingly bitter sectionalism between north and south was drawing the country closer to civil war. Artisans were giving way to a new generation of factory owners seeking new ways to increase markets and profits. European immigrants were flocking to the United States by the thousands to seek jobs in New York, Philadelphia, Boston, and other major cities. The poor and working classes bore the brunt of society's corruption and violence and had to face the daily threat of injury and death on the job as well.

In the late nineteenth century many new industries, like bicycle manufacturing, were created. The bicycle manufacturers in the nineteenth century became the automobile companies of the twentieth century. Since industries were new, safety techniques were primitive. It is doubtful the early capitalists would have been able to implement engineering techniques even if any had been available. This was the age of exploitation, of profit, whatever the cost in wasted resources or in wasted workers. There were always new bodies to replace the broken ones—new bodies that streamed into the swollen cities from the countryside or from Europe.[6]

Workplace accidents were common; workers' compensation was unknown. It was in Germany during the 1880s that Chancellor Otto von Bismarck instituted workers' compensation and security insurance paid for by the employers. Bismarck sought to undercut the socialists by demonstrating to the German working class that its government was in favor of social reform. Soon, most other countries in Europe followed Bismarck's lead.

The United States held itself aloof. Here, an injured employee had to sue unless the employer volunteered to pay. Few employers volunteered compensation because, during the late nineteenth century, they could hardly lose in the courts due to legal precedents that had been set. (A legal precedent is a judicial decision that may be used as a standard in subsequent similar cases.) Precedents had limited both the employer's personal liability and its liability under the venerable doctrine of *respondeat superior*, which held the employer was not responsible for an injury to a third party caused by an employee. In other words, if you were injured by one of my employees while he was at work for me, I would not be responsible for his actions. Since the employee was probably destitute, there was no legal remedy available.

The other significant precedents included the following:

1. The fellow-servant doctrine held that an employer was not responsible if an employee's injury resulted from the negligence of a co-worker. If you were injured because a co-worker caused the accident, the employer would not be responsible.
2. The assumption-of-risk concept held that an injured employee presumably knew of and accepted the risks associated with the job before accepting the position.
3. The contributory negligence argument held that if the employee was responsible or even partially responsible for an accident, the employer was not liable for the accident.

The legal obstacles to employee compensation for on-the-job injuries have been slowly replaced by new and more equitable rules. However, the old laws die hard. New York City attempted to revive the fellow-servant rule in a case where a fire fighter's injury was caused by another fire fighter at the scene of a fire. In 1982, the fellow-servant rule was declared unconstitutional in New York State. In Chapter 4 this is more thoroughly discussed along with other aspects of worker's compensation.

SAFETY IN AMERICA

Fire was an ever-present threat to industry in the nineteenth century, when fire-resistant and fireproof building materials were unknown. Industry utilized open fires, candles, and whale oil lamps. Fires were fought using a method called the "bucket brigade." A line of persons was formed from the fire to a source of water, a bucket filled with water was then passed from person to person, and the person closest to the fire dumped the water on the fire.

In 1736, Benjamin Franklin organized the first volunteer fire department in the colonies, The Union Fire Company of Philadelphia. In 1752, Franklin also founded a fire insurance company, The Philadelphia Contributorship. Company engineers periodically inspected policyholders' homes and shops for fire hazards.

During the early 1800s, volunteer fire companies frequently proved themselves as ineffective as the old bucket brigades at extinguishing big fires. On one occasion, competing volunteer fire companies fought each other while buildings were reduced to ashes. Major fires also drove several insurance companies into bankruptcy, so factories with high concentrations of flammables, like sawmills and textile mills, became uninsurable. This is still true today to a certain extent.

Zechariah Allen, owner of a Rhode Island cotton factory, established an insurance company for textile factories. He hired inspectors to recommend fire safety improvements for policyholders. Allen's own factory was a model of fire safety. Allen used only high-quality construction materials, installed fire walls, and placed fire-fighting equipment throughout the plant.[6]

Though industrial safety was often ignored during the 1800s, industrial fire protection was getting a healthy push from the factory insurance companies of New England. The National Board of Fire Underwriters was formed in 1866. Most insurance companies employed inspectors to evaluate risks and help customers improve their protection.

William B. Whiting, an inspector for the Insurance Company of Providence, collected exhaustive records on the causes of fires during the 1860s. In 1878 Edward Atkinson, a successor, drew on Whiting's records to develop production equipment and techniques that would be less hazardous. Atkinson also worked with the Massachusetts Institute of Technology to develop fire safety standards for lighting and lubricating oils.

In its early days, railroad travel was hazardous for both passengers and crew. Deaths and serious injuries were common. Most accidents could be attributed to the lack of a braking system. This was memorialized in the "Ballad of the Old 97."

> He was going round the bend at 90 miles an hour
> When the whistle turned into a scream
> They found him in the wreck with his hand still
> on the throttle
> Burned to death by the steam.

Before the advent of the air brake, brakemen applied brakes to each car on signal from the engineer. This made stopping for an emergency next to impossible. In 1869 George Westinghouse developed a brake based on compressed air which could be controlled by the engineer. This made railroad travel vastly safer for riders and crew and was adopted by railroads in the United States, Canada, and around the world. Additionally, in 1893 a bill was passed by Congress which banned the dangerous "link-and-pin" method of coupling railroad cars.

SAFETY LAWS IN THE UNITED STATES

Coal mining and railroading had replaced textiles as the leading industries, and both went through bloody labor troubles in the 1870s. Safety-related issues played a part, though not a central one, in both industries. During the late 1860s, miners and railroaders were intensely concerned about the hazards of their jobs and had begun to agitate for better conditions. Frustration over the slow progress doubtlessly contributed to the labor unrest of the 1870s.

Anthracite mining in eastern Pennsylvania was a notoriously dangerous occupation; methane gas, rock falls, and fire were constant threats. Canaries and mice were commonly used as gas and oxygen deficiency detectors since they were apparently very sensitive to lack of oxygen. The Davy Safety Lamp, invented in the early nineteenth century, helped ease some of the danger of sudden explosion. In one 7-year period, unmarred by major disasters, 566 miners were killed and 1,651 injured in Schuykill County, Pennsylvania.

Miners in Schuykill County, heart of the anthracite country, pushed hard for mine safety legislation throughout the mid-1860s. Finally, in 1869, the Pennsylvania legislature passed a mine safety act for "better regulation and ventilation of the mines." It authorized one inspector whose authority was limited to Schuykill County. That same year 109 miners died in an explosion in nearby Luzerne County, and the lawmakers were shaken by the public outcry into extending inspection to all anthracite-producing counties.[8]

In 1869 the Massachusetts legislature established the first State Bureau of Labor Statistics; its mission was to examine wages, prices, and working conditions. Throughout the 1870s, other states followed Massachusetts' lead.

The first factory inspection law was passed by the Massachusetts legislature in 1877. The law required machine guarding, guarding of floor openings and shafts, good housekeeping and fire safety. Inspectors checked these provisions and for compliance with child labor laws as well. Inspector Rufus Wade reported in 1879: "In the matter providing for speedy and safe egress in case of fire or panic, large numbers of manufacturing establishments were found deficient." He made similar observations about child labor and work hour violations.[6]

Throughout the 1870s the Knights of Labor, the largest national union, demanded bureaus of labor statistics in every state and in its 1878 convention demanded that Congress "adopt measures providing for the health and safety of those engaged in mining, manufacturing, and building pursuits." For the first time in U.S. history, labor demanded federal occupational safety and health legislation. Labor's dream of a federal safety law had to be abandoned, but progress continued on the state front. Eight industrial states followed the lead of Massachusetts during the 1870s and passed safety inspection laws. While most were vague and enforcement was weak, they were a start in the right direction. A number of laws affecting safety and health had been passed in response to specific situations. Safety and health legislation was piecemeal, only affecting certain industries and certain problems. Safety legislation until the passage of the Occupational Safety and Health Act of 1970 (PL91-597) has been reactive and not preventive.

In 1905 Upton Sinclair described the unsanitary conditions in the slaughterhouses of Chicago in lurid details.[9] The Hearst newspapers serialized the book nationwide. As a result of this exposé, the Meat Inspection Act of 1906 and the Pure Food and Drug Act of 1906 were passed.

BEGINNING OF THE SAFETY MOVEMENT

Four different groups were advancing the safety movement:

1. Organized labor
2. Voluntary safety organizations
3. Government
4. Industry

At that time the primary goal of labor unions was the safety and health of members. The Bridge, Structural, and Ornamental Iron Workers Union promoted uniform safety standards on all construction projects. In 1890, the United Mine Workers was organized to obtain shorter hours, better pay, and improved safety conditions.

Voluntary safety organizations were just beginning in the late nineteenth century. The first was the American Public Health Association, founded in 1885. It was concerned with all aspects of health, including epidemiology (the study and control of the frequency of disease occurrence—see Chapter 5) and occupational health and safety. Shortly after, the National Fire Protection Association and Underwriters Laboratories were founded, both of which have played an active role in setting safety standards in the United States.

Although many workers were being killed or maimed on the job, employers remained indifferent, finally necessitating government intervention to protect the worker. Equally important, public opinion aroused legislatures and government officials. New safety laws were enacted, as they had been after the book by Upton Sinclair and the fire at the Triangle Shirtwaist Company, in response to public sentiment.

Only when it became apparent to management that accidents cost money in the form of lower productivity did employers begin to take safety seriously. Then heavy industries, like the steel industry, in which workplace accidents occurred frequently, began to set safety standards.

INDUSTRIAL ACCIDENT PREVENTION

In 1931, H. W. Heinrich published a book on industrial safety in which he maintained that accidents result from unsafe actions and unsafe conditions.[3] He contended that people cause far more accidents than unsafe conditions do and that accidents result from a sequence of the following events:

1. Ancestry of social environment
2. Fault of a person
3. Unsafe act or condition
4. Accident
5. Injury

It was a pragmatic approach to a complex subject and provided a conceptual framework on which to build occupational safety and health practices. This was the beginning effort to develop the theoretical framework which is required if safety is to be viewed as a true profession. Heinrich's theory of accident causation was the beginning of safety theory.

DISASTERS—INCURRED AND NATURAL

In the last 10 years, a number of major disasters have occurred, most of them involving great loss of life (see Table 2.1). Good safety programs could have prevented some; others, if not preventable, could have had ameliorated impacts if adequate safety measures had been in effect.

Natural disasters in the 10-year period from 1970 to 1980 affected over 300 million persons, according to the World Health Organization, and substantially more than a million persons perished as a result of these disasters.

One Hundred Million Guinea Pigs

In 1933, the book *One Hundred Million Guinea Pigs* reported that industry was putting untested synthetic chemicals into everything the public bought, using people as guinea pigs to determine whether the new products were safe or not.[5] The more reputable industrial firms, which had a countrywide market and were vulnerable to bad publicity, actually did make some advanced tests of safety, by the standard of the day. They had no federal agency to assist them in interpreting the results of their safety tests. These large firms claimed that the conditions that the book deprecated were due to small "fly-by-night, kitchen table, and garage" factories which had no investment in their good names and could afford the risk of

TABLE 2.1
Major Disasters, 1972–1982

Fires and Explosions

1974 Sao Paulo, Brazil: Fire in 25-story bank building—227 killed.

1974 Nypro (U.K.) Ltd., at Flixborough: Explosion of cyclohexane—28 killed. Minimum injuries only because it was Saturday.

1978 Southgate, Kentucky: Fire in supper club—164 killed. The causes were delayed alarm, rapid fire spread, overcrowding, and lack of automatic sprinkler protection.

1977 Explosion of two grain elevators—54 killed. One in Louisiana and one in Texas within 5 days of each other.

1980 Las Vegas, Nevada: Fire in Grand Hotel—85 killed. Improper electric wiring, rapid fire and smoke development, failure to extinguish fire in early stage, improperly protected vertical openings, improperly enclosed exit stairs, and improper distribution of smoke by air handling system.

Mine Disasters

1972 Kellogg, Idaho: Fire at mine in Sunshine Silver Mine—91 killed.

1972 Wankie, Rhodesia: Mine explosion—427 miners killed.

1975 India: Coal mine explosion and flood—431 miners killed.

Aircraft Disasters

1979 Ukraine, U.S.S.R.: Two jetliners collide—173 killed.

1980 Riydah, Saudi Arabia: Saudi airliner lands in flames—301 killed.

1982 Washington, D.C.: Air Florida Boeing 727 crashes into Potomac River bridge during snowstorm—78 killed.

customer complaints. They could simply vanish if their product caused trouble and start again somewhere else under a new name. As usual, neither side of the argument was completely right. The book was upsetting to many people but did nothing to improve conditions.

Ginger Jake Paralysis

Several hundred people were paralyzed in the legs for life in the 1920s, most of them around the Los Angeles and St. Louis area, due to triortho-creasyl phosphate (TOCP). It was during our prohibition era. Some of our ingenious fellow citizens had found that Jamaica ginger extract USP, available in every drugstore, would make a fine substitute for whiskey or gin and was safer than bootleg liquor. The supply of USP extract became exhausted. An ingenious chemist discovered that when low-grade, but perfectly safe, extract was mixed with the right amount of TOCP, it would pass the USP tests of specific gravity, color, and refractive index. Several hundred gallons were made, sold to drugstores in Los Angeles and St. Louis, and then to the local thirsty, who imbibed the concoction. The result was paralysis, which eventually became known as "Ginger Jake paralysis." The same ester is used as a nonflammable hydraulic fluid in artillery, and its mistaken use for cooking oil has paralyzed French soldiers in Africa.

The Day of Saint Anthony's Fire

In the Middle Ages, especially in Germany, there was a period marked by a strange phenomenon of mass hysterical dancing. Men, women, and children engaged in unplanned, continuous dancing, ending only with exhaustion or death. Entire towns, dropping all work, joined in. Several centuries later, on August 16, 1951, a similar event occurred. In the town of Point-Saint Espirit, France, the bakers received a lot of moldy flour which was baked into bread. Moldy flour contains an ergot similar in effect to lysergic acid diethylamide (LSD). LSD is a potent hallucinogen which causes vivid hallucinations and delusions. The more bread a person ate, the stronger the effect. Almost the entire town was affected, and hundreds of respectable townspeople became psychotic (insane) that summer day. Many persons jumped from upper-story windows. Others jumped into the river, screaming that they were covered with snakes. Animals went crazy. Many persons had to be committed to psychiatric hospitals. Most people subsequently recovered, but a few were still hospitalized several years later. This story would have been great fiction, but it was true.

The *Titanic*

The British luxury passenger liner *Titanic* sank on April 14–15, 1912, while on its maiden voyage. The accident occurred about 95 miles south of Newfoundland after the *Titanic* struck an iceberg. The ship was considered unsinkable; however, 1,513 persons lost their lives. As a result of the *Titanic*'s sinking, in 1913 the International Convention for Safety of Life at Sea was called in London. This meeting set up a number of safety requirements. These included: every ship must have lifeboat spaces for all persons; lifeboat drills must be conducted during each voyage; ships must maintain a 24-hour-a-day radio watch (another liner was less than 20 miles away, but its radio operator was not on duty); and, finally, the International Ice Patrol, which the meeting established, was to warn ships of icebergs in the shipping lanes. North Atlantic ships are now warned twice daily of the position of icebergs. Not since 1959 has there been an iceberg collision.

SAFETY-RELATED ACTS

The Pure Food and Drug Act of 1906

Until 1906 there were no federal laws regulating the safety of goods, drugs, or cosmetics. A few states had pharmacy laws, but all they required was that a record be kept of who purchased "statutory poisons," defined as "a substance, 60 grains [3.6 grams] or a teaspoon of which may endanger the life of an adult person." The act aimed to prevent short weight, adulteration with nonnutritive substances, reliance on preservatives instead of on sanitation, and filth in foods. Filth was defined as rot, mold fragments, rodent hairs, and excreta, although one of these in moderate quantities was considered not actually in itself harmful! Chemicals, by implication, appear to have been regarded as filth. In 1906 we had nothing like what today we call "food technology." There was no reason to use chemicals in food, other than those which had been traditionally used for hundreds of years and generally regarded as safe (GRAS).

Walsh-Healy Public Contracts Act

Franklin D. Roosevelt and his Secretary of Labor Frances Perkins attempted to make government contractors model employers. "No top Government official worked more ardently to develop legislation for underpaid workers and exploited child laborers than Secretary of Labor Frances Perkins."[*] Their effort resulted in the Walsh-Healy Public Contracts Act of 1936. The act required most government contractors to adopt an 8-hour day and a 40-hour week and to pay the prevailing minimum wage. It eliminated use of child labor on government contracts by establishing minimum ages (16 for males and 18 for females). The Walsh-Healy Act was the first legislation which included occupational health standards. The threshold limit values (TLVs) from many substances were included in the health requirements under this act.

Fair Labor Standards Act

One of the most far-reaching pieces of social legislation passed by the U.S. Congress was the Fair Labor Standards Act of 1938. It was principally social legislation and only peripherally safety legislation. In its final form, the act applied to about one-fifth of the U.S. labor force. It banned child labor; set minimum wages at 25 cents per hour for both men and women; and set the maximum work week at 44 hours. However, the 44-hour limit was not

---

[*]Grossman,[2] p. 23.

absolute: employers who wanted employees to work longer had to pay time and a half for overtime. Employment of all children under 16 was prohibited, and children under 18 were banned from hazardous industries.

Occupational Safety and Health Act

The Occupational Safety and Health Act required that the workplace be made safe and healthy. It stipulated that no longer could the worker be made to trade life or limb for employment. The law set the minimum standards for safety in all workplaces in the United States. See Chapter 3 for a complete discussion of this act.

Toxic Substances Control Act of 1976

The Toxic Substances Control Act of 1976, which became effective in 1977, gave the Environmental Protection Agency jurisdiction over the safety testing of all industrial chemicals. This was the first time violations of a safety law could result in both civil penalties and criminal prosecution.

SUGGESTED LEARNING EXPERIENCES

1. If you were going to write your version of the history of the safety movement, what would your focus be? Explain.

2. Examine the legislative history of a safety law. Examine the House and Senate hearings, the *Congressional Record*, the law as passed, the conference committee report, and the final version of the law.

3. Examine the politics of safety while taking into account the current climate.

4. Compare the history of the safety movement with either the history of nursing or the history of the consumer movement.

5. Discuss the following statement in relation to the safety movement: A society that fails to study history is condemned to repeat it.

6. There were many names, such as Hippocrates, Pliny the Elder, and Agricola, mentioned in this chapter of history. Find the original sources, and examine their statements regarding safety. What did you learn?

7. If you were President of the United States, what safety legislation would you introduce to Congress?

8. Frances Perkins was very important in the history of the safety movement. Select another secretary of labor, and discuss what that person did regarding safety.

9. Prepare a poster that would illustrate the history of the safety movement.

10. Many substances have been designated "GRAS." Select one of these and determine by your own investigation the reasons for its thus being designated.

REFERENCES

1. Fuller, J. G. *The Day of St. Anthony's Fire*. New York: Macmillan, 1968.
2. Grossman, J. Fair Labor Standards Act of 1938: Maximum struggle for a minimum wage. *Monthly Labor Review*, 1978, **101**, 22–30. (June)
3. Heinrich, H. W. *Industrial Accident Prevention* (4th ed.). New York: McGraw-Hill, 1959.
4. Herrick, R. The historical perspective. In N. O. Sax (Ed.), *Dangerous Properties of Industrial Materials* (5th ed.). New York: Van Nostrand-Reinhold, 1979, pp 1–39.
5. Kallet, A., and Schlink, F. J. *One Hundred Million Guinea Pigs: Dangers in Everyday Food, Drugs and Cosmetics*. New York: Arno Press, 1933.
6. McCall, B. *Safety First—at Last!* New York: Vantage Press, 1975.
7. Moore, T. *Poetical Works with a Memoir*. Boston: Houghton Mifflin, 1881.
8. National Safety Council. *Accident Prevention Manual for Industrial Operations* (7th ed.), Chicago, 1977.
9. Sinclair, U. *The Jungle*. New York: New American Library, 1980.

3 OCCUPATIONAL SAFETY AND HEALTH ACT

A piece of legislation does not arrive *de novo* out of a vacuum. It is built upon some real or implied need and upon the legislation that has preceded it. Thus need for the Occupational Safety and Health Act (OSHA) existed as long as there have been workers. Ramazzini in the seventeenth century discussed safety in one of the first documents on occupational medicine.

Timing of legislation is critical. The late 1960s was the appropriate time for OSHA to emerge. However, the act was based upon literally dozens of earlier pieces of legislation.

The historical antecedent of OSHA was the creation of the Interstate Commerce Commission in 1887. This commission was established following the deaths of a large number of railroad workers. In 1913 the Department of Labor was created for the specific purpose of improving working conditions. In 1933 President Roosevelt appointed Frances Perkins as Secretary of Labor. During her tenure three laws were enacted that directly affected job safety: the Social Security Act of 1935, the Walsh-Healy Public Contracts Act of 1936, and the Fair Labor Standards Act of 1938. These laws banned child labor, banned government contract work under hazardous conditions, and funded industrial health programs.

Several other laws were passed protecting groups of workers. The Service Contracts Act of 1965 and the Federal Construction Safety and Health Act of 1967 protected workers of government contractors; the Coal Mine Safety and Health Act of 1969 protected coal miners; and the Metal and Non-metallic Mine Safety Act of 1966 protected other miners. By the late 1960s employees of federal contractors and miners were protected.

LEGISLATIVE HISTORY OF OSHA[3]

In his message to Congress in January 1968 President Lyndon B. Johnson called for enactment of comprehensive job safety and health legislation. He cited as inadequate voluntary standards and the several laws concerning specific safety issues. He noted that they were poorly enforced and that there was a shortage of safety and health professionals. Research in occupational safety and health was negligible.

Congressional hearings began in February 1968. Secretary of Labor Willard Wirtz was the first to testify. He cited two casualty lists: number killed in Vietnam and number killed at work every year. Wirtz also claimed that about 75% of the teenagers now entering the work force would suffer some type of partially disabling injury at work. Wirtz felt that the main issue was "whether or not the Congress would stop the bloodshed at work."[2] As expected, labor testified for the legislation, industry against the legislation.

In November 1968 Richard Nixon was elected President. After his inauguration in January 1969, he called upon his staff to review his campaign promises and report how they were going to implement them. The White House asked Under secretary of Labor James D. Hodgson to prepare a bill, which the President presented to Congress in August 1969. Hearings were again conducted not only in Washington but also around the country. The Chamber of Commerce, which had fought vigorously against the Johnson bill, supported the new one. It was introduced into the U.S. Senate by Harrison Williams of New Jersey and into the U.S. House of Representatives by Harley Steigers. Thus it is often referred to as the Williams-Steigers Act. The Committee on Education and Labor in the House of Representatives and the Committee on Labor and Public Welfare in the Senate conducted the hearings. After the hearings were reported out by the respective committees, the House considered and passed its version of the bill on November 23–24 and the Senate considered and passed its version on October 13 and November 16–17. The separate bills were referred to a conference committee. Legislation on the final version of the bill was completed on December 16–17, and it was again passed by both houses of Congress on December 29, 1970. President Richard Nixon signed into law the Williams-Steigers Occupational Safety and Health Act. As can be seen in this chapter, the act gave the federal government the authority to determine, publish, and enforce safety and health standards for most workers. It was a long and sometimes bitter struggle from the first safety legislation in 1969 through the piecemeal laws and similar measures submitted by President Johnson to the Occupational Safety and Health Act of 1970.

PURPOSE OF THE LAW

Public law 91-596, which is officially known as the Occupational Safety and Health Act of 1970 and unofficially known as the Williams-Steigers Act, states its purpose quite forcefully in its preamble. It is to assure safe and healthful working conditions for working men and women. This was to be accomplished as follows:

1. By authorizing enforcement of standards developed under the act
2. By assisting and encouraging the states in their efforts to assure safe and healthful working conditions
3. By providing for research, information, education, and training in the field of occupational safety and health

To Preserve Human Resources

The purpose of the Occupational Safety and Health Act is to preserve human resources by which is meant workers. Just as money is capital, equipment and property are tangible resources. A healthy, safe work force is a human resource. This was to be accomplished as follows:

1. By reducing or eliminating hazards
2. By instituting a new or improving an existing safety program
3. By giving both labor and management separate yet interdependent safety responsibilities and activities
4. By authorizing the secretary of labor to set mandatory occupational safety and health standards
5. By establishing an occupational safety and health review commission to make legal judgments concerning the act
6. By supporting research in occupational safety and health
7. By establishing causal connections between work and illnesses
8. By providing training programs in occupational safety and health
9. By developing and publishing occupational safety and health standards
10. By enforcement
11. By providing a nationwide reporting system for occupational safety and health

Job Safety and Health Protection

Every employer must display in each location or establishment a poster (see Figure 3.1) that explains the protection and obligations of employees. States which have approved occupational safety and health plans, such as Connecticut, may require that a state poster also be displayed.

It applies to work performed in:

1. Any state
2. District of Columbia

job safety and health protection

The Occupational Safety and Health Act of 1970 provides job safety and health protection for workers through the promotion of safe and healthful working conditions throughout the Nation. Requirements of the Act include the following:

Employers: Each employer shall furnish to each of his employees employment and a place of employment free from recognized hazards that are causing or are likely to cause death or serious harm to his employees; and shall comply with occupational safety and health standards issued under the Act.

Employees: Each employee shall comply with all occupational safety and health standards, rules, regulations and orders issued under the Act that apply to his own actions and conduct on the job.

The Occupational Safety and Health Administration (OSHA) of the Department of Labor has the primary responsibility for administering the Act. OSHA issues occupational safety and health standards, and its Compliance Safety and Health Officers conduct jobsite inspections to ensure compliance with the Act.

Inspection: The Act requires that a representative of the employer and a representative authorized by the employees be given an opportunity to accompany the OSHA inspector for the purpose of aiding the inspection.

Where there is no authorized employee representative, the OSHA Compliance Officer must consult with a reasonable number of employees concerning safety and health conditions in the workplace.

Complaint: Employees or their representatives have the right to file a complaint with the nearest OSHA office requesting an inspection if they believe unsafe or unhealthful conditions exist in their workplace. OSHA will withhold, on request, names of employees complaining.

The Act provides that employees may not be discharged or discriminated against in any way for filing safety and health complaints or otherwise exercising their rights under the Act.

An employee who believes he has been discriminated against may file a complaint with the nearest OSHA office within 30 days of the alleged discrimination.

Citation: If upon inspection OSHA believes an employer has violated the Act, a citation alleging such violations will be issued to the employer. Each citation will specify a time period within which the alleged violation must be corrected.

The OSHA citation must be prominently displayed at or near the place of alleged violation for three days, or until it is corrected, whichever is later, to warn employees of dangers that may exist there.

Proposed Penalty: The Act provides for mandatory penalties against employers of up to $1,000 for each serious violation and for optional penalties of up to $1,000 for each nonserious violation. Penalties of up to $1,000 per day may be proposed for failure to correct violations within the proposed time period. Also, any employer who willfully or repeatedly violates the Act may be assessed penalties of up to $10,000 for each such violation.

Criminal penalties are also provided for in the Act. Any willful violation resulting in death of an employee, upon conviction, is punishable by a fine of not more than $10,000 or by imprisonment for not more that six months, or by both. Conviction of an employer after a first conviction doubles these maximum penalties.

Voluntary Activity: While providing penalties for violations, the Act also encourages efforts by labor and management, before an OSHA inspection, to reduce injuries and illnesses arising out of employment.

The Department of Labor encourages employers and employees to reduce workplace hazards voluntarily and to develop and improve safety and health programs in all workplaces and industries.

Such cooperative action would initially focus on the identification and elimination of hazards that could cause death, injury, or illness to employees and supervisors. There are many public and private organizations that can provide information and assistance in this effort, if requested.

More Information: Additional information and copies of the Act, specific OSHA safety and health standards, and other applicable regulations may be obtained from your employer or from the nearest OSHA Regional Office in the following locations:

Atlanta, Georgia
Boston, Massachusetts
Chicago, Illinois
Dallas, Texas
Denver, Colorado
Kansas City, Missouri
New York, New York
Philadelphia, Pennsylvania
San Francisco, California
Seattle, Washington

Telephone numbers for these offices, and additional Area Office locations, are listed in the telephone directory under the United States Department of Labor in the United States Government listing.

Washington, D.C.
1977
OSHA 2203

Ray Marshall
Ray Marshall
Secretary of Labor

U. S. Department of Labor
Occupational Safety and Health Administration

FIGURE 3.1. Job safety and health protection. (This poster is required to be displayed at every workplace in the United States.)

3. Puerto Rico
4. Virgin Islands
5. Canal Zone

All other territories are under federal government jurisdiction.

Exclusions

The major exclusion to this law is employees who are covered by the Atomic Energy Act of 1954, Section 274, as amended, and workers covered by the Mine, Health, and Safety Act.

THE GENERAL DUTY CLAUSE

The general duty clause from Section 5 of the act seems self-explanatory, but is it? "Each employer shall furnish to each of his employees employment which is free from recognized hazards that are likely to cause death or serious physical harm to employees. . . . Each employer shall comply with occupational safety and health standards promulgated under this Act." These statements appear straightforward, but there are a number of ambiguous words. Can you determine which words could cause trouble?

THE ROLE OF THE DEPARTMENT OF LABOR

The role of the Department of Labor is to publish OSHA standards, to enter factories to make inspections, to prescribe regulations for record keeping, to establish and supervise programs for the education and training of employee and employer personnel, to develop and maintain safety and health statistics, and to make grants to the states by identifying needs, developing state plans, monitoring state plans, and canceling plans.

Developing State Plans

The act requires the Department of Labor to encourage the states to develop and operate their own job safety and health programs, provided they "are or will be at least as effective in providing safe and healthful employment."*

Any state desiring to assume responsibility for developing and enforcing its own job safety and health program may submit a plan to the secretary of labor. The plan must meet a number of requirements, including

*Pub. L. 91-596, Sec. 19c(2).

assurances that the designated state agency will have the legal authority to enforce safety and health standards.

THE ROLE OF THE DEPARTMENT OF HEALTH AND HUMAN SERVICES

The role of the Department of Health and Human Services consists of:

Research
Developing criteria for dealing with toxic materials and hazards
Making toxicity determinations on request by employer or employee group
Providing OSHA personnel
Training OSHA personnel
Establishing the National Institute for Occupational Safety and Health which is responsible for
    Developing OSHA health standards
    Education and training
    Health hazard investigation
    Making an annual report to the speaker of the house
    Publishing an annual list of all known toxic substances

NATIONAL ADVISORY COMMITTEE

A 12-person National Advisory Committee on Occupational Safety and Health is provided to advise, consult, and make recommendations to the secretary of labor and to the secretary of health and human services on matters relating to the act. The committee is required to be composed of representatives of management, labor, public health, and occupational safety and health profession.

EMPLOYEE'S RIGHTS

Employees are provided certain rights under OSHA:

1. The right to request the secretary of labor in writing to have safety and health inspections made in the plant, work establishment, or job site

2. The right to have a representative accompany compliance officers when an inspection is made of the plant, work establishment, or job site

3. The right to have employers maintain accurate records of employee exposures to potentially toxic materials or harmful physical agents and to have access to such records regarding his or her own exposures

4. The right to have dangerous substances identified by labeling or posting in the plant or at the job site

5. The right to be informed promptly by the employer of exposure to any toxic materials or harmful physical agents that are present in concentrations or at levels in excess of those prescribed by an applicable standard

6. The right to have citations for employer violations found by the compliance officer posted prominently at the work site

7. The right to have the employee's name withheld from the employer, upon request to OSHA, if a written, signed complaint was filed[3]

EMPLOYER'S RIGHTS

The law also provides specific rights for the employer:

1. The right to be advised by OSHA personnel of the reason for an inspection

2. The right to request and receive proper identification of *OSHA personnel prior to an inspection*

3. The right to participate in the walk-around inspection with the compliance officer and in opening and closing conferences with the officer

4. The right to file a notice of contest with the OSHA assistant director (in the employer's region) within 15 working days of receipt of a citation and notice of penalty, if the employer disagrees with the citation or the penalty proposed

5. The right to apply to OSHA for a temporary variance from a standard if the employer is unable to comply because of the unavailability of materials, equipment, or personnel to make changes within the required time

6. The right to apply to OSHA for a permanent variance from a standard if the employer can prove the facilities or methods of operation provide protection that is at least as effective as that required by the standard

7. The right to be assured of the confidentiality of any trade secrets observed by an OSHA compliance officer during an inspection[3]

STANDARDS

To carry out the purposes of the act, the secretary of labor is required to develop and adopt mandatory safety and health standards. These must be observed by employers covered by the act.

The "occupational safety and health standard is a rule which requires conditions, or the adoption or use of one or more practices, means, methods, operations, or processes reasonably necessary or appropriate, to provide safe or healthful employment and places of employment."

Three types of standards are authorized: (1) interim standards, (2) emergency temporary standards, and (3) permanent standards.

The secretary of labor was required to issue interim standards within two years after the act's enactment. They were made up of national consensus standards and established federal standards. Consensus standards are issued by the American National Standards Institute (ANSI) when they are developed and approved according to its procedures. Such standards provide an opportunity for diverse views to be considered. Representatives of industry, labor, insurance companies, and professional organizations reach an agreement on the contents of a standard. The ANSI standards are largely intended to establish the best practice for the subject covered. They are not intended to have the force of law, although some ANSI standards were published as regulations by governmental agencies before OSHA. Established federal standards are any occupational safety and health standards set up by an agency of the United States in effect or contained in an act of Congress when OSHA was enacted. These include standards established under the following acts:

1. Walsh-Healy Public Contracts Act
2. Maritime Safety Act
3. Service Contract Act of 1965
4. Construction Safety Act
5. Arts and Humanities Acts
6. Longshoremen's and Harbor Workers' Compensation Act

As OSHA develops effective safety and health standards, standards issued under the previously listed acts will be superseded. On May 29, 1971, the first interim standards package (Title 29, Chapter 17, Part 1910—more commonly known as 29 CFR 1910 General Industry Standard) was issued and became effective on August 27, 1971. Emergency temporary standards take effect immediately upon publication in the *Federal Register*, without regard to the rule-making provisions of the Administrative Procedure Act.

INSPECTIONS

An OSHA inspector is allowed to enter and inspect any workplace and its environs during regular business hours. He or she may inspect and investigate any pertinent conditions; inspect any equipment, structures, and machines; and finally, question any person privately who is relevant to the inspection.

Advance Notice

The act prohibits giving advance notice of inspection under criminal penalty. There are a few limited circumstances when advance notice may be given:

1. In cases of apparent imminent danger, so the employer can eliminate or reduce the hazard as soon as possible
2. In cases where special preparations are necessary for an inspection
3. In cases where the inspection will be performed after business hours
4. In cases when it would be necessary to assure the presence of representatives of employer, employee, or other special personnel needed to aid the inspection
5. In cases when the area director finds that by giving advance notice the inspection would be enhanced

Priorities

The priorities for an OSHA inspection have been set by order of precedence:

1. *Imminent Danger.* ". . . to restrain any conditions or practices in any place of employment which are such that a danger exists which could reasonably be expected to cause death or serious physical harm immediately or before the imminence of such danger can be eliminated through the enforcement procedures otherwise provided by this Act."*
2. *Catastrophes or Fatal Accidents.* Catastrophes are considered to be incidents resulting in hospitalization of five or more employees.

*Pub. L. 91-596, Sec. 132.

Employers must notify the Department of Labor in the event of any catastrophe or fatality.

3. **OSHA Target Industries.**   The purpose of this approach is to concentrate efforts on the industries with the highest recorded injury rates.

4. **Valid Employee Complaints.**   "Any employees or representative of employees who believe that a violation of a safety and health standard exists . . . may request an inspection."*

5. **Reinspections.**   Inspections are normally conducted on establishments cited for alleged serious violations to determine whether or not hazards have been corrected.

Duties of Compliance Officer

The primary aim of compliance officers is to enforce the standards issued under this act, for which they must have thorough knowledge of the standards. The principal standard that compliance officers use is the General Industry Standard (29 CFR 1910), which will be discussed in detail.

Another task is to review the administrative requirements of the act. These include ascertaining the following:

1. Has the employer posted the notice (Figure 3.1) informing employees of their rights and obligations?

2. Has the employer complied with the record-keeping requirements? Is the log up-to-date?

3. Has the employer given advance notice to employee representatives?

At the end of an inspection, compliance officers are required to conduct a closing conference with the employer. They discuss their observations and recommendations. If safety and health violations are found, applicable sections of the standards should be provided to the employer. The inspection officer does not issue any citations or penalties.

VIOLATIONS

In the course of an inspection, compliance officers evaluate whether the employer was complying with the standards published under the act and whether the employer complied with the general duty clause. They prepare a report for the OSHA area director, who decides what citations will be issued. Violations are presented in order of increasing severity.

*Pub. L. 91-596, Sec. 8f(1).

1. *De minimis.*   A violation of a standard which really has no immediate relationship to occupational safety and health. This might include a safety sign which has graffiti in it.

2. *Nonserious.*   A hazard directly related to occupational safety and health but which would not cause serious physical injury. Many housekeeping hazards, like uncleanliness and clutter, are nonserious violations.

3. *Serious.*   A hazard which has a direct relationship to occupational safety and health and can cause serious injury or death. Serious hazards might include a scaffold without a guardrail or the use of a portable electric tool in a wet environment. The employer knew or should have known about such hazards.

4. *Willful.*   A hazard which has a direct relationship to occupational safety and health and can cause serious injury or death. In addition, the employer was aware of the hazard and made no attempt to correct it.

5. *Repeat.*   A hazard present upon a previous inspection and which is still present upon reinspections.

CITATIONS

The citation (Figure 3.2) or a copy of it must be posted by the employer or safety engineer at or near the place where the violation occurred. This is to make employees aware of the hazard. The citation must remain posted for three working days or until the violation is corrected, whichever is longer. The citation must be posted even if it is being contested.

PENALTIES

The act provides for both civil and criminal penalties, which are shown in Table 3.1. Most violations result in the assessment of a penalty. Penalties must be paid within 15 working days after the employer receives the citation and notification of penalties. If the penalty, the citation, or both are being contested, payment does not have to be remitted until a final decision is made on the appeal.

SUGGESTED LEARNING EXPERIENCES

1. Interview a safety officer in a plant, having the person detail a description of the OSHA visit and the reaction to it.

¹ ISSUANCE DATE	² OSHA NUMBER
	579

³ REGION	⁴ AREA	⁵ PAGE
01	0280	1 OF 1

PENALTIES ARE DUE WITHIN 15 DAYS OF RECEIPT OF THIS NOTIFICATION UNLESS CONTESTED (See enclosed Booklet)

This Section May Be Detached Before Posting

⁶ TYPE OF VIOLATION(S)	CITATION NO.
SERIOUS	1

INSPECTION DATE:

INSPECTION SITE:

TO:

THE LAW REQUIRES that a copy of this Citation be posted immediately in a prominent place at or near the location of the violation(s) cited below. The Citation must remain posted until the violations cited below have been corrected, or for 3 working days (excluding weekends and Federal holidays) whichever is longer.

This citation describes violations of the Occupational Safety and Health Act of 1970. The penalty(ies) listed below are based on these violations. You must correct the violations referred to in this citation by the dates listed below and pay the penalties proposed, unless within 15 working days excluding weekends and Federal holidays) from your receipt of this citation and penalty you mail a notice of contest to the U.S. Department of Labor Area Office at the address shown above. (See the enclosed booklet which outlines your responsibilities and courses of action and should be ead in conjunction with this form.)

ITEM NUMBER STANDARD, REGULATION OR SECTION OF THE ACT VIOLATED; DESCRIPTION	⁹ DATE BY WHICH VIOLATION MUST BE CORRECTED	¹⁰ PENALTY
The violations described in this citation are alleged to have occurred on or about the day the inspection was made unless otherwise indicated within the description given below.		
1 29 CFR 1910.23(a)(8): Floor hole(s), into which persons could accidentally walk, were not guarded by standard railings with standard toeboards on all exposed sides or by floor hole covers of standard strength and construction: Ambit Steel Precipitator Overhead Platform – Floor holes in grating in front of precipitators #1, #4 and #5 were not guarded or covered.	1/14/80	$300.00

AREA DIRECTOR.

$300.00
¹¹ TOTAL PENALTY FOR THIS CITATION

Make check or Money Order Payable To: "DOL-OSHA" Indicate OSHA No on Remittance

CITATION AND NOTIFICATION OF PENALTY   OSHA-2 REV. 5/76

TABLE 3.1
Penalties for OSHA Violations

Violation	Penalty required	Maximum civil penalty ($)	Maximum criminal penalty
De minimus	No		
Nonserious	No	1,000	
Serious	No	1,000	
Willful	No	10,000	
Willful resulting in death	Yes	10,000	6 months
Willful resulting in death, 2nd conviction	Yes	20,000	12 months
Repeated	No	10,000	
Falsifying records	Yes	10,000	6 months
Not posting	No	1,000	
Assaulting compliance officer	Yes	5,000	3 years

2. As a safety officer, what would you wish to have prepared for an OSHA visit?

3. Debate: OSHA should or should not be abolished.

4. In your own workplace, if some of your employees are disgruntled about some of OSHA's requirements and compliances, how would you ideally handle this? After you have developed a plan, assess to see how realistic your plan seems.

5. What do you see ahead for OSHA in the next five years? Be sure to include your rationale.

6. Stage a short one-act play detailing the development of OSHA.

7. What kind of employee protection would you want to have in your workplace?

8. Playing a managerial role, give a critique of OSHA.

9. What positive changes would you propose for OSHA over the next 10 years?

10. Develop a priority list you would have if you had a safety position in a concern that had not complied or was not complying with the current safety and health laws and regulations.

FIGURE 3.2.   Citation and notification of penalty. (This is an example of an actual OSHA citation for workplace violation).

APPENDIX

Public Law 91-596
91st Congress, S. 2193
December 29, 1970

An Act

84 STAT. 1590

To assure safe and healthful working conditions for working men and women;
by authorizing enforcement of the standards developed under the Act; by
assisting and encouraging the States in their efforts to assure safe and healthful working conditions; by providing for research, information, education, and
training in the field of occupational safety and health; and for other purposes.

*Be it enacted by the Senate and House of Representatives of the
United States of America in Congress assembled,* That this Act may
be cited as the "Occupational Safety and Health Act of 1970".

Occupational
Safety and
Health Act of
1970.

CONGRESSIONAL FINDINGS AND PURPOSE

SEC. (2) The Congress finds that personal injuries and illnesses arising out of work situations impose a substantial burden upon, and are
a hindrance to, interstate commerce in terms of lost production, wage
loss, medical expenses, and disability compensation payments.

(b) The Congress declares it to be its purpose and policy, through
the exercise of its powers to regulate commerce among the several
States and with foreign nations and to provide for the general welfare,
to assure so far as possible every working man and woman in the
Nation safe and healthful working conditions and to preserve our
human resources—

    (1) by encouraging employers and employees in their efforts
to reduce the number of occupational safety and health hazards
at their places of employment, and to stimulate employers and
employees to institute new and to perfect existing programs for
providing safe and healthful working conditions;

    (2) by providing that employers and employees have separate
but dependent responsibilities and rights with respect to achieving safe and healthful working conditions;

    (3) by authorizing the Secretary of Labor to set mandatory
occupational safety and health standards applicable to businesses
affecting interstate commerce, and by creating an Occupational
Safety and Health Review Commission for carrying out adjudi-
catory functions under the Act;

    (4) by building upon advances already made through employer
and employee initiative for providing safe and healthful working
conditions;

    (5) by providing for research in the field of occupational
safety and health, including the psychological factors involved,
and by developing innovative methods, techniques, and
approaches for dealing with occupational safety and health
problems;

    (6) by exploring ways to discover latent diseases, establishing
causal connections between diseases and work in environmental
conditions, and conducting other research relating to health problems, in recognition of the fact that occupational health standards
present problems often different from those involved in occupational safety;

    (7) by providing medical criteria which will assure insofar as
practicable that no employee will suffer diminished health, functional capacity, or life expectancy as a result of his work
experience;

    (8) by providing for training programs to increase the num-
ber and competence of personnel engaged in the field of occupational safety and health;

(9) by providing for the development and promulgation of occupational safety and health standards;

(10) by providing an effective enforcement program which shall include a prohibition against giving advance notice of any inspection and sanctions for any individual violating this prohibition;

(11) by encouraging the States to assume the fullest responsibility for the administration and enforcement of their occupational safety and health laws by providing grants to the States to assist in identifying their needs and responsibilities in the area of occupational safety and health, to develop plans in accordance with the provisions of this Act, to improve the administration and enforcement of State occupational safety and health laws, and to conduct experimental and demonstration projects in connection therewith;

(12) by providing for appropriate reporting procedures with respect to occupational safety and health which procedures will help achieve the objectives of this Act and accurately describe the nature of the occupational safety and health problem;

(13) by encouraging joint labor-management efforts to reduce injuries and disease arising out of employment.

DEFINITIONS

SEC. 3. For the purposes of this Act—

(1) The term "Secretary" mean the Secretary of Labor.

(2) The term "Commission" means the Occupational Safety and Health Review Commission established under this Act.

(3) The term "commerce" means trade, traffic, commerce, transportation, or communication among the several States, or between a State and any place outside thereof, or within the District of Columbia, or a possession of the United States (other than the Trust Territory of the Pacific Islands), or between points in the same State but through a point outside thereof.

(4) The term "person" means one or more individuals, partnerships, associations, corporations, business trusts, legal representatives, or any organized group of persons.

(5) The term "employer" means a person engaged in a business affecting commerce who has employees, but does not include the United States or any State or political subdivision of a State.

(6) The term "employee" means an employee of an employer who is employed in a business of his employer which affects commerce.

(7) The term "State" includes a State of the United States, the District of Columbia, Puerto Rico, the Virgin Islands, American Samoa, Guam, and the Trust Territory of the Pacific Islands.

(8) The term "occupational safety and health standard" means a standard which requires conditions, or the adoption or use of one or more practices, means, methods, operations, or processes, reasonably necessary or appropriate to provide safe or healthful employment and places of employment.

(9) The term "national consensus standard" means any occupational safety and health standard or modification thereof which (1), has been adopted and promulgated by a nationally recognized standards-producing organization under procedures whereby it can be determined by the Secretary that persons interested

51

and affected by the scope or provisions of the standard have reached substantial agreement on its adoption, (2) was formulated in a manner which afforded an opportunity for diverse views to be considered and (3) has been designated as such a standard by the Secretary, after consultation with other appropriate Federal agencies.

(10) The term "established Federal standard" means any operative occupational safety and health standard established by any agency of the United States and presently in effect, or contained in any Act of Congress in force on the date of enactment of this Act.

(11) The term "Committee" means the National Advisory Committee on Occupational Safety and Health established under this Act.

(12) The term "Director" means the Director of the National Institute for Occupational Safety and Health.

(13) The term "Institute" means the National Institute for Occupational Safety and Health established under this Act.

(14) The term "Workmen's Compensation Commission" means the National Commission on State Workmen's Compensation Laws established under this Act.

APPLICABILITY OF THIS ACT

Sec. 4. (a) This Act shall apply with respect to employment performed in a workplace in a State, the District of Columbia, the Commonwealth of Puerto Rico, the Virgin Islands, American Samoa, Guam, the Trust Territory of the Pacific Islands, Wake Island, Outer Continental Shelf lands defined in the Outer Continental Shelf Lands Act, Johnston Island, and the Canal Zone. The Secretary of the Interior shall, by regulation, provide for judicial enforcement of this Act by the courts established for areas in which there are no United States district courts having jurisdiction.   67 Stat. 462. 43 USC 1331 note.

(b) (1) Nothing in this Act shall apply to working conditions of employees with respect to which other Federal agencies, and State agencies acting under section 274 of the Atomic Energy Act of 1954, as amended (42 U.S.C. 2021), exercise statutory authority to prescribe or enforce standards or regulations affecting occupational safety or health.   73 Stat. 688.

(2) The safety and health standards promulgated under the Act of June 30, 1936, commonly known as the Walsh-Healey Act (41 U.S.C. 35 et seq.), the Service Contract Act of 1965 (41 U.S.C. 351 et seq.), Public Law 91-54, Act of August 9, 1969 (40 U.S.C. 333), Public Law 85-742, Act of August 23, 1958 (33 U.S.C. 941), and the National Foundation on Arts and Humanities Act (20 U.S.C. 951 et seq.) are superseded on the effective date of corresponding standards, promulgated under this Act, which are determined by the Secretary to be more effective. Standards issued under the laws listed in this paragraph and in effect on or after the effective date of this Act shall be deemed to be occupational safety and health standards issued under this Act, as well as under such other Acts.   49 Stat. 2036. 79 Stat. 1034. 83 Stat. 96. 72 Stat. 835. 79 Stat. 845; Ante, p. 443.

(3) The Secretary shall, within three years after the effective date of this Act, report to the Congress his recommendations for legislation to avoid unnecessary duplication and to achieve coordination between this Act and other Federal laws.   Report to Congress.

(4) Nothing in this Act shall be construed to supersede or in any manner affect any workmen's compensation law or to enlarge or diminish or affect in any other manner the common law or statutory rights, duties, or liabilities of employers and employees under any law with respect to injuries, diseases, or death of employees arising out of, or in the course of, employment.

DUTIES

SEC. 5. (a) Each employer—

(1) shall furnish to each of his employees employment and a place of employment which are free from recognized hazards that are causing or are likely to cause death or serious physical harm to his employees;

(2) shall comply with occupational safety and health standards promulgated under this Act.

(b) Each employee shall comply with occupational safety and health standards and all rules, regulations, and orders issued pursuant to this Act which are applicable to his own actions and conduct.

OCCUPATIONAL SAFETY AND HEALTH STANDARDS

80 Stat. 381;
81 Stat. 195.
5 USC 500.

SEC. 6. (a) Without regard to chapter 5 of title 5, United States Code, or to the other subsections of this section, the Secretary shall, as soon as practicable during the period beginning with the effective date of this Act and ending two years after such date, by rule promulgate as an occupational safety or health standard any national consensus standard, and any established Federal standard, unless he determines that the promulgation of such a standard would not result in improved safety or health for specifically designated employees. In the event of conflict among any such standards, the Secretary shall promulgate the standard which assures the greatest protection of the safety or health of the affected employees.

(b) The Secretary may by rule promulgate, modify, or revoke any occupational safety or health standard in the following manner:

(1) Whenever the Secretary, upon the basis of information submitted to him in writing by an interested person, a representative of any organization of employers or employees, a nationally recognized standards-producing organization, the Secretary of Health, Education, and Welfare, the National Institute for Occupational Safety and Health, or a State or political subdivision, or on the basis of information developed by the Secretary or otherwise available to him, determines that a rule should be promulgated in order to serve the objectives of this Act, the Secretary may request the recommendations of an advisory committee appointed under section 7 of this Act. The Secretary shall provide such an advisory committee with any proposals of his own or of the Secretary of Health, Education, and Welfare, together with all pertinent factual information developed by the Secretary or the Secretary of Health, Education, and Welfare, or otherwise available, including the results of research, demonstrations, and experiments. An advisory committee shall submit to the Secretary its recommendations regarding the rule to be promulgated within ninety days from the date of its appointment or within such longer or shorter period as may be prescribed by the Secretary, but in no event for a period which is longer than two hundred and seventy days.

Advisory committee, recommendations.

(2) The Secretary shall publish a proposed rule promulgating, modifying, or revoking an occupational safety or health standard in the Federal Register and shall afford interested persons a period of thirty days after publication to submit written data or comments. Where an advisory committee is appointed and the Secretary determines that a rule should be issued, he shall publish the proposed rule within sixty days after the submission of the advisory committee's recommendations or the expiration of the period prescribed by the Secretary for such submission.

(3) On or before the last day of the period provided for the submission of written data or comments under paragraph (2), any interested person may file with the Secretary written objections to the proposed rule, stating the grounds therefor and requesting a public hearing on such objections. Within thirty days after the last day for filing such objections, the Secretary shall publish in the Federal Register a notice specifying the occupational safety or health standard to which objections have been filed and a hearing requested, and specifying a time and place for such hearing.

(4) Within sixty days after the expiration of the period provided for the submission of written data or comments under paragraph (2), or within sixty days after the completion of any hearing held under paragraph (3), the Secretary shall issue a rule promulgating, modifying, or revoking an occupational safety or health standard or make a determination that a rule should not be issued. Such a rule may contain a provision delaying its effective date for such period (not in excess of ninety days) as the Secretary determines may be necessary to insure that affected employers and employees will be informed of the existence of the standard and of its terms and that employers affected are given an opportunity to familiarize themselves and their employees with the existence of the requirements of the standard.

(5) The Secretary, in promulgating standards dealing with toxic materials or harmful physical agents under this subsection, shall set the standard which most adequately assures, to the extent feasible, on the basis of the best available evidence, that no employee will suffer material impairment of health or functional capacity even if such employee has regular exposure to the hazard dealt with by such standard for the period of his working life. Development of standards under this subsection shall be based upon research, demonstrations, experiments, and such other information as may be appropriate. In addition to the attainment of the highest degree of health and safety protection for the employee, other considerations shall be the latest available scientific data in the field, the feasibility of the standards, and experience gained under this and other health and safety laws. Whenever practicable, the standard promulgated shall be expressed in terms of objective criteria and of the performance desired.

(6) (A) Any employer may apply to the Secretary for a temporary order granting a variance from a standard or any provision thereof promulgated under this section. Such temporary order shall be granted only if the employer files an application which meets the requirements of clause (B) and establishes that (i) he is unable to comply with a standard by its effective date because of unavailability of professional or technical personnel or of materials and equipment needed to come into compliance with the standard or because necessary construction or alteration of facilities cannot be completed by the effective date, (ii) he is taking all available steps to safeguard his employees against the hazards covered by the standard, and (iii) he has an effective program for coming into compliance with the standard as quickly as

Publication in Federal Register.

Hearing, notice.

Publication in Federal Register.

Toxic materials.

Temporary variance order.

practicable. Any temporary order issued under this paragraph shall prescribe the practices, means, methods, operations, and processes which the employer must adopt and use while the order is in effect and state in detail his program for coming into compliance with the standard. Such a temporary order may be granted only after notice to employees and an opportunity for a hearing: *Provided*, That the Secretary may issue one interim order to be effective until a decision is made on the basis of the hearing. No temporary order may be in effect for longer than the period needed by the employer to achieve compliance with the standard or one year, whichever is shorter, except that such an order may be renewed not more than twice (I) so long as the requirements of this paragraph are met and (II) if an application for renewal is filed at least 90 days prior to the expiration date of the order. No interim renewal of an order may remain in effect for longer than 180 days.

Notice, hearing.

Renewal.

Time limitation.

(B) An application for a temporary order under this paragraph (6) shall contain:

(i) a specification of the standard or portion thereof from which the employer seeks a variance,

(ii) a representation by the employer, supported by representations from qualified persons having firsthand knowledge of the facts represented, that he is unable to comply with the standard or portion thereof and a detailed statement of the reasons therefor,

(iii) a statement of the steps he has taken and will take (with specific dates) to protect employees against the hazard covered by the standard,

(iv) a statement of when he expects to be able to comply with the standard and what steps he has taken and what steps he will take (with dates specified) to come into compliance with the standard, and

(v) a certification that he has informed his employees of the application by giving a copy thereof to their authorized representative, posting a statement giving a summary of the application and specifying where a copy may be examined at the place or places where notices to employees are normally posted, and by other appropriate means.

A description of how employees have been informed shall be contained in the certification. The information to employees shall also inform them of their right to petition the Secretary for a hearing.

(C) The Secretary is authorized to grant a variance from any standard or portion thereof whenever he determines, or the Secretary of Health, Education, and Welfare certifies, that such variance is necessary to permit an employer to participate in an experiment approved by him or the Secretary of Health, Education, and Welfare designed to demonstrate or validate new and improved techniques to safeguard the health or safety of workers.

Labels, etc.

(7) Any standard promulgated under this subsection shall prescribe the use of labels or other appropriate forms of warning as are necessary to insure that employees are apprised of all hazards to which they are exposed, relevant symptoms and appropriate emergency treatment, and proper conditions and precautions of safe use or exposure.

Protective equipment, etc.

Where appropriate, such standard shall also prescribe suitable protective equipment and control or technological procedures to be used in connection with such hazards and shall provide for monitoring or measuring employee exposure at such locations and intervals, and in such manner as may be necessary for the protection of employees. In addition, where appropriate, any such standard shall prescribe the type and frequency of medical examinations or other tests which shall be made available, by the employer or at his cost, to employees exposed to such hazards in order to most effectively determine whether the health of such employees is adversely affected by such exposure. In the event such medical examinations are in the nature of research, as determined by the Secretary of Health, Education, and Welfare, such examinations may be furnished at the expense of the Secretary of Health, Education, and Welfare. The results of such examinations or tests shall be furnished only to the Secretary or the Secretary of Health, Education, and Welfare, and, at the request of the employee, to his physician. The Secretary, in consultation with the Secretary of Health, Education, and Welfare, may by rule promulgated pursuant to section 553 of title 5, United States Code, make appropriate modifications in the foregoing requirements relating to the use of labels or other forms of warning, monitoring or measuring, and medical examinations, as may be warranted by experience, information, or medical or technological developments acquired subsequent to the promulgation of the relevant standard.

(8) Whenever a rule promulgated by the Secretary differs substantially from an existing national consensus standard, the Secretary shall, at the same time, publish in the Federal Register a statement of the reasons why the rule as adopted will better effectuate the purposes of this Act than the national consensus standard.

(c)(1) The Secretary shall provide, without regard to the requirements of chapter 5, title 5, United States Code, for an emergency temporary standard to take immediate effect upon publication in the Federal Register if he determines (A) that employees are exposed to grave danger from exposure to substances or agents determined to be toxic or physically harmful or from new hazards, and (B) that such emergency standard is necessary to protect employees from such danger.

(2) Such standard shall be effective until superseded by a standard promulgated in accordance with the procedures prescribed in paragraph (3) of this subsection.

(3) Upon publication of such standard in the Federal Register the Secretary shall commence a proceeding in accordance with section 6(b) of this Act, and the standard as published shall also serve as a proposed rule for the proceeding. The Secretary shall promulgate a standard under this paragraph no later than six months after publication of the emergency standard as provided in paragraph (2) of this subsection.

(d) Any affected employer may apply to the Secretary for a rule or order for a variance from a standard promulgated under this section. Affected employees shall be given notice of each such application and an opportunity to participate in a hearing. The Secretary shall issue such rule or order if he determines on the record, after opportunity for an inspection where appropriate and a hearing, that the proponent of the variance has demonstrated by a preponderance of the evidence that the conditions, practices, means, methods, operations, or processes used or proposed to be used by an employer will provide employment and places of employment to his employees which are as safe and healthful as those which would prevail if he complied with the standard. The rule or order so issued shall prescribe the conditions the employer must maintain, and the practices, means, methods, operations, and processes which he must adopt and utilize to the extent they

Medical examinations.

80 Stat. 383.

Publication in Federal Register.

Temporary standard. Publication in Federal Register. 80 Stat. 381; 81 Stat. 195. 5 USC 500.

Time limitation.

Variance rule.

differ from the standard in question. Such a rule or order may be modified or revoked upon application by an employer, employees, or by the Secretary on his own motion, in the manner prescribed for its issuance under this subsection at any time after six months from its issuance.

Publication in Federal Register.

(e) Whenever the Secretary promulgates any standard, makes any rule, order, or decision, grants any exemption or extension of time, or compromises, mitigates, or settles any penalty assessed under this Act, he shall include a statement of the reasons for such action, which shall be published in the Federal Register.

Petition for judicial review.

(f) Any person who may be adversely affected by a standard issued under this section may at any time prior to the sixtieth day after such standard is promulgated file a petition challenging the validity of such standard with the United States court of appeals for the circuit wherein such person resides or has his principal place of business, for a judicial review of such standard. A copy of the petition shall be forthwith transmitted by the clerk of the court to the Secretary. The filing of such petition shall not, unless otherwise ordered by the court, operate as a stay of the standard. The determinations of the Secretary shall be conclusive if supported by substantial evidence in the record considered as a whole.

(g) In determining the priority for establishing standards under this section, the Secretary shall give due regard to the urgency of the need for mandatory safety and health standards for particular industries, trades, crafts, occupations, businesses, workplaces or work environments. The Secretary shall also give due regard to the recommendations of the Secretary of Health, Education, and Welfare regarding the need for mandatory standards in determining the priority for establishing such standards.

ADVISORY COMMITTEES; ADMINISTRATION

Establishment; membership.

SEC. 7. (a) (1) There is hereby established a National Advisory Committee on Occupational Safety and Health consisting of twelve members appointed by the Secretary, four of whom are to be designated by the Secretary of Health, Education, and Welfare, without

80 Stat. 378.
5 USC 101.

regard to the provisions of title 5, United States Code, governing appointments in the competitive service, and composed of representatives of management, labor, occupational safety and occupational health professions, and of the public. The Secretary shall designate one of the public members as Chairman. The members shall be selected upon the basis of their experience and competence in the field of occupational safety and health.

(2) The Committee shall advise, consult with, and make recommendations to the Secretary and the Secretary of Health, Education, and Welfare on matters relating to the administration of the Act. The Committee shall hold no fewer than two meetings during each calendar year. All meetings of the Committee shall be open to the public

Public transcript.

and a transcript shall be kept and made available for public inspection.

(3) The members of the Committee shall be compensated in accordance with the provisions of section 3109 of title 5, United States

80 Stat. 416.

Code.

(4) The Secretary shall furnish to the Committee an executive secretary and such secretarial, clerical, and other services as are deemed necessary to the conduct of its business.

(b) An advisory committee may be appointed by the Secretary to assist him in his standard-setting functions under section 6 of this Act. Each such committee shall consist of not more than fifteen members and shall include as a member one or more designees of the Secretary of Health, Education, and Welfare, and shall include among its members an equal number of persons qualified by experience and affiliation to present the viewpoint of the employers involved, and of persons similarly qualified to present the viewpoint of the workers involved, as well as one or more representatives of health and safety agencies of the States. An advisory committee may also include such other persons as the Secretary may appoint who are qualified by knowledge and experience to make a useful contribution to the work of such committee, including one or more representatives of professional organizations of technicians or professionals specializing in occupational safety or health, and one or more representatives of nationally recognized standards-producing organizations, but the number of persons so appointed to any such advisory committee shall not exceed the number appointed to such committee as representatives of Federal and State agencies. Persons appointed to advisory committees from private life shall be compensated in the same manner as consultants or experts under section 3109 of title 5, United States Code. The Secretary shall pay to any State which is the employer of a member of such a committee who is a representative of the health or safety agency of that State, reimbursement sufficient to cover the actual cost to the State resulting from such representative's membership on such committee. Any meeting of such committee shall be open to the public and an accurate record shall be kept and made available to the public. No member of such committee (other than representatives of employers and employees) shall have an economic interest in any proposed rule.

80 Stat. 416.

Recordkeeping.

(c) In carrying out his responsibilities under this Act, the Secretary is authorized to—

(1) use, with the consent of any Federal agency, the services, facilities, and personnel of such agency, with or without reimbursement, and with the consent of any State or political subdivision thereof, accept and use the services, facilities, and personnel of any agency of such State or subdivision with reimbursement; and

(2) employ experts and consultants or organizations thereof as authorized by section 3109 of title 5, United States Code, except that contracts for such employment may be renewed annually; compensate individuals so employed at rates not in excess of the rate specified at the time of service for grade GS-18 under section 5332 of title 5, United States Code, including traveltime, and allow them while away from their homes or regular places of business, travel expenses (including per diem in lieu of subsistence) as authorized by section 5703 of title 5, United States Code, for persons in the Government service employed intermittently, while so employed.

Ante, p. 198-1.

80 Stat. 499;
83 Stat. 190.

INSPECTIONS, INVESTIGATIONS, AND RECORDKEEPING

SEC. 8. (a) In order to carry out the purposes of this Act, the Secretary, upon presenting appropriate credentials to the owner, operator, or agent in charge, is authorized—

(1) to enter without delay and at reasonable times any factory, plant, establishment, construction site, or other area, workplace or environment where work is performed by an employee of an employer; and

(2) to inspect and investigate during regular working hours and at other reasonable times, and within reasonable limits and in a reasonable manner, any such place of employment and all pertinent conditions, structures, machines, apparatus, devices, equipment, and materials therein, and to question privately any such employer, owner, operator, agent or employee.

(b) In making his inspections and investigations under this Act the Secretary may require the attendance and testimony of witnesses and the production of evidence under oath. Witnesses shall be paid the same fees and mileage that are paid witnesses in the courts of the United States. In case of a contumacy, failure, or refusal of any person to obey such an order, any district court of the United States or the United States courts of any territory or possession, within the jurisdiction of which such person is found, or resides or transacts business, upon the application by the Secretary, shall have jurisdiction to issue to such person an order requiring such person to appear to produce evidence if, as, and when so ordered, and to give testimony relating to the matter under investigation or in question, and any failure to obey such order of the court may be punished by said court as a contempt thereof.

(c) (1) Each employer shall make, keep and preserve, and make available to the Secretary or the Secretary of Health, Education, and Welfare, such records regarding his activities relating to this Act as the Secretary, in cooperation with the Secretary of Health, Education, and Welfare, may prescribe by regulation as necessary or appropriate for the enforcement of this Act or for developing information regarding the causes and prevention of occupational accidents and illnesses. In order to carry out the provisions of this paragraph such regulations may include provisions requiring employers to conduct periodic inspections. The Secretary shall also issue regulations requiring that employers, through posting of notices or other appropriate means, keep their employees informed of their protections and obligations under this Act, including the provisions of applicable standards.

(2) The Secretary, in cooperation with the Secretary of Health, Education, and Welfare, shall prescribe regulations requiring employers to maintain accurate records of, and to make periodic reports on, work-related deaths, injuries and illnesses other than minor injuries requiring only first aid treatment and which do not involve medical treatment, loss of consciousness, restriction of work or motion, or transfer to another job.

(3) The Secretary, in cooperation with the Secretary of Health, Education, and Welfare, shall issue regulations requiring employers to maintain accurate records of employee exposures to potentially toxic materials or harmful physical agents which are required to be monitored or measured under section 6. Such regulations shall provide employees or their representatives with an opportunity to observe such monitoring or measuring, and to have access to the records thereof. Such regulations shall also make appropriate provision for each employee or former employee to have access to such records as will indicate his own exposure to toxic materials or harmful physical agents. Each employer shall promptly notify any employee who has been or is being exposed to toxic materials or harmful physical agents in concentrations or at levels which exceed those prescribed by an applicable occupational safety and health standard promulgated under section 6, and shall inform any employee who is being thus exposed of the corrective action being taken.

84 STAT. 1600

(d) Any information obtained by the Secretary, the Secretary of Health, Education, and Welfare, or a State agency under this Act shall be obtained with a minimum burden upon employers, especially those operating small businesses. Unnecessary duplication of efforts in obtaining information shall be reduced to the maximum extent feasible.

(e) Subject to regulations issued by the Secretary, a representative of the employer and a representative authorized by his employees shall be given an opportunity to accompany the Secretary or his authorized representative during the physical inspection of any workplace under subsection (a) for the purpose of aiding such inspection. Where there is no authorized employee representative, the Secretary or his authorized representative shall consult with a reasonable number of employees concerning matters of health and safety in the workplace.

(f)(1) Any employees or representative of employees who believe that a violation of a safety or health standard exists that threatens physical harm, or that an imminent danger exists, may request an inspection by giving notice to the Secretary or his authorized representative of such violation or danger. Any such notice shall be reduced to writing, shall set forth with reasonable particularity the grounds for the notice, and shall be signed by the employees or representative of employees, and a copy shall be provided the employer or his agent no later than at the time of inspection, except that, upon the request of the person giving such notice, his name and the names of individual employees referred to therein shall not appear in such copy or on any record published, released, or made available pursuant to subsection (g) of this section. If upon receipt of such notification the Secretary determines there are reasonable grounds to believe that such violation or danger exists, he shall make a special inspection in accordance with the provisions of this section as soon as practicable, to determine if such violation or danger exists. If the Secretary determines there are no reasonable grounds to believe that a violation or danger exists he shall notify the employees or representative of the employees in writing of such determination.

(2) Prior to or during any inspection of a workplace, any employees or representative of employees employed in such workplace may notify the Secretary or any representative of the Secretary responsible for conducting the inspection, in writing, of any violation of this Act which they have reason to believe exists in such workplace. The Secretary shall, by regulation, establish procedures for informal review of any refusal by a representative of the Secretary to issue a citation with respect to any such alleged violation and shall furnish the employees or representative of employees requesting such review a written statement of the reasons for the Secretary's final disposition of the case.

(g)(1) The Secretary and Secretary of Health, Education, and Welfare are authorized to compile, analyze, and publish, either in summary or detailed form, all reports or information obtained under this section.

Reports, publication.

(2) The Secretary and the Secretary of Health, Education, and Welfare shall each prescribe such rules and regulations as he may deem necessary to carry out their responsibilities under this Act, including rules and regulations dealing with the inspection of an employer's establishment.

Rules and regulations.

84 STAT. 1601

CITATIONS

SEC. 9. (a) If, upon inspection or investigation, the Secretary or his authorized representative believes that an employer has violated a requirement of section 5 of this Act, of any standard, rule or order promulgated pursuant to section 6 of this Act, or of any regulations prescribed pursuant to this Act, he shall with reasonable promptness issue a citation to the employer. Each citation shall be in writing and shall describe with particularity the nature of the violation, including a reference to the provision of the Act, standard, rule, regulation, or order alleged to have been violated. In addition, the citation shall fix a reasonable time for the abatement of the violation. The Secretary may prescribe procedures for the issuance of a notice in lieu of a citation with respect to de minimis violations which have no direct or immediate relationship to safety or health.

(b) Each citation issued under this section, or a copy or copies thereof, shall be prominently posted, as prescribed in regulations issued by the Secretary, at or near each place a violation referred to in the citation occurred.

Limitation.
(c) No citation may be issued under this section after the expiration of six months following the occurrence of any violation.

PROCEDURE FOR ENFORCEMENT

SEC. 10. (a) If, after an inspection or investigation, the Secretary issues a citation under section 9(a), he shall, within a reasonable time after the termination of such inspection or investigation, notify the employer by certified mail of the penalty, if any, proposed to be assessed under section 17 and that the employer has fifteen working days within which to notify the Secretary that he wishes to contest the citation or proposed assessment of penalty. If, within fifteen working days from the receipt of the notice issued by the Secretary the employer fails to notify the Secretary that he intends to contest the citation or proposed assessment of penalty, and no notice is filed by any employee or representative of employees under subsection (c) within such time, the citation and the assessment, as proposed, shall be deemed a final order of the Commission and not subject to review by any court or agency.

(b) If the Secretary has reason to believe that an employer has failed to correct a violation for which a citation has been issued within the period permitted for its correction (which period shall not begin to run until the entry of a final order by the Commission in the case of any review proceedings under this section initiated by the employer in good faith and not solely for delay or avoidance of penalties), the Secretary shall notify the employer by certified mail of such failure and of the penalty proposed to be assessed under section 17 by reason of such failure, and that the employer has fifteen working days within which to notify the Secretary that he wishes to contest the Secretary's notification or the proposed assessment of penalty. If, within fifteen working days from the receipt of notification issued by the Secretary, the employer fails to notify the Secretary that he intends to contest the notification or proposed assessment of penalty, the notification and assessment, as proposed, shall be deemed a final order of the Commission and not subject to review by any court or agency.

(c) If an employer notifies the Secretary that he intends to contest a citation issued under section 9(a) or notification issued under subsection (a) or (b) of this section, or if, within fifteen working days

84 STAT. 1602

of the issuance of a citation under section 9(a), any employee or representative of employees files a notice with the Secretary alleging that the period of time fixed in the citation for the abatement of the violation is unreasonable, the Secretary shall immediately advise the Commission of such notification, and the Commission shall afford an opportunity for a hearing (in accordance with section 554 of title 5, United States Code, but without regard to subsection (a)(3) of such section). The Commission shall thereafter issue an order, based on findings of fact, affirming, modifying, or vacating the Secretary's citation or proposed penalty, or directing other appropriate relief, and such order shall become final thirty days after its issuance. Upon a showing by an employer of a good faith effort to comply with the abatement requirements of a citation, and that abatement has not been completed because of factors beyond his reasonable control, the Secretary, after an opportunity for a hearing as provided in this subsection, shall issue an order affirming or modifying the abatement requirements in such citation. The rules of procedure prescribed by the Commission shall provide affected employees or representatives of affected employees an opportunity to participate as parties to hearings under this subsection.

80 Stat. 384.

JUDICIAL REVIEW

SEC. 11. (a) Any person adversely affected or aggrieved by an order of the Commission issued under subsection (c) of section 10 may obtain a review of such order in any United States court of appeals for the circuit in which the violation is alleged to have occurred or where the employer has its principal office, or in the Court of Appeals for the District of Columbia Circuit, by filing in such court within sixty days following the issuance of such order a written petition praying that the order be modified or set aside. A copy of such petition shall be forthwith transmitted by the clerk of the court to the Commission and to the other parties, and thereupon the Commission shall file in the court the record in the proceeding as provided in section 2112 of title 28, United States Code. Upon such filing, the court shall have jurisdiction of the proceeding and of the question determined therein, and shall have power to grant such temporary relief or restraining order as it deems just and proper, and to make and enter upon the pleadings, testimony, and proceedings set forth in such record a decree affirming, modifying, or setting aside in whole or in part, the order of the Commission and enforcing the same to the extent that such order is affirmed or modified. The commencement of proceedings under this subsection shall not, unless ordered by the court, operate as a stay of the order of the Commission. No objection that has not been urged before the Commission shall be considered by the court, unless the failure or neglect to urge such objection shall be excused because of extraordinary circumstances. The findings of the Commission with respect to questions of fact, if supported by substantial evidence on the record considered as a whole, shall be conclusive. If any party shall apply to the court for leave to adduce additional evidence and shall show to the satisfaction of the court that such additional evidence is material and that there were reasonable grounds for the failure to adduce such evidence in the hearing before the Commission, the court may order such additional evidence to be taken before the Commission and to be made a part of the record. The Commission may modify its findings as to the facts, or make new findings, by reason of additional evidence so taken and filed, and it shall file such modified or new findings, which findings with respect to questions of fact, if supported by substantial evi-

72 Stat. 941;
80 Stat. 1323.

dence on the record considered as a whole, shall be conclusive, and its recommendations, if any, for the modification or setting aside of its original order. Upon the filing of the record with it, the jurisdiction of the court shall be exclusive and its judgment and decree shall be final, except that the same shall be subject to review by the Supreme Court of the United States, as provided in section 1254 of title 28,

62 Stat. 928.

United States Code. Petitions filed under this subsection shall be heard expeditiously.

(b) The Secretary may also obtain review or enforcement of any final order of the Commission by filing a petition for such relief in the United States court of appeals for the circuit in which the alleged violation occurred or in which the employer has its principal office, and the provisions of subsection (a) shall govern such proceedings to the extent applicable. If no petition for review, as provided in subsection (a), is filed within sixty days after service of the Commission's order, the Commission's findings of fact and order shall be conclusive in connection with any petition for enforcement which is filed by the Secretary after the expiration of such sixty-day period. In any such case, as well as in the case of a noncontested citation or notification by the Secretary which has become a final order of the Commission under subsection (a) or (b) of section 10, the clerk of the court, unless otherwise ordered by the court, shall forthwith enter a decree enforcing the order and shall transmit a copy of such decree to the Secretary and the employer named in the petition. In any contempt proceeding brought to enforce a decree of a court of appeals entered pursuant to this subsection or subsection (a), the court of appeals may assess the penalties provided in section 17, in addition to invoking any other available remedies.

(c) (1) No person shall discharge or in any manner discriminate against any employee because such employee has filed any complaint or instituted or caused to be instituted any proceeding under or related to this Act or has testified or is about to testify in any such proceeding or because of the exercise by such employee on behalf of himself or others of any right afforded by this Act.

(2) Any employee who believes that he has been discharged or otherwise discriminated against by any person in violation of this subsection may, within thirty days after such violation occurs, file a complaint with the Secretary alleging such discrimination. Upon receipt of such complaint, the Secretary shall cause such investigation to be made as he deems appropriate. If upon such investigation, the Secretary determines that the provisions of this subsection have been violated, he shall bring an action in any appropriate United States district court against such person. In any such action the United States district courts shall have jurisdiction, for cause shown to restrain violations of paragraph (1) of this subsection and order all appropriate relief including rehiring or reinstatement of the employee to his former position with back pay.

(3) Within 90 days of the receipt of a complaint filed under this subsection the Secretary shall notify the complainant of his determination under paragraph 2 of this subsection.

THE OCCUPATIONAL SAFETY AND HEALTH REVIEW COMMISSION

Establishment; membership.

SEC. 12. (a) The Occupational Safety and Health Review Commission is hereby established. The Commission shall be composed of three members who shall be appointed by the President, by and with the advice and consent of the Senate, from among persons who by reason

of training, education, or experience are qualified to carry out the
functions of the Commission under this Act. The President shall designate one of the members of the Commission to serve as Chairman.

(b) The terms of members of the Commission shall be six years **Terms.**
except that (1) the members of the Commission first taking office shall
serve, as designated by the President at the time of appointment, one
for a term of two years, one for a term of four years, and one for a
term of six years, and (2) a vacancy caused by the death, resignation,
or removal of a member prior to the expiration of the term for which
he was appointed shall be filled only for the remainder of such
unexpired term. A member of the Commission may be removed by the
President for inefficiency, neglect of duty, or malfeasance in office.

(c) (1) Section 5314 of title 5, United States Code, is amended by **80 Stat. 460.**
adding at the end thereof the following new paragraph:

"(57) Chairman, Occupational Safety and Health Review
Commission."

(2) Section 5315 of title 5, United States Code, is amended by add- **Ante, p. 776.**
ing at the end thereof the following new paragraph:

"(94) Members, Occupational Safety and Health Review
Commission."

(d) The principal office of the Commission shall be in the District **Location.**
of Columbia. Whenever the Commission deems that the convenience
of the public or of the parties may be promoted, or delay or expense
may be minimized, it may hold hearings or conduct other proceedings
at any other place.

(e) The Chairman shall be responsible on behalf of the Commission
for the administrative operations of the Commission and shall appoint
such hearing examiners and other employees as he deems necessary
to assist in the performance of the Commission's functions and to
fix their compensation in accordance with the provisions of chapter
51 and subchapter III of chapter 53 of title 5, United States Code, **5 USC 5101,**
relating to classification and General Schedule pay rates: *Provided,* **5331.**
That assignment, removal and compensation of hearing examiners **Ante, p. 198-1.**
shall be in accordance with sections 3105, 3344, 5362, and 7521 of title 5,
United States Code.

(f) For the purpose of carrying out its functions under this Act, two **Quorum.**
members of the Commission shall constitute a quorum and official
action can be taken only on the affirmative vote of at least two
members.

(g) Every official act of the Commission shall be entered of record, **Public records.**
and its hearings and records shall be open to the public. The Commission is authorized to make such rules as are necessary for the orderly
transaction of its proceedings. Unless the Commission has adopted a
different rule, its proceedings shall be in accordance with the Federal
Rules of Civil Procedure. **28 USC app.**

(h) The Commission may order testimony to be taken by deposition
in any proceedings pending before it at any state of such proceeding.
Any person may be compelled to appear and depose, and to produce
books, papers, or documents, in the same manner as witnesses may be
compelled to appear and testify and produce like documentary
evidence before the Commission. Witnesses whose depositions are taken
under this subsection, and the persons taking such depositions, shall be
entitled to the same fees as are paid for like services in the courts of
the United States.

(i) For the purpose of any proceeding before the Commission, the
provisions of section 11 of the National Labor Relations Act (29
U.S.C. 161) are hereby made applicable to the jurisdiction and powers **61 Stat. 150;**
of the Commission. **Ante, p. 930.**

84 STAT. 1605

Report.

(j) A hearing examiner appointed by the Commission shall hear, and make a determination upon, any proceeding instituted before the Commission and any motion in connection therewith, assigned to such hearing examiner by the Chairman of the Commission, and shall make a report of any such determination which constitutes his final disposition of the proceedings. The report of the hearing examiner shall become the final order of the Commission within thirty days after such report by the hearing examiner, unless within such period any Commission member has directed that such report shall be reviewed by the Commission.

(k) Except as otherwise provided in this Act, the hearing examiners shall be subject to the laws governing employees in the classified civil service, except that appointments shall be made without regard to section 5108 of title 5, United States Code. Each hearing examiner shall receive compensation at a rate not less than that prescribed for GS-16 under section 5332 of title 5, United States Code.

80 Stat. 453.

Ante, p. 198-1.

PROCEDURES TO COUNTERACT IMMINENT DANGERS

SEC. 13. (a) The United States district courts shall have jurisdiction, upon petition of the Secretary, to restrain any conditions or practices in any place of employment which are such that a danger exists which could reasonably be expected to cause death or serious physical harm immediately or before the imminence of such danger can be eliminated through the enforcement procedures otherwise provided by this Act. Any order issued under this section may require such steps to be taken as may be necessary to avoid, correct, or remove such imminent danger and prohibit the employment or presence of any individual in locations or under conditions where such imminent danger exists, except individuals whose presence is necessary to avoid, correct, or remove such imminent danger or to maintain the capacity of a continuous process operation to resume normal operations without a complete cessation of operations, or where a cessation of operations is necessary, to permit such to be accomplished in a safe and orderly manner.

(b) Upon the filing of any such petition the district court shall have jurisdiction to grant such injunctive relief or temporary restraining order pending the outcome of an enforcement proceeding pursuant to this Act. The proceeding shall be as provided by Rule 65 of the Federal Rules, Civil Procedure, except that no temporary restraining order issued without notice shall be effective for a period longer than five days.

28 USC app.

(c) Whenever and as soon as an inspector concludes that conditions or practices described in subsection (a) exist in any place of employment, he shall inform the affected employees and employers of the danger and that he is recommending to the Secretary that relief be sought.

(d) If the Secretary arbitrarily or capriciously fails to seek relief under this section, any employee who may be injured by reason of such failure, or the representative of such employees, might bring an action against the Secretary in the United States district court for the district in which the imminent danger is alleged to exist or the employer has its principal office, or for the District of Columbia, for a writ of mandamus to compel the Secretary to seek such an order and for such further relief as may be appropriate.

65

REPRESENTATION IN CIVIL LITIGATION

SEC. 14. Except as provided in section 518(a) of title 28, United States Code, relating to litigation before the Supreme Court, the 80 Stat. 613. Solicitor of Labor may appear for and represent the Secretary in any civil litigation brought under this Act but all such litigation shall be subject to the direction and control of the Attorney General.

CONFIDENTIALITY OF TRADE SECRETS

SEC. 15. All information reported to or otherwise obtained by the Secretary or his representative in connection with any inspection or proceeding under this Act which contains or which might reveal a trade secret referred to in section 1905 of title 18 of the United States Code shall be considered confidential for the purpose of that section, 62 Stat. 791. except that such information may be disclosed to other officers or employees concerned with carrying out this Act or when relevant in any proceeding under this Act. In any such proceeding the Secretary, the Commission, or the court shall issue such orders as may be appropriate to protect the confidentiality of trade secrets.

VARIATIONS, TOLERANCES, AND EXEMPTIONS

SEC. 16. The Secretary, on the record, after notice and opportunity for a hearing may provide such reasonable limitations and may make such rules and regulations allowing reasonable variations, tolerances, and exemptions to and from any or all provisions of this Act as he may find necessary and proper to avoid serious impairment of the national defense. Such action shall not be in effect for more than six months without notification to affected employees and an opportunity being afforded for a hearing.

PENALTIES

SEC. 17. (a) Any employer who willfully or repeatedly violates the requirements of section 5 of this Act, any standard, rule, or order promulgated pursuant to section 6 of this Act, or regulations prescribed pursuant to this Act, may be assessed a civil penalty of not more than $10,000 for each violation.

(b) Any employer who has received a citation for a serious violation of the requirements of section 5 of this Act, of any standard, rule, or order promulgated pursuant to section 6 of this Act, or of any regulations prescribed pursuant to this Act, shall be assessed a civil penalty of up to $1,000 for each such violation.

(c) Any employer who has received a citation for a violation of the requirements of section 5 of this Act, of any standard, rule, or order promulgated pursuant to section 6 of this Act, or of regulations prescribed pursuant to this Act, and such violation is specifically determined not to be of a serious nature, may be assessed a civil penalty of up to $1,000 for each such violation.

(d) Any employer who fails to correct a violation for which a citation has been issued under section 9(a) within the period permitted for its correction (which period shall not begin to run until the date of the final order of the Commission in the case of any review proceeding under section 10 initiated by the employer in good faith and not solely for delay or avoidance of penalties), may be assessed a civil penalty of not more than $1,000 for each day during which such failure or violation continues.

66

(e) Any employer who willfully violates any standard, rule, or order promulgated pursuant to section 6 of this Act, or of any regulations prescribed pursuant to this Act, and that violation caused death to any employee, shall, upon conviction, be punished by a fine of not more than $10,000 or by imprisonment for not more than six months, or by both; except that if the conviction is for a violation committed after a first conviction of such person, punishment shall be by a fine of not more than $20,000 or by imprisonment for not more than one year, or by both.

(f) Any person who gives advance notice of any inspection to be conducted under this Act, without authority from the Secretary or his designees, shall, upon conviction, be punished by a fine of not more than $1,000 or by imprisonment for not more than six months, or by both.

(g) Whoever knowingly makes any false statement, representation, or certification in any application, record, report, plan, or other document filed or required to be maintained pursuant to this Act shall, upon conviction, be punished by a fine of not more than $10,000, or by imprisonment for not more than six months, or by both.

65 Stat. 721;
79 Stat. 234.

(h)(1) Section 1114 of title 18, United States Code, is hereby amended by striking out "designated by the Secretary of Health, Education, and Welfare to conduct investigations, or inspections under the Federal Food, Drug, and Cosmetic Act" and inserting in lieu thereof "or of the Department of Labor assigned to perform investigative, inspection, or law enforcement functions".

62 Stat. 756.

(2) Notwithstanding the provisions of sections 1111 and 1114 of title 18, United States Code, whoever, in violation of the provisions of section 1114 of such title, kills a person while engaged in or on account of the performance of investigative, inspection, or law enforcement functions added to such section 1114 by paragraph (1) of this subsection, and who would otherwise be subject to the penalty provisions of such section 1111, shall be punished by imprisonment for any term of years or for life.

(i) Any employer who violates any of the posting requirements, as prescribed under the provisions of this Act, shall be assessed a civil penalty of up to $1,000 for each violation.

(j) The Commission shall have authority to assess all civil penalties provided in this section, giving due consideration to the appropriateness of the penalty with respect to the size of the business of the employer being charged, the gravity of the violation, the good faith of the employer, and the history of previous violations.

(k) For purposes of this section, a serious violation shall be deemed to exist in a place of employment if there is a substantial probability that death or serious physical harm could result from a condition which exists, or from one or more practices, means, methods, operations, or processes which have been adopted or are in use, in such place of employment unless the employer did not, and could not with the exercise of reasonable diligence, know of the presence of the violation.

(l) Civil penalties owed under this Act shall be paid to the Secretary for deposit into the Treasury of the United States and shall accrue to the United States and may be recovered in a civil action in the name of the United States brought in the United States district court for the district where the violation is alleged to have occurred or where the employer has its principal office.

84 STAT. 1608

STATE JURISDICTION AND STATE PLANS

SEC. 18. (a) Nothing in this Act shall prevent any State agency or court from asserting jurisdiction under State law over any occupational safety or health issue with respect to which no standard is in effect under section 6.

(b) Any State which, at any time, desires to assume responsibility for development and enforcement therein of occupational safety and health standards relating to any occupational safety or health issue with respect to which a Federal standard has been promulgated under section 6 shall submit a State plan for the development of such standards and their enforcement.

(c) The Secretary shall approve the plan submitted by a State under subsection (b), or any modification thereof, if such plan in his judgment—

(1) designates a State agency or agencies as the agency or agencies responsible for administering the plan throughout the State,

(2) provides for the development and enforcement of safety and health standards relating to one or more safety or health issues, which standards (and the enforcement of which standards) are or will be at least as effective in providing safe and healthful employment and places of employment as the standards promulgated under section 6 which relate to the same issues, and which standards, when applicable to products which are distributed or used in interstate commerce, are required by compelling local conditions and do not unduly burden interstate commerce,

(3) provides for a right of entry and inspection of all work-places subject to the Act which is at least as effective as that provided in section 8, and includes a prohibition on advance notice of inspections,

(4) contains satisfactory assurances that such agency or agencies have or will have the legal authority and qualified personnel necessary for the enforcement of such standards,

(5) gives satisfactory assurances that such State will devote adequate funds to the administration and enforcement of such standards,

(6) contains satisfactory assurances that such State will, to the extent permitted by its law, establish and maintain an effective and comprehensive occupational safety and health program applicable to all employees of public agencies of the State and its political subdivisions, which program is as effective as the standards contained in an approved plan,

(7) requires employers in the State to make reports to the Secretary in the same manner and to the same extent as if the plan were not in effect, and

(8) provides that the State agency will make such reports to the Secretary in such form and containing such information, as the Secretary shall from time to time require.

(d) If the Secretary rejects a plan submitted under subsection (b), he shall afford the State submitting the plan due notice and opportunity for a hearing before so doing.   *Notice of hearing.*

(e) After the Secretary approves a State plan submitted under subsection (b), he may, but shall not be required to, exercise his authority under sections 8, 9, 10, 13, and 17 with respect to comparable standards promulgated under section 6, for the period specified in the next sentence. The Secretary may exercise the authority referred to above until he determines, on the basis of actual operations under the

State plan, that the criteria set forth in subsection (c) are being applied, but he shall not make such determination for at least three years after the plan's approval under subsection (c). Upon making the determination referred to in the preceding sentence, the provisions of sections 5(a)(2), 8 (except for the purpose of carrying out subsection (f) of this section), 9, 10, 13, and 17, and standards promulgated under section 6 of this Act, shall not apply with respect to any occupational safety or health issues covered under the plan, but the Secretary may retain jurisdiction under the above provisions in any proceeding commenced under section 9 or 10 before the date of determination.

Continuing evaluation.

(f) The Secretary shall, on the basis of reports submitted by the State agency and his own inspections make a continuing evaluation of the manner in which each State having a plan approved under this section is carrying out such plan. Whenever the Secretary finds, after affording due notice and opportunity for a hearing, that in the administration of the State plan there is a failure to comply substantially with any provision of the State plan (or any assurance contained therein), he shall notify the State agency of his withdrawal of approval of such plan and upon receipt of such notice such plan shall cease to be in effect, but the State may retain jurisdiction in any case commenced before the withdrawal of the plan in order to enforce standards under the plan whenever the issues involved do not relate to the reasons for the withdrawal of the plan.

Plan rejection, review.

(g) The State may obtain a review of a decision of the Secretary withdrawing approval of or rejecting its plan by the United States court of appeals for the circuit in which the State is located by filing in such court within thirty days following receipt of notice of such decision a petition to modify or set aside in whole or in part the action of the Secretary. A copy of such petition shall forthwith be served upon the Secretary, and thereupon the Secretary shall certify and file in the court the record upon which the decision complained of was issued as provided in section 2112 of title 28, United States Code.

72 Stat. 941; 80 Stat. 1323.

Unless the court finds that the Secretary's decision in rejecting a proposed State plan or withdrawing his approval of such a plan is not supported by substantial evidence the court shall affirm the Secretary's decision. The judgment of the court shall be subject to review by the Supreme Court of the United States upon certiorari or certification as provided in section 1254 of title 28, United States Code.

62 Stat. 928.

(h) The Secretary may enter into an agreement with a State under which the State will be permitted to continue to enforce one or more occupational health and safety standards in effect in such State until final action is taken by the Secretary with respect to a plan submitted by a State under subsection (b) of this section, or two years from the date of enactment of this Act, whichever is earlier.

FEDERAL AGENCY SAFETY PROGRAMS AND RESPONSIBILITIES

Sec. 19. (a) It shall be the responsibility of the head of each Federal agency to establish and maintain an effective and comprehensive occupational safety and health program which is consistent with the standards promulgated under section 6. The head of each agency shall (after consultation with representatives of the employees thereof)—

(1) provide safe and healthful places and conditions of employment, consistent with the standards set under section 6;

(2) acquire, maintain, and require the use of safety equipment, personal protective equipment, and devices reasonably necessary to protect employees;

69

84 STAT. 1610

(3) keep adequate records of all occupational accidents and ill- Recordkeeping.
nesses for proper evaluation and necessary corrective action;

(4) consult with the Secretary with regard to the adequacy as
to form and content of records kept pursuant to subsection (a)(3)
of this section; and

(5) make an annual report to the Secretary with respect to Annual report.
occupational accidents and injuries and the agency's program
under this section. Such report shall include any report submitted
under section 7902(e)(2) of title 5, United States Code. 80 Stat. 530.

(b) The Secretary shall report to the President a summary or digest Report to
of reports submitted to him under subsection (a)(5) of this section, President.
together with his evaluations of and recommendations derived from
such reports. The President shall transmit annually to the Senate and Report to
the House of Representatives a report of the activities of Federal Congress.
agencies under this section.

(c) Section 7902(c)(1) of title 5, United States Code, is amended
by inserting after "agencies" the following: "and of labor organizations representing employees".

(d) The Secretary shall have access to records and reports kept Records, etc.;
and filed by Federal agencies pursuant to subsections (a) (3) and (5) availability.
of this section unless those records and reports are specifically required
by Executive order to be kept secret in the interest of the national
defense or foreign policy, in which case the Secretary shall have access
to such information as will not jeopardize national defense or foreign
policy.

RESEARCH AND RELATED ACTIVITIES

SEC. 20. (a)(1) The Secretary of Health, Education, and Welfare,
after consultation with the Secretary and with other appropriate
Federal departments or agencies, shall conduct (directly or by grants
or contracts) research, experiments, and demonstrations relating to
occupational safety and health, including studies of psychological
factors involved, and relating to innovative methods, techniques, and
approaches for dealing with occupational safety and health problems.

(2) The Secretary of Health, Education, and Welfare shall from
time to time consult with the Secretary in order to develop specific
plans for such research, demonstrations, and experiments as are necessary to produce criteria, including criteria identifying toxic substances, enabling the Secretary to meet his responsibility for the
formulation of safety and health standards under this Act; and the
Secretary of Health, Education, and Welfare, on the basis of such
research, demonstrations, and experiments and any other information
available to him, shall develop and publish at least annually such
criteria as will effectuate the purposes of this Act.

(3) The Secretary of Health, Education, and Welfare, on the basis
of such research, demonstrations, and experiments, and any other
information available to him, shall develop criteria dealing with toxic
materials and harmful physical agents and substances which will
describe exposure levels that are safe for various periods of employment, including but not limited to the exposure levels at which no
employee will suffer impaired health or functional capacities or
diminished life expectancy as a result of his work experience.

(4) The Secretary of Health, Education, and Welfare shall also
conduct special research, experiments, and demonstrations relating
to occupational safety and health as are necessary to explore new
problems, including those created by new technology in occupational
safety and health, which may require ameliorative action beyond that

4 STAT. 1611

which is otherwise provided for in the operating provisions of this Act. The Secretary of Health, Education, and Welfare shall also conduct research into the motivational and behavioral factors relating to the field of occupational safety and health.

Toxic substances, records.

(5) The Secretary of Health, Education, and Welfare, in order to comply with his responsibilities under paragraph (2), and in order to develop needed information regarding potentially toxic substances or harmful physical agents, may prescribe regulations requiring employers to measure, record, and make reports on the exposure of employees to substances or physical agents which the Secretary of Health, Education, and Welfare reasonably believes may endanger the health or safety of employees. The Secretary of Health, Education, and Welfare also is authorized to establish such programs of medical examinations and tests as may be necessary for determining the incidence of occupational illnesses and the susceptibility of employees to such illnesses. Nothing in this or any other provision of this Act shall be deemed to authorize or require medical examination, immunization, or treatment for those who object thereto on religious grounds, except where such is necessary for the protection of the health or safety of others. Upon the request of any employer who is required to measure and record exposure of employees to substances or physical agents as provided under this subsection, the Secretary of Health, Education, and Welfare shall furnish full financial or other assistance to such employer for the purpose of defraying any additional expense incurred by him in carrying out the measuring and recording as provided in this subsection.

Medical examinations.

Toxic substances, publication.

(6) The Secretary of Health, Education, and Welfare shall publish within six months of enactment of this Act and thereafter as needed but at least annually a list of all known toxic substances by generic family or other useful grouping, and the concentrations at which such toxicity is known to occur. He shall determine following a written request by any employer or authorized representative of employees, specifying with reasonable particularity the grounds on which the request is made, whether any substance normally found in the place of employment has potentially toxic effects in such concentrations as used or found; and shall submit such determination both to employers and affected employees as soon as possible. If the Secretary of Health, Education, and Welfare determines that any substance is potentially toxic at the concentrations in which it is used or found in a place of employment, and such substance is not covered by an occupational safety or health standard promulgated under section 6, the Secretary of Health, Education, and Welfare shall immediately submit such determination to the Secretary, together with all pertinent criteria.

Annual studies.

(7) Within two years of enactment of this Act, and annually thereafter the Secretary of Health, Education, and Welfare shall conduct and publish industrywide studies of the effect of chronic or low-level exposure to industrial materials, processes, and stresses on the potential for illness, disease, or loss of functional capacity in aging adults.

Inspections.

(b) The Secretary of Health, Education, and Welfare is authorized to make inspections and question employers and employees as provided in section 8 of this Act in order to carry out his functions and responsibilities under this section.

Contract authority.

(c) The Secretary is authorized to enter into contracts, agreements, or other arrangements with appropriate public agencies or private organizations for the purpose of conducting studies relating to his responsibilities under this Act. In carrying out his responsibilities under this subsection, the Secretary shall cooperate with the Secretary of Health, Education, and Welfare in order to avoid any duplication of efforts under this section.

(d) Information obtained by the Secretary and the Secretary of Health, Education, and Welfare under this section shall be disseminated by the Secretary to employers and employees and organizations thereof.

(e) The functions of the Secretary of Health, Education, and Welfare under this Act shall, to the extent feasible, be delegated to the Director of the National Institute for Occupational Safety and Health established by section 22 of this Act.

Delegation of functions.

TRAINING AND EMPLOYEE EDUCATION

SEC. 21. (a) The Secretary of Health, Education, and Welfare, after consultation with the Secretary and with other appropriate Federal departments and agencies, shall conduct, directly or by grants or contracts (1) education programs to provide an adequate supply of qualified personnel to carry out the purposes of this Act, and (2) informational programs on the importance of and proper use of adequate safety and health equipment.

(b) The Secretary is also authorized to conduct, directly or by grants or contracts, short-term training of personnel engaged in work related to his responsibilities under this Act.

(c) The Secretary, in consultation with the Secretary of Health, Education, and Welfare, shall (1) provide for the establishment and supervision of programs for the education and training of employers and employees in the recognition, avoidance, and prevention of unsafe or unhealthful working conditions in employments covered by this Act, and (2) consult with and advise employers and employees, and organizations representing employers and employees as to effective means of preventing occupational injuries and illnesses.

NATIONAL INSTITUTE FOR OCCUPATIONAL SAFETY AND HEALTH

SEC. 22. (a) It is the purpose of this section to establish a National Institute for Occupational Safety and Health in the Department of Health, Education, and Welfare in order to carry out the policy set forth in section 2 of this Act and to perform the functions of the Secretary of Health, Education, and Welfare under sections 20 and 21 of this Act.

Establishment.

(b) There is hereby established in the Department of Health, Education, and Welfare a National Institute for Occupational Safety and Health. The Institute shall be headed by a Director who shall be appointed by the Secretary of Health, Education, and Welfare, and who shall serve for a term of six years unless previously removed by the Secretary of Health, Education, and Welfare.

Director, appointment, term.

(c) The Institute is authorized to—

(1) develop and establish recommended occupational safety and health standards; and

(2) perform all functions of the Secretary of Health, Education, and Welfare under sections 20 and 21 of this Act.

(d) Upon his own initiative, or upon the request of the Secretary or the Secretary of Health, Education, and Welfare, the Director is authorized (1) to conduct such research and experimental programs as he determines are necessary for the development of criteria for new and improved occupational safety and health standards, and (2) after consideration of the results of such research and experimental programs make recommendations concerning new or improved occupational safety and health standards. Any occupational safety and health standard recommended pursuant to this section shall immediately be forwarded to the Secretary of Labor, and to the Secretary of Health, Education, and Welfare.

(e) In addition to any authority vested in the Institute by other provisions of this section, the Director, in carrying out the functions of the Institute, is authorized to—

(1) prescribe such regulations as he deems necessary governing the manner in which its functions shall be carried out;

(2) receive money and other property donated, bequeathed, or devised, without condition or restriction other than that it be used for the purposes of the Institute and to use, sell, or otherwise dispose of such property for the purpose of carrying out its functions;

(3) receive (and use, sell, or otherwise dispose of, in accordance with paragraph (2)), money and other property donated, bequeathed, or devised to the Institute with a condition or restriction, including a condition that the Institute use other funds of the Institute for the purposes of the gift;

(4) in accordance with the civil service laws, appoint and fix the compensation of such personnel as may be necessary to carry out the provisions of this section;

(5) obtain the services of experts and consultants in accordance with the provisions of section 3109 of title 5, United States Code;

80 Stat. 416.

(6) accept and utilize the services of voluntary and noncompensated personnel and reimburse them for travel expenses, including per diem, as authorized by section 5703 of title 5, United States Code;

83 Stat. 190.

(7) enter into contracts, grants or other arrangements, or modifications thereof to carry out the provisions of this section, and such contracts or modifications thereof may be entered into without performance or other bonds, and without regard to section 3709 of the Revised Statutes, as amended (41 U.S.C. 5), or any other provision of law relating to competitive bidding;

(8) make advance, progress, and other payments which the Director deems necessary under this title without regard to the provisions of section 3648 of the Revised Statutes, as amended (31 U.S.C. 529); and

(9) make other necessary expenditures.

Annual report to HEW, President, and Congress.

(f) The Director shall submit to the Secretary of Health, Education, and Welfare, to the President, and to the Congress an annual report of the operations of the Institute under this Act, which shall include a detailed statement of all private and public funds received and expended by it, and such recommendations as he deems appropriate.

GRANTS TO THE STATES

SEC. 23. (a) The Secretary is authorized, during the fiscal year ending June 30, 1971, and the two succeeding fiscal years, to make grants to the States which have designated a State agency under section 18 to assist them—

(1) in identifying their needs and responsibilities in the area of occupational safety and health,

(2) in developing State plans under section 18, or

(3) in developing plans for—

(A) establishing systems for the collection of information concerning the nature and frequency of occupational injuries and diseases;

(B) increasing the expertise and enforcement capabilities of their personnel engaged in occupational safety and health programs; or

(C) otherwise improving the administration and enforcement of State occupational safety and health laws, including standards thereunder, consistent with the objectives of this Act.

(b) The Secretary is authorized, during the fiscal year ending June 30, 1971, and the two succeeding fiscal years, to make grants to the States for experimental and demonstration projects consistent with the objectives set forth in subsection (a) of this section.

(c) The Governor of the State shall designate the appropriate State agency for receipt of any grant made by the Secretary under this section.

(d) Any State agency designated by the Governor of the State desiring a grant under this section shall submit an application therefor to the Secretary.

(e) The Secretary shall review the application, and shall, after consultation with the Secretary of Health, Education, and Welfare, approve or reject such application.

(f) The Federal share for each State grant under subsection (a) or (b) of this section may not exceed 90 per centum of the total cost of the application. In the event the Federal share for all States under either such subsection is not the same, the differences among the States shall be established on the basis of objective criteria.

(g) The Secretary is authorized to make grants to the States to assist them in administering and enforcing programs for occupational safety and health contained in State plans approved by the Secretary pursuant to section 18 of this Act. The Federal share for each State grant under this subsection may not exceed 50 per centum of the total cost to the State of such a program. The last sentence of subsection (f) shall be applicable in determining the Federal share under this subsection.

(h) Prior to June 30, 1973, the Secretary shall, after consultation with the Secretary of Health, Education, and Welfare, transmit a report to the President and to the Congress, describing the experience under the grant programs authorized by this section and making any recommendations he may deem appropriate.

Report to President and Congress.

STATISTICS

SEC. 24. (a) In order to further the purposes of this Act, the Secretary, in consultation with the Secretary of Health, Education, and Welfare, shall develop and maintain an effective program of collection, compilation, and analysis of occupational safety and health statistics. Such program may cover all employments whether or not subject to any other provisions of this Act but shall not cover employments excluded by section 4 of the Act. The Secretary shall compile accurate statistics on work injuries and illnesses which shall include all disabling, serious, or significant injuries and illnesses, whether or not involving loss of time from work, other than minor injuries requiring only first aid treatment and which do not involve medical treatment, loss of consciousness, restriction of work or motion, or transfer to another job.

34 STAT. 1615

(b) To carry out his duties under subsection (a) of this section, the Secretary may—

(1) promote, encourage, or directly engage in programs of studies, information and communication concerning occupational safety and health statistics;

(2) make grants to States or political subdivisions thereof in order to assist them in developing and administering programs dealing with occupational safety and health statistics; and

(3) arrange, through grants or contracts, for the conduct of such research and investigations as give promise of furthering the objectives of this section.

(c) The Federal share for each grant under subsection (b) of this section may be up to 50 per centum of the State's total cost.

(d) The Secretary may, with the consent of any State or political subdivision thereof, accept and use the services, facilities, and employees of the agencies of such State or political subdivision, with or without reimbursement, in order to assist him in carrying out his functions under this section.

Reports.

(e) On the basis of the records made and kept pursuant to section 8(c) of this Act, employers shall file such reports with the Secretary as he shall prescribe by regulation, as necessary to carry out his functions under this Act.

(f) Agreements between the Department of Labor and States pertaining to the collection of occupational safety and health statistics already in effect on the effective date of this Act shall remain in effect until superseded by grants or contracts made under this Act.

AUDITS

SEC. 25. (a) Each recipient of a grant under this Act shall keep such records as the Secretary or the Secretary of Health, Education, and Welfare shall prescribe, including records which fully disclose the amount and disposition by such recipient of the proceeds of such grant, the total cost of the project or undertaking in connection with which such grant is made or used, and the amount of that portion of the cost of the project or undertaking supplied by other sources, and such other records as will facilitate an effective audit.

(b) The Secretary or the Secretary of Health, Education, and Welfare, and the Comptroller General of the United States, or any of their duly authorized representatives, shall have access for the purpose of audit and examination to any books, documents, papers, and records of the recipients of any grant under this Act that are pertinent to any such grant.

ANNUAL REPORT

SEC. 26. Within one hundred and twenty days following the convening of each regular session of each Congress, the Secretary and the Secretary of Health, Education, and Welfare shall each prepare and submit to the President for transmittal to the Congress a report upon the subject matter of this Act, the progress toward achievement of the purpose of this Act, the needs and requirements in the field of occupational safety and health, and any other relevant information. Such reports shall include information regarding occupational safety and health standards, and criteria for such standards, developed during the preceding year; evaluation of standards and criteria previously developed under this Act, defining areas of emphasis for new criteria and standards; an evaluation of the degree of observance of applicable occupational safety and health standards, and a summary

75

of inspection and enforcement activity undertaken; analysis and evaluation of research activities for which results have been obtained under governmental and nongovernmental sponsorship; an analysis of major occupational diseases; evaluation of available control and measurement technology for hazards for which standards or criteria have been developed during the preceding year; description of cooperative efforts undertaken between Government agencies and other interested parties in the implementation of this Act during the preceding year; a progress report on the development of an adequate supply of trained manpower in the field of occupational safety and health, including estimates of future needs and the efforts being made by Government and others to meet those needs; listing of all toxic substances in industrial usage for which labeling requirements, criteria, or standards have not yet been established; and such recommendations for additional legislation as are deemed necessary to protect the safety and health of the worker and improve the administration of this Act.

NATIONAL COMMISSION ON STATE WORKMEN'S COMPENSATION LAWS

SEC. 27. (a) (1) The Congress hereby finds and declares that—
(A) the vast majority of American workers, and their families, are dependent on workmen's compensation for their basic economic security in the event such workers suffer disabling injury or death in the course of their employment; and that the full protection of American workers from job-related injury or death requires an adequate, prompt, and equitable system of workmen's compensation as well as an effective program of occupational health and safety regulation; and
(B) in recent years serious questions have been raised concerning the fairness and adequacy of present workmen's compensation laws in the light of the growth of the economy, the changing nature of the labor force, increases in medical knowledge, changes in the hazards associated with various types of employment, new technology creating new risks to health and safety, and increases in the general level of wages and the cost of living.
(2) The purpose of this section is to authorize an effective study and objective evaluation of State workmen's compensation laws in order to determine if such laws provide an adequate, prompt, and equitable system of compensation for injury or death arising out of or in the course of employment.
(b) There is hereby established a National Commission on State Workmen's Compensation Laws. *Establishment.*
(c) (1) The Workmen's Compensation Commission shall be composed of fifteen members to be appointed by the President from among members of State workmen's compensation boards, representatives of insurance carriers, business, labor, members of the medical profession having experience in industrial medicine or in workmen's compensation cases, educators having special expertise in the field of workmen's compensation, and representatives of the general public. The Secretary, the Secretary of Commerce, and the Secretary of Health, Education, and Welfare shall be ex officio members of the Workmen's Compensation Commission: *Membership.*
(2) Any vacancy in the Workmen's Compensation Commission shall not affect its powers.
(3) The President shall designate one of the members to serve as Chairman and one to serve as Vice Chairman of the Workmen's Compensation Commission.

76

Quorum.

(4) Eight members of the Workmen's Compensation Commission shall constitute a quorum.

Study.

(d)(1) The Workmen's Compensation Commission shall undertake a comprehensive study and evaluation of State workmen's compensation laws in order to determine if such laws provide an adequate, prompt, and equitable system of compensation. Such study and evaluation shall include, without being limited to, the following subjects: (A) the amount and duration of permanent and temporary disability benefits and the criteria for determining the maximum limitations thereon, (B) the amount and duration of medical benefits and provisions insuring adequate medical care and free choice of physician, (C) the extent of coverage of workers, including exemptions based on numbers or type of employment, (D) standards for determining which injuries or diseases should be deemed compensable, (E) rehabilitation, (F) coverage under second or subsequent injury funds, (G) time limits on filing claims, (H) waiting periods, (I) compulsory or elective coverage, (J) administration, (K) legal expenses, (L) the feasibility and desirability of a uniform system of reporting information concerning job-related injuries and diseases and the operation of workmen's compensation laws, (M) the resolution of conflict of laws, extraterritoriality and similar problems arising from claims with multistate aspects, (N) the extent to which private insurance carriers are excluded from supplying workmen's compensation coverage and the desirability of such exclusionary practices, to the extent they are found to exist, (O) the relationship between workmen's compensation on the one hand, and old-age, disability, and survivors insurance and other types of insurance, public or private, on the other hand, (P) methods of implementing the recommendations of the Commission.

Report to
President
and Congress.

(2) The Workmen's Compensation Commission shall transmit to the President and to the Congress not later than July 31, 1972, a final report containing a detailed statement of the findings and conclusions of the Commission, together with such recommendations as it deems advisable.

Hearings.

(e)(1) The Workmen's Compensation Commission or, on the authorization of the Workmen's Compensation Commission, any subcommittee or members thereof, may, for the purpose of carrying out the provisions of this title, hold such hearings, take such testimony, and sit and act at such times and places as the Workmen's Compensation Commission deems advisable. Any member authorized by the Workmen's Compensation Commission may administer oaths or affirmations to witnesses appearing before the Workmen's Compensation Commission or any subcommittee or members thereof.

(2) Each department, agency, and instrumentality of the executive branch of the Government, including independent agencies, is authorized and directed to furnish to the Workmen's Compensation Commission, upon request made by the Chairman or Vice Chairman, such information as the Workmen's Compensation Commission deems necessary to carry out its functions under this section.

(f) Subject to such rules and regulations as may be adopted by the Workmen's Compensation Commission, the Chairman shall have the power to—

(1) appoint and fix the compensation of an executive director, and such additional staff personnel as he deems necessary, without regard to the provisions of title 5, United States Code, governing appointments in the competitive service, and without regard to the provisions of chapter 51 and subchapter III of chapter 53 of such title relating to classification and General Schedule

80 Stat. 378.
5 USC 101.

5 USC 5101,
5331.

84 STAT. 1618

pay rates, but at rates not in excess of the maximum rate for GS-18 of the General Schedule under section 5332 of such title, and — Ante, p. 198-1.

(2) procure temporary and intermittent services to the same extent as is authorized by section 3109 of title 5, United States Code.

80 Stat. 416.

(g) The Workmen's Compensation Commission is authorized to enter into contracts with Federal or State agencies, private firms, institutions, and individuals for the conduct of research or surveys, the preparation of reports, and other activities necessary to the discharge of its duties. — Contract authorization.

(h) Members of the Workmen's Compensation Commission shall receive compensation for each day they are engaged in the performance of their duties as members of the Workmen's Compensation Commission at the daily rate prescribed for GS-18 under section 5332 of title 5, United States Code, and shall be entitled to reimbursement for travel, subsistence, and other necessary expenses incurred by them in the performance of their duties as members of the Workmen's Compensation Commission. — Compensation; travel expenses.

(i) There are hereby authorized to be appropriated such sums as may be necessary to carry out the provisions of this section. — Appropriation.

(j) On the ninetieth day after the date of submission of its final report to the President, the Workmen's Compensation Commission shall cease to exist. — Termination.

ECONOMIC ASSISTANCE TO SMALL BUSINESSES

SEC. 28. (a) Section 7(b) of the Small Business Act, as amended, is amended— — 72 Stat. 387; 83 Stat. 802. 15 USC 636.

(1) by striking out the period at the end of "paragraph (5)" and inserting in lieu thereof "; and"; and

(2) by adding after paragraph (5) a new paragraph as follows:

"(6) to make such loans (either directly or in cooperation with banks or other lending institutions through agreements to participate on an immediate or deferred basis) as the Administration may determine to be necessary or appropriate to assist any small business concern in effecting additions to or alterations in the equipment, facilities, or methods of operation of such business in order to comply with the applicable standards promulgated pursuant to section 6 of the Occupational Safety and Health Act of 1970 or standards adopted by a State pursuant to a plan approved under section 18 of the Occupational Safety and Health Act of 1970, if the Administration determines that such concern is likely to suffer substantial economic injury without assistance under this paragraph."

(b) The third sentence of section 7(b) of he Small Business Act, as amended, is amended by striking out "or (5)" after "paragraph (3)" and inserting a comma followed by "(5) or (6)".

(c) Section 4(c)(1) of the Small Business Act, as amended, is amended by inserting "7(b)(6)," after "7(b)(5),". — 80 Stat. 132. 15 USC 633.

(d) Loans may also be made or guaranteed for the purposes set forth in section 7(b)(6) of the Small Business Act, as amended, pursuant to the provisions of section 202 of the Public Works and Economic Development Act of 1965, as amended. — 79 Stat. 556. 42 USC 3142.

ADDITIONAL ASSISTANT SECRETARY OF LABOR

SEC. 29. (a) Section 2 of the Act of April 17, 1946 (60 Stat. 91) as amended (29 U.S.C. 553) is amended by— — 75 Stat. 338.

78

84 STAT. 1619

(1) striking out "four" in the first sentence of such section and inserting in lieu thereof "five"; and

(2) adding at the end thereof the following new sentence, "One of such Assistant Secretaries shall be an Assistant Secretary of Labor for Occupational Safety and Health.".

80 Stat. 462.

(b) Paragraph (20) of section 5315 of title 5, United States Code, is amended by striking out "(4)" and inserting in lieu thereof "(5)".

ADDITIONAL POSITIONS

SEC. 30. Section 5108(c) of title 5, United States Code, is amended by—

(1) striking out the word "and" at the end of paragraph (8);

(2) striking out the period at the end of paragraph (9) and inserting in lieu thereof a semicolon and the word "and"; and

(3) by adding immediately after paragraph (9) the following new paragraph:

"(10)(A) the Secretary of Labor, subject to the standards and procedures prescribed by this chapter, may place an additional twenty-five positions in the Department of Labor in GS–16, 17, and 18 for the purposes of carrying out his responsibilities under the Occupational Safety and Health Act of 1970;

"(B) the Occupational Safety and Health Review Commission, subject to the standards and procedures prescribed by this chapter, may place ten positions in GS–16, 17, and 18 in carrying out its functions under the Occupational Safety and Health Act of 1970."

EMERGENCY LOCATOR BEACONS

72 Stat. 775.
49 USC 1421.

SEC. 31. Section 601 of the Federal Aviation Act of 1958 is amended by inserting at the end thereof a new subsection as follows:

"EMERGENCY LOCATOR BEACONS

"(d)(1) Except with respect to aircraft described in paragraph (2) of this subsection, minimum standards pursuant to this section shall include a requirement that emergency locator beacons shall be installed—

"(A) on any fixed-wing, powered aircraft for use in air commerce the manufacture of which is completed, or which is imported into the United States, after one year following the date of enactment of this subsection; and

"(B) on any fixed-wing, powered aircraft used in air commerce after three years following such date.

"(2) The provisions of this subsection shall not apply to jet-powered aircraft; aircraft used in air transportation (other than air taxis and charter aircraft); military aircraft; aircraft used solely for training purposes not involving flights more than twenty miles from its base; and aircraft used for the aerial application of chemicals."

SEPARABILITY

SEC. 32. If any provision of this Act, or the application of such provision to any person or circumstance, shall be held invalid, the remainder of this Act, or the application of such provision to persons or circumstances other than those as to which it is held invalid, shall not be affected thereby.

December 29, 1970   - 31 -   Pub. Law 91-596

84 STAT. 1620

APPROPRIATIONS

SEC. 33. There are authorized to be appropriated to carry out this Act for each fiscal year such sums as the Congress shall deem necessary.

EFFECTIVE DATE

SEC. 34. This Act shall take effect one hundred and twenty days after the date of its enactment.

Approved December 29, 1970.

○

LEGISLATIVE HISTORY:

HOUSE REPORTS: No. 91-1291 accompanying H.R. 16785 (Comm. on
               Education and Labor) and No. 91-1765 (Comm. of
               Conference).
SENATE REPORT No. 91-1282 (Comm. on Labor and Public Welfare).
CONGRESSIONAL RECORD, Vol. 116 (1970):
     Oct. 13, Nov. 16, 17, considered and passed Senate.
     Nov. 23, 24, considered and passed House, amended, in lieu
              of H.R. 16785.
     Dec. 16, Senate agreed to conference report.
     Dec. 17, House agreed to conference report.

REFERENCES

1. Occupational Safety and Health Administration. *All about OSHA* (Rev. ed.). Washington, D.C.: Government Printing Office, 1980.
2. U.S. Congress. Select Subcommittee on Labor of the Committee on Education and Labor. Hearings on H.R. 14816, 90th Congress, 2nd Session, 1968.
3. U.S. Congress. *Legislative History of the Occupation and Health Act of 1970.* Report prepared by the Subcommittee on Labor, June 1971.

4 WORKERS' COMPENSATION

In the United States, with a work force of 104 million,[9] there are approximately 5.4 million workers injured seriously enough to lose time from the job and 11.3 million workers injured on the job. In 1980 there were 13,000 fatalities and 45 million person-days of work lost. In 1977 payments of workers' compensation claims reached $8.57 billion, and employers spent $14 billion to insure or to self-insure workers. This amounted to 1.71% of their payroll.[5] The overall cost of accidents in the United States in 1980 was $83.2 billion.[21]

Safety engineers are frequently involved with workers' compensation claims and/or the incidents which precipitate these claims. The annual number of claims is *very substantial* and often requires the safety engineer to expend considerable effort. A reasonable estimate of just the paperwork involved in administering these claims might amount to 2.8 million person-days of work annually. This would be sufficient to keep 1,400 safety engineers employed full time. So, if you want a more meaningful career in safety engineering than processing workers' compensation claims, accident prevention and health promotion may be your major goal as a safety engineer.

The policies, procedures, and benefits for workers' compensation are constantly changing. In 1979 alone, nearly 200 laws affecting almost every aspect of workers' compensation were passed in the 50 states. Therefore, it is essential for the safety engineer not only to understand current policies, procedures, and benefits thoroughly but to keep abreast of changes in them. One way to do this is to follow the changes published in the *Federal Register* and the *Analysis of Workers' Compensation Laws*[5] published by the U.S. Chamber of Commerce.

BEGINNINGS OF WORKERS' COMPENSATION

The history of compensation for injury dates from earliest recorded times. In the Old Testament there is a reference to helping those who were injured:

> . . . and if men strive together and one smite another with a stone, or with his fist, and he die not, but keepth his bed, and if he rise again, and walk abroad upon his staff, then shall he that smote him be quit, only he shall pay for the loss of his time and shall cause him to be thoroughly healed (Exod. 21:18).

The first principles of compensation appear to have been developed by the Salic-Francs (the northern tribes of the Merovingian Empire). About A.D. 500 ideas of compensation were further developed by the Anglo-Saxons and finally codified in laws by King Canute, the Danish-born king of England, from A.D. 1016 to 1035. Canute stated principles of compensation for particular injuries, and his schedule of compensation appears almost unchanged to the present time.[4]

In England relationships between employer and employee (master and servant) were governed by common law. "Common law" refers to rules evolved as a result of judicial decisions, rather than to statutes enacted by a legislative body. Each decision builds upon prior ones in an attempt to fully explicate some law or judicial decision. For instance, in defining "in the course of employment," one decision may expand the meaning to include accidents in the company cafeteria, while another includes the company parking lot. So by the legal precedence after these decisions, "in the course of employment" means any accident which occurs from the time a person enters the company parking lot at the beginning of the shift until leaving it at the end of the shift. A recent decision expanded this concept even further. If a company pays for sports uniforms for members of a team, then employees are covered by workers' compensation for injuries while engaged in sporting events when wearing the uniforms even if unrelated to their work.

The industrial revolution, which began in the eighteenth century in England, radically changed employer–employee relationships. Whereas before the industrial revolution a master might have had four or five workers or apprentices working with their hands, an industrial employer might employ hundreds of people working at machinery which was often very dangerous. However, the precedents set by common law during the pre–industrial revolution period did not change. Under this system, an injured worker must still prove the case in court to receive compensation, just as it is done today to collect for injuries sustained in an automobile accident.

Several precedents set under common law often precluded an impaired worker from collecting compensation for a work-related injury.

1. *Assumption of risk.*   An employee who accepted a position automatically accepted any risks associated with the job. The worker knew and understood the hazards and could take a job elsewhere if he or she did not care to assume those risks.

2. *No contributory negligence.*   An employee had to prove that he or she in *no* way contributed to an accident by negligence. For instance, if the ladder being used had a bad rung and the employee had an accident because the rung broke, the employee contributed to the accident because he could see the broken rung and did not take appropriate precautions or actions.

3. *Fellow-servant rule.*   An employee's claim could also be disqualified if a fellow employee caused the accident. For example, if a worker was burned with a welding torch held by another employee, the employer was not liable for compensation because the accident was caused by a fellow worker. An even more ridiculous application of the fellow-worker rule happened to railroad workers. A worker did not properly couple two freight cars; after 150 miles the train uncoupled and several railroad personnel were injured. However, the courts ruled that the injured workers were not eligible for compensation because a fellow worker was negligent. Because of precedents like these, very few disabled workers ever received any compensation for their injuries. They were often forced to accept charity or public welfare in order to survive.[17]

Formal laws to protect the worker were enacted in Europe in the eighteenth century, and the first workers' compensation law was passed in Germany in 1884. This law was compulsory: the employer was required to insure all workers, while the government managed the nonprofit insurance fund. In England, Parliament enacted a workers' compensation plan in 1897. The British plan was optional, in that the employer could either accept or reject coverage, but if it were rejected, the employer lost the right to use the common-law defenses discussed earlier.

In the United States, it was estimated that only 1 in 8 persons severely crippled by job-related injuries received any compensation at all. Only rarely did a person who was not seriously injured receive any compensation whatsoever. The socioeconomic impact of large numbers of impoverished disabled workers or their widows depending upon welfare prompted journalists, like Upton Sinclair, to draw public attention to the situation.[22]

In 1902 Maryland passed a law providing for a cooperative accident insurance fund, which represented the first legislation embodying to any degree the principle of workers' compensation. However, the scope of the act was very restricted. Benefits, which were very minimal, were provided only for fatal accidents. Within three years, the courts declared the act unconstitutional. In 1908 Massachusetts passed legislation which authorized the establishment of private plans of compensation. This law never became important because it had very little impact.

In 1907 Congress enacted the first workers' compensation law applicable to certain federal employees. President Theodore Roosevelt had previously urged the compensation law's passage in his message to Congress. He pointed out that the burden of an accident fell upon the disabled person, the spouse, and children. However, in 1908 the law was declared unconstitutional, but Congress enacted a second law that overcame the criticism of the courts. Though utterly inadequate, it was the first real workers' compensation act in the United States.

An initial effort on the state level was conducted by New York in 1909 when a special commission was appointed to examine the possibility of changing the adversary approach to workers' compensation. In 1909 the commission recommended a workers' compensation law which embodied a concept of no-fault liability. This law was enacted by the New York State Legislature on January 25, 1910. The law was soon tested in the courts in *Ives* v. *Buffalo Railroad Company*. The courts declared the law unconstitutional and "plainly revolutionary" and the doctrine of liability without fault to be in conflict with the laws of the United States.[20]

Collecting workers' compensation was one concern of businesspeople and the legislators in the early twentieth century. Some businesspeople equated compensation benefits with charity—being paid for not working. Investigations confirmed that disabled workers rarely recovered lost income and that even when they did, awards were inequitable and won only after long delays. The whole system was costly. It offered an incentive for safety programs; it engendered hostility between employer and employee; and compensation for death and disability was hopelessly inadequate.

The issue that confronted lawmakers was how to structure workers' compensation programs. A critical prerequisite was to find a reliable way of assessing loss of function in an individual. Two schools of thought evolved:

1. Reimbursement for loss of wage-earning capacity only
2. The "whole-man" theory, which recommended compensating the injured worker not only for loss of wage-earning capacity but also for the degree of physical impairment, that is, the percentage to which the worker was less than a "whole man" as a result of the injury[6]

The problem was complicated somewhat because the definition of earning capacity had been modified by a number of well-recognized legal exceptions, presumptions, and fictions. The best-known and most extensive was the schedule award that provided automatic compensation for losses, or loss of use of, limbs or members, whether the claimant demonstrated loss of earnings or not. An arbitrary number of weeks' disability was chosen as an approximation of lost wages for a particular injury.

The whole-man theory arose because although an injury may result in permanent impairment, all injuries do not interfere with earning capacity. For example, although the loss of one's reproductive capability would undoubtedly draw a financial award at common law, it would not bring workers' compensation as it does not affect earning capacity. For the same reason, loss of the sense of taste or smell is noncompensable. Another reason for the whole-man theory was that most impairments were from injuries for which schedule awards were not made. These included sexual or reproductive dysfunction, back problems, and psychiatric disability.

In 1911, nine states passed workers' compensation acts—Wisconsin, New Jersey, California, Washington, Kansas, New Hampshire, Ohio, Illinois, and Massachusetts—but it was not until 1948 that the last of the 48 states, Mississippi, passed a law compensating workers for injuries. This law became effective on January 1, 1949.

THE NATURE AND PURPOSE OF WORKERS' COMPENSATION

Workers' compensation is a form of social security. It is similar to social security disability insurance or unemployment compensation in that it provides a certain measure of income protection to the worker who is unable to work because of job-related disability or to the family if the worker is killed. The object of workers' compensation is to provide enough continuous income for the injured worker and the family to maintain their economic and social status while the worker is temporarily or permanently disabled.

Workers' compensation is not intended for the recovery of damages for pain and suffering caused by the workplace injury. Workers' compensation is an employment benefit for the majority of workers. In 1977, 87.8% of all employees were covered by workers' compensation.[5] The right of a worker to sue his or her employer for damages in court was surrendered in return for guaranteed protection under workers' compensation, regardless of who was at fault in the injury. It is quite similar to no-fault automobile insurance, which is available in many states where the automobile owner's insurance policy provides coverage regardless of blame in an automobile collision.

The objectives of workers' compensation are:

1.   Prompt and competent medical treatment
2.   Replacement of income
3.   Rehabilitation of the injured person

4. Return to employment of the injured person
5. Accident prevention and control

DISABILITIES AS A RESULT OF EMPLOYMENT

Disability refers to all physical and psychological impairments that a person suffers within a specific social and environmental context.[16] A disability determination for workers' compensation may be thought of as a quantitative diagnosis in terms of percent of disability as opposed to the medical or nursing diagnosis.

Most workers' compensation laws contain one seemingly universal statement, which reads "arising out of and in the course of employment." This statement appears to set the basic conditions for receiving workers' compensation benefits. Although the statement specifically implies a relationship between injury or disease and employment, it is sufficiently ambiguous to make it difficult to determine whether a person should receive benefits.

Analyzing the phrase may assist the safety engineer in making a preliminary determination as to whether the employee's claim appears valid. The phrase "arising out of" seems to imply that there is a cause-and-effect relationship between the injury and the job.

The fact that an employee suffered a "nervous breakdown" or had a heart attack on the job is often not enough evidence to support a claim of "arising out of." In the first case, if the person was locked in a control room for eight hours a day with no contact with other people or the outside at all, it might imply a situation called "sensory deprivation" that is well known to affect most people psychologically and cause them to hallucinate and/or have delusions, both of which are associated with a "nervous breakdown." In the second instance, severe stress on the job is known to be a factor contributing to myocardial infarctions, otherwise known as "heart attacks." However, a person's personality, life-style, and physical condition may also contribute to coronary disease. Two types of personality have been identified in relation to coronary heart disease:

1. Type *B* personality is associated with the cool, calm person who is not easily upset.
2. Type *A* personality, often associated with heart attacks, is usually tense and hard-driving.

The phrase "in the course of employment" usually means from the time an employee arrives at the employer's premises until the time of leaving at the end of the shift. This definition is deceptively simple. Many legal decisions have enlarged the meaning of "employment," while others have often raised unanswered questions. Is the employer's parking lot considered part of the premises? Does coverage extend to a restaurant if a dining area is not provided at the work site? What if the work site is 60 miles from the nearest bank where a worker can cash a paycheck? What if an employee is attending a convention in a distant city, goes out in the evening, has a few drinks, is injured in an accident, and becomes permanently disabled? Is this "in the course of employment"? Would it make any difference if the employee was not intoxicated?

Although many claims are quite straightforward, others must be settled by litigation. Who helps decide whether the worker needs legal help?

Example: Mr. Cardillo worked on a contract at Grand Coulee Dam and lived 100 miles from the job site. Under the union contract the contractor agreed to pay travel expenses or $15 per day. Mr. Cardillo stayed at the job site and traveled home every weekend. Mr. Cardillo was killed in an auto collision on his way to work. The court (*Cardillo* v. *Liberty Mutual,* 330 U.S. 469) decided that compensation should be awarded to his widow.

The reasoning for the award was that the employer was required to provide transportation. It was irrelevant whether the employer furnished a vehicle of its own or paid for the expense of transportation. The safety engineer needs to be well educated in compensation law to communicate with the attorneys, physicians, employees, employers, and state compensation board involved in a workers' case.

Types of Disabilities

The four major types of disablement are defined as follows:

Temporary total disability.   An injury or impairment other than death which temporarily incapacitates a person totally from any gainful employment

Temporary partial disability.   Any injury or impairment which results in a temporary loss of use of any member (hand, arm, or leg) or bodily function or part of a member or function

Permanent partial disability.   Any injury or impairment which results in the complete loss of or loss of use of any member or function or part of a member or function of the body

Permanent total disability.   Any injury or impairment other than death which totally incapacitates a person from any gainful employment

Let us examine the scope of the problem for a medium-sized manufacturing plant with about 250 employees and a payroll of $5 million a year. The cost of workers' compensation will be at least $85,000 a year if the facility has a reasonably good accident record. In a typical year there will be approximately 2,000 person-hours per employee, or 500,000 hours total. There will be approximately 20.4 workplace injuries among the 250 employees, and of these injuries, 8.25 will be serious enough to result in filing a workers' compensation claim. There will also be a chance of 1 in 100 that this firm will suffer an industrial fatality during the year. In addition to the direct cost of workers' compensation claims, there will be substantial indirect costs, including such factors as nonproductive time of other workers near an accident scene, overtime required as a result of an accident, and cost of obtaining a replacement. Accident prevention is cost-effective in and of itself, but in addition there are other special costs to the organization which include the suffering of the injured worker, the consequent lowering of morale among other workers, the training of a new worker, and the decreased productivity of at least one worker while the new worker is being trained and oriented.

WORKERS' COMPENSATION TODAY

Today, in every state, the District of Columbia, and Puerto Rico, workers' compensation is the principal economic security program for work-related disability. The U.S. Congress has enacted two laws: (1) the Federal Employees' Compensation Act, which provides workers' compensation for all government employees, and (2) the Longshoremen's and Harbor Workers' Compensation Act, which covers all maritime workers (except in Puerto Rico). Employees in the ten provinces of Canada, the Yukon, and Northwest Territories are covered by workers' compensation laws administered by their respective governments. Labour Canada (the Canadian equivalent of the U.S. Department of Labor) administers the Government Employees' Compensation Act for Canadian federal employees.

Workers' compensation is compulsory throughout the United States and Canada, except in three states: New Jersey, South Carolina, and Texas. In these states a company may either accept or reject the Workmens' Compensation Act; however, if it rejects the act, it loses the three common-law defenses—assumption of risk, contributory negligence, and fellow-servant rule (discussed previously in this chapter).

All workers' compensation laws are based upon the *fault-free convention:* Regardless of fault, an employee receives a portion of wages if incapacitated by injury or diseases that "arise out of and in the course of this employment." However, because each state or province has individual laws, benefits vary widely from one jurisdiction to another.

Workers' compensation has six basic objectives:

1. To provide sure, prompt, and reasonable income and medical benefits to work-accident victims and/or income benefits to their dependents, regardless of fault

2. To provide a recourse remedy and thereby reduce court delays, costs, and workloads arising from personal injury litigation

3. To relieve public and private charities of the financial drain for uncompensated industrial accidents

4. To eliminate payments of fees to lawyers and witnesses as well as time-consuming trials and appeals

5. To encourage employer interest in safety and rehabilitation with appropriate experience-rating mechanisms

6. To promote the study of the causes of accidents[5]

Typical Compensation Provisions

In order to meet these objectives, most compensation programs include the following provisions:

1. The employee is automatically entitled to benefits when suffering personal injury from an accident "arising out of and in the course of employment."

2. Negligence and fault are immaterial: The employee's contributory negligence does not affect his or her rights, nor does the employer's complete freedom from fault reduce its liability.

3. Coverage is limited to employees as distinguished from independent contractors.

4. Benefits include hospital and medical expenses and wages, usually one-half to two-thirds the average weekly wage. Death benefits for dependents are provided.

5. The employee and dependents, in exchange for assured benefits, forfeit their common-law right to sue the employer for damages for any injury covered by the act.

6. Administration is usually in the hands of commissions, and as far as possible, rules of procedure, evidence, and conflicts of law are relinquished to achieve the beneficial purposes of legislation. (Petty squabbles are thus avoided.)

7. The employer is required to secure its liability through private insurance, state fund insurance (in some states), or "self-insurance."[5]

The basic purpose of compensation is to tide the injured worker over the period of financial difficulty that may result from lost earning power. It is not intended for the recovery of damages.[18] In addition to workers' compensation, employees are also obtaining protection from private or collectively bargained disability plans and benefit programs and, more significantly, from social security.

Every workers' compensation law has a "schedule," or list of certain permanently disabling conditions, showing the benefits payable for each. The schedule ratings are based on presumed relationships between impairments and earning capacity. These schedules are used to award benefits and eliminate the need to determine the exact degree of disability in each individual case. Because the schedules are simple, they are popular with administrators and legislators.

Is the claimant entitled to benefits because of actual loss of wages or because the injury threatens future economic loss? Quite apart from loss of wages, should the worker be indemnified for sustaining physical, anatomical, or functional loss, with concomitant personal suffering? Despite 50 years of experience with compensation for permanent disability, there is still no consistent concept of what exactly is being compensated and, consequently, no accurate measure for determining the extent of the injured worker's loss. The problem is twofold: (1) many injuries defy disability rating and (2) neither physical nor economic recovery can be predicted. These difficulties, in turn, erect barriers to rehabilitation and create a legal vicious circle. Since temporary benefits are minimal, workers press for permanent disability ratings in minor cases and thus keep their healing benefits until they get a new job. Where there is real permanent disability and a dispute over the size of the claim, the injured worker may actually avoid a job in order not to lose the award—an award which looks more real in a time of anxiety and insecurity than the dubious possibility of rehabilitation in a tenuous job. This is quite different from "malingering." The need for lawyers merely reinforces this behavior. Insurance carriers resist claims, or at least minimize awards, whenever the liability cannot be measured, convinced of the need to do so because of the workers' incentive to maximize claims. Employers, annoyed by controversial cases, often conclude that it is better to fire the employee.

Consequently, the inept handling of permanent disability cases has completely subverted the primary purpose of screening the injured worker, and everyone involved is forced to do things that most discourage rehabilitation. The injured worker does not want to be rehabilitated; the insurance carrier does not want to recognize the claim; and the employer does not want to retain the disabled worker.

In this jungle of litigation is the daily parade of lawyers with their medical experts, all attempting to fix damages in dollars and cents.[1] This system results in significant interruption of medical care, impedes physical rehabilitation, and completely disregards vocational rehabilitation. The injured worker is awarded a sum of money to divide between lawyer, physician, and himself or herself and a disability which the worker alone keeps.[10]

Workers' Compensation Schedules and Disability Evaluation

A study by Collins of permanent disability cases in California examined the difficulty of classification. In an analysis of over 30,000 cases, 78% of the major permanent disability cases could not be classified according to a loss schedule. The 22% that could be classified involved additional physical impairments which were unable to be rated.[7]

In 1949, a study of the Massachusetts Workmen's Compensation Law by a special commission found that the compensation schedules for injury would affect only 4.5% of those injured. In 1944, the last year for which accurate figures were provided, of 36,470 injury cases, there were 1,239 for which there were scheduled compensations.[8] In another study, Goulston found that only about 35% of the finger injuries would be ratable on most workers' compensation schedules.[13] These studies indicate that disability schedules are not effective in providing compensation to disabled workers, especially those with hand impairments.

The variety of hand injuries is very large, ranging from cuts (laceration) to amputation of the entire hand at the wrist. With four fingers, thumb, palm, web space, and wrist, the number of possible combinations of injuries becomes a problem. Since many hand injuries involve multiple structures, categorizing hand injuries in disability schedules is difficult. Leaming, Walder, and Brainthwaite, showed nine different categories of lacerations. Out of 7,875 diagnosable hand injuries, only 519 (6.6%) were classified in present schedules.[15]

The primary aim in evaluating permanent physical impairment and disability is to identify activities precluded by the handicap. At least three factors must be considered before an evaluation can be made with any degree of accuracy: (1) pain; (2) anxiety neurosis and psychoneurotic behavior; and (3) preexisting depreciation of body tissues.[16]

Presumably, determining the relative permanence of disability involves appraising how the patient's present and projected ability to earn a living is affected by impairment in relation to nonmedical factors, such as age, sex, education, and economic and social environment. However, because nonmedical factors are so difficult to measure, permanent impairment is in fact the only criterion for judging permanent disability, which is more often the case than is readily acknowledged.[14] This approach places all patients at approximately the same level, whether man or woman, left- or right-handed, 25 to 65 years old, college graduate or a fourth-grade dropout, famous violinist or stevedore.

The National Commission on State Worker's Compensation Laws was created by the Occupational Safety and Health Act. This presidential commission was directed to undertake a comprehensive study and evaluation of state workers' compensation laws. The commission had five major objectives:

1. Broader coverage for employees for work-related injuries and diseases
2. Better protection against interruption of income
3. Sufficient medical care and rehabilitation

4. Effective system for delivery of services and benefits
5. Enhancement of safety programs

The commission reported the following results:

> The inescapable conclusion is that state workers' compensation laws in general are inadequate and inequitable. While several states have good programs, and while medical care and some other aspects of workers' compensation are commendable in most states, the strong points are too often matched by the weak.[19]

The present status of how these objectives were met is shown in Table 4.1.

Extent of Coverage

Although we consider workers' compensation to be a right of all workers, only 87.8 percent of all employees were covered in 1977.[5] The largest uncovered groups were agricultural workers, domestic workers, and casual workers (day laborers). Every state has specific exceptions. Michigan excludes professional athletes who earn twice as much as the state's average wage. This means that members of the Detroit Tigers (baseball), Detroit Lions (football), and Detroit Redwings (hockey) are not covered by workers' compensation.

Medical Care

Medical care for occupational injuries is virtually unlimited. However, when agreeing to a cash settlement of the claim, a worker generally forfeits further medical coverage. Medical care for occupationally related diseases is sometimes more restricted. Certain diseases that are usually job-related include byssinosis (a chronic industrial disease associated with the inhalation of cotton dust over a long time and characterized by chronic bronchitis) and mesothelioma (a cancerous tumor associated with inhalation of asbestos). Some states limit claims for occupational diseases in the following ways:

1. Limit the diseases which are covered.
2. Limit the time in which to file a claim. This is quite critical for diseases like cancer, which may take many years to develop after the person is exposed to a hazard.
3. In a large number of states, disability payments are deducted from death benefits the heirs might receive.

TABLE 4.1
Compliance with Each of 16 Recommended Standards[a]

	No. of States Meeting Recommended Standards		
	December 1966	January 1972	Present
Compulsory law	27	31	47
No numerical exception	23	27	39
Farm employment covered	10	17	
Full coverage of occupational diseases	30	41	50
Rehabilitation division in workers' compensation agency	17	22	
Maintenance benefits during rehabilitation	19	27	
Full medical care for accidental injuries	39	41	50
Full medical care for occupational diseases	28	36	47
Workers' compensation agency has authority to supervise medical aid	23	26	
Worker's initial choice of physician	21	25	
Broad second-injury fund	16	20	
Adequate time to file occupational disease claims	21	24	
Waiting period of not more than 3 days with retroactive benefits after 2 weeks	7	10	22
Death benefits during widowhood	11	15	27
Permanent total disability benefits for period of disability	30	31	40
Temporary total disability benefits to equal two-thirds of state average wage	3	10	20
Total standards met	325	403	
Total possible standards	800	800	800

SOURCE: U.S. Department of Labor, Employment Standards Administration, Office of Workmen's Compensation Programs, Division of Workmen's Compensation Standards.

[a] For explanation of the standards, see U.S. Department of Labor, Employment Standards Administration, "State Workmen's Compensation Laws: A Comparison of Major Provisions with Recommended Standards."

Maximum Monetary Benefits

The monetary benefits provided for the disabled worker include cash benefits to replace income while disabled and cash payments for scheduled (specific) injuries. The benefits vary from state to state and province to province. Cash income benefits are generally placed at $66\frac{2}{3}\%$ in the United States and 75% in Canada of the employee's wages. However, these payments are limited by a maximum weekly payment. Some maximum weekly payments in certain states are shown in Table 4.2.

TABLE 4.2
Maximum Weekly Income Benefits for Total Disability in 1980

State/Province	Amount	Limit
Alabama	136.00	300 weeks
Alaska	650.00	Disability
Connecticut	261.00	Disability
Massachusetts	227.31	$45,000
Tennessee	107.00	400 weeks* $42,800
Texas	119.00	401 weeks
Virginia	199.00	500 weeks
Federal employees	722.78	Life or disability
British Columbia	293.42	Life or disability
Ontario	266.85	Life or disability
Nova Scotia	215.75	Life or disability

*After 400 weeks the benefit payment is reduce to $15 per week.

The safety engineer should strongly recommend to employees that they carry supplementary insurance when in such states as Georgia and Tennessee where workers' compensation benefits are poor.

Percent of Salary

In addition to the maximum benefits which may be paid to a worker, the actual percent of the salary is quite important. Since many workers will not receive the maximum, the National Commission on State Workmen's Compensation Laws has suggested that benefits be placed at $66\frac{2}{3}\%$ of wages. This amount is based upon the notion that a worker's salary is reduced by federal income tax. Further, since the worker does not have to go to work, certain job-related expenses, like transportation, meals, and clothes, are reduced. How does this work out in reality for a number of employees?

Scheduled Injuries

Income for scheduled injuries consists of payments for specific injuries that involve loss or loss of use of certain body members. In many states, the amount paid is a specific number of weeks of presumed disability multiplied by the weekly benefit rate. The injuries for which there are scheduled benefits include:

1. Arm at shoulder
2. Hand
3. Thumb
4. First finger (index)
5. Second finger (middle)
6. Third finger (ring finger)
7. Fourth finger (little finger)
8. Leg at hip
9. Foot
10. Great toe
11. Other toes
12. One eye
13. Hearing, one ear
14. Hearing, both ears

See Table 4.3 for a comparison of benefits.

Rehabilitation

Rehabilitation is the process of reconditioning a worker for employment and/or living after a debilitating injury or disease. This is a basic objective of workers' compensation. However, it is rarely accomplished because employers do not want to pay for it and physicians are concerned only

TABLE 4.3
Comparison of Income Benefits for Scheduled Injuries

	Federal Employment Compensation Act	Highest State	Lowest State
Arm at shoulder	225,507	105,957 (Ill.)	6,750 (Mass.)
Hand	176,358	81,070 (Pa.)	5,250 (Mass.)
Leg at hip	208,161	99,220 (Pa.)	6,000 (Mass.)
One eye	115,645	66,920 (Pa.)	8,000 (Mass.)
Hearing, both ears	144,556	62,920 (Pa.)	8,000 (Mass. & N. Dak.)

SOURCE: Data derived from Analysis of Workers' Compensation Laws, U.S. Chamber of Commerce, 1980

with the medical aspects. Rehabilitation is so important that Mississippi will generously allow the disabled worker up to $10 a week while undergoing rehabilitation.

A TYPICAL WORKERS' COMPENSATION PLAN

Most every workers' compensation law uses a "schedule" of certain permanently disabling conditions for computing benefits for permanent impairments. These are based upon a supposed relationship between impairment and earning capacity. They are used directly or as a benchmark to set benefits and eliminate the necessity of determining the exact degree of disability in individual cases. Their popularity, particularly with administrators and legislators, is considerable. Only Canada, its provinces and territories, and the states of Florida and Nevada assess disability by other approaches.

One approach has been to set specific dollar values for particular impairments. In some cases, this is in addition to the amount the person received while recovering in temporary total disability. In other cases, the temporary total disability paid during the period the person is unable to work is deducted from the amount for a particular impairment. Total amputation of the fourth (little) finger may result in a payment of $500, while amputation of the thumb may result in a payment of $2,000. The federal government has used this approach; however, its benefits are approximately 15 times greater than that awarded by the lowest-paying state under workers' compensation. In the dollar approach, the rater examines a picture of the hand which shows the values of amputation of the various digits, and the disabled worker is awarded compensation on that basis. No allowance is made for any impairment except amputation. A variation of this approach allows for assessment of impairment other than amputation up to the amount set for amputation. However, there are no guides to set how one goes about making these types of ratings.

A second approach used by some states (for example, Oklahoma) values each finger in terms of a week's disability. Specific awards are determined for loss of use of each of the fingers and thumb. A specified award is made for the loss of use of the hand as a whole. The value for total loss of the thumb is 60 weeks' disability, which is equivalent to amputation of the whole thumb at the metacarpophalangeal joint. The value for loss of use of the fourth (little) finger is 15 weeks of disability. Under this approach, evaluation of permanent physical impairment to a part of the hand cannot exceed the specified award for the total loss of that part. Basically, the difference between this approach (the one used by Oklahoma) and the one described previously (the one used by the federal government) is that impairment is given in terms of weeks of disability, rather than a specific dollar value.[3]

An Alternative to a Schedule

No doubt, in many cases, the schedules serve their purposes; but the shortcomings of the schedules are serious and are becoming increasingly so. The difficulties start with the very concept of permanent disability benefits, which is the premise that following healing, there is a disabling condition remaining more or less static and unchangeable for the worker's lifetime. Many of the ways in which an accident affects the body or a disease manifests itself permit neither concise definition nor objective measurement.

It is clear that following healing, many (if not most) permanent disability cases have symptoms which defy objective definition and measurement. Determination of the extent of permanent disability, or whether disability exists at all, depends on the degree to which a compensation law (or its interpretation) recognizes these symptoms.

Is the claimant entitled to benefits because of suffering an actual loss of wages or, rather, because the injury threatens a future economic loss? Or quite apart from loss of wages, should the claimant be indemnified because of having sustained a physical, an anatomical, or a functional loss with concomitant personal suffering? Despite 50 years of compensation experience with permanent disability, there exists today no consistent concept of what is being compensated in the United States and, therefore, it necessarily follows, no effective measures for determining the extent of the injured worker's loss.

American Medical Association

The American Medical Association has developed a schedule of permanent impairments which were published in a series from 1958 to 1970 as supplements to the *Journal of the American Medical Association*. For example, hand impairment is evaluated in terms of amputations and loss of range of motion.[2] Under this approach, a totally disabled person is one who is approximately 100% disabled, whereas a person who has an amputation of the dominant thumb is 22% disabled. Restricted range of motion is also assessed. This approach described and assessed amputation and loss of range of motion but did not consider the behaviorial consequences of the damage.

The guide sets up a schedule by which one can evaluate any anatomic damage or loss of function after the person has achieved the maximum medical recovery possible. Rating of permanent impairment is an appraisal of the anatomic or functional loss as it affects a person's effectiveness in tasks of daily living. These losses may affect "self-care, communication of normal living, posture, ambulation, elevation, traveling, and nonspecialized hand activities." Competent disability evaluation, according to the *American Medical Association Guide*, requires a complete examination of the impaired subsystem, objective functional measurement, and avoidance of subjective factors.[2]

The guide argues that evaluation of permanent impairment is a function which physicians alone are competent to perform, as the guide states:

> Evaluation of permanent impairment defines the scope of medical responsibility and, therefore, represents the physician's role in the evaluation of permanent disability. Evaluation of permanent impairment is an appraisal of the nature and extent of the patient's illness or injury as it affects his personal efficiency in the activities of daily living."

How is it reasonably possible to do this without taking into consideration age, sex, or the other factors mentioned? The reason for wanting to establish an evaluation of permanent impairment solely on the basis of an accurate "objective measurement of function and avoidance of subjective impressions and factors, as the patient's age, sex, or employability,[11] also is not understood by the guide. The fallacy of this approach, of course, lies in its twofold attempt to make something scientific and empirical out of disability or impairment evaluation, though neither is capable of true scientific evaluation. As we shall see later, the obsessive self-defeating thirst for objectivity in the medical legal field is minimized, while the relatively meager tools of x-ray and electroencephalogram are overemphasized. The aim for scientific certainty, rather than reasonable probability, has resulted in a compulsive search for objectivity that from a legal point of view is unnecessary and from a medical point of view is fruitless and frustrating.

Now, just what is it that this American Medical Association "scientific" guide to evaluation of permanent impairment does? A careful examination of the "guide" discloses that what the tables do is to set up percentages of impairment based on three things: (1) loss of motion, (2) ankylosis, and/or (3) amputation of a joint.

Veterans Administration Schedule

The Veterans Administration has developed an extensive schedule of impairments of all varieties. This guide is a rating schedule for the medical evaluation of disability resulting from all types of diseases and injuries encountered as a result of or incident to military service. The evaluator looks up the impairment from this list, and that is the percentage of disability assigned to the impaired person. The major difference between the Veterans Administration approach and the aforementioned workers' compensation plans is that in the former approach, the impaired person receives a monthly pension, rather than a fixed award. A second difference is that the impairment can be reassessed from time to time under the Veterans Administration system. As can be seen from these alternatives, another approach is needed.

New Approach

Work is progressing on an alternative to a schedule. Gloss and Wardle have developed the Hand Disability Rating Scale which is a first attempt to assess disability in a reliable and valid manner.[12] This rating scale, which will become part of a proposed system for assessing all impairments, was developed in the following way.

Ten persons, consisting of hand surgeons and of specialists in occupational medicine, physical therapy, and occupational therapy, who work with hand-injured people, were interviewed concerning the effects of hand injuries on an individual. Four persons with hand injuries were interviewed concerning the effect of a hand impairment on activities of daily living. Records of hand-injury cases at the Department of Labor and Industry, state of North Carolina, were reviewed for the effect of hand injuries on work performance. A list of effects of hand impairment were developed. These effects were then sorted into several categories, each of which provided the basis for the development of a subscale. The total list was then reviewed by a specialist in occupational medicine, a hand surgeon, and a physical therapist. Their amendments were incorporated into the instrument. The revised instrument, prior to use, was then reviewed by three hand surgeons.

The final Hand Disability Rating Scale consisted of 64 items (Figure 4.1). These items represented the basic effects of the impairment on the individual's functioning. The items were grouped into 10 subscales. The 10 subscales measured the effect of impairment upon an area of functioning, such as activities of daily living or work impairment. These subscales were assigned equal value and were summed to provide a total disability score. Following construction of the Hand Disability Rating Scale (HDRS), comparative analyses were conducted with the HDRS, the American Medical Association (AMA) Rating Schedule, and the Veterans Administration (VA) Rating Schedule to determine which instrument provided the most stable and valid assessment of hand impairment.[11] For that purpose a sample of 120 patients with permanent impairment of the hands was assessed on the three measures by two raters. In addition, each patient was administered tests of finger dexterity, hand strength, hand–eye coordination, and hand rate of manipulation.

The interrater reliabilities based upon the correlation between the two raters were .75 for the AMA schedule, .84 for the VA schedule, and .98 for the HDRS. The internal consistency reliability of the HDRS was .87. For the injured-hand psychomotor test, scores were more highly correlated with the HDRS than either the AMA or VA scales in all cases. Multiple correlations between the subscales of the HDRS and the psychomotor test scores were statistically significant in all but one case. A factor analysis of the HDRS items comprised of 12 factors was computed, 9 of which corresponded with various HDRS subscales.

I. PHYSICAL IMPAIRMENT

Anatomic Damage

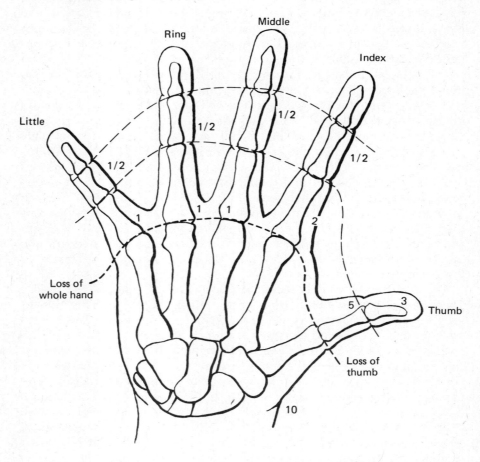

	Points
Amputation of entire hand at wrist = 20 points	_____
Amputation of all fingers except thumb = 10 points	_____
Amputation of Thumb = 5 points	_____
At IP Joint = 3 points	_____
Entire thumb and understructure = 10 points	_____
Amputation of Index finger = 2 points	_____
Dip Joint = 1 point	_____
Pip Joint = ½ point	_____
Total points I	_____

FIGURE 4.1. Hand Disability Rating Scale (*from Gloss, 1982*).

II. PHYSIOLOGICAL DAMAGE

A. Muscle Impairment

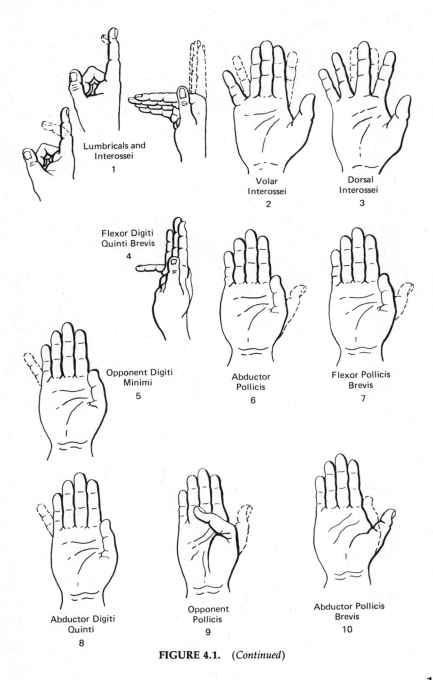

Lumbricals and Interossei
1

Volar Interossei
2

Dorsal Interossei
3

Flexor Digiti Quinti Brevis
4

Opponent Digiti Minimi
5

Abductor Pollicis
6

Flexor Pollicis Brevis
7

Abductor Digiti Quinti
8

Opponent Pollicis
9

Abductor Pollicis Brevis
10

FIGURE 4.1. (*Continued*)

	None	Impairment
1. Lumbricals and Interossei		
2. Volar Interossei		
3. Dorsal Interossei		
4. Flexor Digiti Quinti Brevis		
5. Opponeas Digiti Minimi		
6. Adductor Pollicis		
7. Flexor Pollicis Brevis		
8. Abductor Digiti Quinti		
9. Opponeas Pollicis		
10. Abductor Pollicis Brevis		

Points
_____

0 = no muscle damage or
no muscle damage in addition to muscle loss due to amputation. Slight muscle weakness which does not interfere with normal functioning.

_____

1 = Limited muscle damage which affects one of the motions shown in Figure 2.

_____

2 = Extensive muscle damage which affects more than one of the muscular motions in Figure 2.

_____

(Maximum score 2 points)

_____

Flexor Carpi
Ulnaris
1

Flexor Carpi Radialis
2

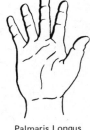

Palmaris Longus
3

Flexor Digitonum
Profundus
4

Flexor Digitonum
Sublimis
5

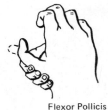

Flexor Pollicis
Longus
6

FIGURE 4.1. (*Continued*)

B. *Tendon Involvement*

	Damage		
	None	Some	Much
1. Flexor Carni Ulnaris	____	____	____
2. Flexor Carni Radialis	____	____	____
3. Palmaris Longus	____	____	____
4. Flexor Digitorum Profundus	____	____	____
5. Flexor Digitorum Sublimis	____	____	____
6. Flexor Pollicis Longus	____	____	____
7. Extensor Pollicis Longus	____	____	____
8. Extensor Pollicis Brevis	____	____	____
9. Abductor Pollicis Longus	____	____	____
10. Extensor Carni and Radialis Brevis and Longus	____	____	____
11. Extensor Indices Proprius	____	____	____
12. Extensor Carni Ulnaris	____	____	____
13. Extensor Digitorum Communis	____	____	____

Points
_____

0 = No tendon damage or no additional tendon damage due to amputation. Slight weakness in flexor digitorum tendon which does not interfere with most normal function.

1 = Limited tendon damage that affects functioning of one finger except thumb or index finger thumb relationship.

2 = Tendon damage that affects more than one finger, or affects thumb and index finger.

Maximum score 2 points   _____

C. *Scar Tissue or Adhesions*

0 points = little scar tissue or few adhesions that do not affect hand functioning.

1 point = active and passive joint movement will show some degree of limitation.

2 points = scar tissue affects movement of tendon and joints and is likely to continually break down.

Maximum score 2 points   _____

D. *Sensory Involvement*

Points
_____

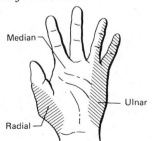
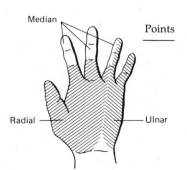

FIGURE 4.1. *(Continued)*

0 points = Disappearance of over-response and recovery of two-point discrimination within the autonomic zone.

1 point = Superficial pain and tactile sense in the autonomic zone or deep cutaneous pain in the autonomic zone.

2 points = No sensation

Pain

0 points = No pain

1 point = The pain annoys but is forgotten during activity.

2 points = The pain interferes or prevents activity

(Total possible points for Section II is 10)

Total points II _____

Points _____

III. RANGE OF MOTION

Only the injured digit will be rated. Amputations will not affect range of motion. In cases of arthritis every part will be rated. Degree of which movement if any has been lost will be measured by a goniometer. These ratings will be recorded after each joint reduction of range for each finger equals total score.

1. Thumb: MP Ext____ MP Flex____ 1P Ext____ 1P Flex____ Abd____ = 3 points _____

2. Index: MP Ext____ MP Flex____ PIP Ext____ PIP Flex____ DIP Ext____ DIP Flex____ = 2 points _____

3. Long MP Ext____ MP Flex____ PIP Ext____ PIP Flex____ DIP Ext____ PIP Flex____ = 1 point _____

4. Fing: MP Ext____ MP Flex____ PIP Ext____ DIP Flex____ DIP Ext____ DIP Flex____ = 1 point _____

5. Fifth: MP Ext____ MP Flex____ PIP Ext____ PIP Flex____ DIP Ext____ DIP Flex____ = 1 point _____

6. Wrist: Ext____ Flex____ = 2 points _____

(Total possible points for III is 10)

Total points III _____

Points _____

IV. DISFIGUREMENT

A. *Scars*

0 points = small scar (less than 1 inch palmar or dorsal sides) not affecting function.

1 point = moderate scars (1 to 2 inches on palmar or dorsal side)

2 points = large scars—all scars larger than moderate

$3\frac{1}{3}$ points = extensive scars or multiple wide scars _____

B. *Missing Members*

0 points = single digit with less missing than from pip joint

1 point = multiple digits from pip joint, single digit

FIGURE 4.1. (*Continued*)

2 points = more than one digit except thumb

3½ points = thumb, or all fingers except thumb _____

C. *General Appearance*

Is the injury immediately noticeable?   Yes or no _____

Is the injury disfiguring?   Yes or no _____

Is the sight of the injury unpleasant?   Yes or no _____

(Total possible points in IV is 10 points)

Total points IV _____

Points _____

V. FUNCTIONAL IMPAIRMENT

A. *Coordination*

"The smooth flow of movement in the execution of motor tasks (North Carolina State University, 1971, p. 4).

Does injury make coordination difficult?

Does patient have to rely on trick motions to accomplish a two-hand task such as tying shoes?

0 points = no difficulty

1 point = moderate difficulty

2 points = marked difficulty _____

B. *Hand Strength*

Since the injury or onset of disease, has there been a reduction of hand strength?

0 points = none

½ point = moderate reduction

1 point = marked reduction _____

C. *Pinch Strength Measures*

In three-point prehension, the object is grasped with the tips of the thumb, index finger, and middle finger. Palmar prehension, also involves the thumb, index and middle fingers but the object is grasped with pads of thumb and fingers rather than the tips. Lateral prehension involves the thumb and index finger only: the object is grasped between the pad of the thumb and the radial side of the index finger between the distal and proximal interphalangeal joints.

Pinch strength

0 points = no reduction in strength

½ point = moderate reduction in strength

1 point = marked reduction in strength _____

D. *Endurance*

Lasting quality, duration.

Can you continue to use your hands for extended periods of time? Like before the injury or disease?

0 points = no reduction in use

FIGURE 4.1.   (*Continued*)

1 point = moderate reduction in use

2 points = marked reduction in use   _____

E. **Sensory Recovery**

The amount of sensory process available after injury is stablized.

0 points = full sensation or no deficit. This also applies to amputation

1 point = an over-response or superficial pain

2 points = no protective reaction   _____

F. **Age**

Age___

0 points = age 20 to 30

1 point = age 31 to 50

2 points = ages 50 and over   _____

G. **Dexterity**

"Skill or adroitness in using the hands or body."

In comparison with your hand prior to the accident or in comparison with your other hand, have your skills been reduced? How?

_____

_____

0 points = no difference between hands in time it takes to complete
task

1 point = less than 30 seconds difference in time it takes to complete task

2 points = over 1 minute difference in time between the two hands   _____

(Total possible points for V is 10)

Total points V   _____

Points

_____

VI. **WORK IMPAIRMENT**

A. **Restrictions on Work**

How much can you lift with your injured hand?

_____

1. Lifting

0 points = no difficulty

½ point = unable or difficulty lifting objects weighting less than
those lifted with injured hand in everyday living.

½ point = unable or difficulty lifting objects with injured hand
more than those lifted in everyday living.   _____

2. Pushing or pulling—difficulty in pushing or pulling an object
such as a lawnmower with injured hand.

Do you have any difficulty in pushing or pulling objects? If
yes, what?

_____

0 points = no difficulty

1 point = difficulty   _____

FIGURE 4.1. (*Continued*)

3. Gripping—ability to tightly hold an object grasping a finger, a door handle.

Do you have any difficulty in gripping or grasping object with your injured hand? If yes, what?

_____

0 points = no difficulty
1 point = difficulty

_____

4. Holding—ability to handle an object or carrying coffee cup, etc.

Do you have any difficulty in holding or carrying an object with your injured hand? If yes, what?

_____

0 points = no difficulty
1 point = difficulty

_____

5. Manipulating = difficulty in buttoning buttons using the injured hand.

Do you have difficulty in manipulating objects using your injured hand? If yes, explain.

_____

0 points = no difficulty
1 point = difficulty

_____

6. Climbing—ability climbing a ladder.

Do you have difficulty in climbing as a result of injury to your hand? If yes, How?

_____

0 points = no difficulty
1 point = unable to climb ladder holding on with two hands

_____

7. Writing. Able to write with pen and pencil if preferred writing hand was injured.

Has the hand with which you write been injured?

If yes, how has this injury affected your writing?

0 points = no difficulty or injured hand is not the preferred writing hand
1 point = difficulty

_____

8. Machinery hazard. Does work around machinery and/or equipment present a danger to you because of difficulty in operating? If yes, How?

_____

0 points = no hazard
1 point = hazard

_____

9. Work breaks. Do you receive adequate work breaks for the kind of work you do? If no, explain.

_____

0 points = yes
1 point = no

_____

FIGURE 4.1. (*Continued*)

10. Allergic reactions such as contact dermatitis. Do you have contact dermatitis? If yes, how does it affect your work?

_____

_____

If yes, when did you acquire it?_____

Is it persistent?

0 points = no

1 point = yes _____

(Total possible points for Section VI is 10)

Total points VI _____

Points _____

VII. INTANGIBLE WORK FACTORS

A. *No Opportunity for Rehabilitation*

The person lives at a distance from the rehabilitation center from which commuting is not possible, or if person lives a great distance and housing is not available at rehabilitation center. No place at work where he/she can have an easier job at the same pay

or

Occupation is such that rehabilitation will not help (Tool and die maker, surgeon, violinist).

0 points = rehabilitation is no problem

5 points = no rehabilitation because of distance, housing, easier job, occupation _____

B. *Employer's Cooperation*

0 points = total employer cooperation. If necessary the employer will re-employ worker in other positions until recovery completed and place in.position based upon recovered function

3 points = the employer will permit employee to return to work only in same position and after recovery is completed.

5 points = absence of employer's cooperation. The employer will not retain injured workers. _____

(Maximum number of points for VII is 10)

Total points VII _____

Points _____

VIII. PSYCHOLOGICAL IMPAIRMENT

A. *Meaning to Person*

What does this injury mean to you?

_____

_____

What has having this injury meant to you?

_____

_____

0 points = nothing

FIGURE 4.1. (*Continued*)

1 point = some adjustments had to be made
2 points = many adjustments had to be made   _____

B. *Person's Motivation to Be Rehabilitated*

What does he see is his motivation to be rehabilitated?

_____

What plans has he made?

_____

What plans will he make?

_____

What does he see himself doing?

_____

0 points = started rehabilitation or made plans
1 point = intends to be rehabilitated
2 points = has no intention to be rehabilitated   _____

C. *Person's Need*

What have been the psychological effects of his injury to him?

_____

To his family?

_____

0 points = none
1 point = 1
2 points = many (depressed)   _____

D. *Secondary Gains*

What has he gained from this injury?

_____

What has happened since the injury that he didn't expect?

_____

Has this injury made him an invalid?

_____

0 points = minimal
1 point = some dependent behavior such as being taken care of
           for a short period of time
2 points = invalid (either physically or emotionally)   _____

E. *Self-Concept*

What has happened to his thinking?

_____

FIGURE 4.1. *(Continued)*

Does he think he is less of a person?

_____

_____

Less of a man/woman?

_____

_____

What can't he/she do that he/she could do before the injury?

_____

_____

What would he/she like to change the most?

_____

_____

What about himself/herself would he/she like to change the most?

_____

_____

0 points = no change
1 point = some change
2 points = poor self-concept   _____

           (Total possible points in VIII is 10)   _____

           Total points VIII

                                                                       Points

                                                                       _____

IX. LIMITATION ON ACTIVITIES OF DAILY LIVING

A. Self

1. Toileting. Any way he/she needs help? In what specific areas?

   _____

   _____

   0 points = no interference
   ½ point = interference   _____

2. Eating. Does he/she need meat cut?
   Anything he/she can't handle such as glass, silverware, nut cracker?

   0 points = no interference
   ½ point = interference   _____

3. Sexual. Any difficulties in this area since the injury?

   0 points = no difficulty since the injury
   ½ point = new difficulties   _____

4. Miscellaneous. Any problems in taking care of himself/herself that he/she sees. (Setting clock, opening and closing windows, turning knobs.)

   0 points = no problems
   ½ point = problems   _____

5. Dressing. Problem with zippers, buttons, pants, slipping garments over the head, tieing neckties, tieing shoes, putting panty hose on?

   0 points = no difficulty
   ½ point = difficulty   _____

FIGURE 4.1.   (*Continued*)

6. Safety. Has patient lost protective mechanisms?

0 points = no

½ point = yes   _____

B. Home Environment

1. Food preparation. Cooking meals, setting table, serving dinner.

0 points = no difficulty

½ point = difficulty   _____

2. Food purchase, storage and use. Putting groceries on shelf and in refrigerator, repackaging.

0 points = no difficulty

½ point = difficulty   _____

3. Food clean-up. Washing dishes and putting them away. Cleaning-up.

0 points = no difficulty

½ point = difficulty   _____

4. Housecleaning. Dusting, vacuuming, making beds, sweeping, mopping.

0 points = no difficulty

½ point = difficulty   _____

5. Telephoning. Any problem holding receiver? Any problem dialing telephone? Any problem answering telephone?

0 points = no difficulty

½ point = difficulty   _____

6. Home maintenance. Any jobs no longer can do? Difficulty using hammer, screws and bolts, wrench, drill, lawnmower, raking, hoeing, cutting hedge, weeding, climbing ladders, painting, planting, gardening, carrying furniture, moving furniture, carrying heavy boxes.

0 points = no difficulty

½ point = difficulty   _____

7. Shopping. Any difficulty getting packages off shelf, carrying boxes, using vending machines?

0 points = no difficulty

½ point = difficulty   _____

8. Laundry. Washing clothers, carrying baskets, folding clothers, ironing, putting clothers on hanger, putting clothes away.

0 points = no difficulty

½ point = difficulty   _____

9. Clothing maintenance. Any difficulty hanging clothes? Any difficulty sewing by hand? Any difficulty sewing by machine? Any difficulty mending? Any difficulty doing handiwork? What changes has injury made in this area?

_____

_____

0 points = no difficulty

½ point = difficulty   _____

FIGURE 4.1.   (*Continued*)

10. Family
    a. Child care. Diapering, dressing, feeding, playing with.

0 points = no additional difficulty since the injury or no children

$\frac{1}{2}$ point = additional difficulty since the injury   _____

    b. Social interaction. Any change in other's reaction to you?

_____

_____

Any change in who asks whom out when?

_____

_____

In how you feel about being around other people, any change?

_____

_____

Any difficulty communicating with others because of problems in letter writing? Typing?

_____

_____

0 points = no difficulty

$\frac{1}{2}$ point = difficulty because of the injury   _____

    c. Finance. How has this been altered?

0 points = no change

$\frac{1}{2}$ point = change has occurred because of the injury   _____

    d. Entertainment. What changes have happened here with you?

_____

_____

What changes have happened with others?

_____

_____

What did you used to do that you can't do now?

_____

_____

What would you like to change?

_____

_____

0 points = no change

$\frac{1}{2}$ point = changes have occurred since the injury   _____

    e. Change. What has been the biggest change in your family since the injury?

Positive_____

Negative_____

0 points = no change

$\frac{1}{2}$ point = change   _____

                (Maximum points possible for IX is 10)

                Total points IX   _____

                Sum of points in I–IX   _____

Points
_____

FIGURE 4.1. (*Continued*)

X. DOMINANCE

A. Dominant (if injury does not affect activities of daily living or work requirements or is less than twenty percent total disability then do not rate dominance: If meets above criteria, have patient respond to which is his preferred hand. Sum of total points in I–IX (from above)

_____ × 0.10 = _____

B. Nondominant = injured hand is not the preferred hand = 0 points.

Total points X   _____

TOTAL   _____

FIGURE 4.1.   (*Continued*)

The results of this study were encouraging. The HDRS showed both substantial stability and significant relationship with the various tests of hand function. Further, the HDRS fared better by those standards than either the AMA or VA schedules. Subsequent revisions of the HDRS are in process, and the scale is being expanded to assess impairments of the upper and lower extremities.

FILING A WORKERS' COMPENSATION CLAIM

The process of filing a workers' compensation claim appears to be quite the same between jurisdictions. In each case the safety person or the person who administers first aid files a form with the agency that administers workers' compensation. The extent of the impairment is verified by a statement of the attending physician.

In this section, the process of filing a claim in two jurisdictions will be examined: the state of Oregon and the province of Ontario.

Workers' compensation in Oregon is administered by the Workers' Compensation Department. Workers' compensation is a no-fault insurance program. The employer assumes responsibility for all work-related accidents and occupational diseases. The employer pays for all necessary medical care, for loss of wages during periods of temporary disability, and for permanent disability. The employer may assume its own risks (self-insurer) if it is large enough or insure itself either through the state-owned fund or through private insurance companies. There are only two states that provide all of their employees workers' compensation. In return for the fact that the employer pays all costs, the employees give up their right to sue for damages. A worker is entitled to benefits if a compensable injury has been suffered. A worker who feels that he or she has suffered a compensable injury reports to the employer as soon as possible. The safety engineer fills out a State of Oregon Worker's and Employer's Report of Occupational Injury or Disease (see Figure 4.2). All employers or their insurers are required to notify the Workers' Compensation Department of

STATE OF OREGON
WORKER'S AND EMPLOYER'S
REPORT OF OCCUPATIONAL
INJURY OR DISEASE

CLAIM NO.

WORKER'S NOTICE TO EMPLOYER

| 1. FIRST NAME, MIDDLE INITIAL, LAST NAME | 2. TELEPHONE | 3. DATE OF BIRTH / / | 4. SEX | 5. SOCIAL SECURITY NUMBER / / | DO NOT WRITE IN THIS COLUMN |

6. STREET AND NUMBER 7. ADDRESS WHERE INJURED OR EXPOSED TO DISEASE (STREET, CITY)

8. CITY STATE ZIP CODE | 9. COUNTY IN WHICH INJURY OCCURRED | 10. ON EMPLOYER'S PREMISES? ☐ YES ☐ NO | 11. HOSPITALIZED AS INPATIENT? ☐ Yes ☐ No (IF EMERGENCY ROOM ONLY, MARK "NO")

12. FULL NAME AND ADDRESS OF ATTENDING PHYSICIAN | 13. NAME AND CITY OF HOSPITAL

14. DATE AND HOUR OF INJURY OR DATE OF DIAGNOSIS OF OCCUPATIONAL DISEASE / / A.M. P.M. | 15. NATURE OF INJURY OR DISEASE (EXAMPLES: CUT. BRUISE. POISONING. ETC.) | 16. PART OF BODY AFFECTED ☐ LEFT ☐ RIGHT

17. DESCRIBE ACCIDENT FULLY. DESCRIBE WHAT YOU WERE DOING WHEN INJURED. IF YOU FELL, STATE IF IT OCCURRED INDOORS OR OUTDOORS. IF YOU WERE STRUCK, NAME THE OBJECT WHICH STRUCK YOU. WERE YOU LIFTING, PULLING, PUSHING, OR CARRYING? IF MACHINERY WAS INVOLVED, NAME MACHINERY AND DESCRIBE ITS FUNCTION. NAME CHEMICAL INVOLVED IF APPROPRIATE.

18. NAMES OF WITNESSES | 19. I AUTHORIZE RELEASE OF MEDICAL INFORMATION. WORKER'S SIGNATURE | 20. DATE SIGNED

EMPLOYER'S ACKNOWLEDGMENT OF NOTICE

21. EMPLOYER'S NAME | 22. MAILING ADDRESS | ZIP CODE | 23. WCD NUMBER

24. DATE EMPLOYER FIRST KNEW OF INJURY / / | 25. NATURE OF BUSINESS (MFG, SHOES, TRUCKING FOR HIRE, LOG HAULING, RETAIL GROCERY, ETC.) | 26. ADDRESS OF EMPLOYER FACILITY IF DIFFERENT FROM EMPLOYER'S MAILING ADDRESS

27. WORKER'S OCCUPATION | 28. DEPARTMENT IN WHICH WORKER IS REGULARLY EMPLOYED | 29. HOW LONG WITH PRESENT EMPLOYER? | 30. NAMES OF OTHER WORKERS INJURED IN THIS ACCIDENT. (IF NONE, WRITE NONE)

31. SIGNATURE OF AUTHORIZED EMPLOYER REPRESENTATIVE | 32. TITLE | 33. TELEPHONE | 34. DATE SIGNED / /

INSURER USE ONLY

35. CLAIM IS | (A) ☐ ACCEPTED (B) ☐ DENIED (C) ☐ DEFERRED | AS | (D) ☐ DISABLING (E) ☐ NONDISABLING | (F) ☐ OCCUPATIONAL DISEASE (G) ☐ INJURY | DATE _____ / /
SIGNATURE OF INSURER'S REPRESENTATIVE

EMPLOYER—PLEASE ANSWER ALL QUESTIONS

36. IF THE ACCIDENT WAS CAUSED BY ANYONE BESIDES WORKER, GIVE NAME AND ADDRESS | 37. WAS ACCIDENT CAUSED BY FAILURE OF A MACHINE OR PRODUCT? ☐ YES ☐ NO

38. DESCRIBE ACCIDENT FULLY. WHAT WAS EMPLOYEE DOING WHEN INJURED? IF EMPLOYEE FELL, DID IT OCCUR INDOORS OR OUTDOORS? IF EMPLOYEE WAS STRUCK, NAME THE OBJECT WHICH STRUCK HIM (HER). WAS EMPLOYEE LIFTING, PULLING, PUSHING OR CARRYING? | 39. CHECK THE APPROPRIATE BOX(S) IF THE NONFATAL INJURY OR DIAGNOSED OCCUPATIONAL ILLNESS RESULTED IN (A) ☐ LOSS OF CONSCIOUSNESS (B) ☐ RESTRICTION OF WORK OR MOTION

40. NAMES AND ADDRESSES OF WITNESSES

41. WHAT MACHINE, SUBSTANCE, CHEMICAL, OR OBJECT WAS MOST CLOSELY CONNECTED WITH INJURY OR EXPOSURE? | 42. WERE MECHANICAL GUARDS OR OTHER SAFEGUARDS PROVIDED? ☐ YES ☐ NO | 43. DID WORKER RETURN TO HIS/HER NEXT SCHEDULED SHIFT AFTER THE ACCIDENT? ☐ YES ☐ NO

44. IN WHICH STATE WAS WORKER HIRED? | 45. IN WHICH STATE(S) WAS WORKER HIRED TO WORK? | 46. FATALITY? ☐ YES ☐ NO DATE / / | 47. TIME WORKER LEFT WORK HOUR AM PM | 48. DATE RETURNED TO WORK / /

49. NUMBER OF DAYS WORKED PER WEEK ☐ 3 OR LESS ☐ 4 ☐ 5 ☐ 6 ☐ 7 | 50. NAME SCHEDULED DAYS OFF | 51. WORKING SHIFT FROM AM PM TO AM PM | 52. NUMBER OF HOURS WORKED PER SHIFT

53. DOES WORKER RECEIVE TIPS? ☐ YES AMOUNT REPORTED TO ☐ NO EMPLOYER PER WEEK $ | 54. WAGE (INCLUDING BOARD AND ROOM) $ ☐ HR. ☐ DAY ☐ WK. ☐ MO. | 55. AVERAGE WAGE PER DAY IF PIECEWORK

56. UNDER WHAT CLASS CODE OF YOUR POLICY WERE WORKER'S WAGES REPORTED? | 57. DID INJURY HAPPEN DURING COURSE OF EMPLOYMENT? ☐ YES ☐ NO ☐ UNKNOWN | 58. IS INJURED WORKER A CORPORATE OFFICER, PARTNER SOLE PROPRIETOR? ☐ YES ☐ NO | 59. IS EMPLOYER AN ☐ CORPORATION ☐ OTHER ☐ INDIVIDUAL ☐ PARTNERSHIP

60. IF YOU DOUBT VALIDITY OF CLAIM STATE REASON | 61. EMPLOYER'S SIGNATURE | 62. DATE SIGNED / /

FORM 436-801 REV. 12-77

WORKER: COMPLETE ITEMS 1 THRU 20.
EMPLOYER: COMPLETE BALANCE OF THIS REPORT EXCEPT ITEM 35.
YOU SEND PAGES 1 THRU 3, WITH CARBON INTACT TO YOUR COMPENSATION CARRIER.
YOU KEEP PAGE 4.
YOU GIVE PAGE 5 TO THE WORKER.
(NOTE: COMPLETING THE SHADED ITEMS SATISFIES OSHA FORM 101 RECORD KEEPING REQUIREMENTS).

PAGE 1

FIGURE 4.2. State of Oregon Worker's and Employer's Report of Occupational Injury or Disease (*courtesy Workers' Compensation Department, State of Oregon*).

114

STATE OF OREGON
WORKER'S AND EMPLOYER'S
REPORT OF OCCUPATIONAL
INJURY OR DISEASE

WORKER'S NOTICE TO EMPLOYER

1. FIRST NAME, MIDDLE INITIAL, LAST NAME	2. TELEPHONE	3. DATE OF BIRTH	4. SEX	5. SOCIAL SECURITY NUMBER

6. STREET AND NUMBER	7. ADDRESS WHERE INJURED OR EXPOSED TO DISEASE (STREET, CITY)

8. CITY STATE ZIP CODE	9. COUNTY IN WHICH INJURY OCCURRED	10. ON EMPLOYER'S PREMISES? ☐ YES ☐ NO	11. HOSPITALIZED AS INPATIENT? ☐ Yes ☐ No (IF EMERGENCY ROOM ONLY, MARK "NO")

12. FULL NAME AND ADDRESS OF ATTENDING PHYSICIAN	13. NAME AND CITY OF HOSPITAL

14. DATE AND HOUR OF INJURY OR DATE OF DIAGNOSIS OF OCCUPATIONAL DISEASE A.M. P.M.	15. NATURE OF INJURY OR DISEASE (EXAMPLES: CUT, BRUISE, POISONING, ETC.)	16. PART OF BODY AFFECTED ☐ LEFT ☐ RIGHT

17. DESCRIBE ACCIDENT FULLY. DESCRIBE WHAT YOU WERE DOING WHEN INJURED. IF YOU FELL, STATE IF IT OCCURRED INDOORS OR OUTDOORS. IF YOU WERE STRUCK, NAME THE OBJECT WHICH STRUCK YOU. WERE YOU LIFTING, PULLING, PUSHING, OR CARRYING? IF MACHINERY WAS INVOLVED, NAME MACHINERY AND DESCRIBE ITS FUNCTION. NAME CHEMICAL INVOLVED IF APPROPRIATE.

18. NAMES OF WITNESSES	19. I AUTHORIZE RELEASE OF MEDICAL INFORMATION. WORKER'S SIGNATURE	20. DATE SIGNED

EMPLOYER'S ACKNOWLEDGMENT OF NOTICE

21. EMPLOYER'S NAME	22. MAILING ADDRESS	ZIP CODE 23. WCD NUMBER

24. DATE EMPLOYER FIRST KNEW OF INJURY	25. NATURE OF BUSINESS (MFG, SHOES, TRUCKING FOR HIRE, LOG HAULING, RETAIL GROCERY, ETC.)	26. ADDRESS OF EMPLOYER FACILITY IF DIFFERENT FROM EMPLOYER'S MAILING ADDRESS

27. WORKER'S OCCUPATION	28. DEPARTMENT IN WHICH WORKER IS REGULARLY EMPLOYED	29. HOW LONG WITH PRESENT EMPLOYER?	30. NAMES OF OTHER WORKERS INJURED IN THIS ACCIDENT. (IF NONE, WRITE NONE)

31. SIGNATURE OF AUTHORIZED EMPLOYER REPRESENTATIVE	32. TITLE	33. TELEPHONE	34. DATE SIGNED

THIS COPY TO BE SIGNED BY EMPLOYER AND RETURNED TO WORKER

READ THE FOLLOWING INSTRUCTIONS AND INFORMATION CAREFULLY.

WORKER'S RECEIPT

NOTICE TO WORKER

After you have completed the "Worker's Notice to the Employer", lines 1 through 20, your employer completed lines 21 through 34 and returned this page to you as acknowledgement of your claim. THIS IS A RECEIPT. Keep it as your record.

Your employer will submit the claim for you.

You will receive written notice from your employer's insurer of the action taken on your claim. If your claim is deferred, the insurer must notify you of the acceptance or denial within 60 days from the date your employer had notice of your claim. If denied, the reason for the denial will be included.

If questions regarding your claim are not resolved by your employer or his insurance company, contact the Workers' Compensation Department.

WORKERS' COMPENSATION DEPARTMENT
LABOR & INDUSTRIES BUILDING
Salem, Oregon 97310
CALL SALEM: 378-3302 or
TOLL FREE: 1-800-452-0288

FORM 436-801
REV. 12-77

PAGE

5

FIGURE 4.2. (*Continued*)

Form 82
2/79

IN ALL CASES OF INJURY

The EMPLOYER shall:

1 Make sure that first aid is given immediately, in accordance with the Regulations.

2 Record the first-aid treatment or advice given to the employee.

3 Complete and give to the employee a Treatment Memorandum (Form 156) if medical aid* is needed.

4 Provide immediate transportation to a hospital, a doctor's office, or the employee's home, if necessary.

5 Submit to the WCB an Employer's Report of Accidental Injury/Industrial Disease (Form 7) and any other information that may be requested.

The EMPLOYEE shall:

1 Promptly obtain first aid.

2 Notify the employer immediately of any injury requiring medical aid* and obtain from the employer a completed Treatment Memorandum (Form 156) to take to the doctor or the hospital.

3 Choose a doctor or other qualified practitioner, with the understanding that a change of doctor cannot be made without the permission of the WCB.

4 Complete and promptly return all report forms received from the WCB.

*Medical aid includes medical, surgical, optometrical, and dental aid; the services of osteopaths, chiropractors, and chiropodists; hospital and skilled nursing care; and the provision and maintenance of artificial members and appliances made necessary as a result of the injury.

In an emergency, a doctor or a hospital staff member may notify the WCB of an employee's injury.

Employers are required by the Workmen's Compensation Act to keep this form posted in a conspicuous place in full view of all their employees.

Workmen's Compensation Board

2 Bloor Street East
Toronto, Ontario
M4W 3C3

Claims Enquiry
Telephone
(416) 965-8851

DANS TOUS LES CAS D'ACCIDENTS

L'EMPLOYEUR DOIT:

1 S'assurer que les premiers secours se donnent sans délai, conformément aux règlements.
2 Noter dans un registre les premiers secours ou les conseils donnés à l'employé.
3 Remplir une note pour traitement (formulaire 156) et la remettre à l'employé s'il a besoin d'assistance médicale.*
4 Assurer le transport immédiat de l'employé chez le médecin, à l'hôpital ou à son domicile, le cas échéant.
5 Soumettre à la Commission des accidents du travail, un rapport de blessure accidentelle/maladie professionnelle (formulaire 7) et tous autres renseignements qui pourraient être demandés.

L'EMPLOYÉ DOIT:

1 Obtenir les premiers secours sans attendre.
2 Signaler immédiatement à l'employeur toute blessure qui requiert une assistance médicale* et obtenir de lui une note pour traitement (formulaire 156), remplie en bonne et due forme, à remettre au médecin ou à l'hôpital.
3 Choisir un médecin ou un autre praticien qualifié, étant entendu qu'il ne peut changer de médecin sans l'autorisation de la Commission des accidents du travail.
4 Remplir et remettre au plus tôt tous les formulaires reçus de la Commission des accidents du travail.

L'assistance médicale comprend les soins médicaux, chirurgicaux, optométriques et dentaires, les services d'ostéopathes, de chiropracteurs et de pédiatres; les soins hospitaliers ainsi que les services infirmiers dispensés par des personnes diplômées, la fourniture et l'entretien de membres artificiels et d'appareils nécessités par suite d'une blessure.

En cas d'urgence, le médecin ou un membre du personnel hospitalier peut signaler la blessure de l'employé à la Commission des accidents du travail.

Les employeurs sont tenus par la Loi sur les accidents du travail de placer cet avis en évidence dans un endroit fréquenté par tous les employés.

IN TUTTI I CASI D'INFORTUNIO

IL DATORE DI LAVORO DOVRÀ:

1 Assicurarsi che il pronto soccorso sia prestato immediatamente, in conformità con le regole.
2 Registrare il pronto soccorso oppure gli avvisi dati all'operaio.
3 Riempire e consegnare al lavoratore un Memorandum di Cura (Modulo 156) qualora questi abbia bisogno di assistenza medica.*
4 Provvedere immediatamente per il trasporto all'ospedale, al dottore oppure a casa, se necessario.
5 Inviare al WCB la relazione sull'infortunio o sulla malattia industriale (Modulo 7) e qualsiasi altra informazione che potrebbe essere richiesta.

L'OPERAIO DOVRÀ:

1 Ottenere subito il pronto soccorso.
2 Notificare immediatamente al datore di lavoro qualsiasi infortunio per cui ha bisogno di assistenza medica* e ottenere dal medesimo un Memorandum di Cura (Modulo 156), compilarlo e consegnarlo al dottore oppure all'ospedale.
3 Scegliere un dottore o un altro professionista qualificato, tenendo presente che non può cambiare dottore non è ammesso senza il permesso del WCB.
4 Compilare e spedire prontamente tutti i moduli che si ricevono dal WCB.

L'assistenza medica include: chirurgia, ottometria e assistenza dentistica, cure di osteopatia, di chiropratica e di pedicuristi; cure in ospedale ed assistenza infermieristica qualificata, nonché rifornimento e mantenimento di membri ed apparecchi artificiali necessari a causa dell'infortunio.

In caso di emergenza, il dottore o il personale dell'ospedale possono comunicare all'WCB l'infortunio dell'operaio.

I datori di lavoro sono obbligati dal decreto del WCB di tenere affisso, in luogo visibile a tutti i loro operai, questo avviso (Modulo).

EM TODOS OS CASOS DE ACIDENTE

O PATRÃO DEVE:

1 Prestar a primeira assistência imediatamente, de acordo com os regulamentos.
2 Registrar a primeira assistência prestada ou as directrizes dadas ao empregado.
3 Completar e dar ao empregado o Treatment Memorandum (Form 156) se for necessário assistência médica.*
4 Dar transporte imediato para o hospital, consultório médico ou para casa do empregado, se for necessário.
5 Enviar para a WCB a primeira comunicação do acidente (Employer's Report of Accidental Injury/Industrial Disease Form 7) e qualquer outra informação que possa ser pedida.

O EMPREGADO DEVE:

1 Obter a primeira assistência médica imediatamente.
2 Informar o patrão imediatamente de qualquer acidente que requeira assistência médica*, e obter do patrão, devidamente preenchida o (Treatment Memorandum Form 156) para ser entregue ao médico ou no hospital.
3 Escolher um médico ou outra pessoa qualificada, e tomar nota que a mudança de médico não pode ser feita sem a autorização da WCB.
4 Completar e devolver imediatamente todos os formulários recebidos da WCB.

Assistência médica inclui médicos, médicos cirurgiões, oculistas, e dentistas. Inclui ainda os serviços osteopáticos, de "chiropractors", de calistas, no hospital e de enfermagem; o fornecimento e manutenção de membros artificiais e utensílios necessários, usados em resultado do acidente.

Numa emergência um médico ou um membro do hospital pode notificar a WCB de qualquer acidente acordido com um empregado.

Os patrões são obrigados pela lei da WCB a afixar este cartaz em local próprio e que possa ser visto por todos os seus empregados.

ΕΙΣ ΟΛΑΣ ΤΑΣ ΠΕΡΙΠΤΩΣΕΙΣ ΑΤΥΧΗΜΑΤΩΝ

Ο ΕΡΓΟΔΟΤΗΣ ΘΑ ΠΡΕΠΗ:

1 Να βεβαιωθή ότι πρώτες βοήθειες παρέχονται αμέσως και σύμφως προς τους κανονισμούς.
2 Να σημειώση τις πρώτες βοήθειες ή συμβουλές που εδόθησαν στον εργάτη.
3 Να συμπληρώση και να δώση στον εργάτη το έντυπο που καλείται (Treatment Memorandum Form 156) εάν χρειάζεται ιατρική βοήθεια.*
4 Να εξασφαλίση άμεσον μεταφοράν σε νοσοκομείο ή εις ιατρικό γραφείο ή στο σπίτι του εργάτου εάν υπάρξη ανάγκη.
5 Να παρουσιάση στο WCB μια αναφορά εργοδότου το οποίον καλείται, και κάθε άλλη αναφορά που μπορεί να είναι αναγκαία. (Employer's Report of Accidental Injury/Industrial Disease Form 7)

Ο ΕΡΓΑΤΗΣ ΘΑ ΠΡΕΠΗ:

1 Να λάβη αμέσως πρώτες βοήθειες.
2 Να ειδοποιήση τον εργοδότη αμέσως σχετικά με κάθε τραύμα που απαιτεί ιατρική βοήθεια* και να πάρη από τον εργοδότη συμπληρωμένο το έντυπο που καλείται (Treatment Memorandum Form 156) το οποίο θα δώση στο γιατρό ή στο νοσοκομείο.
3 Να διαλέξη ένα γιατρό ή άλλο κατάλληλο ιατρικό πρόσωπο με την επίγνωση ότι αλλαγή γιατρού δεν μπορεί να γίνη χωρίς την άδεια του WCB.
4 Να συμπληρώση και να επιστρέψη αμέσως όλα τα έντυπα που μπορεί να λάβη από το WCB.

Ο όρος ιατρική βοήθεια περιλαμβάνει ιατρική, χειρουργική, οπτομετρική και οδοντιατρική βοήθεια, τας υπηρεσίας από χειροπρακτορας, φυσιοθεραπευτάς και ποδιάτρους, νοσοκομειακή θεραπευτική και νοσοκομειακή νοσηλεία, το εφοδιασμό και συντήρηση τεχνητών μελών του σώματος και άλλα απαραίτητα, λόγω του ατυχήματος.

Εις περίπτωσιν εκτάκτου ανάγκης, κάποιος γιατρός ή κάποιο μέλος, από το προσωπικό του νοσοκομείου μπορεί να ειδοποιήση το WCB σχετικά με τον τραυματισμό του εργάτου.

Οι εργοδόται είναι υποχρεωμένοι από τους νόμους του WCB να τοιχοκολλήσουν το παρόν έντυπο σε κάποιο κατάλληλο, σημείο ώστε να είναι ορατό σε όλους τους εργάτας.

FIGURE 4.3. Form 82 of Workmen's Compensation (*courtesy Workmen's Compensation Board, Toronto, Ontario*).

every claim for a disabling compensable injury within 21 days of the date the employer had notice or knowledge of the injury.

Workers' compensation in Ontario is administered by the Workmen's Compensation board which was established in 1915. The present act provides compensation, medical aid, rehabilitation, and pensions for Ontario employees disabled on the job. All medical costs are paid by the Workmen's Compensation Board. Employees who are disabled and who are incapable of earning full wages beyond the accident day receive weekly compensation payments totaling 75% of their average gross earnings. Maximum compensation in 1980 was $320.19, which is tax-free.

All employers must have posted a copy of Form 82 (see Figure 4.3) which advises employees of their rights and benefits under the Workmen's Compensation Act. All employers must have a first aid box and a qualified attendant who has a certificate from St. John Ambulance. In order to encourage and assist employers to have their employees trained in first aid, the Workmen's Compensation Board will pay for the training of up to two employees per shift. When an accident occurs, the following sequence takes place:

1. After an employee is provided first aid, the attendant is required to complete and give to the employee a "Treatment Memorandum" (see Figure 4.4), which authorizes additional treatment by whomever the employee chooses, if necessary.

2. The employer must then provide immediate transportation to a hospital, doctor, or employee's home.

3. The safety engineer files an "Employer's Report of Accidental Injury or Industrial Disease" (see Figure 4.5) with the Workmen's Compensation Board which establishes the claim.

4. The physician who sees the injured worker must file a "Doctor's First Report" (see Figure 4.6).

5. Under the law, an employer must report any accident involving lost time or medical aid within three days of learning about it. Employers do not have to report minor injuries that need first aid only, but they are required to keep a record of these accidents. The information required to fill out the form is shown in Figure 4.7.

6. The employee must fill out an Employee's Report of Accidental Injury or Industrial Disease (see Figure 4.8).

Cost of Workplace Injuries

Workers' compensation benefits have been rapidly increasing, and there is no sign that this trend will slow down or even catch its breath. Workers' compensation has been increasing at a rate of over $3 billion a year, and

Treatment Memorandum

Mr. Mrs. Miss	Last Name	Social Insurance Number

First Name(s)

Address

Doctor/Hospital

The above claims to have been injured in our employ on

_____ 19_____ and requires medical aid.

We are sending a report to The Workmen's Compensation Board, Ontario.

Firm

Address

Official	Date

> The injured employee has the initial choice of doctor, but may not change doctors without permission of The Workmen's Compensation Board, Ontario.

Doctor

If it appears that the injured employee will be disabled from earning full wages on any day beyond the day of accident, please submit a Doctor's First Report, Form 8, to The Workmen's Compensation Board.

Delay in completion may delay payment of compensation.

The Workmen's Compensation Board supports early vocational rehabilitation. If your patient is disabled immediate action is recommended to ensure that appropriate rehabilitation measures are instituted. Many employers accommodate their injured employees advantageously by minor modifications to their normal jobs or by transfer to other occupations more suited to their current temporary disabilities. To assist the employer and the Board in planning such measures, the Board urges that you discuss this matter with your patient and co-operate with the employer's medical staff or responsible representative in implementing a program which is reasonable and appropriate for the injured employee.

When submitting your account please indicate that you have received this form.

0156(09/79)

FIGURE 4.4. Treatment Memorandum (*courtesy Workmen's Compensation Board, Toronto, Ontario*).

Employer's Report of Accidental Injury or Industrial Disease

Workmen's Compensation Board
2 Bloor Street East
Toronto, Ontario
M4W 3C3

● Please see reverse for further details.

Shaded areas are for W.C.B. use only.

Claim No.

Employer Identification

Firm Name

Firm No. | Rate No. | Phone No.

Address

City/Town | Province | Postal Code

Plant, dept., or worksite where employed

Employee Reference No.

Miner's Certificate No.

Employee Identification

Last Name | First Name | Sex

Marital Status | Area Code | Phone No.

Address (no., street, apt.)

City/Town | Province | Postal Code

Date of Birth
day | month | year

Date of Employment
day | month | year

Occupation at time of the injury and years of experience in that occupation | Years Exp.

Language Spoken if not English

Ⓑ Name and address of attending physician(s)

Social Insurance No.

Date and hour of accidental injury
day | month | year | time   m

Ⓒ Date and hour reported to employer
day | month | year | time   m

Ⓓ History of Accidental Injury or Industrial disease

1. What happened to cause the injury?

2. Explain what the employee was doing and the effort involved.

3. Identify the size, weight and type of equipment or materials involved.

4. Describe injury, part of body involved and specify left or right side.

5. Where did the accident occur?

6. What conditions contributed to the accident?

7. Give the names and addresses of witnesses or persons having knowledge of the injury.

Claim Information

No Yes

(E) 1. Is the injured person an owner, spouse of the employer, (sub) contractor or executive of the business?

2. Did the accident happen outside Ontario? If yes, state Canadian province or country._____

3. Was anyone not in your employ totally or partially responsible for the accident?

(F) 4. Do you have any reason to doubt the history of injury?

No Yes

5. At the time of injury, was the employee doing work other than for the purpose of the employer's business?

6. Was there any serious and wilful misconduct involved? (G)

7. To your knowledge, has the employee had a previous similar disability? (H)

8. Do you have any information that the employee could have returned to work earlier?

Earnings and Lost Time Information

(I) Will the employee be totally or partially disabled beyond the day of injury?

□ No □ Yes—*Complete this section.*

1. If employed less than one week, enter normal one day's earnings. $_____

or

(J) 2. If employed for one or more weeks prior to accident, enter earnings up to four weeks.

From			To			Gross Earnings
day	month		day	month		

(K) Lost time without earnings, give dates and reason

Earnings for last day worked | Normal earnings for last day worked

(N) Authorized Signature | Official Title

Date and hour last worked | Date and hour returned to work

day month year time | day month year time

Normal working hours on last day worked | Estimate length of time off work

From ___ m To ___ m

	S	M	T	W	T	F	S	Total

(L) Enter employee's normal working days by : F = full day H = half day and total weekly pay hours.

From ___ day ___ month ___ year ___ time

To ___ day ___ month ___ year ___ time

If the employee worked after the first layoff, please enter dates.

(M) Enter particulars of any payments, allowances or benefits made, or to be made, for the period of disability.

Status	Injury	Sts	Injury	Sts	Injury	Sts	App.
	No. El.	Adj.	Vol.	Cst.	T.C.	Sig.	Mult.

0007 (8/80)

FIGURE 4.5. Employer's report of accidental injury or industrial disease (*courtesy Workmen's Compensation Board, Toronto, Ontario*).

Guide to Completing Employer's Report of Accidental Injury or Industrial Disease

When to Fill Out This Form

Accidents requiring first aid only do not have to be reported to the Board, but the Workmen's Compensation Act requires that you keep a record of details.

Claims may be submitted for eye glasses, dentures and artificial appliances damaged in an accident while being worn.

Section 117 of the Act requires that you file a report within three days of learning of an occupational injury or disease that disables an employee or requires medical aid. This form must be completed and sent to the Workmen's Compensation Board, Box 588, Postal Station F, Toronto, Ontario M4Y 2S4. Please make a copy for your own records.

Please type or print clearly in ink. If all of the information is not immediately available to you, please send what you have and the rest later. If additional space, please attach a separate letter.

Employer Identification

(A) If you intend to have your own identification number for this claim, such as the employee's payroll number to be shown on future correspondence or enquiries, please enter in the space provided. For mining companies and contractors doing mine work, please also report the employee's miner's certificate number.

Employee Identification

(B) If the employee has difficulty communicating in English and may require an interpreter, please state the language spoken.

History of Accidental Injury or Industrial Disease

(C) Please state the date and hour that the injury was first reported to an employer representative, such as first aid, immediate supervisor, or time office.

(D) The history of accidental injury or industrial disease should clearly describe an accident, circumstances surrounding the onset of pain in the apparent absence of an accident, or the events leading up to the industrial disease. Please answer the following questions explicitly to avoid additional enquiries.

1. Describe anything unusual that may have caused the injury. Examples: employee slipped, tripped, fell, struck wrist.
2. What was the employee doing when the symptoms were noticed? Was an awkward position, repetitive motion or physical exertion involved?
3. State the size and weight of any objects handled. Specify the tools, equipment, machinery, chemicals, and materials involved.
4. Describe the type of injury, all parts of the body affected and when applicable specify right or left side. Examples: cut right hand, low back pain, rash to both feet.
5. Where specifically did the accident occur? Examples: company parking lot, machine shop, Kingston construction site, Highway 400 near Barrie.
6. What conditions contributed to the accident? Examples: faulty equipment, icy parking lot, oily machine shop floor, littered worksite, slippery roads, ladder not tied down.
7. Were there any eye witnesses to the accident or others having knowledge of the history of injury as reported by the employee? If so, please state their names and addresses.

Claim Information

(E) Employers, owners, partners, independent operators, and their spouses, or an executive officer of a business, must have personal coverage to be considered an employee for the purposes of compensation. An executive officer is anyone holding the position of Chairman, Vice-Chairman of the Board of Directors, President, Vice-President, Secretary, Treasurer, or Director in a limited liability company, or General Manager or Manager designated an officer by by-law or resolution of the Directors.

For additional information about the above including contractors and sub-contractors, please refer to the current booklet titled "General Information and Guide For Completing the Employer's Statement of Payroll" or consult the office nearest you.

(F) Your explanation of any doubts about the history of injury should take into account statements given by all of the witnesses.

(G) Serious and wilful misconduct is the deliberate disobedience of expressed order, or the breach of a law or rule which is enforced, well known to and designed for the safety of the employees. A thoughtless act does not constitute serious and wilful misconduct.

(H) List any claim numbers for a similar disability if such are immediately available. Do not delay submission of this form to obtain them.

Earnings and Lost Time Information

(I) This section must be completed if the employee was (will be) totally or partially disabled beyond the day of injury whether or not full wages are paid. Full or partial advances are to be stated in the appropriate area.

(J) Enter the employee's gross earnings including overtime, bonuses, and additional benefits such as tips, gratuities, free meals, and free accommodation. State the number of free meals or number of days of free room and board during each pay period. Do not include vacation pay or temporary expenses for out of town jobs.

If the employee worked from one to four weeks for your company prior to the accident, enter the actual pay periods worked including partial weeks. If the employee worked more than four weeks, enter the four complete weeks prior to the accident. If the employee is paid bi-weekly, semi-monthly or monthly, include the equivalent up to four weeks. For each pay period, enter the gross wages. If employee is on annual salary, indicate the amount and that it applies for one year.

(K) If the employee lost any time without pay during any of the reported pay periods, please state the dates off work and the reason. Examples: illness, holidays, strike, lack of work, weather conditions.

(L) Enter the normal days with F for a full day or H for a half day worked plus the total number of hours per week for which the employee is paid. Please state to two (2) decimal places. Examples: state 40 hours as 40 and 40¼ hours as 40.25.

For **rotating shift workers**, print **shift** across the boxes and state weekly total representing an average number of hours per week for which the employee is paid. When the employee returns to work, complete the **Employer's Subsequent Statement** (Form 9) stating the total number of shifts lost plus the number of pay hours per shift.

(M) If the employee will receive any benefits from your company or any other insurance plan for the period of disablement, state the amount of benefits and the dates covered.

Authorized Signature

(N) This report must be signed by an authorized representative of your company. A partner or an executive officer of the company (except a sole owner) may *not* sign the report of his/her own injury.

If you require further assistance, or more compensation information, please telephone the office nearest you.

Hamilton	(416) 523-1800	North Bay	(705) 472-5200	Sault Ste. Marie	(705) 942-3002	Timmins	(705) 267-64
Kingston	(613) 544-9682	Ottawa	(613) 238-7851	Sudbury	(705) 675-9301	Toronto	(416) 965-88
Kitchener/Waterloo	(519) 576-4130	St. Catharines	(416) 937-2020	Thunder Bay	(807) 623-4545	Windsor	(519) 256-34
London	(519) 433-2331						

If you are not in the local calling area, check your telephone directory for the toll-free telephone number.

CLAIMS SERVICES DIVISION

FIGURE 4.5 *(Continued)*

FIRST MEDICAL REPORT
FOR WORKERS' COMPENSATION CLAIMS

ND ORIGINAL AND DUPLICATE
THE INSURER, PLEASE!

| | FOR WCD USE ONLY | | | | | – | | | |

rer Claim No.   Emp. No. — FOR WCD ONLY   LOC.

(1) Time of Injury ☐ AM ☐ PM   Date of Injury /   /

Worker's Legal Name (First, Middle Initial, Last)   Date of Birth /   /   Male ☐ Female ☐   Last Date Worked (2)

Worker's Address   City   State   Zip   Worker's Tel. No. (3)

Social Security Number — — (4)

Occupation (5)

Hospitalized as Inpatient? If yes, give hosp. name: ☐ Yes ☐ No. (6)

THE WORKERS' COMP. DEPT. ASKS YOU TO SEND THIS FORM PROMPTLY AND DIRECTLY TO THE INSURER PROPERLY ADDRESSED. IF ANY QUESTION, CALL TOLL-FREE NUMBER SHOWN AT THE RIGHT. THIS REPORT IS DELINQUENT IF HELD MORE THAN 72 HOURS.

Employer's Telephone Number (7)

Employer's Business Name (be specific)   Address   City   State (8)

Was Body Part Injured Before? If yes, describe: ☐ Yes ☐ No (9)

WORKER'S STATEMENT OF CAUSE AND NATURE OF INJURY OR EXPOSURE   Give name of your private health insurance company.

When signed this authorizes release of medical information and becomes NOTICE OF CLAIM.

Signature of Worker (10)

DESCRIBE COMPLAINTS (11)

NATURE AND LOCATION OF INJURY OR EXPOSURE (12)

Is Condition Work Related? If "no", explain ☐ Yes ☐ No ☐ Undetermined (13)

Released for Work? ☐ Yes ☐ No   IF YES, GIVE DATE /   /   Regular Modified (Give Limitations) Check One: ☐ ☐ (14)

R ATTENTION HOSP. EMERG. ROOM (1) COMPLETE ABOVE SECTIONS (2) ATTACH "ER" REPORT (3) SEND TO INSURER

X-Rays? If Yes, Give Findings. ☐ Yes ☐ No (15)

DIAGNOSIS (16)

FIRST TREATMENT
Time of Day ☐ AM ☐ PM   Date: /   Type of Treatment: (17)

Date of Next Treatment /   /   Estimate Length of Further Treatment ___ Months and/or ___ Weeks   Medically Stationary? ☐ Yes ☐ No   Will Injury Cause Permanent Impairment? ☐ Yes ☐ Undetermined ☐ No (18)

If Case Referred to another Doctor, Give Name and Address: (19)

REMARKS: (20)

Type Name of Physician and Degree   Address   Telephone Number (21)

Medical Aid Account Number   Date /   /   Doctor's Signature (22)

436-827 (8-79) STATE OF OREGON • Workers' Compensation Department • Labor and Industries Building • Salem, OR 97310

Insurer's Copy

INSURER

If you do not know the name and address of the insurer, contact the employer. If the employer cannot be reached, call Workers' Compensation Department Toll Free and Ask for "EMPLOYER'S COVERAGE". From Portland 229-5700 From Eugene 686-7500 From Salem 378-4954 Direct From All other areas 1-800-452-7813

FIGURE 4.6. Doctor's first report.

there is no light at the end of the tunnel. In 1980 the Alliance of American Insurers reported substantial losses in a number of states. The losses per dollar of premium were $0.45 in Kentucky, $0.33 in Maine, $0.30 in Wyoming, and $0.20 in Rhode Island. For the federal government, disability retirements more than tripled in the period from 1955 to 1974. A total of 339,436 disability pensions were granted during this period, ranging from 8,244 granted in 1955 to 30,015 granted in 1974.

Real Costs

Employer Identification

A. If you intend to have your own identification number for this claim, such as the employee's payroll number to be shown on future correspondence or enquiries, please enter in the space provided. For mining companies and contractors doing mine work, please also report the employee's miner's certificate number.

Employee Identification

B. If the employee has difficulty communicating in English and may require an interpreter, please state the language spoken.

History of Accidental Injury or Industrial Disease

C. Please state the date and hour that the injury was first reported to an employer representative, such as first aid, immediate supervisor, or time office.

D. The history of accidental injury or industrial disease should clearly describe an accident, circumstances surrounding the onset of pain in the apparent absence of an accident, or the events leading up to the industrial disease. Please answer the following questions explicitly to avoid additional enquiries.

1. Describe anything unusual that may have caused the injury. Examples: employee slipped, tripped, fell, struck wrist.

2. What was the employee doing when the symptoms were noticed? Was an awkward position, repetitive motion or physical exertion involved?

3. State the size and weight of any objects handled. Specify the tools, equipment, machinery, chemicals, and materials involved.

4. Describe the type of injury, all parts of the body affected and when applicable specify right or left side. Examples: cut right hand, low back pain, rash to both feet.

5. Where specifically did the accident occur? Examples: company parking lot, machine shop, Kingston construction site, Highway 400 near Barrier.

6. What conditions contributed to the accident? Examples: faulty equipment, icy parking lot, oily machine shop floor, littered worksite, slippery roads, ladder not tied down.

7. Were there any eye witnesses to the accident or others having knowledge of the history of injury as reported by the employee? If so, please state their names and addresses.

FIGURE 4.7   Guide to completing employer's report (courtesy Workmen's Compensation Board, Toronto, Ontario).

Claim Information

E. Employees, owners, partners, independent operators, and their spouses, or an executive officer of a business, must have personal coverage to be considered an employee for the purposes of compensation. An executive officer is anyone holding the position of Chairman, Vice-Chairman of the Board of Directors, President, Vice-President, Secretary, Treasurer, or Director in a limited liability company, or General Manager or Manager designated an officer by by-law or resolution of the Directors.

For additional information about the above including contractors and subcontractors, please refer to the current booklet titled "General Information and Guide for Completing the Employer's Statement of Payroll, or phone (416) 965-8650.

F. Your explanation of any doubts about the history of injury should take into account all statements given by witnesses.

G. Serious and willful misconduct is the deliberate disobedience of an expressed order, or the breach of a law or rule which is enforced, well known to and designed for the safety of the employees. A thoughtless act does not constitute serious and willful misconduct.

H. List any claim numbers for a similar disability if such are immediately available. Do not delay submission of this form to obtain them.

Earnings and Lost-Time Information

I. This section must be completed if the employee was (will be) totally or partially disabled beyond the day of injury whether or not full wages are paid. Full or partial advances are to be stated in the appropriate area.

J. Enter the employee's gross earnings including overtime, bonuses, and additional benefits such as tips, gratuities, free meals, and free accommodation. State the number of free meals, and free accommodation. State the number of free meals or number of days of free room and board during each pay period. Do not include vacation pay or temporary expenses for out of town jobs. If the employee worked from one to four weeks for your company prior to the accident, enter the actual pay periods worked including partial weeks. If the employee worked more than four weeks, enter the four complete weeks prior to the accident. If the employee is paid bi-weekly, semi-monthly or monthly, include the equivalent of up to four weeks. For each pay period, enter the gross wages. If the employee is on salary, indicate the amount and that it applies for one year.

K. If the employee lost any time without pay during any of the reported pay periods, please state the dates off work and the reason. Examples: illness, holidays, strike, lack of work, weather conditions.

L. Enter the normal days with F for a full day or H for a half day worked plus the total number of hours per week for which the employee is paid. Please state to two (2) decimal places. Examples: state 40 hours as 40 and 40¼ hours as 40.25.

For rotating shift workers, print shift across the boxes and state a weekly total representing an average number of hours per week for which the employee is paid. When the employee returns to work, complete the Employer's Subsequent Statement (Form 9) stating the total number of shifts lost plus the number of pay hours per shift.

M. If the employee will receive any benefits from your company or any other insurance plan for the period of disablement, state the amount of benefits and the dates covered.

Authorized Signature

N. This report must be signed by an authorized representative of your company. A partner or any executive officer of the company (except a sole owner) may not sign the report of his/her own injury.

FIGURE 4.7. (Continued)

125

Workmen's Compensation Board
2 Bloor Street East
Toronto, Ontario
M4W 3C3

COMPLETE AND RETURN THIS REPORT AT ONCE

CLAIM NO.

EMPLOYEE'S NAME & ADDRESS

PLEASE <u>PRINT</u> YOUR FULL NAME & ADDRESS AND SOCIAL INSURANCE NUMBER IN THE AREA PROVIDED BELOW IF IT DOES <u>NOT</u> AGREE AS SHOWN ON THIS FORM.

ACCIDENT DATE

Last Name

First Name & Middle Initial

Street Address

City/Town

Province

SOCIAL INSURANCE NO.

Postal Code

Social Insurance No. ___ ___

Phone No.

AGE_____ WEIGHT _____ HEIGHT _____ FT _____ INS —SEX _____ MARRIED _____

LANGUAGE PREFERRED

OCCUPATION _____

1
A. GIVE THE DATE AND HOUR OF THE ACCIDENT.
B. GIVE THE DATE AND HOUR YOU FIRST LAY OFF WORK.
C. GIVE THE DATE YOU FIRST REPORTED ACCIDENT TO YOUR EMPLOYER, FOREMAN OR OTHER OFFICIAL.
D. NAME THE PERSON TO WHOM REPORT WAS MADE
E. IN WHAT CITY, TOWN OR PLACE DID THE ACCIDENT HAPPEN?
F. DID IT HAPPEN ON EMPLOYER'S PREMISES? SPECIFY EXACTLY WHERE.

19 ___ AT ___ O'CLOCK ___ M.
19 ___ AT ___ O'CLOCK ___ M.
19 ___ AT. ___ O'CLOCK ___ M.
POSITION
PROVINCE OF

2
A. WAS THE WORK YOU WERE DOING FOR THE PURPOSE OF YOUR EMPLOYER'S BUSINESS?
B. WAS IT PART OF YOUR REGULAR WORK WITH HIM? _____

IN THE SPACE BELOW TELL US EXACTLY WHAT OCCURRED TO CAUSE YOUR DISABILITY. IF YOUR DISABILITY IS A **STRAIN** or **HERNIA**, YOUR HISTORY MUST CONTAIN THIS ADDITIONAL INFORMATION, (A) **DESCRIPTION, SIZE** AND **WEIGHT** OF ANY OBJECT BEING HANDLED (B) THE **PURPOSE** FOR WHICH THIS OBJECT WAS BEING HANDLED. (C) THE NAME OF HELPER, IF BEING ASSISTED. (D) DESCRIBE **POSITION** of **BODY, ARMS** AND **LEGS** AT EXACT TIME DISABILITY WAS FIRST NOTICED. (E) DESCRIBE WHAT YOU FELT AND NAME EXACT PART OF BODY AFFECTED.
NOTE — A FULL AND CLEAR DESCRIPTION IS NECESSARY IN ORDER TO AVOID FURTHER CORRESPONDENCE AND DELAY

3

4 DESCRIBE INJURY, IF APPLICABLE, STATE WHETHER RIGHT OR LEFT SIDE.

5 HOW LONG HAD YOU BEEN DOING THIS TYPE OF WORK FOR THIS EMPLOYER?

6
A. HAVE YOU HAD A SIMILAR DISABILITY BEFORE?
B. DO YOU HAVE OR HAVE YOU HAD ANY PHYSICAL DEFECT, MUTILA-TION, OR CHRONIC DISEASE?
(IF YOUR ANSWER IS "YES" TO EITHER QUESTION, GIVE PARTICULARS AND DATE OF LAST TREATMENT.)

0006(03/81)

SEE ALSO QUESTIONS ON BACK OF FORM

0006

FIGURE 4.8. Employee's report of accidental injury or industrial disease (*courtesy Workmen's Compensation Board, Toronto, Ontario*).

126

7	IF YOUR INJURY IS A HERNIA - A. GIVE EXACT TIME AND DATE YOU FIRST FELT PAIN OR OTHER SENSATION. B. GIVE EXACT TIME AND DATE YOU FIRST NOTICED A SWELLING	
8	WERE THERE ANY WITNESSES TO YOUR ACCIDENT OR DID YOU COMPLAIN TO ANYONE AT THE TIME ? GIVE NAMES AND ADDRESSES, TWO IF POSSIBLE	
9	IF YOU DELAYED REPORTING TO YOUR EMPLOYER, FOREMAN OR OTHER OFFICIAL, GIVE REASON FOR DELAY	
10	WAS ANYONE OTHER THAN YOUR EMPLOYER OR FELLOW EMPLOYEES TO BLAME FOR OR INVOLVED IN THE ACCIDENT? IF SO, GIVE NAME AND ADDRESS.	
11	A. DESCRIBE FIRST TREATMENT (FIRST AID, IF GIVEN), STATING WHEN RECEIVED AND GIVE NAME OF PERSON GIVING SAME B. NAME AND ADDRESS OF ATTENDING DOCTOR. C. NAME AND ADDRESS OF DENTIST, IF TEETH INJURED. D. IF TREATED AT HOSPITAL (OTHER THAN X-RAY) NAME HOSPITAL	
12	A. ARE YOU WORKING OR ABLE TO WORK? ☐ YES ☐ NO B. DID YOU WORK FOR ANY EMPLOYER BETWEEN THE FIRST DAY OFF AND NOW? IF YES, SHOW DATES ALSO NAME AND ADDRESS OF EMPLOYER. C. IF YOU CONTINUED AT WORK AFTER INJURY, GIVE DATES AND DESCRIBE WORK PERFORMED	ON WHAT DATE WERE YOU FIRST ABLE TO WORK? _____ _____ _____ FROM _____ TO _____
13	A. ARE YOU OWNER OR PARTNER IN THE BUSINESS OR A CONTRACTOR, SUBCONTRACTOR OR RELATIVE OF THE EMPLOYER? B. DO YOU HOLD ANY OF THE FOLLOWING OFFICES?	NO ☐ IF YES, STATE WHICH _____ NO ☐ PRESIDENT ☐ VICE PRESIDENT ☐ DIRECTOR ☐ SECRETARY ☐ TREASURER ☐

DO NOT ANSWER THESE QUESTIONS UNLESS YOUR DISABLEMENT LASTS BEYOND THE DAY OF ACCIDENT

SHOW SEPARATELY FOR EACH WEEK YOUR **GROSS** WAGES AND LOST TIME FOR THE FOUR WEEKS OR PAY PERIODS PRECEDING ACCIDENT INCLUDE BONUSES, TIPS, ETC., BUT DO **NOT** ADD OR INCLUDE VALUE OF VACATION PAY STAMPS OR VACATION PAY PERCENTAGE. **IF YOU RECEIVED FREE MEALS, OR FULL BOARD** (3 MEALS AND BED) **IN ADDITION TO WAGES, SHOW NUMBER OF MEALS, OR NUMBER OF DAYS BOARD IN EACH OF PAY PERIODS BELOW ***. IF YOU WORKED ELSEWHERE DUE TO LACK OF WORK IN EMPLOYMENT WHERE ACCIDENT OCCURRED, PLEASE ADVISE

14			OTHER ALLOWANCES *		LOST TIME				
	WEEK	WAGES	MEALS	BOARD	HOLIDAYS WITHOUT PAY	SICKNESS WITHOUT PAY	LACK OF WORK	OTHER	REASON
	FROM ____ TO ____	$ ____	____ MLS	____ DAYS	____ DAYS	____ DAYS	____ DAYS	____ DAYS	____
	FROM ____ TO ____	$ ____	____ MLS	____ DAYS	____ DAYS	____ DAYS	____ DAYS	____ DAYS	____
	FROM ____ TO ____	$ ____	____ MLS	____ DAYS	____ DAYS	____ DAYS	____ DAYS	____ DAYS	____
	FROM ____ TO ____	$ ____	____ MLS	____ DAYS	____ DAYS	____ DAYS	____ DAYS	____ DAYS	____

15	A. INDICATE YOUR NORMAL WORKING DAYS BY PLACING 'X' IN SPACES UNDER DAYS OF THE WEEK.

SUN	MON	TUE	WED	THU	FRI	SAT

B. WHAT WOULD HAVE BEEN YOUR NORMAL WORKING HOURS ON THE DAY OF LAY OFF? FROM _____ M. TO _____ M.

16	A. HAVE YOU BEEN PAID OR WILL YOU BE PAID ANYTHING BY YOUR EMPLOYER FOR THE PERIOD OF YOUR DISABILITY? IF SO, GIVE PARTICULARS. B. WHAT WERE, OR WILL YOU BE PAID FOR THE DAY YOU LAY OFF?	_____ TOTAL AMOUNT $ _____ $ _____ C. WHAT IS YOUR USUAL DAY'S PAY $ _____
17	ARE YOU RECEIVING OR ENTITLED TO RECEIVE ANY OTHER PAYMENT OR ALLOWANCE FOR YOUR DISABILITY? IF SO, GIVE PARTICULARS.	_____ TOTAL PER WEEK $ _____
18	HAVE YOU HAD OTHER ACCIDENTS OR DISABILITIES WHICH WERE REPORTED TO THE BOARD? IF SO, GIVE DATES AND NATURE OF INJURY.	_____ 19 _____ _____ 19 _____
19	THIS IS FORM 6. HAVE YOU PREVIOUSLY SENT TO THE BOARD A FORM 6 FOR THIS ACCIDENT?	

I DECLARE ALL THE ABOVE IS TRUE AND CORRECT AND I CLAIM COMPENSATION AND/OR MEDICAL AID.

SIGNED THIS _____ DAY OF _____ 19 _____ EMPLOYEE SIGN HERE _____

FIGURE 4.8. (*Continued*)

The real cost includes indirect costs. Indirect costs are many times the direct (or compensation) costs. Safety experts used to think the ratio was 4 to 1. But recent studies indicate the indirect costs are at least 9 times the direct costs. Indirect costs include such factors as lost production time, damage to plant and equipment, depressed morale, loss of skills, and retraining. Thus the "real cost" of industrial injuries in the United States and Canada is several times the millions per year that is paid out in workers' compensation claims.

SUGGESTED LEARNING EXPERIENCES

1. Discuss the history of workers' compensation from its inception to the present. What are some of the common themes? What issues are different?

2. Give a critique of workers' compensation laws in two different states. How are they similar? different?

3. As an employee, what improvements in workers' compensation would you like enacted?

4. What should a safety engineer know about workers' compensation?

5. Examine the AMA *Guide*.[1] Rate the following impairments: disarticulation of the arm at the shoulder, amputation of the middle finger and the thumb, and ankylosis of the knee at an unfavored position.

6. Malingering! What is it and what do you do about it?

7. Assess the Hand Disability Rating Scale. What improvements would you make?

8. Estimate the cost of safety for a shipbuilder that has 175 workers and is bidding for a new construction job?

9. Debate: Today workers' compensation is or is not adequate for workplace injuries and occupational illnesses.

10. There are five objectives of workers' compensation. How well have these objectives been met?

REFERENCES

1. Aiken, A. P. Basic requirements for an adequate compensation system. *Rhode Island Medical Journal*, 1955, **43**, 503–509.

2. American Medical Association. *Guide to the Evaluation of Permanent Impairment*. Chicago, 1972.

3. Bell, J. P. Rating of permanent physical impairment to the hand. *Oklahoma State Medical Association Journal,* 1967, **60,** 325–326.

4. Bertelsen, A., and Capener, N. Fingers, compensation and King Canute. *Journal of Bone and Joint Surgery, British Volume,* 1960, **42B,** 390–392.

5. Chamber of Commerce of the United States. *Analysis of Workers' Compensation Laws.* Washington, DC: Chamber of Commerce, 1981.

6. Cheit, E. F. *Injury and Recovery in the Course of Employment.* New York: Wiley, 1961.

7. Collins, I. *Permanent Disability Rating Schedule.* Sacramento, Calif.: Industrial Accident Commission, 1947.

8. Commonwealth of Massachusetts. *Report of the Special Commission Established to Make an Investigation and Study of Subjects of Benefits Payable under the Workmen's Compensation Law in Case of Certain Injuries.* Senate Report No. 427, 1949.

9. Current labor statistics. *Monthly Labor Review,* 1981, **104,** 67–104. (Dec.)

10. Felton, J. S. Workmen's compensation and rehabilitation. Efforts towards social maturity. *Rehabilitation Literature,* 1962, **23,** 230–234.

11. Gloss, D. S., and Wardle, M. G. Reliability and validity of the American Medical Association Guide for the Evaluation of Permanent Impairment. *Journal of the American Medical Association,* 1982, **248,** 2292–2296.

12. Gloss, D. S., and Wardle, M. G. *The Hand Disability Rating Scale; A Guide to the Evaluation of Permanent Impairment of the Hand.* Unpublished report, MDHR Associates, 1982.

13. Goulston, E. Industrial finger injuries. *Medical Journal of Australia,* 1972, **2,** 530–532.

14. Kessler, H. L. *Accidental Injuries* (2nd ed.). Philadelphia: Lea & Febiger, 1941.

15. Leaming, D. B., Walder, D. N., and Brainthwaite, F. The treatment of hands. *British Journal of Surgery,* 1960, **48,** 247–270.

16. McBride, E. D. *Disability Evaluation.* Philadelphia: Lippincott, 1963.

17. McCall, B. *Safety First at Last.* New York: Vantage Press, 1975.

18. Manning, G. C. *Disability and the Law.* Baltimore: Williams & Wilkins, 1962.

19. National Commission on States' Workmen's Compensation Laws. Report. Washington, D.C.: GPO, 1972.

20. National Safety Council. *Accident Prevention Manual for Industrial Operations* (7th ed.) Chicago, 1977.

21. National Safety Council. *Accident Facts.* Chicago, 1981.

22. Sinclair, U. *The Jungle.* New York: New American Library, 1980.

5 SAFETY AND THE RELATED PROFESSIONS

Responsibility for safety is not limited to a single profession. Several professions directly protect workers' safety and health; these are occupational medicine, industrial nursing, industrial hygiene, and, of course, safety engineering. Other professions which have a less direct relationship to occupational health and safety are fire protection, security, hazardous materials control, and mechanical engineering.

Still other professions support safety by systematically analyzing specific problems and developing and testing hypotheses. They are anthropometrics, biomechanics, epidemiology, ergonomics, health physics, and toxicology. Researchers establish the norms and data bases from which the safety profession may take action. With so many professions providing direct and indirect safety and health care for the worker, a certain amount of role confusion arises.

This chapter describes the relationship between safety engineering and related professions and suggests ways to improve cooperation and effectiveness.

THE SAFETY PROFESSION

Safety Engineer

The role of the safety engineer is what this entire book is about. This person is a professional who is committed to making the workplace as safe as possible. Principal responsibilities include ensuring a safe workplace,

	Industrial hygiene	Toxicology	Health physics	Ergonomics	Epidemiology	Anthropometrics	Mechanical eng.	Hazard material control	Security	Fire protection	Industrial nursing	Occupational med.	Safety eng.
Accident prevention													X
Accident investigation				X	X								X
Human error				X									X
Workers' compensation	X										X	X	X
Treatment of injuries and illnesses											X	X	
Workplace layout	X			X			X	X	X	X			
Health promotion												X	
Emergency preparation									X	X			X
Safety training	X												X
OSHA log											X		X
Hazardous materials		X						X		X			
Ergonomics	X			X		X							
Fire alarms									X	X			
Ventilation	X						X	X					
Temperature	X			X			X						
Illumination	X			X									X
Noise	X			X									X
Radiation			X										

FIGURE 5.1   Matrix illustrating overlapping responsibilities of the safety-related professions.

accident investigation, loss control and risk management, and safety training. They are discussed thoroughly in the specific chapters dealing with those topics. Safety engineering does not perform its work in a vacuum. It collaborates and cooperates with a number of other professions to make the workplace safe and healthy. Some of these relationships are shown in Figure 5.1. This figure shows that safety engineering shares many functions with other occupational safety and health specialties. For instance, occupational medicine, industrial nursing, and safety engineering all have functions in handling workers' compensation claims.

The safety engineer must have a bachelor of science degree to enter the field of safety. The background for the degree should include a variety of courses in engineering, psychology, communications, mathematics, and science. The safety professional should obtain certification by the Board of Certified Safety Professionals. In addition, the safety professional should obtain further training in ergonomics, management, public health, and engineering. The safety professional should be a member of the American Society of Safety Engineers and should routinely read *Professional Safety*, *National Safety News*, *Occupational Health and Safety*, and the *Journal of Occupational Accidents*.

PROFESSIONALS DIRECTLY RESPONSIBLE FOR OCCUPATIONAL SAFETY AND HEALTH

Occupational Medicine

Occupational medicine is concerned with industrial and preventive medicine.[13] The physician has learned to detect a cause–effect relationship between an illness and an agent at the workplace, that is, to diagnose and trace the onset of an illness to exposure to a specific harmful agent.

The fundamental responsibility of occupational physicians is to monitor the health and safety of employees. They consult with other professionals concerned with occupational safety and health. They are familiar with the health hazards peculiar to the industry or industries they serve. Occupational physicians are typically more concerned with health hazards and with medical treatment. Safety engineers are typically more concerned with physical hazards and with prevention. They are knowledgeable about most occupational diseases, such as byssinosis, silicosis, and asbestosis.

An outbreak of dermatitis will usually affect the industrial hygienist, the safety engineer, and the occupational physician. The industrial hygienist will be concerned with finding the cause of the outbreak, the safety engineer with the personal protective equipment needed and the workers' compensation claims which might need to be filed. The occupational physician will be concerned with the medical management of the cases, together with finding the substance which produced the observed symptoms.

Occupational physicians have a degree in medicine—an M.D., of course—plus residency in a medical speciality and a postgraduate degree in public health, environmental health, toxicology, or nuclear medicine.

Like all doctors, they must keep abreast of developments in a rapidly growing field. They will routinely read the *Journal of Occupational Medicine, Industrial Medicine and Surgery*, and the *American Journal of Industrial Medicine*. The occupational physician is probably a member of the American Association of Occupational Physicians.

Occupational physicians must have a broad, thorough educational and training base on which to build because they are apt to encounter situations demanding instant decisions, such as immediate treatment of people after an industrial spill, or the failure of a scaffolding, or an industrial fire. These professionals are in a position to improve the general attitude toward safety, depending on their commitment to workplace safety.

Occupational Health Nursing

Like the occupational physician, the occupational health nurse is concerned with employee health care. This encompasses activities such as health education, physical fitness regimes, preretirement programs, and the mobilization of workers to take responsibility for their own health. Examples of the last are changing a diet to prevent cardiovascular disease, obesity, or malnutrition; changing exercise patterns to prevent problems arising from a sedentary life-style; and changing attitudes toward the use of alcohol and drugs to prevent the health problems associated with their use.[14] Brenny[5] states that the 10 core functions of occupational health nursing include the following:

1. Administration of the nursing service
2. Health assessments
3. Care for workers with occupational illness or injury
4. Care for workers with nonoccupational illness or injury
5. Counseling and referral techniques
6. Employee health education services
7. Collaboration and consultation with other health care professionals
8. Use of community health services
9. Participation in plant safety inspections
10. Participation in special health programs, such as programs to reduce stress or to control alcoholism

The Occupational Safety and Health Act of 1970, with its stated purposes of assuring safe and healthful working conditions for all workers, has helped to bring the need for a comprehensive occupational health program into sharp focus. The occupational health nurse, as a member of the occupational health team, has the responsibility to participate in a colleague relationship with others in planning, implementing, and evaluating the total occupational health program and to take a leading role in identifying the nursing needs of the workers and in developing the nursing program to meet these needs.[14]

The occupational health nurse is often seen as principally concerned with the health and welfare of the work force. The nurse tends to many nonoccupational illnesses. The occupational health nurse is typically the first health care person to see an accident victim and to render first aid when applicable. The occupational health nurse understands the nurse practice limits and knows when to refer the patient to the occupational physician and/or to transfer the patient to a medical treatment facility. The nurse often maintains the OSHA log and knows which injuries and illnesses are recordable. Occupational health nurses, because of their early interventions, have a critical role in occupational safety and health.

The occupational health nurse is a registered nurse and holds a license to practice in one of the 50 states or 10 provinces. The occupational nurse usually has a minimum of a bachelor of science degree in nursing from an accredited school of nursing. The occupational nurse generally reads *Occupational Health Nursing, Nursing Outlook,* and the *American Journal of Nursing.* The occupational nurse is usually a member of the American Association of Occupational Nurses.

Industrial Hygienist

Industrial hygiene is a science dedicated to the recognition, evaluation, and control of those environmental factors or stresses, arising in or from the workplace, which may cause sickness, impaired health and well-being, or significant discomfort and inefficiency among workers or among the citizens of the community. Industrial hygienists generally have training and a degree in science, engineering, or medicine. In addition, they should have a master's in public health (M.P.H.) or a master of science in hygiene in industrial hygiene. Industrial hygienists should be certified by the American Board of Industrial Hygiene.[1] In addition, they should read the *American Industrial Hygiene Journal, Archives of Environmental Health,* and *Dangerous Properties of Industrial Materials* to keep abreast of current information in industrial hygiene. The industrial hygienist will probably belong to the American Industrial Hygiene Association.

Industrial hygienists are responsible for: (1) the recognition of environmental factors and stresses that occur in the workplace and their effects on persons; (2) the evaluation and measurement of these factors and stresses which impair a person's health; and (3) the techniques necessary to control or minimize such factors and stresses.[13] Industrial hygienists have nine core functions which include the following:

1. Administer the industrial hygiene program.
2. Examine the workplace environment and its environs:
   a. Study work operations and processes and obtain full details of the nature of work, materials and equipment used, products and by-products, number and sex of employees, and hours of work.
   b. Take appropriate measurements to determine the magnitude of exposure or nuisance to workers and the public.
3. Interpret the results of the examination of the work environment and environs in terms of ability to impair health and cause community nuisance and/or damage; present specific conclusions to appropriate interested parties, such as management and health officials.
4. Make specific recommendations for control measures.
5. Prepare and interpret rules, regulations, standards, and procedures for the healthful conduct of work.
6. Present expert industrial hygiene testimony when required.

7. Teach workers and the public in prevention of occupational disease and community nuisance.
8. Conduct epidemiologic studies of workers and industries to discover causes of occupational disease.
9. Conduct research to advance knowledge concerning the effects of occupation upon health and the means of preventing occupational health impairment, community air pollution, noise, nuisance, and related problems.

Industrial hygienists bridge the gap between manufacturing areas and the medical department. Their training enables them to apply the chemical, engineering, physical, biological, or medical knowledge needed to prevent occupational disease, which may range from deafness to heat stroke, from simple skin rash to cancer. Special techniques and knowledge enable them to plan for and assess environmental health problems and to provide the plant physician with the health background of the employee's job.[13]

Many occupational diseases have symptoms similar to nonoccupational ailments. Industrial hygienists can point out to the physician dangerous tasks within the plant. Such guidance enables the physician to correlate the patient's condition and complaints with potential health hazards of the job. When physicians know what to look for, they can, when necessary, ask for specific biochemical tests to find out whether the normal operation of the patient's bodily functions has been altered.[13]

PROFESSIONS RELATING INDIRECTLY TO OCCUPATIONAL SAFETY AND HEALTH

Fire Protection

Fire protection means constant surveillance. It requires the careful handling of hazardous or volatile materials and prompt reporting to the fire protection professional of anything amiss. Understanding how to handle flammable materials is critical to safety. Fire professionals must know auto ignition temperatures, flash points, and other vital information about industrial chemicals. They publish instructions to guide operating personnel in the selection, use, and handling of such substances. They would probably read *Fire Prevention, Fire Technology, Fire Engineering*, and *Fire Journal*.

The fire professional's domain is the entire workplace environment. It is imperative, considering the potential seriousness of consequences, not only to identify hazards but to implement rigorous precautionary measures and see that they are followed by everyone. These may include prohibiting smoking in certain areas, requiring that protective clothing be worn, using least hazrdous solvents, providing proper ventilation, and establishing regular monitoring of hazardous situations by the fire professional.

Security

Security is imperative for the maintenance of both the workplace and the worker. Security is the protection by whatever means necessary of the workplace materials to ensure they are not hampered or tampered with in any way; it is also the protection of the employee's belongings, that is, the condition the employee leaves his or her office in should be the same condition it will be in on the following day.

The scope of activity of the security personnel consists of doing whatever is necessary to protect the worker, the workplace environment, and the materials therein.

Hazardous Materials Control

Hazardous materials control is usually associated with environmental health. This profession is concerned with identification and control of any and all substances considered hazardous to the worker. A hazardous materials control plan must be completed by every facility using toxic substances. This plan will document the handling procedures and control of every hazardous material from the time it is received until it is disposed.

Figure 5.2 shows a typical warning label, and Figure 5.3 is a hazardous materials data sheet which is required by Federal Standard 313A. This form must be completed for each hazardous material used in a facility. Compliance with both is necessary so that anyone associated with a hazardous material is informed by a clear, informative, and distinctive label of what precautions are necessary. The information sheet contains all the information that anyone might need about a substance and is completed by a knowledgeable person on acceptance of the material. This is especially important in terms of materials handling and transportation. Professionals in hazardous materials control would probably read *Environmental Science and Technology* and *Dangerous Properties of Industrial Materials Report*.

Mechanical Engineering

Mechanical engineering is one of the engineering disciplines. A mechanical engineer usually has a bachelor's degree in mechanical engineering (B.S.M.E.) and should be a registered professional engineer (P.E.). Mechanical engineering is usually involved in the physical properties of a design and the environmental factors which affect this design. Mechanical engineers often specialize in heating, ventilating, and air conditioning (HVAC). They are also concerned with noise and vibration.

HAZARDOUS WASTE

FEDERAL LAW PROHIBITS IMPROPER DISPOSAL

IF FOUND, CONTACT THE NEAREST POLICE, OR
PUBLIC SAFETY AUTHORITY, OR THE
U.S. ENVIRONMENTAL PROTECTION AGENCY

PROPER D.O.T.
SHIPPING NAME _____ UN OR NA# _____

GENERATOR INFORMATION:

NAME _____

ADDRESS _____

CITY _____ STATE _____ ZIP_____

EPA
ID NO. _____

EPA
WASTE NO. _____

ACCUMULATION
START DATE _____

MANIFEST
DOCUMENT NO. _____

HANDLE WITH CARE!
CONTAINS HAZARDOUS OR TOXIC WASTES

STYLE WM-6

© LABELMASTER, CHICAGO, IL. 60626

FIGURE 5.2. Illustration of a hazardous material label (*copyright © from Labelmaster, Chicago, IL 60626*).

Noise and Vibration. Noise is unwanted sound. Noise causes major problems in many industrial activities, and it is the combined job of the mechanical engineer and safety engineer to control it. Noise will be discussed in detail in Chapter 14. Some of the techniques used to control noise include using acoustical absorbing materials, using barriers, soundproofing work stations, and designing quieter processes.

Vibration is the periodic motion of an object. This motion may make any piece of machinery shake. Mechanical engineers reduce or eliminate vibration by securing or attaching a vibrating piece of equipment to a solid object or by using a material to absorb vibration.

U.S. DEPARTMENT OF LABOR
Occupational Safety and Health Administration

Form Approved
OMB No. 44-R1387

MATERIAL SAFETY DATA SHEET

Required under USDL Safety and Health Regulations for Ship Repairing,
Shipbuilding, and Shipbreaking (29 CFR 1915, 1916, 1917)

SECTION I

MANUFACTURER'S NAME	EMERGENCY TELEPHONE NO.
ADDRESS (Number, Street, City, State, and ZIP Code)	

CHEMICAL NAME AND SYNONYMS	TRADE NAME AND SYNONYMS
CHEMICAL FAMILY	FORMULA

SECTION II - HAZARDOUS INGREDIENTS

PAINTS, PRESERVATIVES, & SOLVENTS	%	TLV (Units)	ALLOYS AND METALLIC COATINGS	%	TLV (Units)
PIGMENTS			BASE METAL		
CATALYST			ALLOYS		
VEHICLE			METALLIC COATINGS		
SOLVENTS			FILLER METAL PLUS COATING OR CORE FLUX		
ADDITIVES			OTHERS		
OTHERS					

HAZARDOUS MIXTURES OF OTHER LIQUIDS, SOLIDS, OR GASES	%	TLV (Units)

SECTION III - PHYSICAL DATA

BOILING POINT (°F.)	SPECIFIC GRAVITY (H₂O=1)
VAPOR PRESSURE (mm Hg.)	PERCENT, VOLATILE BY VOLUME (%)
VAPOR DENSITY (AIR=1)	EVAPORATION RATE (_____ =1)
SOLUBILITY IN WATER	
APPEARANCE AND ODOR	

SECTION IV - FIRE AND EXPLOSION HAZARD DATA

FLASH POINT (Method used)	FLAMMABLE LIMITS	Lel	Uel
EXTINGUISHING MEDIA			
SPECIAL FIRE FIGHTING PROCEDURES			
UNUSUAL FIRE AND EXPLOSION HAZARDS			

FIGURE 5.3. This data sheet must be completed for each hazardous substance located on the premises of any concern (*U.S. Department of Labor, Occupational Safety and Health Administration*).

SECTION V - HEALTH HAZARD DATA

THRESHOLD LIMIT VALUE

EFFECTS OF OVEREXPOSURE

EMERGENCY AND FIRST AID PROCEDURES

SECTION VI - REACTIVITY DATA

STABILITY	UNSTABLE		CONDITIONS TO AVOID	
	STABLE			

INCOMPATABILITY *(Materials to avoid)*

HAZARDOUS DECOMPOSITION PRODUCTS

HAZARDOUS POLYMERIZATION	MAY OCCUR		CONDITIONS TO AVOID	
	WILL NOT OCCUR			

SECTION VII - SPILL OR LEAK PROCEDURES

STEPS TO BE TAKEN IN CASE MATERIAL IS RELEASED OR SPILLED

WASTE DISPOSAL METHOD

SECTION VIII - SPECIAL PROTECTION INFORMATION

RESPIRATORY PROTECTION *(Specify type)*

VENTILATION	LOCAL EXHAUST		SPECIAL
	MECHANICAL *(General)*		OTHER

PROTECTIVE GLOVES	EYE PROTECTION

OTHER PROTECTIVE EQUIPMENT

SECTION IX - SPECIAL PRECAUTIONS

PRECAUTIONS TO BE TAKEN IN HANDLING AND STORING

OTHER PRECAUTIONS

FIGURE 5.3. *(Continued)*

140

Heating, Ventilating, and Air Conditioning.   The ultimate goal of heating, ventilation, and air conditioning (HVAC) is to create a climate of work in which true thermal comfort prevails. However, this seldom is achievable.

Heat.   Belding has discussed several ways to reduce heat stress.[3] These are decreasing the physical work of the task, modifying the thermal environment, considering the thermal conditions of the rest area, and requiring appropriate clothing (conventional work clothing, aluminized reflective clothing, or thermally conditioned clothing). See Figure 5.4 for an illustration of desired ambient temperatures for various work conditions. See Table 5.1 for a checklist for controlling heat stress and strain.

Cold.   Protection against cold temperatures uses the same principles as protection against heat, except that with heat humans have the adaptive mechanism of sweating while with cold they can only shiver or have their skin blood vessels constrict. According to Belding, the adaptation of humans to cold has been based on their ingenuity in providing themselves with insulative clothing and heated shelter. One approach to assess the

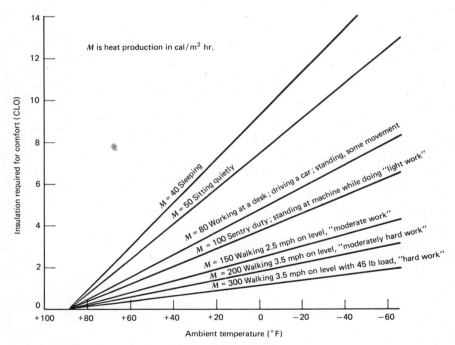

FIGURE 5.4.   Ambient temperatures for various work activities (*from NIOSH*).

TABLE 5.1
Checklist for Controlling Heat Stress and Strain

Item	Actions for Consideration
Components of Heat Stress	
M. Body heat production of task	Reduce physical demands of the work; powered assistance for heavy tasks
R. Radiative load	Interpose line-of-sight barrier
	Furnace wall insulation
	Metallic reflecting screen
	Heat reflective clothing
	Cover exposed parts of body
C. Convective load	If air temperature above 35°C (95°F)
	Reduce air temperature
	Reduce airspeed across skin
	Wear clothing
Emax, maximum evaporative cooling by sweating	Increase by
	Decreasing humidity
	Increasing airspeed
Acute heat exposure R, C, and Emax	Air- or fluid-conditioned clothing
	Vortex tube
Duration and timing	Shorten duration each exposure
	More frequent better than long to exhaustion
Exposure limit	Self-limited, based on formal indoctrination of workers and foremen and on signs and symptoms of overstrain
Recovery	Air-conditioned space nearby
Individual fitness for work in heat	Determine by medical evaluation, primarily of cardiovascular status
	Careful break-in of unacclimatized workers
	Water intake at frequent intervals
	Non–job-related fatigue or mild illness may temporarily contraindicate exposure (e.g., low-grade infection, diarrhea, sleepless night)

SOURCE: Belding, H. S. Control of exposure to heat and cold. In NIOSH, *The Industrial Environment—Its Evaluation and Control*. Washington, D.C.: GPO, 1973.

impact of cold is by using windchill, which is the measure of perceived cold (see Figure 5.5).

The insulating value of clothing depends on the thickness of air which it effectively immobilizes, not on specific insulation of the materials themselves. Thus an equal thickness of steel wool and eiderdown will provide about the same insulation. Cold weather outfits should be multilayered so that they can be adjusted for weather and variables of work load. The outer layer and perhaps a secondary layer should be of wind resistant fabric. Lightness of garments is achieved through use of resilient, low-density

WIND-CHILL CHART											
Estimated Wind Speed MPH	ACTUAL THERMOMETER READING °F.										
	50	40	30	20	10	0	−10	−20	−30	−40	−50	−60
	EQUIVALENT TEMPERATURE °F.										
Calm	50	40	30	20	10	0	−10	−20	−30	−40	−50	−60
5	48	37	27	16	6	−5	−15	−26	−36	−47	−57	−68
10	40	28	16	4	−9	−21	−33	−46	−58	−70	−83	−95
15	36	22	9	−5	−18	−36	−45	−58	−72	−85	−99	−112
20	32	18	4	−10	−25	−39	−53	−67	−82	−96	−110	−124
25	30	16	0	−15	−29	−44	−59	−74	−88	−104	−118	−133
30	28	13	−2	−18	−33	−48	−63	−79	−94	−109	−125	−140
35	27	11	−4	−20	−35	−49	−67	−82	−98	−113	−129	−145
40	26	10	−6	−21	−37	−53	−69	−85	−100	−116	−132	−148

LITTLE DANGER FOR PROPERLY CLOTHED PERSON	INCREASING DANGER Flesh May Freeze Within One Minute	GREAT DANGER Flesh May Freeze Within 30 Seconds

FIGURE 5.5.   Wind-Chill Chart (*used with permission from Concern, Nov./Dec., 1981, p. 5*).

materials (quilted fibers, pile, loosely woven wool or synthetic). Looseness of fit and easily adjustable closures will provide help in modifying the insulation to meet variable needs for work and rest.[3]

Ventilation. The term "general ventilation" suggests that a room or an entire building is flushed by supplying and exhausting large volumes of air throughout the area. Properly used, general ventilation can be very effective for the removal of large volumes of heated air or for the removal of low concentrations of nontoxic or low-toxicity contaminants. General, or dilution, ventilation is achieved by either natural or mechanical means. Often the best overall result is obtained with a combination of mechanical and natural air supply and exhaust. Ventilation is an important engineering control technique available for improving or maintaining the quality of the air in the occupational work environment. Broadly defined, ventilation is a method of controlling the environment with airflow. In industrial ventilation the airflow may be used for one or a combination of the following reasons:

1. Heating or cooling
2. Removing a contaminant
3. Diluting the concentration of a contaminant
4. Supplying makeup air

These basic uses of industrial ventilation can be divided into three major applications:

1. The prevention of fire and explosions
2. The control of atmospheric contamination to healthful levels
3. The control of heat and humidity for comfort[11]

The control of potentially hazardous airborne contaminants by ventilation can be accomplished by either one or both of two methods: (1) diluting the concentration of the contaminant before it reaches the worker's breathing zone by mixing it with uncontaminated air or (2) capturing and removing the contaminant at or near its source or point of generation, thus preventing the release of the contaminant into the workroom. The first of these methods is termed general ventilation or "dilution ventilation"; the second is called "local exhaust ventilation." Dilution ventilation does not reduce or eliminate the total amount of hazardous material released into the workroom air; local exhaust ventilation is the preferred and more economical method for contaminant control.[11]

PROFESSIONS WHICH HAVE DEVELOPED INFORMATION USED IN OCCUPATIONAL SAFETY

Epidemiology

Epidemiology is the branch of medical science that deals with the incidence, prevalence, distribution, and control of diseases. The area of epidemiology which describes the distribution of health status in terms of age, sex, race, and geography might be considered an extension of demography (the study of population) to health and disease. The second area involves an explanation of the patterns of disease in terms of causes. Although many disciplines seek to learn about the determinants of disease, the special contribution of epidemiology is its application of the statistics of frequency and distribution of disease.[10]

Epidemiology measures the frequency of illnesses and deaths in specific population groups and the relationship between exposure and incidence of disease. Thus studies of illness in groups of workers have made it possible to relate some diseases to substances with which the workers were in contact. Epidemiologic studies point to suspected and probable associations but do not prove cause–effect relationships. For example, epidemiologic studies of coal miners demonstrated that prolonged exposure to coal mine dust could produce pneumoconiosis (black lung). Other studies have shown the relationship between workers' illness and exposure to sugar cane dust (bagassosis), cotton dust (byssinosis), silica dust (silicosis), and various fibrous silicates (asbestosis).

Epidemiology has an excellent record of identifying human carcinogens. With few exceptions, most industrial carcinogens were first identified in humans. For example, it was known as early as 1775 that coal tar caused cancer in humans, but not until 1915 did an animal study confirm this knowledge.[16] Epidemiologic studies have revealed the carcinogenic properties of substances and chemicals such as arsenic, vinyl chloride, ionizing radiation, and other agents.

The most important goal of epidemiology is to identify causal mechanisms. Such knowledge can provide clues to developing preventive measures against diseases not currently preventable. This aim encompasses a number of subsidiary objectives:

1. Developing hypotheses that explain patterns of disease distribution in terms of specific human characteristics or experiences
2. Testing such hypotheses through specially designed studies
3. Testing the validity of the concepts on which disease control programs are based, through the use of epidemiologic data collected in conjunction with the programs
4. Aiding in the classification of ill persons into groups that appear to have etiologic factors in common

Even if the etiologic factors are not fully identified, similarity of epidemiologic behavior may point to etiologic similarity even of clinically distinct entities. Conversely, differences in the epidemiologic distributions of subgroups of a clinical entity may suggest that such subgroups be regarded as separate disease entities for purposes of etiologic investigation.[10]

It must be emphasized that the conclusion of an epidemiologic study is seldom a simple "yes" or "no." With so many intangibles in observational studies, epidemiologic studies need to be replicated; evaluated in terms of the design, data quality, and analysis done; and viewed in conjunction with knowledge from other biomedical disciplines.[16] Epidemiologic studies are based on records accumulated over time, and since the responsibility of maintaining these records usually lies with individuals other than epidemiologists, it would benefit safety engineers to be familiar with epidemiology, especially in terms of the data required and their sources. It is crucial for safety engineers to seek advice and support from epidemiologists.[16]

Biomechanics

Biomechanics is the branch of engineering devoted to improving the person-machine–task relationship in an effort to reduce operator discomfort and fatigue.[13] Biomechanics is a very effective tool in the prevention of excessive work stress. Cumulative effects of excessive physical stress on the worker can, in a rather insidious and subtle manner, result in physical debilities such as low back pain, tenosynovitis (inflammation of the tendon sheaths), and bursitis (inflammation of the bursa (the sac between a tendon and a bone, especially on the shoulder or elbow). These are often a cause of workers' compensation claims.

Cases of excessive fatigue and discomfort are, frequently, forerunners of soreness and pain. By exerting a strong distracting influence on a worker, these stresses can render him or her more prone to major accidents. Discomfort and fatigue will tend to make the worker less capable of maintaining the proper vigilence for safe task performance.[12]

Ergonomics

Ergonomics is the application of human biological science in conjunction with the engineering sciences to achieve the optimum mutual adjustment of humans and their work, the benefits being measured in terms of human efficiency and well-being. Other names for ergonomics are human engineering and human factors.

The ergonomics approach goes beyond productivity, health, and safety. It includes consideration of the total physiological and psychological demands of the job upon the worker.

In a broad sense, the benefits which can be expected from designing work systems to minimize physical stress on workers are as follows:

1. More efficient operation
2. Fewer accidents
3. Lower cost of operation
4. Reduced training time
5. More effective use of personnel

The human body can endure considerable discomfort and stress and can perform many awkward and unnatural movements for a limited period of time. When unnatural conditions or motions continue for prolonged periods, the physiological limitations of the worker may be exceeded. To ensure a high level of performance on a continuing basis, work systems must be tailored to human capacities and limitations.

The Person–Machine Model. Human engineers look upon the person and the machine as interacting components operating in an overall system. A simplified model includes the basic interactions between the person and the machine. The human operator is an organic sensor, a data processor, and a controller located between the displays and controls of a machine. An input of some kind is transformed by the machine into a signal which is displayed. The information so displayed is sensed by the operator, processed mentally, and translated into control responses. The control actions in turn alter the behavior of the machine to produce an output and further changes in the displays, thus causing the whole cycle to be repeated.

A person driving an automobile is a good example of such a system. The driver reacts to inputs from the speedometer and other displays on the dashboard, inputs from the road and outside environment, noise from the engine, feedback to his or her muscles from the steering wheel, and other stimuli. From these inputs the driver makes decisions to take certain control actions. These control actions affect the movements of the automobile, which in turn furnish new and different inputs to the driver.

This model of a simple person–machine system provides a framework for some of the things with which human engineers are concerned. Some important areas of work in human engineering are the allocation of functions between persons and machines; task analysis; the design of information displays, controls, panel layouts, and workplaces; design for maintainability; and the work environment.

Functional Allocation.   The ergonomist often determines the functions between persons and machines, that is, to "divide up a job" or decide who should do what in a system. When we ask about the allocation of functions between persons and machines, we ask two essential questions: Which functions of the system should be assigned to persons and which to machines? What kinds of things can and should human operators do in person–machine systems? Decisions of this kind are ideally made after a careful consideration of the kinds of things that a person can do better than machines (for example, perceiving, responding to low probability emergency situations) and those things that machines can do better than people (for example, computing, handling large amounts of information). These decisions are also ideally made early in the design process because they affect all the later design in the system (see Table 5.2).

Task Analysis.   This entails drawing up a very detailed and explicit list of the functions that persons will be doing in the completed system. The task analysis provides data essential for making decisions about selection standards, training requirements, work loads, the numbers of people who will be needed in the system, and the design of equipment to support the operator in his or her task. See Figure 5.6 for an example of task analysis.

Display Design.   Displays are those devices by which machines communicate information to persons. The first problem is the choice of the sensory channel to use for displaying information. Information may be displayed visually (by a dial, gauge, video screen, or printout) or aurally (by a buzzer, bell, gong, or voice). Some kinds of information are better displayed to one particular sense channel, other kinds of information to another.

TABLE 5.2
Capabilities: Persons versus Machines

Person	Machine
Can reason and make decisions inductively.	Has no inductive capability.
Can follow a random and variable strategy.	Always follows the preprogrammed strategy.
Can improvise and exercise judgment based on memory, experience, education, and reasoning.	Is better at routine functions.
Can make judgments and take action when preset procedures are impossible.	Programs for all conceivable situations, such as emergencies, and corrective or alternative actions are impracticable.
Can adapt performance since a person learns by experience, education, and reasoning.	Cannot learn facts and capabilities other than those it is programmed to learn.
Has high ability to reason out ambiguities and vague statements and information.	Is highly limited if input lacks clarity.
Can interpret an input signal accurately even in the presence of distraction, high noise level, jamming, or masking.	Can have performance degraded by interference, so it may fail entirely.
Can fill in lacking portions to supplement superficial training.	Pertinent facts and programming must be present and complete for accomplishment of function.
Can undertake new programs without extensive or precise programming.	Reprogramming must be as complete and precise as initial programming.
Can sometimes overcome effects of failure of one part of the nervous system through use of other parts.	Electronic systems will sometimes fail completely if only a single circuit element fails.
Is self-maintaining, or requires comparatively little care.	Maintenance is always required and increases with system complexity.
Is small and light in weight for all functions that can be performed, and requires little power.	Equivalent capabilities are generally heavier and bigger and have high power and cooling requirements.
Is in good supply and inexpensive for most functions.	Complexity and supply is limited by cost and production time.
Can override preset procedures and plans if necessary or preferable.	Can accomplish only preprogrammed actions within their designed capabilities.
Can add reliability to system performance by the ability to make repairs on associated equipment.	Generally has no repair capabilities.

TABLE 5.2 (*Continued*)

Person	Machine
Can detect and sometimes correct own mistakes.	Machines make few mistakes once their programs have been checked out. Programs frequently have self-check routines.
Can sometimes tolerate overloads without complete failure; in other cases, performance deteriorates slowly.	Even small overloads can cause complete breakdown or disruption of operations.
Has high performance flexibility.	Performs only tasks for which it was built and programmed.
Performance can be degraded by fatigue, boredom, or diurnal cycling.	Performance will be degraded only by lack of calibration or maintenance.
Long repetitive tasks will impair performance.	Performs repetitive or precise tasks well.
Can refuse to perform even when capable of doing so.	Will always respond to proper instructions except when there is a malfunction.
Can detect low-probability events impracticable in machine systems.	Many unexpected events cannot be handled adequately because of the size and complexity of the equipment required.
Can exert comparatively small force. Generally cannot execute a large force smoothly for extended periods.	Can exert large forces smoothly and precisely for almost any period of time.
Is not adapted to high-speed search of voluminous information.	Searching of voluminous information is a basic function of computers.
Is interested in personal survival.	Lacks consciousness of personal existence.
Is emotional in relation with others and in stress situations.	Has no personal relations or emotions.
Performance may deteriorate with work-cycle duration.	Performance is impaired relatively little with long work cycle if maintenance has been adequate.
Great individual differences can take place in performances by different personnel.	Only very minor differences in performances will take place by similar types of machines.
Has certain sensing abilities machines do not have: smell and taste.	Range of abilities generally extends outside human limits for those abilities it does have: can see into infrared and ultraviolet.
Quickly saturates capacity for accomplishing diversified functions.	Can be designed to accomplish a large number of functions at once. Ability to do each rapidly increases its ability to do many sequentially. Can frequently do many simultaneously.

149

Task Analysis

System	Lawnmower	Subsystem	Starting		Sheet 1 of 1	
Function	Operator	Equipment Cont/Disp	Equipment Location	Gross/Sub Task Description	Task Crit	Time (Sec)
Safety check		Engine on–off switch	On handle	1. Push switch to OFF.		5
				2. Check that switch is off.		1
				3. Bend down.		1
				4. Lift mower to 45° angle.		3
				5. Check that blades are free.		10
				6. Lower mower to horizontal.		2
				7. Stand up.		2
Starting engine		Engine on–off choke control		8. Push switch to ON.		1
				9. Pull out choke.		2
				10. Grasp starting cord.		5
				11. Pull starting; if engine starts, continue; if not reverse cord.		
				12. If engine fails to start, return to step 10 and repeat.		
		Choke control		13. If engine starts push in.		2

FIGURE 5.6. A task analysis for starting a gasoline-driven lawn mower.

Control Design.   Controls are devices that enable a person to communicate and transmit information to machines. Controls have to be selected so as to match human actions and the functions required of the machine. Some important factors that influence the design of controls are the control–display ratio, the compatibility of direction–movement relationships, safeguards against accidental activation, and control coding.

Panel Layout.   Displays and controls are typically located on control panels of equipment. It is important to place displays and controls in convenient locations.

Workplace Layout.   The topics that have to be considered here are workplace dimensions, the location of controls and displays, seat and panel design, the design of doors and accesses for easy entry and exit, and protective devices such as seat belts and restraining harnesses that may be required for certain situations.

Design for Maintainability.   As machines become more complex, problems of repair and maintenance become more critical. Human engineering design for maintainability includes such things as the design of efficient fault-finding strategies, the location of units for easy access, the design of auxiliary tools and test equipment, and the production of simple, easy-to-use maintenance manuals.

Work Environment.   Some of the important factors influencing the work environment that are studied by the human engineer include lighting; noise; vibration; acceleration and zero-gravity effects; variations in temperature, humidity, and barometric pressure; noxious gases, fumes, and other contaminants; and radiation.

Ergonomics has much information that the safety engineer needs to know in order to make the most judicial decisions relating to the workplace environment. One of the best-known safety engineers has continually maintained that if a safety engineer could only take one course it should be one in ergonomics.

Anthropometry

Anthropometry is the theory and practice of taking bodily dimensions with suitable instruments and the orderly treatment of the resulting data. As a method of distinguishing the bodily characteristics of individuals of different races and stages of development (physical growth), it has long been used by anthropologists. It is largely in the last few decades of the twentieth century that anthropometry has been used by those concerned in charting the diverse dimensions of persons in relation to the technological environment in which, increasingly, they are required to live and work.

For ergonomics, industrial medicine, and general industrial requirements, the basic need is to relate the dimensions of a person to his or her equipment so that the latter can be used with maximum advantage, efficiency, ease, and safety. This is important in ensuring that the individual attains optimum performance, that the unnecessary onset of fatigue will be avoided, and that other kinds of physiological stress will be minimized and the likelihood of accidents thereby reduced.

Function and structure are two aspects of work suitability. Measures of physical work capacities represent the functional aspects of occupational requirements for employment. However, the structural characteristics of women are equally important. Machine tools and equipment have been designed primarily for men to operate. Control panels were designed for male operators. Now that adult female workers are also employed, changes may have to be made in the design of new equipment and older equipment may have to be retrofitted to meet the female structure.[17]

Work Physiology

Work physiology is the study of the physiological requirements of work. It measures the physical work capacity of men and women. How much work is a person capable of performing? Workers in this field typically measure how much oxygen the body can use, how much capacity a person's lungs have, and what the capacity of the heart is. A second measure of work is the amount of energy consumed while working. This measure is the calorie. Some work, like sleeping, only requires 1 calorie per minute, while other work, like running a marathon, requires much more work. Physiologists have used the calorie to determine the extent of work. They have classified work as sedentary, light, moderate, heavy, very heavy, and ultraheavy. This data is very important in determining employment in occupations requiring hard physical work. See Table 5-3 for the caloric equivalent of work.

Wardle and Gloss published results of a study concerning women performing strenuous work.[17] The principal conclusion of this research indicates that a woman of average stature and weight can perform many jobs requiring hard physical work. Women should not be compared with men in selection for positions requiring light, moderate, heavy, or very heavy work. Rather, they should be compared with the occupational demands of the position.[8]

The fear that women will get hurt working in nontraditional, male-dominated jobs is not uncommon. The U.S. Department of Labor's Bureau of Labor Statistics (BLS) has shown that the relative frequency and kinds of injury occurring to both women and men in the same occupations and industries were similar. This seems to indicate that work activity, not the worker, is a more important determinant of injuries. The data also seemed to show that women in traditionally male-dominated jobs will suffer the

TABLE 5.3
Grades for Evaluating Intensity of Industrial Work (kcal/min)

Work Grade	65-kg man	55-kg woman
Sedentary	0–2	0–1.4
Light	2.0–4.9	1.5–3.4
Moderate	5.0–7.4	3.0–5.4
Heavy	7.5–9.9	5.5–7.4
Very heavy	10.0–12.4	7.5–9.9
Unduly heavy	12.5+	10.0+

SOURCE: Wardle, M.G., and Gloss, D. S. Women's capacities to perform strenuous work. *Women and Health*, 1980, **3**, 5–15 (Summer).

same kinds of injuries and generally with the same relative frequency as their male counterparts. Also, men in traditionally female-dominated occupations will suffer injuries common to their female counterparts with the same frequency. Women workers are no more likely to put their backs "out of whack" by lifting than are their male co-workers.

Health Physics

Health physics is the study of radiation exposures. One of the major concerns of health physics is that of determining the lowest radiation exposure possible and understanding what the risks of radiation are. In essence, health physics is for maximum radiation protection. This is most obvious in the form of the lead apron which technicians wear when taking x-rays and pregnant women wear when having x-rays taken; it is the reason why those taking the x-rays move outside the room while the film is being exposed.

Initially, the focus was on somatic effects (Hiroshima), then on genetic effects (effects of radiation on the unborn child and a concern with the radiation effects on future generations). Presently, the focus is also on multiple sources of exposure and how to measure the benefit part of the benefit–risk equation in a pluralistic society where one person's view of benefit is quite different from that of another.[15]

This field of health physics is very important in our advancing technological age in which radiation is used in many forms for many things, from microwave stoves in employees' cafeterias to the nuclear reactor plant which supplies all the electricity for an area. It is a field that we can never know too much about, because of the inherent devastating and damaging long-range effects radiation exposure may have.

Since one of the safety engineer's primary tasks is to make the workplace as safe as possible for all employees, the worker must know what radiation exposure there is and what the possible consequences are and must decide whether he or she wishes to have the exposure. It is the responsibility of the safety engineer to know or consult with a knowledgeable person in this area so every employee may have this information. It is also imperative that the surrounding community be made aware of the radiation levels and what they mean.

Health physics is a growing field and is one that will continue to grow in both depth and magnitude as more is known about radiation and more uses are made of radiation.

Environmental Health

Environmental health is that body of knowledge which studies what the environment consists of, including the pollutants and contaminants in the atmosphere. It is an umbrella discipline under which many of the bodies of knowledge, such as hazardous wastes, industrial pollutants, emission levels, illumination levels, and noise thresholds, are encompassed. This is another area that is growing and becoming of interest to all workers because of the risks therein. This is an area in which the government takes a very active role in setting regulatory standards.

Toxicology

Toxicology is the study of poisons. Environmental toxicology is the nature, distribution, and interaction of foreign chemicals that have become part of the environment. Toxicology includes determination of toxic levels and tolerance levels of a substance and the acceptable boundaries of pollutants in our environment.

One very important aspect of toxicology is the dose–response curve. Substances can differ in the concentrations at which they have any effect at all (potency level), the extent to which an effect is intensified by graded increase in dose (slope), and the maximum effect obtained (ceiling). Paracelsus recognized this property of chemicals over 500 years ago when he observed "only the dose makes a poison."

ROLE CONFUSION IN OCCUPATIONAL SAFETY AND HEALTH

Who is responsible for the health of workers? Is it the employer, the government, the unions, the workers themselves, or the courts? Or is it a combination of some or all of them? Without cooperation, collaboration, and coordination among all parties, employee safety and health will not be given priority, and safety programs will not be developed or implemented.

Even if the law required court orders to employers to provide the safest possible environment, if workers choose not to wear protective equipment or refuse to use machine guards, they can make the workplace unsafe for themselves. Sometimes the most effective policing agent is the union because workers are frequently more loyal to their unions than to the company. Many professionals who have never belonged to unions, may not understand or take seriously the power of the union. Strategically, it is very important for key people in the local unions to be aware of the need for the safety restrictions and to support management in enforcing them. Workers feel the union is responsible for their getting where they are and for the benefits they have, and they will do what the union says. It is imperative that all work together to resolve differences and to make sure safety is equally important to all.

As discussed in this chapter, safety is an interdisciplinary concern, involving people from many professions whose duties and responsibilities often overlap, such as the safety engineer and the industrial hygienist. This can produce role confusion—that is, it is not always clear who is supposed to be doing what. (See Figure 5.1.) Because of this, a brief introduction to role theory and its effect on occupational safety and health is included.

Role Theory

The term "role" is commonly used to refer to both the expected and the actual behaviors associated with a position. Role theory utilizes a conceptual model that predicts how persons will perform in a given role and the types of behaviors that may be expected under a particular set of circumstances.[4] Goffman describes how people change roles with changing circumstances.[9] In *Presentation of the Self in Everyday Life* he eloquently portrays Shakespeare's line "All the world is a stage and all its people players." Safety engineers play a variety of roles with the workers, the foremen, the safety committee, the management, their peers, and their associates.[10]

MULTIPLE ROLES OF SAFETY ENGINEERS

The safety engineer has many roles in an organization, from consultant to researcher to inspector. Not only do his or her responsibilities overlap with each other, but they also overlap with the responsibilities of other safety and health specialists. This can lead to role ambiguity and conflict which may cause stress and poor work performance.

Role ambiguity is vagueness, uncertainty, and lack of agreement on role expectations. It occurs in many occupations. Role ambiguity occurs in safety engineering because the expectation of the job differs greatly from the company policies and procedures.

The role of the safety engineer in controlling hazards is in conflict with his or her role in management to increase return on investment (ROI). When a safety engineer makes a cost trade off and allows a hazard to continue unabated, this is the result of role conflict. Role conflict occurs in many other professions. Board of education members are frequently having a role conflict between providing quality education and increasing taxes. American soldiers in Vietnam had a severe role conflict between being soldiers in an unpopular war and being American citizens who did not want to be in Vietnam in the first place. Workers frequently have a role conflict between the production norms of the peer group and those of management.

Role ambiguity and role conflict both produce stress and anxiety. When stress and anxiety are severe enough, they lead to impaired performance and job dissatisfaction. When stress and anxiety are prolonged and severe, they can cause physical and/or psychological symptoms to occur.[7]

One of the commonest strategies for dealing with role conflict, ambiguity, and strain is partial or total withdrawal from the situation. This means less participation, but more internal comfort. Another way is to figure out a logical way to reduce pressure while at the same time being just as effective.

SUGGESTED LEARNING EXPERIENCES

1. Conduct a seminar on professional, ethical, and legal aspects of a safe and healthy environment.

2. Collect interviews from the various levels of workers in an industrial organization concerning their beliefs about safety and their roles in safety. Is there a relationship?

3. Conduct a seminar with various members representing different related professions, including safety, focusing on their frustrations and concerns in their jobs.

4. Attend an interdisciplinary plant meeting. Know the goals of the group. Observe the communication patterns in the group. Discuss the safety professional's participation in this group. From your observation, what statement would you make regarding his or her power in the group?

5. If you were the key safety professional in an organization and could hire only one other safety professional, what discipline would you prefer and why?

6. As a safety professional in an organization that had six other safety- and health-related professionals, how would you suggest task allocation so there would be no duplication of effort?

7. Prepare a poster depicting the similarities and differences of safety in each of its related professions.

8. In the town in which you live, are health and safety professionals acting in any capacity (elected, voluntary, as consultants to decision-making boards? Why? Why not?

9. If you had to do it over again, would you choose the profession you did? Discuss.

10. In terms of your development as a safety professional, what is a key concept that has meaning to you? Explain.

REFERENCES

1. American Board of Industrial Hygiene. Bulletin. Akron, Ohio, 1980. (Feb.)

2. Barachas, J. D., Berger, P. A., Ciaranello, R. D., and Elliott, G. R. *Psychopharmacology from Theory to Practice.* London and New York: Oxford University Press, 1977.

3. Belding, H. S. Control of exposure to heat and cold. In NIOSH, *The Industrial Environment—Its Evaluation and Control.* Washington, D.C.: GPO, 1973.

4. Biddle, B., and Thomas, E. (Eds.). *Role Theory: Concepts and Research.* New York: Wiley, 1966.

5. Brenny, A. A. Reconsidering core functions: For management's sake. *Occupational Health and Safety,* 1981, **50,** 40–43. (Aug.)

6. Brown, M. S. *Occupational Health Nursing Principles and Practices.* New York: Springer, 1981.

7. Conway, M. W. Theoretical approaches to the study of roles. In *Role Theory Perspectives for Health Professionals.* New York: Appleton-Century-Croft, 1978.

8. Gloss, D. S., and Wardle, M. G. Ergonomics and the working woman. *Occupational Health and Safety,* 1979, **49,** 20–24. (May/June)

9. Goffman, E. *The Presentation of the Self in Everyday Life.* New York: Anchor, 1960.

10. McMahon, B., and Pugh, T. F. *Epidemiology Principles and Methods.* Boston: Little, Brown, 1972.

11. Mutchler, J. E. Principles of ventilation. In NIOSH, *The Industrial Environment—Its Evaluation and Control.* Washington, D.C.: GPO, 1973. pp. 573–582.

12. The occupational nurse: Key member of the medical team. *Occupational Hazards,* 1981, **41,** 111–114.

13. Olishifshi, J. B., and McElroy, F. E. *Fundamentals of Industrial Hygiene.* Chicago: National Safety Council, 1971.

14. Tinkham, C. W. The plant as the patient of the occupational health nurse. *Nursing Clinics of North America,* 1972, **7,** 99–107. (Mar.)

15. Wald, N. Keynote address. *Health Physics Society: Health Physics in the Healing Arts.* Rockville, Md.: Health Service, Food and Drug Administration, 1973.

16. Wang, O. A practical guide for non-epidemiologists. *Occupational Health and Safety,* 1981, **50,** 21–30. (Nov.)

17. Wardle, M. G., and Gloss, D. S. Women's capacities to perform strenuous work. *Women and Health,* 1980, **3,** 5–15. (Summer)

PART 2 HAZARD RECOGNITION AND CONTROL

6 ACCIDENT CONDITIONS AND CONTROLS

Accidents happen. Conditions which can cause hazards will always exist. It is the duty of safety engineers, using both theory and practice, to understand problems and possible consequences and take whatever actions are appropriate to correct a hazardous situation. Identifying and correcting hazardous conditions tests the safety engineer's knowledge, training, and creativity. Measuring and recording accidents and occupational illnesses are important for assessing the situation.[3] This chapter describes unsafe conditions and acts; presents a theory of accident causation; and discusses ergonomics and strategies of hazard control.

Potential hazards and recommended controls are shown in Table 6.1. An unsafe, or hazardous, condition means that certain behavior is likely to result in an accident. For example, working with electric tools in the rain can result in severe shock.

THE UNSAFE CONDITION

An unsafe condition is any condition, which in the right set of circumstances, will lead to an accident. Some potential unsafe conditions which are found in industry include the following:

1. Lifting, pushing, or pulling of objects
2. Falling and flying objects
3. Stationary and moving objects

TABLE 6.1
Examples of Potential Hazards and Recommended Control

Safety Area	Potential Hazard	Recommended Control
Electrical	Wire connections (bare terminals)	Insulated sleeving
	Exposed voltages on contacts and terminals	Barriers or guards of transparent, nonconductive material; should use socket connectors
	Current/voltage warnings	Marked per ANSI Z35.1-1972 format[a]
	Connectors (damage, breakage)	Connectors recessed on chassis or panels
	Power and interconnecting cables	Color coded
	Mismating connections	Separate connectors for AC/DC when a power option is provided for given equipment; positive means, such as keying, to prevent inadvertent reversing or mismatching of electrical and mechanical connections
	Fuse usage in primary distribution	Not permitted; should use circuit breakers
	Fuses	Located in serviceable locations
		Branch fuses located at the module level to confine power supply overloads to the failed module
	Off the shelf	As a minimum, will be UL-approved
Lighting and current	Lightning and other intermediate voltage/current hazards	Grounding rods, lightning arrestors on signal cables; grounding in power cable to a central-point ground stud

TABLE 6.1 (*Continued*)

Safety Area	Potential Hazard	Recommended Control
Chemical	Battery venting	Enclosure constructed to prevent expelling of electrolyte or leakage of gas or other ignition sources into main equipment
	Toxic materials	Use of NIOSH Toxic Substance List to identify and prevent usage of toxic materials
Noise		Design equipment not to exceed the noise limits of MIL-STD-1474B[b] or 29 CFR 1910.95[c]

[a] American National Standards Institute, *Accident Prevention Signs Specifications* (ANSI 235.1-1972). New York: 1972.
[b] U.S. Department of Defense, *Noise Limits for Army Material* (MIL-STD-1474B). Washington, D.C., June 1979.
[c] Occupational Safety and Health Administration, *Code of Federal Regulations*. General Industry Standards (29 CFR 1910). U.S. Department of Labor, Washington, D.C., June 1981.

4. Working surfaces
5. Scaffolds, ladders, vehicles, stairs, and shafts
6. Involuntary or voluntary motions
7. Being caught in, under, or between a moving and a stationary object or two or more moving objects
8. Pressure, vibration, or friction between person and source of injury
9. Electric shock or electrocution
10. Temperature extremes

THE UNSAFE ACT

An unsafe act means carrying out a task under less than safe conditions. Examples are fooling around, not completing a routine checklist, climbing a ladder with broken rungs, and not wearing proper protective equipment. Unsafe acts are responsible for most work-related disabling injuries and deaths.

In 1980 the disabling work injuries in the nation totaled approximately 2,200,000. Of these, about 13,000 were fatal and 80,000 resulted in some permanent impairment. Injuries to the trunk occurred most frequently, with thumb and finger injuries next. These break down into the following categories:

Eyes	110,000
Head (except eyes)	130,000
Arms	200,000
Trunk	640,000
Hands	150,000
Fingers	330,000
Legs	290,000
Feet	110,000
Toes	40,000
General	200,000

The leading causes of death are shown in Table 6.2. Among persons of all ages, accidents are the fourth-leading cause of death. For persons aged 15 to 24 years, accidents claim more lives than all other causes combined, and about 5 times more than the next leading cause of death. Four out of five accident victims in this group are males.[11]

A THEORY OF ACCIDENT CAUSATION

A theory of accident causation based on the sequence of events that culminate in injury has been proposed.[7] A principle utilized in this theory is that the greater the amount of energy available for potential damage, the earlier in the causal sequence prevention is directed. For example, the margin of unneeded protection in crash-padding and safety belts decreases with greater velocity of impact, so that injuries secondary to very high velocity crashes can be completely prevented only by preventing the crashes themselves, rather than by merely properly packaging the vehicle occupants.

STRATEGIES FOR CONTROL

Other strategies for accident control include pinpointing hazardous conditions through job study observations and work practice audits; preventing the accumulation of hazardous materials; reducing the amount of energy accumulation; preventing or modifying the release of energy; separating individuals from energy by time and space; raising the injury threshold; acclimatizing the worker; and controlling behavior.

TABLE 6.2
Leading Causes of All Deaths

	No. of Deaths	D
All ages	1,927,788	
Heart disease	729,510	334
Cancer	396,992	182
Stroke[b]	175,629	80
Accidents	105,561	48
Motor vehicle	52,411	24
Falls	13,690	6
Drowning	7,026	3
Fires, burns	6,163	3
Other	26,271	12
Under 1 year	45,945	1,434
Anoxia	9,556	298
Congenital anomalies	8,404	262
Complications of pregnancy and childbirth	5,544	173
Immaturity	3,677	115
Pneumonia	1,499	47
Accidents	1,262	39
Ingestion of food, object	296	9
Motor vehicle	264	8
Mech. suffocation	242	8
Fires, burns	154	5
Other	306	9
1 to 4 years	8,429	69
Accidents	3,504	29
Motor vehicle	1,287	11
Fires, burns	742	6
Drowning	630[c]	5
Ingestion of food, object	167	1
Falls	121	1
Other	557	5
Congenital anomalies	1,027	8
Cancer	599	5
5 to 14 years	12,030	34
Accidents	6,118	17
Motor vehicle	3,130	9
Drowning	1,010[c]	3
Fires, burns	586	1
Firearms	297	1
Other	1,095	3
Cancer	1,500	4
Congenital anomalies	650	2

TABLE 6.2 (*Continued*)

	No. of Deaths	Death Rate[a]
15 to 24 years	48,500	118
Accidents	26,622	64
Motor vehicle	19,164	46
Drowning	2,180[c]	5
Firearms	581	1
Poison (solid, liquid)	577	1
Other	4,120	11
Homicide	5,443	13
Suicide	5,115	12
25 to 44 years	103,991	179
Accidents	29,024	43
Motor vehicle	14,574	25
Drowning	1,700[c]	3
Poison (solid, liquid)	1,210	2
Falls	1,053	2
Fires, burns	1,044	2
Other	5,443	9
Cancer	16,866	29
Heart disease	14,167	24
45 to 64 years	434,246	990
Heart disease	151,564	346
Cancer	134,115	306
Stroke[b]	21,670	49
Accidents	18,774	43
Motor vehicle	8,048	18
Falls	2,101	5
Fires, burns	1,400	3
Drowning	910[c]	2
Surg. complications	901	2
Other	5,414	13
Cirrhosis of liver	16,449	37
Diabetes mellitus	7,790	18
65 to 74 years	452,259	3,027
Heart disease	183,880	1,231
Cancer	119,623	801
Stroke[b]	36,390	244
Diabetes mellitus	9,629	64
Accidents	9,072	61
Motor vehicle	3,217	22
Falls	1,852	13
Fires, burns	789	5
Surg. complications	783	5
Ingestion of food, object	483	3
Other	1,948	13

TABLE 6.2 (*Continued*)

	No. of Deaths	Death
Pneumonia	9,225	62
Cirrhosis of liver	6,209	42
75 years and over	822,388	9,012
Heart disease	377,322	4,135
Cancer	121,569	1,332
Stroke[b]	113,336	1,242
Pneumonia	33,777	370
Arteriosclerosis	24,046	264
Accidents	15,185	164
Falls	7,830	86
Motor vehicle	2,727	30
Surg. complications	963	10
Fires, burns	918	10
Ingestion of food, object	716	8
Other	2,031	22
Diabetes mellitus	14,748	162
Emphysema	5,949	65

SOURCE: Deaths are for 1978, latest official figures from National Center for Health Statistics, Public Health Service, U.S. Department of Health and Human Services.

[a] Deaths per 100,000 population in each age group. Rates are averages for age groups, not individual ages.
[b] Cerebrovascular disease.
[c] Partly estimated.

Job Study Observations

In job study observations and work practice audits, data collected will pinpoint workplace areas where accidents and unsafe work practices are occurring and will itemize specific workplace behaviors that may result in accidents.

Forms that might be used in accident investigation and auditing are illustrated in Figures 6.1 through 6.4. Chapter 14 discusses this in detail, but it is important enough to highlight here also because the data it provides serves as a basis for controls.

Critical Incident Technique. The critical incident technique for identifying unsafe acts has proved a useful data collection source. Using this technique, a sample of employees are interviewed and asked to describe unsafe practices they have performed or have seen performed. From this information a hazard control program can be developed and the quality of safe

Injury Accident _____
No-Injury Accident _____
Date _____ Name of Injured _____ Dept._____

TIME LOST

1. How much time did other employees lose by talking, watching, or helping at accident? Number of employees _____ X hours = [   ]

2. How much productive time was lost because of damage equipment or loss of reduced output by injured worker?
   Estimate hours = [   ]

3. How much time did injured employee lose for which he was paid on the day of the injury?
   Estimate hours = [   ]

4. Will overtime be necessary?
   Estimate hours = [   ]

5. How much of the supervisors' or other managements' time was lost as a result of this accident?
   Estimate hours = [   ]

6. Were additional costs incurred due to hiring and training or replacement?
   Training Time Estimate hours = [   ]

7. Describe the damage to material or equipment. _____
   _____
   _____

8. If machine and/or operations were idle, can loss of production be made up?
   Yes _____ No _____

9. Any demurrage or other cost involved? Yes _____ No _____

TOTAL ACCIDENT COSTS

   To compute the total costs of the accident, it is necessary to complete the following cost information. Should the supervisor have access to this information, it is advised he complete as much as possible. Safety personnel will develop those costs now known by supervisor. Actual costs can be developed at the time such information is available and this report can be revised accordingly, if desirable.

10. Costs associated with questions 1 through 9 (actual and/or estimated costs).

10-1	$
10-2	$
10-3	$
10-4	$
10-5	$
10-6	$
10-7	$
10-8	$
10-9	$_____

TOTAL

11. Costs associated with giving emergency medical attention, first aid, ambulance costs, etc. (actual and/or estimated costs). $

FIGURE 6.1. Department Supervisor's Accident Cost Report. (Used with permission from National Safety Council).

168

12. Workers Compensation costs (temporary total, permanent partial, permanent total, fatal, etc., costs actual and/or estimated). $

13. Hospital and medical costs (includes hospital, doctor's fees, and other medical costs actual and/or estimated). $

14. Costs associated with placing injured on other work on restricted work when unable to perform regular work (actual and/or estimated costs). $

TOTAL

15. Estimated total cost of accident: GRAND TOTAL $

Name of Supervisor _____

Fill in and send to the safety department not later than day after accident.

FIGURE 6.1.   (*Continued*)

performance measured. This technique is used to identify causal factors for injurious and noninjurious accidents. A study of companies with high and low accident rates concluded that:

1. There is greater management concern and involvement in safety matters, as reflected by the rank and stature of the company's safety officer, regular inclusion of safety issues in plant meeting agenda, and personal inspections of work areas by a top plant official, in some instances on a near daily basis.

2. There are more open, informal communications between workers and management and frequent everyday contacts between workers and supervisors on both safety and other job matters.

3. There are tidier work areas with more orderly plant operations, better ventilation and lighting, and lower noise levels.

4. The work force has more older, married workers with longer job service and less absenteeism and turnover.

5. There is more regard for the use and effectiveness of measures other than suspensions and dismissals in disciplining violations of safety rules.

6. There is greater availability of recreational facilities for worker use during off-hour jobs.

7. Greater efforts are made to involve worker families in campaigns for promoting safety consciousness both on and off the job.

8. There are well-defined selection, placement, and job advancement procedures with opportunities for training in developing new skills.[8]

Class 1 _____
(Permanent partial or temporary total disability)

Class 2 _____
(Temporary partial disability or medical treatment case requiring outside physician's care)

Class 3 _____
(Medical treatment case requiring local dispensary care)

Class 4 _____
(No injury)

Name _____

Date of injury _____ Its nature _____

Department _____ Operation _____ Hourly wage _____

Hourly wage of supervisor $ _____

Average hourly wage of workers in department where injury occurred $ _____

1. Wage cost of time lost by workers who were not injured, if paid by employer $ _____

   a. Number of workers who lost time because they were talking, watching, helping _____ .

      Average amount of time lost per worker _____ hours _____ minutes.

   b. Number of workers who lost time because they lacked equipment damaged in accident or because they needed output or aid of injured worker _____ .

      Average amount of time lost per worker _____ hours _____ minutes.

2. Nature of damage to material or equipment _____

Net cost to repair, replace, or put in order the above material or equipment $ _____

3. Wage cost of time lost by injured worker while being paid by employer (other than workmen's compensation payments) $ _____

   a. Time lost on day of injury for which worker was paid _____ hrs. _____ mins.

   b. Number of subsequent days' absence for which worker was paid _____ days (other than workmen's compensation payments) _____ hours per day.

   c. Number of additional trips for medical attention on employer's time on succeeding days after worker's return to work _____ .

      Average time per trip _____ hrs. _____ mins. Total trip time _____ hrs. _____ mins.

   d. Additional lost time by employee, for which he was paid by company _____ hrs. _____ mins.

FIGURE 6.2. Investigator's Cost Data Sheet. (*used with permission from National Safety Council*).

4. If lost production was made up by overtime work, how much more did the work cost than if it had been done in regular hours? (Cost items: wage rate difference, extra supervision, light, heat, cleaning for overtime) $ _____

5. Cost of supervisor's time required in connection with the accident $ _____

   a. Supervisor's time shown on Dept. Supervisor's Report _____ hrs. _____ mins.

   b. Additional supervisor's time required later _____ hrs. _____ mins.

6. Wage cost due to decreased output of worker after injury if paid old rate $ _____

   a. Total time on light work or at reduced output _____ days _____ hours per day.

   b. Worker's average percentage of normal output during this period _____ %.

7. If injured worker was replaced by new worker, wage cost of learning period $ _____

   a. Time new worker's output was below normal for his own wage _____ days _____ hours per day. His average percentage of normal output during time _____%. His hourly wage $ _____.

   b. Time of supervisor or others for training _____ hrs. Cost per hour $ _____.

8. Medical cost to company (not covered by workmen's compensation insurance) $ _____

9. Cost of time spent by higher supervision on investigation, including local processing of workmen's compensation application forms. (No safety or prevention activities should be included.) $ _____

10. Other cost not covered above (e.g., public liability claims; cost of renting replacement equipment; loss of profit on contracts cancelled or orders lost if accident causes net reduction in total sales; loss of bonuses by company; cost of hiring new employee if the additional hiring expense is significant; cost of excessive spoilage by new employee; demurrage). Explain fully. $ _____

Total uninsured cost (sum of items 1 through 10) $ _____

Name of Company _____

FIGURE 6.2. (*Continued*)

Preventing Accumulation of Hazardous Material

One important strategy for controlling accidents is preventing accumulation of hazardous material. It is an easily understood concept, but the actual execution must be monitored and reinforced or materials will pile up, causing needless accidents. Any accumulation of work materials that add to clutter and create unnecessarily cumbersome work areas is consid-

Area _____   Type Exposure _____
Date _____
Potential Hazard Description _____
_____
Number of Employees Potentially Exposed _____
Desired Level To Be Maintained (TLV) _____
Results Obtained from Study _____
Action Recommended _____

PRELIMINARY HAZARD ASSESSMENT

Equipment _____
Procedure _____
_____
Result/Recommendation _____
_____

HAZARD ASSESSMENT

Equipment _____
_____
Location of Sampling Points _____

Time Schedule for Sampling _____

Procedure _____
_____
_____
_____
_____
_____

FIGURE 6.3.   Hazard assessment design *(used with permission from NIOSH).*

ered unsafe. Good industrial housekeeping is imperative in any accident prevention program and hence in any safety engineering program. The advantages of a clean plant are:

1. Reduction in injuries on the job
2. Increased efficiency, brought about by an easier flow of materials and improved performance
3. Improvement in employees' morale
4. Better community relations, winning the esteem of the community
5. Better labor relations; reduction in complaints by employees
6. Minimization of losses from fires

Reducing the Amount of Energy Accumulation

Reducing energy accumulation is another important accident control strategy necessary to prevent the buildup of thermal energy. This is accom-

Date reported   FOREMAN'S ACCIDENT REVIEW

PLANT NO. 2

Name of injured worker _____ Clock No._____
Date of accident _____
Brief description of alleged accident _____
_____
_____

Indicate below by an "X" whether in your opinion the alleged accident was caused by:

Physical causes
_____Improper guarding?
_____Defective substances or equipment?
_____Hazardous arrangement?
_____Improper illumination?
_____Improper ventilation?
_____Improper dress or apparel?
_____No mechanical cause?
_____Not listed? (Describe briefly) _____

Sometimes the injured person is not directly associated with the causes of an accident. Using an "X" to represent the injured worker and an "O" to represent any other person involved, indicate whether, in your opinion, the accident was caused by:

Unsafe acts
_____Operating without authority?   _____Unsafe loading, placement, mix-
_____Failure to secure or warn?   ing, etc.?
_____Working at unsafe speed?   _____Took unsafe position?
_____Made safety device inoperative?   _____Worked on moving equipment?
_____Used unsafe equipment or hands   _____Teased, abused, distracted, etc.?
      instead of equipment?   _____Did not use safe clothes or per-
_____No unsafe act?   sonal protective equipment?
_____Not listed? (Describe briefly) _____
_____

Personal causes
_____Physical or mental defect?
_____Lack of knowledge or skill?
_____Wrong attitude?
_____Not listed? (Describe briefly) _____

Actions that I have taken to prevent an accident similar to the one reviewed above.
_____
_____

Foreman's signature_____

FIGURE 6.4.   Foreman's Accident Review (*used with permission from National Safety Council*).

plished by understanding the potential heat buildup of specific materials and substances. Good housekeeping may prevent the accumulation of flammable waste materials. Storing flammable substances in proper containers is another means of energy accumulation. Dilution ventilation in battery rooms will prevent hydrogen gas from reaching the lower explosive limit (LEL), which for hydrogen is 40,000 parts per million.

Preventing or Modifying Release of Energy

Another important accident control strategy is preventing or modifying the release of energy. This is the principle behind the use of helmets and shields that spread the impact over as wide an area as possible. This is also the concept that is utilized in the removal of projections on automobile front-dash panels and the addition of padding to these surfaces where there is a possibility of impact with the body.

The force impact is important to consider in hazard control. The rapid deceleration produced when a falling body hits a surface is much more likely to produce injury than the slower deceleration through the use of a fire net.

Separating Individuals from Energy by Time and Space

The use of barriers to decrease the abruptness of impact is a control measure. The barrier is softer than the object impacting with it. Barriers are also used to separate the worker from energy by space. People can also be separated from energy by time by placing a time differential on the release of energy.

Raising the Injury Threshold

Raising the injury threshold is another strategy for accident control. This may be by physical conditioning and training so that the worker's body is at peak fitness and is prepared to function at optimum performance. Regular exercise is essential to maintain muscular strength.

Physical Training.   The importance of consistent physical training cannot be overestimated if physically strenuous work is to be performed. This training should include the following:

1.   Training of the oxygen transporting system, which includes the heart and lungs. This is done by aerobic exercises in which the heart rate is increased beyond 140 beats per minute for young persons. This is called the threshold of training. Running is a good way to perform this exercise.

2. Muscle training, especially of the abdomen and back muscles. Basic calisthenics, which includes sit-ups, push-ups, and toe touches, is good for this purpose.

3. Training aimed at maintaining joint mobility and at the enhancement of the metabolism of the articular cartilage of the joints. This is especially critical for workers doing heavy or strenuous work.

Research shows that in occupations requiring heavy work, accidents are more likely to occur on Monday after a weekend off when the worker has probably not been doing the amount and kind of physical work the job entails. For the worker to be able to sustain high levels of activity for an extended period of time without fatigue, he or she must be able to take in enough oxygen and transport it through the bloodstream at a sufficient rate to meet the oxygen demands of the muscles.[4]

Acclimatization

Acclimatization refers to the body's gradual adjustment to a change in climate or working conditions and its rapid return to a "normal" state when removed from the stressful situation. In addition, to reenter the stressful situation the body must be retrained and the working conditions readjusted to.

An example of appropriate acclimatization would be the steps that would be taken for a worker exposed to a hot environment for the first time. Generally, a two-week program is required which will gradually get the worker and his or her body used to this hot environment. It is suggested that on the first day the worker be exposed to 50% of the work load and time that will eventually be the total required. Each day thereafter, a 10% increase in exposure is scheduled, building up to a 100% total exposure on the sixth day.[6] If the worker has been absent from work for nine or more consecutive days for any reason, a four-day acclimatization period, beginning with a 50% exposure the first day and with daily increments of 20%, will be necessary. If the worker has been ill and away from work for four days, this four-day acclimatization would likewise be needed.[12]

Controlling Behavior

Controlling inappropriate behavior will reduce the chance of an accident. Unacceptable behavior occurs for many reasons, most of which are the result of psychological factors which will be discussed in Chapter 7.

Horseplay. Horseplay is rowdy behavior. This may include playing practical jokes on the victim.

Alcohol.   Accident data is increasingly showing the crucial role that alcohol plays in many accidents, particularly traffic accidents. Approximately 50% of crashes fatal to vehicle occupants, 35% or more of those fatal to adult pedestrians, and 20% of crashes in which vehicle occupants are injured nonfatally result at least in part from prior use of alcoholic beverages.[5]

In industrial accidents, drinking takes its toll and successful strategies must be developed. The cost of alcohol abuse to industry is substantial. Alcohol abuse results in decreased productivity, increased absenteeism, inefficiency, increased number of accidents, increased employee turnover, and increased incidence of actual sickness secondary to the issue of alcohol.[13]

Drugs.   The use of drugs must also be considered in setting up strategies for accident control. Many drugs can make one sleepy or alter one's perceptions. Many drugs interact with one another. The behavioral changes that result from taking a combination of drugs may be quite different from the effects of taking each drug individually.

In some industrial jobs (including those requiring driving) it is essential for safety that certain drugs not be taken, because of possible behavioral or other effects. This requires an agreement that employees (who must take medication incompatible with performance at work) will not lose their jobs but will be given work that they can safely carry out without endangering anyone. The company authorities should know what medications an employee is taking in order to assure that he or she is functioning safely.

ERGONOMICS OF ACCIDENT CONTROL

The objective of ergonomics is to design person–machine systems that operate with minimum error and maximum efficiency. The term "human factors" is also commonly used to describe activity in this area. Ergonomics was discussed in Chapter 5. The ergonomics of accident control is the result of applying the information presented in this chapter in order to make the workplace safe. Table 6.3 shows factors that must be considered in implementing the ergonomics of accident control.

There are some ergonomic safety principles. The first is that the safe way of doing things must always compete with the other, less safe ways, and it is important to maximize the safe way's chances of success. The second principle is that critical points in safety are apt to be connected with disturbances, faults, and other unforeseen and unwanted events occurring during the work process. The third principle is that ensuring comfortable positions for the worker, optimal patterning of required movements, and the best ergonomic layout must take into account not only the normal case in which everything goes well but also any unforeseen happenings, from slight disturbances to emergencies.

One of the basic principles of ergonomics is that the failure of a device should not result in property damage. This "fail-safe principle" is the basis of electric fuses and other automatic fire-protection devices. Another ergonomic principle is that the machine is an extension of the individual, which is why one of its major foci is the appropriate fitting of the worker for the equipment that he or she is to use.

An example of utilizing ergonomic principles is that of seat belt usage. It has been confirmed that the use of safety belts does reduce death and injury in vehicle crashes.[11] To date, no ergonomic/safety campaigns have been successful in promoting the universal use of seat belts, although they are standard equipment on all cars. Consumer resistance or apathy is the key factor in this failure. The installation of a passive restraint, such as air bags, will overcome this apathy.

All workers should be taught the ergonomic principles that apply to their jobs. As an example the following topics should be addressed:

1. The risks to health of unskilled workers.

2. Basic physics, including the principle of levers, the work needed to move a load horizontally, and the work needed to change the direction of motion.

3. The effects on the body. The basic functional anatomy of the spine and the muscles and joints of the trunk and limbs is easily taught to trainees in terms of their practical experience and with reference to basic physics. It is easy to demonstrate muscular activity while muscles are being stretched or lengthened isometrically; how the rib cage is constrained by pushing and carrying; how the shoulder, abdominal, and back muscles contribute to lifting. The increase in intrathoracic and intra-abdominal pressures during materials handling is not difficult to understand, because of the familiarity with breath holding when making an effort.

4. Individual awareness of the body's strengths and weaknesses. The first practical lesson in materials handling which all trainees must learn is how much they can handle comfortably and where, in relation to their bodies, their strengths and weaknesses lie.

5. Learning to avoid the unexpected. There is a need to recognize the physical factors that might contribute to an accident and to eliminate them. For example:

   a. Is the load free to move and not get stuck?

   b. Is it a weight that is comfortable to handle?

   c. Are lifting aids available?

   d. Has the load proper handles to grasp or can they be provided?

   e. Is protective clothing indicated?

   f. Is the area clear of obstructions?

   g. Is the floor clean, dry, and nonslip?

TABLE 6.3
Representative Listing of Factors That Shape Human Performance

Extraindividual

Situational Characteristics	Psychological Stresses	Intraindividual (Organismic) Factors
Temperature, humidity, air quality	Task speed	Previous training/experience
Noise and vibration	Task load	State of current practice or skill
Degree of general cleanliness	High jeopardy risk	
	Threats (of failure, loss of job)	Personality and intelligence variables
Manning parameters	Monotonous, degrading, or meaningless work	Motivation and attitudes
Work hours, work breaks	Long, uneventful vigilance periods	Knowledge of required performance standards
Availability/adequacy of supplies	Conflicts of motive about job performance	Physical condition
Actions by supervisors	Reinforcement absent or negative	Influence of family and other outside persons or agencies
Actions by co-workers and peers	Sensory deprivation	Group identifications
Actions by union representatives	Distractions (noise, glare, movement, flicker, color)	
Rewards, recognition, benefits	Inconsistent cueing	
Organizational structure (e.g., authority, responsibility, communication channels)		

Task and Equipment Characteristics	Physiological Stresses	
Perceptual requirements	Fatigue	
Anticipatory requirements	Pain or discomfort	
Motor requirements (speed, strength, precision)	Hunger or thirst	
	Temperature extremes	
	G-force extremes	
Interpretation and decision making	Atmospheric pressure extremes	
Complexity (information load)	Oxygen insufficiency	
	Vibration	
Long- and short-term	Movement constriction	
Frequency and repetitiveness	Lack of physical exercise	
Continuity (discrete vs. continuous)		
Feedback (knowledge of results)		

TABLE 6.3 (*Continued*)

Extraindividual		Intraindividual
Task and Equipment Characteristics	Physiological Stresses	(Organismic) Factors

Task criticality
Narrowness of task
Team structure
Person–machine inter-
 face factors; design of
 prime equipment, job
 aids, tools, fixtures

Job and Task Instructions

Procedures required
Verbal or written com-
 munications
Cautions and warnings
Work methods
Shop practices

SOURCE: Swain, A. *Estimating Human Error Rates and Their Effects on System Reliability.* Albuquerque, N.M.: Sandia Labs, 1978.

6. Handling skill. The handling tasks chosen for teaching purposes should cover a range of materials, but the emphasis should be on the actual materials handled by the organization concerned. A number of points of general application should be taught. Skill depends on:

 a. Preparation to avoid being caught unaware
 b. Being able to recognize what can be handled comfortably, without help
 c. Keeping the center of mass of the load close to the body when lifting
 d. Not twisting or bending sidewards
 e. Using the legs to get close to the load and to make use of the body weight and the kinetic energy of the body and load
 f. Timing for smooth materials handling

7. Handling aids. Handling aids are available for most materials handling tasks. Often, they can be improvised if they do not exist. Their use must be encouraged, especially with unpaced or incentive jobs.[9]

Lighting

Lighting is another important ergonomic concept to consider in accident control. The aim is to provide sufficient visual contrast for safe operation.

Safety Symbols

Safety symbols are another example of utilizing ergonomic principles for the purpose of controlling accidents. Color is a major factor in safety coding. Yellow is used for marking hazards that could cause slipping, falling, and striking accidents. Yellow also indicates flammable-liquid storage cabinets, materials handling equipment, and radiation hazard areas. Sometimes a black checkerboard pattern or stripe is used with the yellow.[1]

In addition, OSHA specifies danger signs to be an opaque glossy red, black, and white; caution signs to be of a yellow background and the panel, black with yellow letters; safety instruction signs to be white and the panel, green with white letters; and the slow-moving vehicle emblem to be a fluorescent yellow-orange triangle with a dark-red reflective border.[12]

Labeling should indicate handhold positions, cautions against single lifting, or off center of gravity containers. Labels should be printed on all sides of a container, facing in at least two directions so they are visible to the operator at all times.

Accident control is just one example of the value of ergonomics in the repertoire of the systems safety engineer. Armed with the concepts and strategies discussed in this chapter, the safety engineer can make the workplace an increasingly safer area.

SUGGESTED LEARNING EXPERIENCES

1. Illustrate safety symbols that would be appropriate to utilize at a workplace concern of your choice.

2. Discuss the ergonomics of accident control. What advances do you envision happening in this area in the next 20 years? Explain.

3. Discuss invasion of privacy and the place of employment's need to know what medications each employee is taking.

4. Debate whether alcohol support work groups should be mandatory for all employees who have a drinking problem.

5. Discuss what responsibilities you think management should have in relation to providing physical fitness activities for all employees.

6. Select a workplace of your choice that requires worker acclimatization. Describe how you as a safety professional would have this accomplished in that industry. Develop a protocol for the same.

7. Discuss an occupation such as mining which requires a high level of physical conditioning. Would there be any differences for men and women? What criteria would you want to follow to be sure that no one was taking on more strenuous work than his or her body was able to handle?

8. Choose a workplace and develop strategies for control of accidents for that industry. What dominant theme seems to be apparent here?

9. Develop some concepts that seem real to you in terms of accident causation. Is there an underlying framework from which you have developed this model? Explain.

10. Develop a flow chart depicting the different places you as a safety engineer would institute accident controls for a specific industrial operation. Could this flow chart be utilized for a more generalized application to any industry?

REFERENCES

1. American National Standards Institute. *Accident Prevention Signs Specifications* (ANSI Z35.1-1972). New York, 1972.
2. American National Standards Institute. *Method of Recording and Measuring the Off-the-Job Disabling Accidental Injury Experience of Employees* (ANSI Z16.3-1973). New York, 1973.
3. American National Standards Institute. *Recordkeeping for Occupational Injuries and Illnesses* (ANSI Z16.4-1977). New York, 1977.
4. Astrand, P. O., and Rodahl, K. *Textbook of Work Physiology* (2nd ed.). New York: McGraw-Hill, 1977.
5. Brubaker, W. W. Alcoholism in industry. *Occupational Health Nursing*, 1977, **25**(2), 7–10.
6. *Criteria for a Recommended Standard. Occupational Exposure to Hot Environments* (HSM-72-102-69). Washington, D.C.: GPO, 1972.
7. Haddon, W. The prevention of accidents. In D. W. Clark and B. MacMahon (Eds.), *Preventive Medicine*. Boston: Little, Brown, 1967, pp. 591–621.
8. Manuele, F. A. How do you know your hazard control program is effective? *Professional Safety*, 1981, **26**, 18–24 (June).
9. NIOSH. *Industrial Hygiene Engineering and Control*. Cincinnati, Ohio, 1978.
10. National Safety Council. *Accident Facts*. Chicago, 1981.
11. Robertson, L. S., O'Neil, B., and Woxom, C. W. Factors associated with safety belt use. *Journal of Health and Social Behavior*, 1972, **13**, 18–24.
12. Occupational Safety and Health Administration. *Code of Federal Regulations*. General Industry Standards (29 CFR 1910). U.S. Department of Labor, Washington, D.C., June 1981.
13. Whyte, E. L. Coping with alcoholism in industry. *Occupational Health Nursing*, 1977, **25**(7), 9–11.

7 PSYCHOLOGICAL HAZARDS

This chapter deals principally with psychological hazards which play a significant role in accident causation. Psychologically produced problems can adversely affect an otherwise safe environment, creating unsafe conditions in which an accident or occupational illness can occur.

These nonobservable psychological hazards contribute more to accident and injury rates as safety engineering and industrial hygiene reduce accident rates to a point where further reduction is very difficult or economically impractical. At this point, training plays a critical role, by making employees aware of how their own behavior contributes to accidents and injuries. By setting unrealistic goals or by not taking the worker's needs into account, management can set up situations which produce accidents.

PSYCHOLOGICAL STRESS

Psychological hazards are typically manifested in the parts of the body which react to stress and anxiety, such as the head (tension headaches), the stomach (ulcers), and the back (high back pain). There is a delay between exposure to stress and appearance of symptoms. The person is not aware of having been exposed to stress until physical symptoms appear. Except in very rare diseases, such as mesothelioma from asbestos exposure and angiosarcoma from exposure to vinyl chloride, a cause–effect relationship between the disease and the work environment is not uniquely evident. Generally, the relationship of an illness to an occupation is elusive because most occupational diseases are clinically indistinguishable from diseases of nonoccupational origin. It is difficult to determine the extent of job influence because a multiplicity of factors may be involved, including

183

psychological stress, the age of the worker, diet, nutrition, smoking, and general life-style. Seyle has called these stress-related symptoms the General Adaptation Syndrome.[10] He believes the body goes through two stages, which he calls the stage of alarm and the stage of resistance, before symptoms appear. When the resistance fails, then the symptoms appear.

It can be difficult to disentangle the work factors, home factors, and personal factors in less obvious hazards. What is important is to identify the factors, know how to treat them, and know how to prevent them. The worker must be treated as a total human being and not be segmented into parts, each having its own space and not interacting with the others. The aim is to reduce stresses or eliminate them when possible and to enable workers to focus on their work to be as productive as possible and to pay better attention to the job at hand.

Jobs can be stress-producing if they are either too simple or too complex. Examples of job stress-producing factors are low utilization of abilities, lack of participation in decisions about how the job is to be performed, low work complexity, and role ambiguity. These factors have been found to be high among assembly-line workers, forklift truck drivers, and machine operators, but were very low among professors, family physicians, and other professionals. Machine-paced assembly-line workers scored high on boredom and dissatisfaction with their work load. In addition to professors and family physicians, white-collar supervisors ranked as one of the most satisfied occupational groups. Overall, assemblers and relief workers on machine-paced assembly lines registered the highest levels of stress and strain. (On these machines, the work rate was set by the machine.)[12]

Stress levels were found to be affected by job satisfaction, life satisfaction, motivation to work, overall physical health, and self-esteem and by intention to leave the job, absenteeism, depressed mood, and escapist drinking. Nonparticipation in decisions affecting the way the job was structured was also considered a major stress-producing factor.

Professional and technical workers scored highest among the occupational groups in job satisfaction, but they also registered high levels of depression. Machine operators scored lowest for job dissatisfaction and low in perceived health but scored the most psychological health of any group on the mental health measures. Laborers were second lowest in job satisfaction but perceived their health as good and scored well on mental health measures. White-collar workers as a group showed much higher job satisfaction than blue-collar workers but, at the same time, registered slightly higher in depressed mood and slightly poorer in perceived health.[12]

A study was conducted in Tennessee of over 22,000 cases of stress-related health disorders covering 130 occupations. The results showed that 40 occupations had higher-than-expected prevalence rates of stress-related diseases; 70 had the expected rates; and 13 had lower-than-expected rates.

TABLE 7.1
Stress Perceived by Various Occupations

High Stress	Low Stress
Farm owners	College Professors
Managers/administrators	Craftspersons
Office managers	Maids
Foremen	Sewer workers
Secretaries	Checkers/examiners
Assembly-line operators	Stock handlers
Clinical laboratory technicians	Freight handlers
Factory machine operators	Farm laborers
Mine machine operators	Heavy equipment operators
General and construction laborers	Child care workers
Waitress/Waiters	Packers/wrappers
House painters	Personnel/labor relations workers

SOURCE: Excerpted from the April 1981 issue of *Occupational Hazards Magazine*. Copyrighted 1981 by Penton/IPC.

The expected rates were based on morbidity rates developed by the National Center for Health Statistics. Among the 40 in the higher-than-expected category, the 12 with the highest and the lowest rates are shown in Table 7.1.

Although not listed in Table 7.1 two occupations with very high stress are police work and coal mining. This has been documented in a number of books on police work by Wambaugh.[13] The dangers of coal mining, including constant stress for the miners, were discussed in Chapter 2.

NIOSH studied the stress problems of coal miners as part of a larger study of factors motivating the adherence to safe procedures and the use of personal protective equipment. Questionnaires were sent to 486 miners who were compared to blue-collar workers. Miners reported fewer stress conditions than those facing blue-collar workers. Miners reported more participation in decision making about the way their jobs were performed, more utilization of skills, greater equity in wages, and a less variable work load. However, they generally showed more anxiety, depression, and irritation than blue-collar workers.

Machine pacing has also been documented as a cause of stress. Machine-paced assembly workers and their relief persons have been found to have the highest levels of job stress and strain.[12] In the NIOSH study, they complained that their talents and skills were not being realized on the job and that they had little or no input in decisions. They reported frequent feelings of boredom and dissatisfaction with their work load and of anxiety, general irritability, and poor work health. Some of the most frequent health problems associated with machine pacing included muscle spasms, cramps, nervous disorders, and a variety of psychosomatic problems. These included peptic ulcers, coronary heart disease, and essential hypertension (high blood pressure of unknown origin).[12]

Frankenhauser and Gardell as reported by Sheridan studied machine-paced and self-paced work at a Swedish sawmill. They used questionnaires, health examinations, and clinical laboratory tests. The results indicated that machine-paced workers had more feelings of monotony, more general mental strain, and greater exhaustion at the end of the workday, as well as more frequent sick leave requests, than workers in self-paced jobs. Medical examinations revealed a higher incidence of psychosomatic, cardiovascular, and stress disorders in workers with machine-paced tasks as compared with workers in self-paced jobs. Biomedical findings were related to both self-reported feelings of "well-being" and measures of job repetitiousness. Feelings of exhaustion at the end of the shift and the inability of machine-paced workers to relax soon after work were related to increased adrenaline secretion. The authors concluded that lack of control over work pace was an important contributing factor producing "wear and tear" among workers engaged in machine-paced jobs.[12]

Shabecoff explored the relationship between stress and physical symptoms. He reported a study by House, who found that stress was related to the incidence of angina, ulcers, respiratory disease, and dermatitis among workers. House and his colleagues found that the workplace can produce illness in and of itself—such as peptic ulcers and hypertension.[11]

LIFE STRESS

Stress is not only related to workplace activities but is also related to activities of daily living. Unemployment, divorce, and an Internal Revenue Service audit can also cause much stress. Stress has been shown to cause many physical problems, but it is not beyond causing safety problems at work. Holmes and Rahe have developed a Life Stress Scale (Figure 7.1) which can be applied to everyone.[5] The sum of the life change units (LCUs) can be considered an estimate of psychological stress. As can be seen from the scale, different life events are weighted differently. For example, death of a spouse is assigned 63. Loss of a spouse either by death or divorce would cause major changes in one's life.

It would be helpful to have a work stress index based on Holmes and Rahe's scale, detailing the various work stresses and ranking them according to the research done so far. If workers completed both scales and became aware of how much stress they were under, they could take preventive measures.

Life Event	Value
Death of Spouse	100
Divorce	73
Marital separation	65
Jail term	65
Death of close family member	63
Personal injury or illness	53
Marriage	50
Fired from work	47
Marital reconciliation	45
Retirement	45
Change in health of family member	45
Pregnancy	40
Sex difficulties	39
Gain of new family member	39
Business readjustment	39
Change in financial state	38
Death of close friends	37
Change to different line of work	36
Change in number of arguments with spouse	35
Mortgage over $10,000	31
Foreclosure of mortgage or loan	30
Change in responsibilities at work	29
Son or daughter leaving home	29
Trouble with in-laws	29
Outstanding personal achievement	28
Wife begins or stops work	26
Wife begins or stops school	26
Husband begins or stops work	26
Husband begins or stops school	26
Change in living conditions	25
Revision of personal habits	24
Trouble with boss	23
Change in work hours or conditions	20
Change in residence	20
Change in schools	20
Change in recreation	19
Change in church activities	19
Change in social activities	18
Mortgage or loan less than $10,000	17
Change in sleeping habits	16
Change in number of family get-togethers	15
Change in eating habits	15
Vacation	13
Christmas	12
Minor violations of the law	11

FIGURE 7.1. Life events scale (*from Holmes and Rahe*).[5]

To use the Life Stress Scale, check off events which have happened to you within the last two years; then add up the total number of stress units. Holmes and Rahe found that a score of 150 or more was associated with a 50% chance of developing an illness.[5] A score of 300 or more was associated with a 90% chance.

Coronary-Prone Behavior

The Framingham Heart Study was started over 20 years ago to find out what factors are involved in persons who have heart attacks. The researchers started with a large number of normal adults in Framingham, Massachusetts, and evaluated their health periodically. One of their many findings dealt with the personalities of people who have heart attacks. They found two groups of people: Those who were likely to have heart attacks were labeled Type A; those who were not likely to have them, Type B. This has been discussed briefly in Chapter 4.

Behavior manifested by the Type A pattern is characterized by excessive activity, competitiveness, aggressiveness, hostility, impatience, and time urgency that is readily evoked by a variety of challenges in the social and physical environment. Behavior manifested by the Type B pattern is characterized by the relative absence of these behaviors. The Type A pattern significantly increases prediction of the major manifestations of coronary heart disease (CHD).[4]

Another question arising from research on the coronary-prone behavior pattern is when and how such behavior develops. Monasteries are thought of as places where monks spend much time in prayer and reflection. However, some monasteries have much higher incidences of coronary heart disease than others. In a study of 25 monasteries, it was found that those monasteries in which a higher rate of coronary heart disease was found were consistently rated by judges as possessing an environment likely to foster the Type A behavior pattern. The monks in these monasteries also tended to be rated as Type A more often.[9]

WOMEN AND STRESS

As women move from home to the competitive workplace, they are likely to experience the same stress as men. Therefore, the psychological factors due to increased competition will probably cause women to have more accidents in the workplace than they did at home. In a number of studies covering workplaces where both men and women are employed, accident rates are generally the same. To paraphrase an expression of the American Lung Association, as women work like men, they will experience stress like men.

There are a number of stress factors which mainly affect women. Women, in general, do not receive equal pay for equal work. Married female workers still perform the majority of home and child care. Women workers are frequently exposed to sexual harassment. These situations produce substantial stress and, given the right set of circumstances, can cause serious accidents.

As women move into more competitive areas of work, it will be interesting to observe whether they, too, become prone to hypertension, ulcers, alcoholism, and cancer, to determine if they are different from men or if their hormones act as a barrier to some of the stress-related illnesses.

Equal Economic Opportunities

Women are now entering the labor force not as a result of a crisis like World War II but because of powerful economic, demographic, and social forces, together with far-reaching changes in attitudes. The large-scale movement of women into the labor force has created a situation in which human resources may now be fully utilized. However, because of many preconceptions and stereotypes about women workers, equal economic opportunities had to be mandated.[8]

Dual Careers

With increasing numbers of women working, dual-career families will be commonplace. Already it is typical for both partners to be providing income for the family's upkeep. This even seems so for the very wealthy, who want to feel useful and indulge in their wealth. Dual-career families create new sets of problems and stresses. The predominant issue seems to be finding enough hours in the day to function effectively at work, perform the tasks of daily living at home, plus enjoy children, spouse, and social life.[14] In the majority of cases, it is the woman's responsibility to coordinate the home, making sure domestic chores get done. The majority of women cited time pressures as most difficult; their main reasons for working outside the home were to achieve fulfillment, to have a career, and to do meaningful work. Women in dual-career marriages must balance demands to prevent stress and at the same time take advantage of equal economic opportunities.

Sexual Harassment

Sexual harassment by either sex may cause stress to the recipient which can either lead to ill health or may result in a workplace accident. It should not be tolerated and is against the law.[7]

FUTURE SHOCK

There is a future shock ahead for each and every worker. Future shock implies that all our concepts, ideas, and training is likely to become obsolete sometime in the near future. The hazards that confront safety engineers today may not even exist in 10 years. Over this period, their job will not just evolve, but will change without notice by quantum leaps.

Not every worker is flexible. Many employees need a routine and a stability; when these are missing, they experience stress and do not function at maximum performance. The advances in technology can also cause worker stress if workers are afraid that they are going to lose their jobs or that they cannot learn how to use the new equipment or adapt to another set of tasks.

Temporary Organizations

Temporary organizations also possess the capability for producing workplace stress. How does one make long-term commitments in a short-term job? How does one adjust to irregular intakes of money and then to periods of no money? How is the body in this situation able to get on a regular pattern?

OVERTIME AND SHIFT WORK

Similarly, what are the effects of overtime, long shifts (12 hours), and shift work? There is a circadian (24-hour) rhythm that governs biological functioning. Diurnal rhythms control pulse, blood pressure, the cardiopulmonary system, blood composition, endocrine secretions, appetite, elimination, and the wake–sleep cycle. Shift work interrupts these processes and requires that they occur at times when the body is not genetically programmed or environmentally conditioned for them. A number of workers have rotating shifts. This means that for week 1 in a cycle, they work 7:00 A.M. to 3:30 P.M., for week 2 they work 3:00 P.M. to 11:30 P.M., and for week 3 they work 11:00 P.M. to 7:00 A.M. This is extremely difficult for workers and may be hazardous.

A number of studies have shown that the circadian rhythm can be changed, but slowly. In one month the circadian cycle was changed only 4%. It seems that the worker may be physically changing his or her pattern, but the worker's body is not biologically changing. There is disagreement over the extent to which the body, over time, can adapt to changes. Although several studies have found rhythmic adjustment may occur within four days to two weeks, several considerations suggest that such ready adaptation may not be commonplace. A notable minority of shift workers never significantly adjust, biologically, to the alterations imposed on their normal body cycles. Furthermore, any adaptation that may be achieved even among fixed-shift workers is repeatedly undermined by days off, holidays, vacations, and sick leave, when employees revert to normal living schedules. Not surprisingly, problems related to sleep, appetite, and digestion are the most common and persistent complaints for many shift workers.

> Particularly widespread among shift workers is insufficient or poor quality sleep resulting from trouble falling asleep, waking during sleep, and waking up early. Although many of these difficulties are from disruptions in the body's normal diurnal sleep rhythms, sleep during daytime is also often disturbed by excessive and unavoidable light and heat, and by noises from children, housework, telephone calls, and street traffic. . . .

> Lack of adequate sleep and poor quality sleep have been implicated in a number of adverse health and safety consequences, including physical disorders, nervous problems, and deficits in mental and psychomotor performance which can lead to on-the-job accidents. Fatigue is the most widely encountered and upsetting reaction shift workers experience from sleep deprivation. This is particularly true of night and rotating shift workers. . . .

> Loss of appetite and irregular eating habits are a common occurrence among shift workers that may lead to weight loss as well as nutritional deficiency. Although disagreement and lack of evidence predominate the effects of evening and night work on employee safety, there are sound physiological grounds for presuming an increased rate of accidents at night based on laboratory studies of efficiency and errors related to circadian rhythms. Laboratory studies of speed, reaction time, and accuracy show demonstrable deficiency after the evening hours begin. Biologists say it is no coincidence that the human errors which led to the nuclear energy accident at Three Mile Island occurred at 4 A.M. by workers who had been changing shifts every week. Some studies also report that workers on night shifts make more mistakes than day shift employees and that this is particularly true for rotating shift workers.*

One would assume that unexpected overtime would be inefficient because the body would not be prepared for it and functioning would suffer. In 12-hour shifts, once the body becomes adjusted and a pattern established, individual preferences would prevail. Some employees might enjoy having three days off and working four days; other employees may find the fatigue factor is so overwhelming that they cannot adjust, or they simply do not like working long shifts.

*Landry, R. F. Off-beat rhythms and biological variables. *Occupational Health and Safety*, 1981, p. 42.

ILLNESS AS A HAZARD

Illness itself is considered a workplace hazard. It is assumed that illness demands attention and energy and that ill employees will not function at maximum efficiency and will be at greater risk for accidents. This is also true for an early return to work after an illness or accident. Psychological effects of returning to work after an accident or illness include feeling good about oneself, being able to be productive, and having an increase of self-esteem. Employees returning to work should be advised of their physical and mental limitations, if any.

Some employees have been accused of "malingering." Malingering means prolongation of symptoms past a reasonable recovery period and difficulty returning to work. Because of the complexity of the human mind and body and the lack of knowledge of healing rates, mind set, wish to get better, need to be an invalid, pain thresholds, and tolerance levels, it is very difficult to assess malingering and make general statements which apply to all workers. What is important is that persons having a hard time getting back to work be thoroughly evaluated, and assessments done on how safe it is for them to work and their level of functioning. (This is addressed in more detail in Chapter 4.)

Mental Illness

Mental illness is a workplace hazard if it is permitted to get in the way of functioning. However, many people with mental illness are able to function at work and lead productive lives.

Mental illness takes many forms, ranging from mood swings or periods of depression to periods of fantasy divorced from reality. What is important in the workplace is that support be available, and fear and anxiety regarding the job (because an employee is being treated for mental problems) be removed. If an employee assistance program is available, the worker should be encouraged to participate in it. This may ensure that a mental health problem does not become a safety problem.

Normal Effects of Prescription Drugs

Depending on the type of therapy, the behavioral symptoms, and the person's wishes, medications may or may not be used. It is imperative that the employee know the effects of the drugs. Some can cause drowsiness or lower blood pressure, which can lead to dizziness or fainting. Before employees take any medicine (no matter what is being treated), they should be told the reason for taking the drug, how long the drug must be taken, and the dosage.

Because many drugs have side effects, patients should contact the doctor if they observe any behavioral changes which could be related to the medication being taken. These include symptoms such as increased irritability, moodiness, dizziness, general feeling of malaise, nausea, vomiting, drowsiness, edema (swelling), chest pain, headaches, skin rashes, tingling in the extremities, and blurred vision. These side effects can be subtle and the person may be tempted to ignore them. In most cases the symptoms will not disappear until the causative drug is discontinued.

The normal effect of prescription drugs is to alleviate the presenting symptoms so the employee can get on with what needs to be done. The employee needs to be informed of the possible side effects and complications of all drugs he or she is taking and the possible interactive, synergistic, or potentiating effects of drugs, foods, and alcohol. For example, some drugs are known to inhibit each other (interactive); others are known to increase the effect of another drug (potentiating), such as alcohol increasing the effect of sleeping pills and Valium; others act in combination with each other (synergestic), such as phenobarbital and Miltown increasing the effect and action of both. With some drugs certain foods must be eliminated (see Table 7.2). It is very important for the employee to heed

TABLE 7.2
Food and Drugs which Are Contraindicated during Treatment with Monoamine Oxidase Inhibitors[a]

Foods and Beverages	Medications
Meat tenderizers	Demerol
Strong or aged cheeses	Morophine
Sour cream	Dexedrine
Yogurt	L-dopa
Pickles	Methyldopa
Pickled herring and other canned fish	Phenylpropanolamine (in over-the-
Chopped liver	counter cold remedies)
Chicken liver	Tofranil
Avocados	Elavil
Bananas	Aventyl
Raisins	Vivactil
Canned figs	Sinequan
Citrus fruits	
Pineapple	
Broad beans and their pods	
Soy sauce	
Any products made with yeast	
Beer and wine (Chianti and sherry)	
Coffee	
Chocolate	
Chicken	

[a]Monamine oxidase (MOA) inhibitors include Parnate, Nardil, and Marplan, for example.

this in order to avoid initiating a hypotensive crisis, which is signaled by the presence of a headache, sweating, restlessness, palpatations, pallor, chills, a stiff neck, nausea, vomiting, muscle twitching, and chest pain. This is an emergency since it can lead to intracranial hemorrhage and death. The medicine must be immediately discontinued, and a physician notified.

Often consumers are not told about potential side effects when the medicine is prescribed. It is not uncommon to be informed several months late by a nurse or a pharmacist. It is the doctor's responsibility to impart this information, but it is also the consumer's right to ask and be informed.

Over-the-counter drugs also have similar side effects and people must constantly be aware of what these might be. In addition, people must be aware of possible interactive effects with all drugs and alcohol.

DRUG AND ALCOHOL DEPENDENCY

Drug and alcohol dependency is another workplace hazard. Drugs and alcohol alter perceptions, motor control, and judgment and can lead to accidents and myriad problems.

Alcoholism, a problem for all industries, is a particularly terrifying one in industries where the consequences can be so devastating, as when an intoxicated worker has an accident which injures others. For this reason, it is most important that alcoholism is treated and that the treatment is totally supported by management. This is not an easy step; alcohol is being increasingly used to relieve stress and in peer situations where drinking is considered by many an acceptable behavior.

When alcoholism is diagnosed late in its course, alcoholics do not receive appropriate treatment, and unrealistic treatment goals are set as clinical measures of recovery. These problems derive from a combination of sources: society's traditional attitudes toward alcoholism, physicians' attitudes toward alcoholism in their patients, the patient's condition, and the state of the art of treatment. If the attitudes of the physician, the patient, and the employer could be changed, the success of treatment would increase tremendously. The keys to successful recovery from alcoholism are (1) early recognition and diagnosis, which means that treatment options would be available, and (2) whatever the treatment, goals which are conceptually possible for the patient to respond to.[1]

The chief deterrent to early diagnosis of an alcohol problem is our society's dual attitude toward alcohol. Drinking is approved, even encouraged, as an escape from problems, a relaxation technique, and a source of entertainment at parties.

The progression from alcohol abuse to alcoholism is highly individual. For some it is an astonishingly short journey, for others a prolonged migration from health and happiness. The telltale signs of this chronic behavior disorder are (1) undue preoccupation with alcohol and its use, to the detriment of physical and mental health, (2) loss of control when drinking is begun, and (3) a self-destructive attitude in dealing with personal relations and life situations.

Alcoholism is a person's way of dealing with pain—something hurts, and the person may or may not know what. Treatment consists of figuring out some way to make the person hurt less and function better.

Recovery rates for alcoholics who seek treatment are estimated to be somewhere between 60 and 80%. The recovery rate would be even higher if the person were directed toward choosing more appropriate therapies. It is important that the treatment regimen be the patient's choice and that decisions are arrived at through discussion with the physician, therapist, or counselor about the various treatment options. The patient should understand the fundamentals of treatment programs before making a choice.[3]

Alcoholics Anonymous (AA) has proved its effectiveness in helping many alcoholics recover. However, many alcoholics will not seek help from AA, whatever the reason—lack of understanding, personality type, misconceptions, and unacceptable goals.

The alcoholism problem is so widespread that many corporations and government agencies have established programs to deal with this problem. A typical alcoholism plan includes the following:

1. The administration develops and adopts a policy making clear that employees abusing alcohol will be treated confidentially.
2. Referral to the plan is made by the employee, by the supervisor, or by a co-worker.
3. Promotional materials are used to familiarize all employees with the program, to encourage self-referral, and to disseminate information among all employees with respect to alcohol and alcoholism.
4. Thorough training is given to managerial personnel, defining roles telling how to implement the policies and refer troubled employees.
5. Insurance coverage is provided, and its coverages are fully explained to all employees.
6. Provision is made to use community resources such as inpatient facilities, outpatient facilities, consultation services, educational programs, and Alcoholics Anonymous.
7. A counselor is designated.

Some workers look for a magic cure by trying a drug other than alcohol. This has not proved to be an acceptable cure. The aversive reactions of a drug such as Antabuse may provide a negative stimuli initially but have not proved to have long-term effects. The alcoholic stops taking the Antabuse when he or she wishes to drink. Substituting another drug, such as a tranquilizer, is also not effective. Then the person becomes addicted to another substance which can lead to deteriorated performance.

There are many treatment programs available from support groups, to assertiveness training, to relaxation and biofeedback training, to in-depth psychotherapy. The worker will need to choose the program which seems most appropriate and which will be congruent with personal beliefs about treatment. The worker will have to believe in the treatment if it is going to be effective and if he or she is going to be able to follow through with the treatment plan.

Recognizing psychological hazards is an important function of safety engineers, who must develop assessment strategies if workplaces for which they are responsible are to be safe and healthy.

SUGGESTED LEARNING EXPERIENCES

1. Discuss ways you as a safety engineer would proceed with recognizing "nonobservable hazards" in a workplace of your choice.

2. Debate: Occupational stress is an antecedent of occupational illness and industrial accident.

3. How would you as a safety engineer include employees' mental health as a priority in a safety program?

4. What effects on safety do you propose might occur in the occupational safety and health field with the increasing number of dual-career families? What do you envision will happen regarding emotional support, communication among family members, mutual goals, and personal satisfaction?

4. Take one of the occupations with a high prevalence rate for occupational-related stress and illness and develop some preventive strategies for it.

5. In terms of psychological stress prevention, describe (a) an ideal occupation and (b) a typical workday. Does this occupation appeal to you? Why? Why not?

6. Develop a job stress index scale.

7. Develop a grant proposal addressing equal economic opportunities for all.

8. How would you as a safety professional deal with employee attitudes toward emotional problems?

9. Take the life stress scale yourself. What implications for your own health do you see?

10.   Develop a small research project looking at what drugs 25 workers are taking. If possible collect this data. Present your results and your conclusions. Were they what you expected? Discuss.

REFERENCES

1.   Chavetz, M. W. Alcohol and alcoholism. *American Scientist*, 1979, **67**, 293–299.

2.   Dembroski, T. M., MacDougall, J. M., and Shields, J. L. Physiologic reactions to social challenge in persons evidencing the Type A coronary-prone behavior pattern. *Journal of Human Stress*, 1977, **2**, 2–9.

3.   Finn, P. The effects of shift work on the lives of employees. *Monthly Labor Review*, 1981, **104**, 31–35.

4.   Haynes, S. G., Levine, S., Scotch, N., Feinleib, M., and Kannel, W. B. The relationship of psychosocial factors to coronary heart disease in the Framingham study. *American Journal of Epidemiology*, 1978, **107**, 362–383.

5.   Holmes, T. H., and Rahe, R. H. The social readjustment rating scale. *Journal of Psychosomatic Research*, 1967, **2**, 213–218.

6.   Landry, R. F. Off-beat rhythms and biological variables. *Occupational Health and Safety*, 1981, **50**, 40–43. (Apr.)

7.   Ledgerwood, D. E., and Johnson-Dietz, S. Sexual harassment: Implications for employer liability. *Monthly Labor Review*, 1981, **104**, 42–43. (Apr.)

8.   Pifer, A. Women working: Toward a new society. *Urban and Social Change Review*, 1980, **1**, 57–59.

9.   Rowland, K. F., and Sokol, B. A review of research examining the coronary-prone behavior pattern. *Journal of Human Stress*, 1977, **2**, 26–33.

10.   Selye, H. The evolution of the stress concept. *American Scientist*, 1973, **61**, 692–699.

11.   Shabecoff, P. Psychosomatic medicine finds out why work can be sickening. *New York Times*, Feb. 3, 1980, p. 22E.

12.   Sheridan, P. J. NIOSH puts job stress under the microscope. *Occupational Hazards*, 1981, 70–79.

13.   Wambaugh, J. *The Onion Field*. New York: Dell, 1979.

14.   Wardle, M. G. *Dual Careers*. Boston College unpublished report, 1982.

8 INJURY SOURCES

The failure to prevent or control accidents adequately results in injury to workers and damage to the equipment. Many safety engineers have estimated that for every accident, 10 near-misses have occurred. These results have been further supported by airline pilot data from the National Transportation Safety Board.

This chapter examines the incidence of injury by type and severity in a variety of industries, such as manufacturing, health care, and construction.

COMPILING INJURY STATISTICS

Injuries are measured in three ways: in terms of frequency rate, severity rate, and the average days charged for each disabling injury.[3]

Frequency Rate

The disabling-injury frequency rate is one of the most important ways of quantifying injuries. Although different agencies may use different formulas or units for computing the frequency rate, the unit used here is million employee-hour units. On the average an employee in the United States works about 2,000 hours per year. One million employee-hours, then, represents a year's work for 500 employees. In general terms, therefore, a rate of 20.0 may be interpreted as 20 disabling injuries per year for each 500 employees, or 1 disabling injury for each 25 employees. Expressed in another way, if a plant has an injury rate of 20.0 during the year, 1 out of every 25 employees was injured to the extent that at least one full day of work was lost or some permanent impairment was suffered.

Frequency rate (FR) is based on the total number of deaths and on the permanent total, permanent partial, and temporary total disabilities occurring in the period covered by the frequency rate. The rate relates these injuries to the employee-hours worked during this specific time and expresses the number of such injuries in terms of million-hour units by the equation:

$$FR = \frac{\text{no. of disabling injuries} \times 1,000,000}{\text{employee-hours of exposure}}$$

Disabling-Injury Severity Rate

Another important index is called the disabling-injury severity rate. This shows the rate at which days are lost or charged in relation to 1 million employee-hours of work. A severity rate of 500 means that 500 days were lost or charged for every 1 million employee-hours worked. As with the frequency rate, the base may be interpreted more generally in terms of employees, and since 1 million employee-hours represents the yearly experience of about 500 employees, a severity rate of 500 may be interpreted as 500 days lost or charged for each 500 employees, or about one day lost or charged for each employee.

Included in the severity rate are both actual days lost and scheduled charges, with the total heavily weighted by the latter. In most cases in which scheduled charges are assessed, these charges exceed the actual days lost. For example, amputation of the index finger carries a scheduled charge of 400 days, which is used in place of the actual days lost. It is improbable that a disability of this type will cause the worker to lose 400 days.

The severity rate, then, must not be interpreted as showing the number of days lost because of injuries. Actually, it shows the number of days lost or charged, and the charges which represent potential losses of production outweigh the actual days lost or charged at the time of the injury.

This rate is based on the total of all scheduled charges, all deaths, permanent total and permanent partial disabilities, plus the total days of disability from all temporary total injuries occurring during the period for which the severity rate (SR) is computed. The rate relates the days charged to the employee-hours worked during the period and, here again, expresses the loss in terms of million-hour units with the following equation:

$$SR = \frac{\text{total days charged} \times 1,000,000}{\text{employee-hours of exposure}}$$

It is significant that despite OSHA standards, the frequency rate over the last 10 years seems to have edged upward, whereas the severity rate has been slowly going down. For all practical purposes, the severity rate over the last 10 years remained unchanged, but the frequency rate increased by about 55% since 1971. It should be emphasized, however, that these figures are merely guidelines, since they represent statistical data rather than the facts. But they are valuable because they suggest that OSHA standards and inspection procedures are ineffective in decreasing accident and injury rates. Problems with OSHA in relation to injuries are:

1. There are too few OSHA inspectors.
2. Sufficient incentives or reinforcements have not been provided for companies to improve their safety programs.
3. The criminal responsibility of the chief corporate officers in relation to hazardous and toxic materials has not been detailed.

INJURY CLASSIFICATIONS

Permanent Total Disability

Any injury or impairment other than death which totally incapacitates a person from any gainful employment is permanent total disability. The worst hazards arise from machinery—sharp blades, heavy machinery, and parts that intrude. The average cost per injury for 1974 was $29,000. In evaluating the severity of injuries, 6,000 days were charged against fatal and permanent total injuries. This means that if a worker dies from machine-related injuries, the worker's family is compensated at about $4 per day. The average cost for all injuries in this category is slightly over $21,000.

Permanent Partial Disability

This is any injury or impairment which results in the complete loss of or loss of use of any member or function or part of a member or function of the body. The average cost for injuries of this type is almost $26,000 per injury.[2] The cost of serious injury caused by a fall is $39,000.[3] From the employer's point of view, it is more economical if a worker is killed than left with a permanent partial disability. This, by the way, is another way of evaluating problem areas for which solutions must be provided, in an attempt to eliminate or reduce injuries and the terrible cost burden associated with them.

Temporary Total Disability

This is any injury or impairment other than death which temporarily incapacitates a person totally from any gainful employment.

Since the average charge per disabling injury can be computed quickly for any plant, group, or industry for which frequency and severity rates are known (merely by dividing the frequency rate into the severity rate), the average charge per disabling injury for a plant may be compared directly with a similar figure computed for the entire industry of which the plant is a part. For example, the plant with the high frequency and severity rates discussed above had an average charge per injury of 21 days, whereas the average for the entire group of 44 plants was 49 days.[7]

IDENTIFYING KEY ACCIDENT DATA

Key information in determining injury sources is shown in Table 8.1. A chain of events leading to an accident is shown in Figure 8.1. There are at least five points at which preventive actions could have been taken.

In identifying injury sources, it will be important to identify components involved. If, for example, a person attempted to climb a ladder with a defective rung and fell because the rung broke when weight was put on it, the agency-of-accident part would be the defective rung.

The Unsafe Act

The unsafe personal action which caused or resulted in an accident should be designated. It may have been a violation of safe practice, use of an incorrect method, or failure to perform critical actions, wittingly or unwittingly. It may be something the worker should not have done, something which should have been done differently, or failure to do something which should have been done. The person responsible for the unsafe act may or may not have been the person injured. Since the "unsafe act" classification means accidents resulting from mistakes made by people, data tabulations in this category may properly be labeled "accident causes."

For statistical purposes, analyzing an injury or accident report consists of tabulating the answers to the following questions:

1. *Nature of Injury.* What was the injury?
2. *Part of Body.* What part of the body was affected by injury in **1**?
3. *Source of Injury.* What object, substance, exposure, or bodily motion inflicted the injury named in **1**?
4. *Accident Type.* How did the injured person come in contact with the object, substance, or exposure named in **3**, or during what personal movement did the bodily motion named in **3** occur?
5. *Hazardous Condition.* What hazardous physical or environmental condition or circumstance caused or permitted the occurrence of the event named in **4**?
6. *Agency of Accident.* In what object, substance, or part of the premises did the hazardous physical or environmental condition named in **5** exist?

7. *Agency-of-Accident Part.* In what specific part of the agency of accident named in **6** did the hazardous condition named in **5** exist?

8. *Unsafe Act.* What unsafe act of a person caused or permitted the occurrence of the event named in **4**?

Contributing factors should be noted, when they can be determined. Table 8.1 gives typical checklists that can be used for preparing the report. For easier statistical handling of the data, the factors can be individually numbered, and the numbers inserted on the form.

Examples of classifying the key facts in various accidents are presented next.

Accident no. 1. The operator of a circular saw reached over the running saw to pick up a piece of scrap. His hand touched the blade, which was not covered, and his thumb was severely lacerated.

1. Nature of injury—laceration
2. Part of body—thumb
3. Source of injury—circular saw
4. Accident type—struck against
5. Hazardous condition—unguarded
6. Agency of accident—circular saw
7. Agency-of-accident part—blade
8. Unsafe act—clearing moving machine

Accident no. 2. A forklift truck went out of control when one wheel hit a piece of stock lumber which projected into the aisle. The truck ran out of the aisle and struck a machine operator, breaking her leg between the ankle and the knee.

1. Nature of injury—fracture
2. Part of body—lower leg
3. Source of injury—lift truck
4. Accident type—struck by
5. Hazardous condition—improperly placed lumber
6. Agency of accident—truck
7. Agency-of-accident part—none
8. Unsafe act—unsafe placement of material

Accident no. 3. A warehouse employee jumped from the loading platform to the ground instead of using the steps. As he landed, he sprained his ankle.

TABLE 8.1
Checklist for Identifying Key Facts

Nature of Injury	Part of Body	Accident Type	Hazardous Condition	Agency of Accident	Agency-of-Accident Part	
					Unsafe Act	Contributing Factors

Nature of Injury
Foreign body
Cut
Bruises and contusions
Strain and sprain
Fracture
Burns
Amputation
Puncture wound
Hernia
Dermatitis
Ganglion
Abrasions
Others

Part of Body
Head and Neck
Scalp
Eyes
Ears
Mouth, teeth
Neck
Face
Skull
Others
Upper Extremities
Shoulder
Arms (upper)
Forearm
Wrist
Elbow
Hand
Fingers and thumb
Others

Accident Type
Struck against (rough or sharp objects, surfaces; exclusive of falls)
Struck by flying objects
Struck by sliding, falling, or other moving objects
Caught in (on or between)
Fall on same level
Fall to different level
Overexertion (resulting in

Hazardous Condition
Improperly or inadequately guarded
Unguarded
Defective tools, equipment, substances
Unsafe design or construction
Hazardous arrangement
Improper illumination
Improper ventilation
Improper dress
Poor housekeeping
Congested area

Agency of Accident
Machine
Vehicles
Hand tools
Tin and black plate (sheet, stock, or scrap)
Material work handled (other than tin and black plate)
Can and end conveyors
Conveyors (belt, cable, can, dividers, chain, twisters, drops, can elevators)

Unsafe Act
Operating without authority
Failure to warn or secure
Operating at unsafe speed
Making safety devices inoperative
Using defective equipment, materials, tools, or vehicles
Using equipment, tools, materials, or vehicles unsafely
Failure to use

Contributing Factors
Disregard of instructions
Bodily defects
Lack of knowledge, skill
Act of person other than injured
Failure to report to medical department
Others
No contributing factor

Body
Back
Chest
Abdomen
Groin
Others
Lower Ex-
tremities
Hips
Legs
Knee
Ankle
Feet
Toes
Thigh
Others

strain, hernia)
Slip (not a fall)
Contact with
temperature
extremes;
burns
Inhalation, ab-
sorption, in-
gestion,
poisoning
Contact with
electric cur-
rent
Others

Others
No unsafe con-
dition

(chutes, belt,
gravity)
Hoists and
cranes
Elevators (pas-
senger and
freight)
Building (door,
pillar, wall,
window)
Floors or level
surfaces
Stairs, steps, or
platforms
Chemicals
Ladders or scaf-
folds
Electrical ap-
paratus
Boilers, pres-
sure vessels
Others

personal pro-
tective equip-
ment
Failure to use
equipment
provided (ex-
cept personal
protective
equipment)
Unsafe loading,
placing, and
mixing
Unsafe lifting
and carrying
(including in-
secure grip)
Taking an un-
safe position
Adjusting,
clearing jams,
cleaning ma-
chinery in
motion
Distracting,
teasing
Poor house-
keeping
Others
No unsafe act

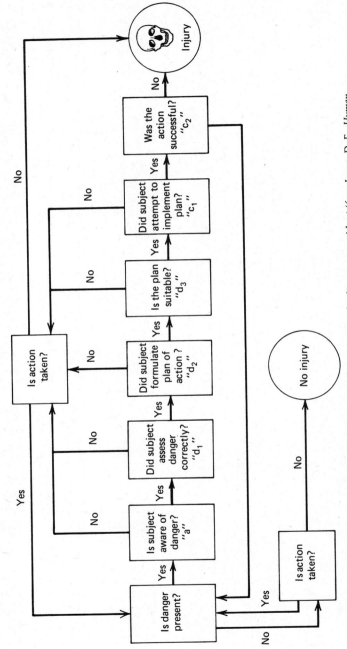

FIGURE 8.1. Illustration of a chain of events leading to an accident (*from James, D. F., Human Factors—Occupational Safety: A Report to the Labour Safety Council of Ontario*).

1. Nature of injury—sprain
2. Part of body—ankle
3. Source of injury—ground
4. Accident type—fall from elevation
5. Hazardous condition—none indicated
6. Agency of accident—none indicated
7. Agency-of-accident part—none indicated
8. Unsafe act—jumping from elevation

Accident no. 4. A laborer working in a trench was suffocated under a mass of earth when the unshored wall of the trench caved in.

1. Nature of injury—asphyxia
2. Part of body—respiratory system
3. Source of injury—earth
4. Accident type—caught under
5. Hazardous condition—lack of shoring
6. Agency of accident—trench
7. Agency-of-accident part—none
8. Unsafe act—none indicated

Accident no. 5. A salesperson stopped her car at an intersection and was waiting for the traffic light to change from red to green. Another car struck her car in the rear. The whiplash effect fractured a vertebra in the salesperson's neck.

1. Nature of injury—fracture
2. Part of body—neck
3. Source of injury—auto
4. Accident type—collision
5. Hazardous condition—traffic hazard
6. Agency of accident—environment
7. Agency-of-accident part—none
8. Unsafe act—none

BASIC ACCIDENT TYPES

The basic accident types include the following: struck against, struck by; fall from elevation; fall on same level; caught in, under, between; rubbed or abraded; bodily reaction; overexertion; contact with electric current; contact with radiation; contact with caustic, toxic, and noxious substances; public transportation accidents; motor vehicle accidents; and drownings. Some of these will be discussed below.

Struck Against

Being struck against can occur from stationary or moving objects and from falling or flying objects. The force of impact is a key factor in this type of injury because of deceleration forces. Also, most of these injuries are totally unexpected, and employees have no chance to defend themselves.

Fall from Above

Falls from above (as opposed to falls on the same level) can occur from scaffolds, walkways, or stairs and into shafts (from the edge of shafts), excavations, or floor openings. This type of accident occurs because stairs were not provided with railings or scaffolds were not provided with guard rails. A shipyard was recently cited and fined by OSHA for not putting a guard rail around missile tube openings on the deck.

Fall on the Same Level

A fall on the same level can be caused by slipping because of poor traction of footgear on the floor. One of the authors slipped recently in an industrial plant because an oil spill from an industrial truck had not been properly removed.

Adhesive friction is a positive grip due to the penetration of one surface into the other. Textured surfaces; gratings; diamond or checkered plate; knurled, dimpled corrugated expanding metal; and serrated, upset, and abrasive surfaces are safest. Although the surface has a positive grip easily evidenced by the impression of the raised surface into the shoe sole, it does not necessarily have slip resistance. In adhesive friction, the raised surfaces are oil-coated and must be classified as having "fluid film friction" in slip resistance (that is, the smaller the sole surface, the greater the risk of slipping). This is why some textured surfaces have less traction than smooth surfaces when both are covered with oil or water.

Because of the number of variables involved, there are no fail-safe methods of preventing slipping, tripping, and falling.

Caught In, Under, or Between Objects

Being caught in, under, or between objects can occur in a stationary or moving meshing object, in two or more moving objects that are not meshing, and in collapsing materials such as earth slides and collapsing buildings.

Rubbed or Abraded

Injury can result from rubbing or abrading by leaning, kneeling, or sitting on nonvibrating objects. These injuries are usually preventable by wearing appropriate protective clothing, such as kneepads. Such injuries can also be caused by constant handling of nonvibrating articles, by rotating objects, and by foreign matter in the eyes. Repeated pressure can result in a blister, usually preventable by using protective clothing, such as padded gloves.

Bodily Reactions

Physical reactions, either voluntary (putting a hand too close to a hot element) or involuntary (having a seizure or momentary loss of memory), can be an injury source.

Overexertion

Overexertion can result from lifting objects that are too heavy to handle without machine assistance or awkward enough to cause loss of balance. The resulting damage is most often a back injury. Overexertion can also occur from pulling, carrying, pushing, or throwing objects that are too heavy if proper body mechanic techniques are not used. Time is also a factor here. For example, the employee's body might not be unduly strained lifting a heavy object and moving it a short distance in a brief amount of time, but severe and long-lasting injury can occur if the worker carries this same object for several minutes.

Contact with Electric Current

Contact with electric current is a serious and frequent cause of workplace injuries. Because of the potential severity of electrical accidents, all employees working with or around high voltages must be thoroughly educated in safety procedures. Consistent monitoring and continuing education must be done.

Contact with Caustic, Toxic, and Noxious Substances

Contact with caustic, toxic, and noxious substances occurs via inhalation through the respiratory system, ingestion through the gastrointestinal system, or absorption via the integumentary system. Contact with the skin is another way that irritants can cause injury.

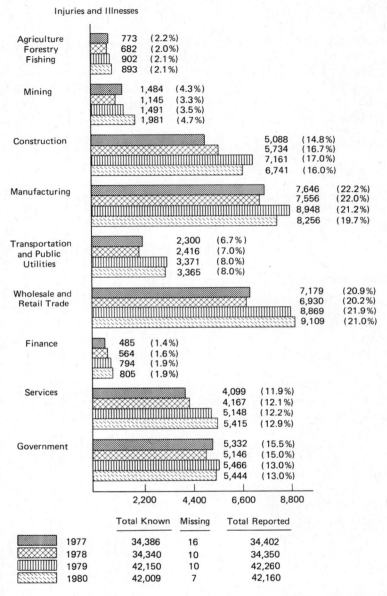

FIGURE 8.2. Work-Related injuries and illnesses and employment by industry, Colorado, 1980 (*used with permission, Colorado Division of Labor*).

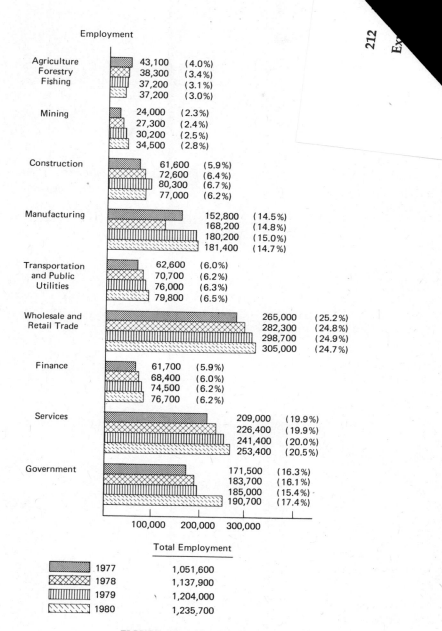

Employment

Agriculture Forestry Fishing	43,100	(4.0%)
	38,300	(3.4%)
	37,200	(3.1%)
	37,200	(3.0%)
Mining	24,000	(2.3%)
	27,300	(2.4%)
	30,200	(2.5%)
	34,500	(2.8%)
Construction	61,600	(5.9%)
	72,600	(6.4%)
	80,300	(6.7%)
	77,000	(6.2%)
Manufacturing	152,800	(14.5%)
	168,200	(14.8%)
	180,200	(15.0%)
	181,400	(14.7%)
Transportation and Public Utilities	62,600	(6.0%)
	70,700	(6.2%)
	76,000	(6.3%)
	79,800	(6.5%)
Wholesale and Retail Trade	265,000	(25.2%)
	282,300	(24.8%)
	298,700	(24.9%)
	305,000	(24.7%)
Finance	61,700	(5.9%)
	68,400	(6.0%)
	74,500	(6.2%)
	76,700	(6.2%)
Services	209,000	(19.9%)
	226,400	(19.9%)
	241,400	(20.0%)
	253,400	(20.5%)
Government	171,500	(16.3%)
	183,700	(16.1%)
	185,000	(15.4%)
	190,700	(17.4%)

100,000   200,000   300,000

Total Employment

	1977	1,051,600
	1978	1,137,900
	1979	1,204,000
	1980	1,235,700

FIGURE 8.2. (*Continued*)

posure to Radiation

Exposure to radiation is increasingly an occupational health hazard. It can occur not just in obvious industries (such as power plants, battery manufacturing, nuclear medicine) but also in industries involved with electromagnetic elements. Radiation is also emitted from cigarette smoking.[2]

Public Transportation

Public transportation accidents, collision and noncollision, include those involving aircraft, ships, boats, street cars, trains, buses, and multipassenger shuttles.

Noncollision accidents occur when a vehicle overturns, goes out of control, or starts or stops abruptly.

Drowning

Drowning is a hazard for employees working around water. This includes occupations such as logging, shipbuilding, and dam construction and any other industry where water is involved.

EMPLOYEE INJURIES

Employee injuries affect any part of the body, but the most common workplace injuries are the head, neck, chest, upper extremities (including hands

	1977		1978		1979		1980	
	#	%	#	%	#	%	#	%
Sprain or strain	14,356	44.7	13,236	41.9	16,645	42.2	15,935	42.9
Cut or puncture	4,798	15.0	4,913	15.5	5,453	13.8	4,813	12.9
Fracture	3,875	12.1	3,944	12.4	4,715	11.9	4,374	11.8
Contusion or bruise	4,540	14.1	3,275	10.3	4,000	10.1	3,112	8.4
Occupational illness	816	2.5	1,887	5.9	2,907	7.4	2,718	7.3
Burn	1113	3.5	1,205	3.7	1,313	3.3	1,350	3.6
Dislocation	937	2.9	1,101	3.4	1,258	3.2	1,236	3.3
Other	1,667	5.2	2,209	6.9	3,204	8.1	3,644	9.8
Total known	32,102	100.0	31,770	100.0	39,495	100.0	37,182	100.0
Unknown	2,129		2,559		2,623		4,834	
Total reported	34,231		34,329		42,118		42,016	

FIGURE 8.3. Nature of injury of all cases reported, Colorado, 1980 (*used with permission, Colorado Division of Labor*).

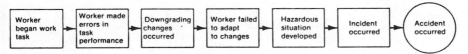

FIGURE 8.4.   Generalized events and causal factors chart (*from James, D. F., Human Factors— Occupational Safety: A Report to the Labour Safety Council of Ontario*).

and wrist), abdomen, genitals, lower extremities, and back. The type of work generally dictates the site of the injury.

All industries have specific hazards, injury types, and occupational illnesses for which employees are at risk. Figures 8.2 through 8.5 and Tables 8.2 through 8.4 show examples of work-related injuries and accidents, some general causal accident factors, and sources and types of workplace hazards.

Occupational illnesses for principle industries are discussed in Chapter 9.

SUGGESTED LEARNING EXPERIENCES

1. Select a workplace of your choice and list the possible injury sources.

2. Develop preventive measures for the above-identified injury sources.

3. Calculate the severity and frequency rate for an industrial injury of your choice.

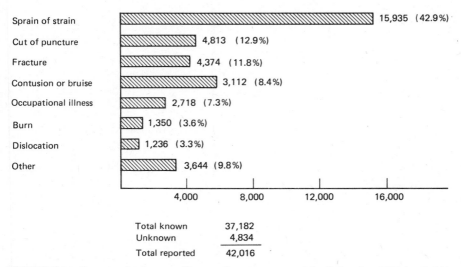

FIGURE 8.5.   Source of injury or illness: all cases reported in Colorado, 1980 (*used with permission, Colorado Division of Labor*).

TABLE 8.2
Work-Related Injuries and Illnesses and Employment by Industry,
Colorado, 1980

Industry	Injuries and Illnesses		Employment		Rate per 100 Workers
	no.	%	no.	%	
Agriculture, forestry, fisheries	893	2.1	37,000	3.0	2.4
Mining	1,981	4.7	34,500	2.8	5.7
Construction	6,741	17.5	77,000	6.2	8.8
Manufacturing	8,256	19.7	181,400	14.7	4.6
Transportation and public utilities	3,365	9.4	79,800	6.5	4.2
Wholesale/retail trade	9,109	21.7	305,000	24.7	3.0
Finance	805	2.0	76,700	6.2	1.1
Services	5,415	18.8	253,400	20.5	2.1
Government	5,444	4.1	190,700	15.4	2.9
Total	42,009	100.0		100.0	3.4
Unknown	7				
Total known	42,016		1,235,700		

SOURCE: Employment data obtained from *Colorado Labor Review*, 1981, **18**(1), 19; Colorado Division of Labor, *Annual Report: Work Related Injuries and Illnesses in Colorado—1980*. Used with permission of Colorado Department of Labor.

4. What forms and procedures would you utilize in determining workplace injury sources.

5. Determine the disability rate for an industrial operation of your choice. What factors would you want to consider?

6. Develop a flow diagram identifying key facts in a hypothetical workplace accident.

7. What factors would you want considered in assessing the slipperiness of floor surfaces? Discuss.

8. Develop a pictorial presentation involving the workplace injury source of overexertion.

9. In assessing your last place of employment, what injury source would

TABLE 8.3
Nature of Injury or Illness by Part of Body Affected, All Cases, Colorado, 1980

Nature of Injury	Head	Eyes	Neck	Upper Extremities	Trunk	Back	Lower Extremities	Multiple Parts	Body System	Other	Total
Burns	2	0	19	88	364	492	374	10	0	0	1,349 (3.6%)
Contusions	107	11	16	660	549	240	1,236	286	0	0	3,105 (8.4%)
Cuts	375	122	7	3,476	63	10	699	53	2	0	4,807 (13.0%)
Dislocation	72	159	2	455	28	6	238	273	0	0	1,233 (3.3%)
Fracture	165	0	28	1,735	546	156	1,629	110	0	0	4,369 (11.8%)
Sprain/strain	5	0	464	1,064	2,517	7,969	3,253	634	0	0	15,906 (42.9%)
Illness	158	110	25	743	197	223	250	137	848	1	2,692 (7.3)
Other	500	542	16	953	132	121	383	884	97	0	3,628 (9.8%)
Total known	1,384 (3.7%)	944 (2.5%)	577 (1.6%)	9,174 (24.7%)	4,396 (11.9%)	9,217 (24.9%)	8,062 (21.7%)	2,387 (6.4%)	947 (2.6%)	1 (0.0%)	37,089
Unknown											4,927
Total Reported											42,016

SOURCE: Colorado Division of Labor, *Annual Report: Work Related Injuries and Illnesses in Colorado—1980*. Used with permission of Colorado Department of Labor. 1981.

215

TABLE 8.4
Nature of Injury or Illness by Source of Injury or Illness, All Cases, Colorado, 1980

Source of Injury or Illness	Nature of Injury or Illness								
	Burns	Contusions	Cuts or Puncture	Dislocation	Fracture	Sprain/ Strain	Illness	Other	Total
Bodily motion	239	1	0	1	197	2,458	155	31	3,082 (8.5%)
Food/food products	9	20	2	118	12	200	124	11	496 (1.4%)
Boxes or containers	177	250	185	15	254	3,410	186	159	4,636 (12.7%)
Buildings/ structures	52	194	129	0	184	450	47	127	1,183 (3.3%)
Chemicals	0	0	10	242	0	1	285	15	553 (1.5%)
Glass (non-containers)	1	1	236	2	5	18	7	15	285 (0.8%)
Machines, pumps	56	249	1,070	30	432	682	93	479	3,091 (8.5%)
Electrical apparatus	12	25	44	30	39	226	32	44	452 (1.2%)

Furniture	37	142	66	3	81	557	48	65	999 (2.7%)
Hand tools	37	118	899	6	125	537	193	89	2,004 (5.5%)
Power tools	9	30	290	11	46	141	45	41	613 (1.7%)
Metal items	78	293	1,035	95	430	1,118	76	461	3,586 (9.9%)
Vehicles	84	570	223	14	486	1,107	100	653	3,174 (8.7%)
Wood items	36	141	147	0	156	579	46	130	1,235 (3.4%)
Working surfaces	325	744	118	0	1,449	2,073	125	509	5,343 (14.7%)
Other	166	363	313	662	423	2,007	971	756	5,661 (15.6%)
Total known	1,318 (3.6%)	3,078 (8.5%)	4,767 (13.1%)	1,229 (3.4%)	4,319 (11.9%)	15,564 (42.8%)	2,533 (7.0%)	3,585 (9.9%)	36,393
Unknown									5,623
Total Reported									42,016

SOURCE: Colorado Division of Labor, Annual Report: Work Related Injuries and Illnesses in Colorado—1980. Used with permission of Colorado Department of Labor. Denver, CO.

be of most concern to you? What preventive measures would you now employ?

10. Examine the Colorado injury and illness data. Take a particular site or type of injury/illness and attempt to determine how to reduce those injuries/illnesses.

REFERENCES

1. Colorado Division of Labor. *Annual Report: Work Related Injuries and illnesses in Colorado—1980.* Denver, CO, 1981.

2. Martell, E. A. Radioactivity in cigarette smoke. *New England Journal of Medicine*, 1982, **30,** 307–310.

3. National Safety Council. *Accident Prevention Manual for Industrial Operations* (7th ed.). Chicago, 1977.

9 INDUSTRIAL HYGIENE

This chapter describes the basic activities of industrial hygiene. It examines environmental controls for ventilation and illumination and the limits to which a person may be used in manual materials handling without resorting to mechanical devices. The chapter discusses hazardous substances and industrial toxicology because nearly every workplace has some substances which can affect the health of workers. An understanding of industrial toxicology will give safety engineers the knowledge they need to know when to limit worker exposure to hazardous substances. The goal of this chapter is not to qualify the reader as an industrial hygienist but to make him or her aware of the types of problems industrial hygienists typically encounter. It will provide some background in the techniques an industrial hygienist uses, some of the special language associated with the field, knowledge as to when to ask for a consultant, and awareness of limitations.

OBJECTIVES OF INDUSTRIAL HYGIENE

Industrial hygienists have two major responsibilities:

1. To determine that the workplace contains nothing harmful to health
2. To apply health standards to the working population[2]

Industrial hygienists must also recognize potential hazards in the work environment, measure the work environment, evaluate the measurements obtained to determine whether a hazard exists, and identify and recommend controls to remove or reduce the hazard exposure of the worker.[15] Industrial hygienists are concerned with preventing hazards in the workplace and removal of existing hazards. The ideal, of course, is to have no hazards at all, but in necessarily hazardous work, like mining, many control measures are available to reduce the accident-causing potential of unavoidable hazards. These measures include substitution of safer materials, processes, or equipment. For example, is there a less toxic or less flammable material that can be used? Is there a better way to do a job, such as using dipping instead of spraying? Is there a better type of equipment for the job, say, using automated equipment instead of manual methods? Can engineering changes, such as machine guards, make the equipment less hazardous?

Brief History of Industrial Hygiene

Until the early twentieth century, worker safety was virtually unknown. The U.S. Public Health Service and the U.S. Bureau of Mines were the first federal agencies to conduct exploratory studies in the mining and steel industries, and these were undertaken as early as 1910. The first state industrial hygiene programs were established in 1913 in the New York and Ohio departments of health.[1]

One of the earliest attempts to link the industrial environment with a specific disease was a study of the frequency of tuberculosis among garment workers. More recently there has been consumer, worker, and union pressure for elimination of workplace accidents, injuries, and diseases linked to the environment. Congress has reacted to these demands by passing three major pieces of legislation: the Metal and Nonmetallic Mine Safety Act of 1966, the Federal Coal Mine Health and Safety Act of 1969, and the Occupational Safety and Health Act of 1970.

The Metal and Nonmetallic Mine Safety Act spells out safety standards; provides for a Safety Board of Review; specifies mandatory reporting of all mine accidents, injuries, and occupational disease; and provides for expanded mine and safety education programs. Although mine conditions improved because of this act, no health or safety representative was on the advisory committee to provide needed input. There was no provision for research in mine safety and health, which meant that no data were gathered to show what was happening in the mines and to evaluate what could be done to make them safer.

The Federal Coal Mine Health and Safety Act of 1969 aimed to attain the highest degree of health protection possible for the miner. It specified mandatory health standards and an advisory committee to study health problems. This act gives the federal government the authority to withdraw miners from any mine and to close it. The act itself reads to

> . . . provide, to the greatest extent possible, that working conditions in each underground coal mine are sufficiently free of respirable dust concentrations in the mine atmosphere to permit each miner the opportunity to work underground during the period of his entire adult working life without incurring any disability from pneumoconiosis or any other occupation-related disease during or at the end of such period.

This is to be accomplished by setting standards for control of dust; mandatory use of respiratory equipment; development of rules for roof support, proper ventilation, grounding of cables, and distribution of underground high voltage; and provision for mandatory medical examinations for the miners at fixed intervals. This act was a major breakthrough in the movement toward achieving a safe and healthy work environment for the miner.[1]

The purpose of the Occupational Safety and Health Act of 1970 (discussed in Chapter 3) is to "assure as far as possible every working man and woman in the nation safe and healthful working conditions and to preserve our human resources."[15] The federal government is empowered to develop and set mandatory safety and health standards for the workplace; it also supports research and education. Many of the tasks for which industrial hygienists are responsible are specified in the act. These include measuring and recording employees' exposure to potentially harmful workplace agents, recording the incidence of occupational disease, and measuring hazardous substances to assure that their concentration levels are below the standards (Threshold Limit Value, TLV, and Maximum Allowable Concentration, MAC) published in the Code of Federal Regulations, especially 29 CFR 1910.

ANALYSIS OF THE WORK ENVIRONMENT

Temperature

One aspect of the workplace environment with which industrial hygiene is concerned is environmental stress caused by temperature. The normal body temperature is not conducive to safe, efficient working conditions. How are these measured? What is optimum, and how can adaptations be made? Table 9.1 shows the implications of an eight-hour day of exposure to various climates, and Tables 9.2 and 9.3 show potential occupational exposures to hot and cold working environments. OSHA requires that temperatures be measured by three methods, dry bulb readings, wet bulb readings, and Globe thermometer readings.

A windchill index has been developed (see Figure 5.5) that shows the temperatures exposed flesh feels on a windy day. People generally do not tolerate cold as well as they do heat. Thermal comfort was discussed previously in Chapter 5.

The effects of temperature stresses on people vary with age (older workers are more sensitive to temperature extremes), sex (women generally have a more difficult time adapting to high temperatures), weight (heavier workers generally have a more difficult time adapting to high temperatures), and physical condition (workers in better physical condition generally can endure more stress). Workers with cardiorespiratory problems

TABLE 9.1
Heat Stress Index Implications of 8-Hour Exposure

Temperature, °C	Effects
−20° to −10°	Mild cold strain.
0° to +20°	No thermal strain.
+20° to +30°	Mild to moderate heat strain. Subtle to substantial decrements in performance may be expected where intellectual forms of work are performed. In heavy work, little decrement is to be expected.
+40° to +60°	Severe heat strain. Involves threat to health unless physically fit. Not suitable for those with cardiovascular or respiratory impairment. Climatization required. Not suitable for tasks requiring sustained mental effort.
+70° to +90°	Very severe heat strain. Personnel must have thorough medical examination and work trial after acclimatization.
+100°	The maximum heat stress that can be tolerated by fit, acclimatized young men.

SOURCE: *Criteria for a Recommended Standard: Occupational Exposure to Hot Environments* (HSM-72-102-69). Washington, D.C.: GPO, 1972.

TABLE 9.2
Workers Exposed to Excessive Heat

Animal-rendering workers
Bakers
Boiler heaters
Cannery workers
Chemical plant operators working near hot containers and furnaces
Cleaners
Coke oven operators
Cooks
Foundry workers
Glass-manufacturing workers
Kiln workers
Miners in deep mines
Outdoor workers during hot weather
Sailors passing hot climatic zones
Shipyard workers when cleaning cargo holds
Smelter workers
Steel and metal forgers
Textile-manufacturing workers (weaving, dyeing)
Tire- (rubber-) manufacturing workers

SOURCE: NIOSH, *A Guide to the Work Relatedness of Disease* (79–116). Washington, D.C.: GPO, 1979.

TABLE 9.3
Workers Exposed to Excessive Cold

Cooling-room workers
Divers
Dry ice workers
Fire fighters
Fishers
Ice makers
Liquified-gas workers
Outdoor workers during cold weather
Packing-house workers
Refrigerated-warehouse workers
Refrigeration workers

SOURCE: NIOSH, *A Guide to the Work Relatedness of Disease* (79–116). Washington, D.C.: GPO, 1979.

usually do not tolerate heat stress well because their circulatory system is not functioning at a maximum level.

The effects of temperature stresses on worker health also vary depending on how they acclimatize to stresses, the length of exposure time, and the temperature. Most people do not acclimatize well to cold, and extremities and respiratory passages in particular must be protected.

Heat stress (see Table 9.4) produces several warning symptoms, which if heeded can be treated. If symptoms are ignored, then heat exhaustion or heat stroke may result.

TABLE 9.4
Heat Stress Symptoms and Interventions

Disorder	Symptom	Treatment
Heat cramps	Painful muscle spasms	Rest
		Replace fluids and salt
Heat syncope	Fainting	Remove from heat
Heat exhaustion	Weakness	Remove from heat
	Pale, cool skin	Replace fluids and salt
	Rapid pulse	Rest
	Decreased blood pressure	
Stroke	Fever, skin hot and dry	Immediate removal from heat
	Red complexion	Bathe to reduce temperature
	Elevated blood pressure	Treat as an emergency
	Rapid pulse	

SOURCE: *Criteria for a Recommended Standard Occupational Exposure to Hot Environments* (HSM-72-102-62). Washington, D.C.: GPO, 1972.

TABLE 9.5
Assessment of Work Load

Type of Work[a]	Average, Kcal/min[b]	Range, Kcal/min
Handwork		
Light	0.4	0.2–1.2
Heavy	0.9	
Work with one arm		
Light	1.5	0.7–2.5
Heavy	1.8	
Work with both arms		
Light	1.5	1.0–3.5
Heavy	2.5	
Work with body		
Light	3.5	2.5–15.0
Moderate	5.0	
Heavy	7.0	
Very heavy	9.0	
A Basic Example		
1. Walking along		2.0 Kcal/min
2. Intermediate value between heavy work with two arms and light work with the body		+ 3.0 Kcal/min
		5.0 Kcal/min
3. Add for basal metabolism		+ 1.0 Kcal/min
Total		6.0 Kcal/min

SOURCE.   NIOSH, *Industrial Hygiene Engineering and Control*. Cincinnati, Ohio, 1978.
[a] Examples of the types of work would be as follows:

Light handwork: writing, hand knitting
Heavy handwork: typewriting
Heavy work with one arm: hammering in nails (shoemaker, upholsterer)
Light work with two arms: filing metal, planing wood, raking a garden
Moderate work with the body: cleaning a floor
Heavy work with the body: railroad-track laying, digging, barking trees

[b] Average values of Kcal/min of metabolic rates during different activities would be as follows for body position and movement: sitting = 0.3; standing = 0.6; walking = 2.0–3.0; and walking uphill requires a 0.8 addition over "walking" per meter (yard) rise.

Specific types of work call for specific limits. For example, where heavy work is required in a cold environment, evaporation of sweat causes further body cooling. Additional layers of clothing are helpful because they provide warmth by conserving warm air. Table 9.5 provides information on how to assess work load, and Table 9.6 shows permissible work load limits.

TABLE 9.6
Permissible Heat Exposure Threshold Limit Values, °C (°F in Parentheses)

Work–Rest Regimen	Work Load		
	Light	Moderate	Heavy
Continuous work	30.0	26.7	25.0
	(86.0)	(80.1)	(77.0)
75% work,			
25% rest, each hour	30.6	28.0	25.9
	(87.1)	(82.4)	(78.6)
50% work,			
50% rest, each hour	31.4	29.4	27.9
	(88.5)	(84.9)	(82.2)
25% work,			
75% rest, each hour	32.2	31.1	30.0
	(90.0)	(88.0)	(86.0)

SOURCE: NIOSH, *Industrial Hygiene Engineering and Control.* Cincinnati, Ohio, 1978.

Humidity

Humidity exposure is also important to consider in evaluating environmental stress. "Relative humidity" refers to the ratio of water vapor in the air to the amount of water the air can hold at a given temperature.

Humidity affects the amount of sweat evaporated. If the humidity is high, then the rate of evaporation of sweat and, thus, the rate of evaporative cooling are lessened, and heat increases in the body. If the air is 100% saturated, then the vapor pressure of the water in the air is at a maximum, and no evaporation of sweat will occur. If the air is 50% saturated and stationary, there will be less evaporation than if the air is 50% saturated but with plenty of air circulation.

The wet bulb thermometer is used for determining humidity along with the psychochromatic chart (see Figure 9.1). In addition, both wet bulb and dry bulb readings are needed to determine relative humidity, dew point, and vapor pressure. (The dew point is the temperature at which the air becomes saturated with a gain or loss in moisture.)

Figure 9-1 shows the effect of humidity and temperature on human comfort.

"Effective temperature" combines temperature, humidity, and air movement into a single value and is a way of measuring relative comfort felt by humans in a particular environment.

Hyperbaric and Hypobaric Environments

There are two types of occupational exposure to these environments. "Hyperbaric" means that air pressure is higher than the pressure at sea level, "hypobaric" means it is lower.

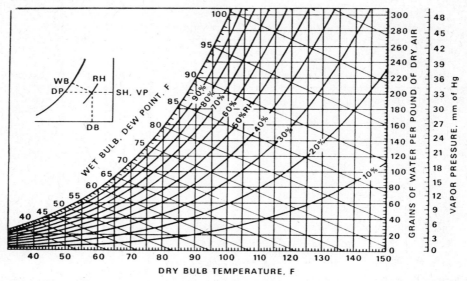

FIGURE 9.1. The effect of temperature and humidity on human comfort (*from NIOSH*).

Among pilots and air crews operating high-performance aircraft at extremely high altitudes (in excess of 30,000 feet), the greatest single hazard is hypoxia (deprivation of oxygen), which at these altitudes results in rapid loss of consciousness. Exposure to these reduced pressures (dysbarism) may also produce symptoms similar to those caused by rapid decompression in divers. The bends, the chokes, neurological disorders, embolism, and ear, sinus, and tooth problems have been experienced by members of air crews. Dysbarism may be complicated by circulatory failure or primary decompression shock evidenced by pallor, profuse sweating, faintness and dizziness, nausea, vomiting, and loss of consciousness. These symptoms are usually relieved by descent from high altitudes.

Potential occupational hazards also exist at much lower altitudes, where the effects of hypoxia are evidenced by impaired judgment and performance and a general feeling of malaise. Acute mountain sickness is also characterized by overwhelming depression, severe headache, nausea, vomiting, and loss of appetite. Virtually all climbers develop one or more symptoms.

Low Altitude. Occupational exposures to low altitudes occur in caisson or tunneling operations by divers. Ears are affected by the change in pressure during compression (descent) or decompression (ascent). In addition, at 4 atmospheres of pressure (14.7 × 4 = 58.8 psi) or more, the gaseous nitrogen in normal air induces a narcotic reaction, producing symptoms such as decreased ability to work and mood changes. Also, inhalation of oxygen at partial pressure exceeding 2 atmospheres (29.4 psi) may result in oxygen poisoning, which causes tingling of the fingers and toes, visual disturbances, hallucinations, confusion, muscle twitching, nausea, and vomiting.

Rapid decompression causes nitrogen bubbles to form in the blood. This causes circulatory damage and local tissue damage characterized by the bends, the chokes, or paralysis. Symptoms of the bends are dull, throbbing pains felt in the joints or deep in the muscles and bones. The chokes is a type of asphyxia affecting the right side of the heart. The earliest symptom is a sensation of substernal (beneath the sternum) discomfort felt during a deep inspiration, which elicits coughing. Paralysis can occur in the lower extremities. All of these decompression symptoms usually disappear rapidly.

Every hyperbaric work environment must be closely monitored because symptoms occur rapidly and are very serious if not immediately treated. People who work in hyperbaric environments must be trained to recognize the symptoms.

Ventilation/Climate Control Systems

Ventilation refers to controlling the air quality and temperature of the workplace. Ventilation systems remove air pollutants and condition the air for worker comfort.

No single type of ventilation system solves all problems.[10] Various types of ventilation include comfort, local exhaust, local supply, makeup air, dilution ventilation, and natural ventilation.

Comfort ventilation moves and conditions the air to assure worker comfort, controls temperature and humidity, and reduces unpleasant odors. Air-conditioning systems in offices and factories are comfort systems.

Local exhaust ventilation removes contaminants generated locally. Air is drawn in at a rate capable of removing contaminants at the source before they are dispersed into the work atmosphere. One example is exhaust hoods over laboratory benches.[6]

Local supply ventilation supplies air to a specified point or operation where it is required and provides spot cooling for workers in hot areas. The push–pull system over an open tank is local supply ventilation.

Makeup air is air supplied to a work environment to compensate for air being exhausted through a local system. Without adequate makeup air, local exhaust systems will not operate efficiently.

Dilution ventilation supplies or exhausts air from a large area in order to dilute, and thus control, pollutants in the total area, as in tunnels to remove carbon monoxide. In general, dilution ventilation is not applicable to highly toxic pollutants. The exhaustion of polluting materials away from a plant through stacks is an example of both natural and dilution ventilation.[13]

TABLE 9.7
Specific Task Illumination Requirements

Work Area or Type of Task	Illumination Levels, Lux[a] (FC)	
	Recommended	Minimum
Assembly, missile component	1,075 (100)	540 (50)
Assembly, general		
Coarse	540 (50)	325 (30)
Medium	810 (75)	540 (50)
Fine	1,075 (100)	810 (75)
Precise	3,230 (300)	2,155 (200)
Benchwork		
Rough	540 (50)	325 (30)
Medium	810 (75)	540 (50)
Fine	1,615 (150)	1,075 (100)
Extrafine	3,230 (300)	2,155 (200)
Business machine operation (calculator, digital, word processor, etc.)	1,075 (100)	540 (50)
Console surface	540 (50)	325 (30)
Corridors	215 (20)	110 (10)
Circuit diagram	1,075 (100)	540 (30)
Dials	540 (50)	325 (30)
Electric equipment testing	540 (50)	325 (30)
Emergency lighting	—	30 (30)
Gauge	540 (50)	325 (30)
Hallways	215 (20)	110 (10)
Inspection tasks, general		
Rough	540 (50)	325 (30)
Medium	1,075 (100)	540 (50)
Fine	2,155 (200)	1,075 (100)
Extrafine	3,230 (300)	2,155 (200)
Machine operation, automatic	540 (50)	325 (30)
Meters	540 (50)	325 (30)
Missiles		
Repair and servicing	1,075 (100)	540 (50)
Storage areas	215 (20)	110 (10)
General inspection	540 (50)	325 (30)
Office work, general	755 (70)	540 (50)
Ordinary seeing tasks	540 (50)	325 (30)
Panels		
Front	540 (50)	325 (30)
Rear	325 (30)	110 (10)
Passageways	215 (20)	110 (10)
Reading		
Large print	325 (30)	110 (10)
Newsprint	540 (50)	325 (30)
Handwritten reports in pencil	755 (70)	540 (50)

TABLE 9.7 (*Continued*)

Work Area or Type of Task	Illumination Levels, Lux[a] (FC)	
	Recommended	Minimum
Small type	755 (70)	540 (50)
Prolonged reading	755 (70)	540 (50)
Recording	755 (70)	540 (50)
Repair work		
General	540 (50)	325 (30)
Instrument	2,155 (200)	1,075 (100)
Scales	540 (50)	325 (30)
Screw fastening	540 (50)	325 (30)
Service areas, general	215 (20)	110 (10)
Stairways	215 (20)	110 (110)
Storage		
Inactive or dead	110 (10)	55 (5)
Live, rough or bulk	110 (10)	55 (5)
Live, medium	325 (30)	215 (20)
Live, fine	540 (50)	325 (30)
Switchboards	540 (50)	325 (30)
Tanks and containers	215 (20)	110 (10)
Testing:		
Rough	540 (50)	325 (30)
Fine	1,075 (100)	540 (50)
Extrafine	2,155 (200)	1,075 (100)
Transcribing and tabulation	1,075 (100)	540 (50)

SOURCE: U.S. Department of Defense, *Human Engineering Design Criteria for Military Systems, Equipment and Facilities* (MIL-STD-1472C). Washington, D.C., 1981.

NOTE: Some unusual inspection tasks may require up to 10,000 lux (1,000 FC).

NOTE: As a guide in determining illumination requirements, the use of a steel scale with $\frac{1}{64}$-in. divisions requires 1,950 lux (180 FC) of light for optimum visibility.

[a] As measured at the task object or 760 mm (30 in.) above the floor.

Illumination

Industrial hygienists are also concerned with the amount of light in the workplace. Research has shown that work performance declines when there is insufficient lighting. Table 9.7 shows recommended illumination levels.

Proper illumination is calculated by the lumen method, which is based on the definition of a footcandle: One footcandle (FC) equals one lumen per square foot (lux); thus footcandles equal lumens striking area. By knowing the initial lumen output of each lamp (published by the lamp manufacturer), the number of lamps installed in the area, and the square feet, one can calculate the lumens per square foot.

ERGONOMICS

Ergonomics is the study of person and machine characteristics in order to optimize the person's living and working environment. Ergonomists conduct research on human capabilities and limitations that relates to work space layout, equipment design, operating procedures, and maintenance.

Several principles relating to person–equipment interface have been enumerated:

Vibrations transmitted at the person–machine interface should not cause somatic reactions such as Reynaud's syndrome.

Moving parts of the body should not be constrained by braces.

Stress concentration on small skin areas or small joints should be avoided.

Ergonomic checklists should always be consulted whenever hand tools are designed, modified, selected, or evaluated.[13]

Person–Machine Interface.   Ergonomics considers the human–machine interface, including speed and difficulty of work and location and type of controls. An ergonomic study of a particular job requires both a functional and a task analysis (see Table 9.8). Then the components of the job can be analyzed and alterations made to create an optimum person–machine interface.

Capacity to Perform Strenuous Work

Ergonomics also addresses physical limitations on performing work because so many industrial accidents occur in relation to lifting. It will be used here to show the kinds of safety guidelines that can be developed.[5] Table 9.9 lists guidelines for lifting. Weight limits established by the military in their guidelines for lifting are listed in Table 9.10. These limits set the maximum amounts that a person is allowed to lift using both hands. Twice the amount for a one-person lift is the permissible limit for a two-person lift, providing the item to be lifted by the two persons is of a convenient configuration. The limits set in Table 9.10 for a 3- to 5-foot lift apply to tasks requiring a person to carry an item up to a distance of 16 feet on a nonrepetitive basis. See Figure 9.2 for examples of safe lifting. Lifting must be studied thoroughly and workers trained extensively if back injuries are to be avoided.[7]

CHEMICALS IN DAILY LIFE

Chemicals are another hazard in the workplace and in the general environment. They are everywhere—in our food and in the air that we breathe.

TABLE 9.8
Conducting a Functional Analysis and a Task Analysis

1. Subdivide the job into the major functions that are necessary to accomplish the desired output. A common set of functions that are useful in the initial analysis of a job is:

   a. Preparation—What steps must the worker take in preparing to perform the desired job?

   b. Observation—What information and data must the worker have in order to perform the job?

   c. Control—What steps must the worker take to control the processes involved? These steps include the mental processes or decisions that must be made to activate given controls.

   d. Physical demands—What physical tasks must the worker perform to accomplish the desired objective?

   e. Termination—What steps must be taken to conclude the job? This includes the cleanup after the job has been completed.

2. Divide the functions into the various tasks that may be performed.

3. Conduct an analysis by asking questions about the individual tasks.

   a. What are the initiating conditions? What is the cause or stimulus that results in the task's being performed in the first place?

   b. What actions must be carried out? What steps must be performed by the individual and/or machine to accomplish the desired result?

   c. What feedback is required to assure that the results of the action are as desired?

   d. What potential errors are possible? What is the cost of the occurrence of these errors in terms of damage to equipment, materials, and/or operator's physical well-being?

   e. What hazards are present that can cause illness or injury to the worker?

   f. What is the required reaction time that is necessary to initiate the task?

   g. What is the time frame in which the task must be completed?

   h. What tools and equipment are required to complete the task or element?

   i. Where is the task or element being performed? What is the physical location and structure of the workplace in which the task or element is performed?

   j. What physical demands are placed upon the worker to perform the task or element?

   k. What skills and knowledge are required of the worker to perform the task or element?

TABLE 9.9
Guidelines for Lifting

The following guidelines for lifting should help workers avoid many incidents, pitfalls, and injuries. Remember the ground rule: Avoid manual material movement, particularly lifting from the floor. Engineer the job to minimize material handling. But if material must be moved by hand, heed the following suggestions:

Be in good physical shape. If you are not used to lifting and vigorous exercise, do not attempt to do difficult lifting tasks.

Think before acting. Place materials conveniently. Have handling aids available. Make sure sufficient space is cleared for action.

Wear safety shoes providing metatarsal protection.

Test the object's weight before handling it. If it appears to be too heavy or bulky, get a mechanical lifting aid, or somebody to help, or both.

Get a good grip on the load; use the palm of the hands; wear gloves.

Get the load close to the body, and pull the load in before lifting. This is the most important lifting rule.

Position your feet so you can get close to the load. Place them far enough apart for stability, and have one foot ahead of the other; let them point in the direction of movement.

Lift primarily by straightening your legs and by slightly unbending your back. Do not twist your back or bend sideways.

Do not perform awkward lifts.

Do not hesitate to get help.

Do not lift at arm's length.

Do not continue heaving when the load is too heavy.

These rules will help to avoid many physical injuries, particularly of the back, but will not prevent all accidents. Additional recommendations include:

Avoid obesity.
Avoid fatigue.
Maintain flexibility.
Don't overdo it.
Maintain overall health.

Items weighing more than the suggested one-person lift should be prominently labeled with weight and lifting instructions. Where mechanical or power lift is required, hoist and lift points should be provided.

SOURCE: NIOSH, *Industrial Hygiene Engineering and Control*. Cincinnati, Ohio, 1978.

232

TABLE 9.10
Weight Limits for Lifting

Height from Ground	Male	Female	Two Persons (Male)
6 ft (1.8 m)	29 lb (13 kg)	20 lb (9 kg)	58 lb (26 kg)
5 ft (1.5 m)	35 lb (16 kg)	24 lb (11 kg)	70 lb (32 kg)
4 ft (1.2 m)	50 lb (23 kg)	34 lb (14 kg)	100 lb (46 kg)
3 ft (0.9 m)	65 lb (29 kg)	43 lb (20 kg)	130 lb (58 kg)
2 ft (0.6 m)	80 lb (36 kg)	53 lb (24 kg)	160 lb (72 kg)
1 ft (0.3 m)	85 lb (39 kg)	56 lb (29 kg)	170 lb (76 kg)

SOURCE: U.S. Department of Defense, *Human Engineering Design Criteria for Military System, Equipment and Facilities* (MIL-STD-1472C). Washington, D.C., 1981.

Additives, once believed safe, are now considered contaminants. There is a growing awareness of chemicals in cosmetics, food, and household cleaning products.

Basic Principles of Toxicology

A poison is defined in the dictionary as a substance which in small quantities can cause illness or death. Commonly, substances like cyanide or arsenic are perceived as poison, while oxygen is not. However, pure oxygen given to newborn babies to start breathing can cause blindness. Two other types of poisons are statutory poison and economic poison. A statutory poison is a substance of which approximately one teaspoonful will endanger a normal adult; 60 grains of the substance will kill an adult. An economic poison was first defined by the United States Congress in the Federal Insecticide, Fungicide, and Rodenticide Act of 1947. By this act pesticides, whether insecticides sprayed on crops or rodenticides used to kill rats and mice, are considered economic poisons. Agent Orange would be an economic poison.

In determining toxicity, lethal dose must be calculated. It is usually written LD_{50}, which is the dose that would yield 50 percent mortality. The toxicity classes range from relatively harmless to extremely toxic. Table 9.11 is a summation of toxicity classes. Toxic materials must be specially labeled to warn the user of the dangers. Figure 9.3 describes precaution labeling for toxic materials.

Safety Standards and Hazardous Materials

Safety standards and hazardous materials levels are based on the lethal dose rate. Federal Standard 313A defines a hazardous material as one which has two or more of the following characteristics:

234

FIGURE 9.2. How to lift safely *(from NIOSH)*.

TABLE 9.11
Toxicity Classes

Toxicity Rating	Descriptive Term	Probable Lethal Dose for One Person
1	Extremely toxic	Taste
2	Highly toxic	4 ml–1 tsp
3	Moderately toxic	30 ml
4	Slightly toxic	500 ml
5	Practically nontoxic	1 liter
6	Relatively harmless	over 1 liter

1. Has a flash point below 200°F (93.3°C), a closed container (a sealed container from which neither liquid nor vapor will escape at ordinary temperatures), or is subject to spontaneous heating or to polymerization with release of large amounts of energy when handled, stored, and shipped without adequate control

2. Has a threshold limit value below 1,000 ppm for gases and vapors, below 500 mg/m^3 for fumes, and below 30 mppcf (million particles per cubic foot) for dusts

3. Will cause 50% fatalities to test animals when a single oral dose is administered in doses of less than 500 mg per kilogram of test animal weight

4. Is a strong oxidizing or reducing agent

5. Causes first-degree burns to skin in short time exposure, or is systematically toxic by skin contact

6. May produce—in the course of normal operations—dusts, gases, fumes, vapors, mists, or smokes with one or more of the above characteristics

7. Produces sensitizing or initiating effects

8. Is radioactive

9. Has special characteristics which in the opinion of the manufacturer could cause harm to personnel if used or stored improperly.[4]

The forms for recording this information are shown in Chapter 5.

Toxic substances enter the body through several routes, including the mouth, respiratory tract, skin, and eyes. The respiratory tract is of prime concern to the industrial hygienist because of the prevalence of airborne hazardous substances and the fact that their levels can be changed or protected against. Eating and smoking are the easiest ways of ingesting hazardous materials. Absorption through the skin is the most common way in industrial environments.[11] Poisons taken in this way may cause dermatitis or may just be absorbed into the body.

PRECAUTIONARY LABEL TEXT

CLASS OF HAZARD	SIGNAL WORD	STATEMENTS OF HAZARD	PRECAUTIONARY MEASURES	INSTRUCTIONS IN CASE OF CONTACT OR EXPOSURE
CONTACT				
Corrosive, Eye	DANGER!	CAUSES (SEVERE) EYE BURNS	Do not get in eyes. Avoid breathing (dust, vapor, mist, gas). Keep container closed. Use with adequate ventilation. Wash thoroughly after handling.	FIRST AID: In case of contact, immediately flush eyes with plenty of water for at least 15 minutes. Call a physician.
Corrosive, Eye and Skin	DANGER!	CAUSES (SEVERE) BURNS	Do not get in eyes, on skin, on clothing. Avoid breathing (dust, vapor, mist, gas). Keep container closed. Use with adequate ventilation. Wash thoroughly after handling.	FIRST AID: In case of contact, immediately flush eyes or skin with plenty of water for at least 15 minutes while removing contaminated clothing and shoes. Call a physician. Wash clothing before re-use. (Discard contaminated shoes.)
Irritant, Eye	WARNING!	CAUSES EYE IRRITATION	Avoid contact with eyes. Wash thoroughly after handling.	FIRST AID: In case of contact, immediately flush eyes with plenty of water for at least 15 minutes. Call a physician.
Irritant, Eye and Skin	WARNING!	CAUSES IRRITATION	Avoid contact with eyes, skin, and clothing. Wash thoroughly after handling.	FIRST AID: In case of contact, immediately flush eyes with plenty of water for at least 15 minutes. Call a physician. Flush skin with water. (Wash clothing before re-use.)
Strong Sensitizer, Skin	WARNING!	MAY CAUSE ALLERGIC SKIN REACTION	Avoid prolonged or repeated contact with skin. Wash thoroughly after handling.	FIRST AID: In case of contact, immediately wash skin with soap and plenty of water.
FLAMMABILITY				
Extremely Flammable Liquid	DANGER!	EXTREMELY FLAMMABLE	Keep away from heat, sparks, and open flame. Keep container closed. Use with adequate ventilation.	
Flammable Liquid	WARNING!	FLAMMABLE	Keep away from heat, sparks, and open flame. Keep container closed. Use with adequate ventilation.	(See Table 2 for selection of appropriate fire-extinguishing statement.)
Flammable Solid	WARNING!	FLAMMABLE	Keep away from heat, sparks, and open flame.	
Combustible Liquid	CAUTION!	COMBUSTIBLE	Keep away from heat and open flame.	
Pyroforic Chemical	DANGER!	EXTREMELY FLAMMABLE CATCHES FIRE IF EXPOSED TO AIR	Keep away from heat, sparks, and open flame. Keep container closed.	
Strong Oxidizer	DANGER!	STRONG OXIDIZER CONTACT WITH OTHER MATERIAL MAY CAUSE FIRE	Keep from contact with clothing and other combustible materials. Do not store near combustible materials. Store in tightly closed container.	Remove and wash contaminated clothing promptly.

FIGURE 9.3. Precaution labeling for toxic materials (*from ANSI, 1977*).

PRECAUTIONARY LABEL TEXT

CLASS OF HAZARD	SIGNAL WORD	STATEMENTS OF HAZARD	PRECAUTIONARY MEASURES	INSTRUCTIONS IN CASE OF CONTACT OR EXPOSURE
INGESTION				
Highly Toxic	DANGER!	MAY BE FATAL IF SWALLOWED	Wash thoroughly after handling.	POISON — Call a Physician FIRST AID: If swallowed, induce vomiting by sticking finger down throat or by giving soapy or strong salty water to drink. Repeat until vomit is clear. Never give anything by mouth to an unconscious person.
Toxic	WARNING!	HARMFUL IF SWALLOWED	Wash thoroughly after handling.	FIRST AID: If swallowed, induce vomiting by sticking finger down throat or by giving soapy or strong salty water to drink. Repeat until vomit is clear. Call a physician. Never give anything by mouth to an unconscious person.
ABSORPTION				
Highly Toxic	DANGER!	MAY BE FATAL IF ABSORBED THROUGH SKIN	Do not get in eyes, on skin, on clothing. Wash thoroughly after handling.	POISON — Call a Physician FIRST AID: In case of contact, immediately flush eyes or skin with plenty of water for at least 15 minutes while removing contaminated clothing and shoes. Wash clothing before re-use. (Discard contaminated shoes.)
Toxic	WARNING!	HARMFUL IF ABSORBED THROUGH SKIN	Avoid contact with eyes, skin, and clothing. Wash thoroughly after handling.	FIRST AID: In case of contact, immediately flush eyes or skin with plenty of water for at least 15 minutes while removing contaminated clothing and shoes. Call a physician. Wash clothing before re-use. (Discard contaminated shoes.)
INHALATION				
Highly Toxic	DANGER!	MAY BE FATAL IF INHALED	Do not breathe (dust, vapor, mist, gas). Keep container closed. Use only with adequate ventilation.	POISON — Call a Physician FIRST AID: If inhaled, remove to fresh air. If not breathing give artificial respiration, preferably mouth-to-mouth. If breathing is difficult, give oxygen.
Toxic	WARNING!	HARMFUL IF INHALED	Avoid breathing (dust, vapor, mist, gas). Keep container closed. Use with adequate ventilation.	FIRST AID: If inhaled, remove to fresh air. If not breathing give artificial respiration, preferably mouth-to-mouth. If breathing is difficult, give oxygen. Call a physician.
Strong Sensitizer, Lungs	WARNING!	MAY CAUSE ALLERGIC RESPIRATORY REACTION	Avoid breathing (dust, vapor, mist, gas). Keep container closed. Use with adequate ventilation.	
Physiologically Inert Vapor or Gas	CAUTION!	(VAPOR) (GAS) REDUCES OXYGEN AVAILABLE FOR BREATHING	Keep container closed. Use with adequate ventilation. Do not enter storage areas unless adequately ventilated.	

FIGURE 9.3. (*Continued*)

Toxic materials can be splashed on eyes; mists, vapors, and gases can also cause a toxic reaction. Eye reactions are usually localized.

RESPONSES TO HAZARDS

Table 9.12 lists responses to hazards. It has been presented in its entirety because it is still relevant today—30 years after it was originally published.

Systemic Effects

A "carcinogen" is a substance that causes cancer. A "teratogen" is a substance that is dangerous only after the sperm has fertilized the egg. Thalidomide is a teratogen which, when taken by pregnant women between the twentieth and thirty-fifth day after conception, almost always causes defects in the embryo. Mutagens change the genetic structure of reproductive cells and can result in injury or death to succeeding generations.

The "transplacental carcinogens" make up a second group of harmful substances. These (DES, for one) cross the placenta, pass into the bloodstream of the fetus, and create conditions for the later development of cancer. Since the cancer takes years to develop, it is difficult to foresee which substances will have this effect.

Teratogens include cadmium, carbaryl, carbon tetrachloride, chromium compound, 4-dimethylaminobenzene, formaldehyde, lead, mercury, oxides of nitrogen, paraquat, and parathion.

Analysis of the Work Environment

If any of these substances are in the workplace, it is necessary to analyze whatever chemicals may be in use. These analyses require knowing the threshold limit values of each substance and the permissible levels that are to be tolerated.

The threshold limit values (TLVs) were developed by the American Conference of Governmental Industrial Hygienists (ACGIH). The values set standards for occupational exposure to airborne substances. They represent "conditions under which it is believed that nearly all workers may be repeatedly exposed without adverse effect."

The TLV is expressed in parts per million (ppm)—that is, parts of vapor or gas per million parts of air by volume. It can be an approximate milligram (mg) of particulate per cubic meter of air (sometimes abbreviated mg/m^3). Concentrations of mineral dust are expressed as millions of particles per cubic foot (mppcf). Vapors and gases are also expressed as the percentage of flammable amounts in air, by volume. The lower flammable limit of hydrogen gas (H_2), for example, is 4%. This would correspond to 40,000 ppm. Thus, one part of hydrogen gas in 99 parts of air—a 1% mixture—wouldn't be a fire hazard (10,000 ppm).

TLVs are offered as guides for the industrial hygienist. They are evaluated every year by ACGIH, and values are changed, if necessary, based on new information. It is the most complete list of toxic hazards currently available. Analysis of chemicals in the workplace must be done thoroughly and completely. Such an analysis will be the basis for determining air content, evaluating what workers have been or are being exposed to, determining safe levels, and applying protective measures.

OCCUPATIONAL HEALTH RELATED TO WOMEN

With more women entering the workforce, greater attention is being paid to contaminant levels. "Women are entering the workforce at a rate of almost two million every year. Today 42.8 million women in the United States work as compared to 18 million in 1950. By 1990, working women are expected to number 54 million."[14] Because of hazards known to pose a threat to women of childbearing age, these women have been barred from certain high-risk jobs. For example:

> In 1975, General Motors of Canada, Ltd., barred women capable of bearing children from areas of high lead exposure.
>
> Women have been barred in recent years from working with polyvinyl chloride (PVC) in a Firestone Tire & Rubber Co. plant in Pottstown, Pa.
>
> Twenty-eight of thirty-seven women transferred out of Bunker Hill Co.'s Kellogg, Idaho, lead smelting plant were placed in jobs with less overtime and fewer seniority rights.
>
> Olin Corporation prohibits women workers of childbearing age in ammunition plants where they could be accidentally exposed to benzene ($C_6 H_6$) and at installations where women could be accidentally exposed to carbon disulfide (CS_2), used in cellophane manufacturing.[14]

This dilemma will increasingly involve all workers because evidence is mounting that men's reproductive systems are also affected. These environmental hazards can damage their reproductive ability and perhaps cause them to transmit defective genes.

Some substances interfere with the ability of men to father children. It has been known for more than 60 years that reproductive abnormalities can result even if only the male parent is exposed to lead. A 1916 study by Thomas Oliver showed that the number of stillborn babies born to wives of house painters was three times the number born to women in general.

TABLE 9.12
Factors Relating to Occupational Injuries and Diseases

I. Classes of Causative Agents
   A. Direct trauma (lacerations, contusions, fractures, etc.)
   B. Temperature
      1. Heat (burns, heat cramps in miners and steel workers)
      2. Cold (frostbite, trenchfoot in soldiers)
   C. Humidity
      1. With heat (heat exhaustion)
      2. With cold (chilblains)
   D. Pressure ("bends" in divers)
   E. Noise (deafness in jet aircraft testers)
   F. Other vibrations (white fingers in compressed-air drill operators)
   G. Radiation
      1. Microwave (cataract or deep burns)
      2. Infrared (cataract in glassblowers, circulatory collapse in metal furnace operators)
      3. Visible (retinal burns from the sun, laser exposures, skin sensitization to light following chemical treatments)
      4. Ultraviolet (sunburn, flash burns of the eyes in welders)
      5. X-rays (dermatitis or leukemia in radiologists)
      6. Gamma rays (deep tissue destruction, blood dyscrasias, cancer)
      7. Particles
         a. Alpha particles (bone cancer from ingested materials)
         b. Beta particles (skin burns in laboratory workers)
         c. Neutrons (cataracts in cyclotron workers)
         d. Other particles (unknown effects)
   H. Agents classified by physical state
      1. Dusts causing pneumoconiosis
         a. Mineral dusts such as silica, silicates, beryllium oxide, (BeO) asbestos, and others
         b. Vegetable dusts such as bagasse or cotton
         c. Coal dust
      2. Metal fumes (zinc shakes in smelters)
      3. Mists (chromic acid CrO_6) ulcers of nasal septum)
      4. Gases such as carbon monoxide (asphyxia in blast furnace operators)
      5. Vapors such as carbon disulfide (C_2S) (central nervous system damage in viscose rayon workers)
      6. Liquids
         a. Aqueous (contact dermatitis in textile dye workers)
         b. Oily (oil folliculitis in machinists)
         c. Organic (systemic poisoning from skin penetration from such liquids as phenylhydrazine)
      7. Solids (damage to exposed skin, eyes, hair, or mucous membranes from corrosive, irritating, or sensitizing materials)
   I. Agents classified by chemical state
      1. Inorganic
         a. Metals (lead poisoning in battery workers)

TABLE 9.12 (*Continued*)

         b. Acids (sulfuric acid bronchitis in chemical workers)
         c. Alkalies (alkali burns from caustics)
         d. Salts (cyanide poisoning from sodium cyanide in case-hardening metals)
      2. Carbon compounds (organic and inorganic)
  J. Infections (anthrax in wool sorters, brucellosis in meat-packers, viral infections in research laboratory workers)
  K. Parasites (hookworm in agricultural workers)
  L. Animal products (dermatitis in meat or fish packers)
  M. Plant products (poison ivy in highway maintenance workers)
  N. Other biologic products (dermatitis in persons manufacturing antibiotics)
  O. Carcinogens (bladder tumor in dye intermediate chemical operators, scrotal cancer in mule skinners)
II. Portal of Entry of Agent into the Body
  A. Respiratory (all airborne agents)
  B. Gastrointestinal (waterborne agents and the proportion of inhaled agents leaving the trachea by ciliary action)
  C. Skin (airborne agents, liquids, often in combination with other portals)
  D. Organs of special sense (eye and radiations, ear and noise)
III. Body System or Organ Showing Major Impairment
  A. Respiratory (silicosis, asbestosis)
  B. Circulatory (nitroglycerine poisoning in explosives workers)
  C. Digestive, including liver (hepatitis from chlorinated hydrocarbons)
  D. Genitourinary (nephrosis from carbon tetrachloride, mercury, arsenic)
  E. Hematopoietic (bone marrow injury from benzol)
  F. Skin, hair, nails (dermatitis in rubber workers, hair loss in thallium handlers)
  G. Central and peripheral nervous (mercury poisoning in hatters)
  H. Eye (cataract)
  I. Ear (deafness)
  J. Organs of locomotion and bony changes in fluoride workers (osteitis from radium dial painting)
  K. Generalized (lead or other heavy-metal poisoning)
IV. Types of Industries and Prominent Hazards
  A. Agriculture, forestry, fishing (trauma)
  B. Mining, quarrying (trauma, pneumoconiosis)
  C. Construction (trauma; effects of heat, cold, and pressure)
  D. Manufacturing (trauma, toxicity from chemicals, pneumoconiosis)
  E. Transportation, communication, utilities (trauma, temperature)
  F. Trade (toxicity from chemicals, trauma)
  G. Finance, insurance, real estate (trauma)
  H. Service industries (acute exposure to chemicals in maintenance workers)
  I. Government (wide variety of hazards since government employees perform all sorts of activities from street repairs to fighting forest fires)

SOURCE: Meigs, J. W. *Factors Relating to Occupational Injuries and Diseases: Teaching Outline.* New Haven, Conn.: Department of Public Health, Yale University, 1952.

Today lead is still found in industrial paints. Another study by Oliver found that of 32 pregnancies in women married to lead workers, there were 11 miscarriages, one stillbirth, and 13 deaths in the first year of life. Only two of the children lived to adulthood. These studies were conducted long ago. Why has it taken so long for us to be concerned about lead in our environment? Another study found that of the 36 substances from which industry reports that it bars women of childbearing age, 21 can also affect men, causing cancer, infertility, or mutation in their sperm.[14]

Exposure of workers to dangerous substances versus concerns about equal rights and equal economic opportunities is still at issue. Many women are reaching for the best possible jobs for which they qualify. They need information concerning the potential hazards in order to make informed decisions. The issue is whether certain age groups should arbitrarily be barred from hazardous work areas. What are the economic implications of such a policy? How can all workers be educated about the risks to themselves and their families and still accept a risky job? Both workers and management have rights and responsibilities. Can this issue be resolved by cooperation, or will it require legislative action?

Much more research is required not only in the effects of chemicals but also in the psychological effects on workers who voluntarily change job goals or are legislated into doing so. How does this affect motivation? What responsibility is placed upon the industrial hygienist in terms of educating the workers? What job change decisions will be seen in the next decade because of these risks?

Many workers must face up to these issues because they are the ones involved with the issues, as for instance with air pollution caused by smoking. However, progress has been slow. It can be perplexing to know when it is appropriate to say, "I prefer you do not smoke here," or "No smoking," or to ignore it. Personal needs and wants can clash with professional responsibilities and obligations. How is such a decision made? What implications will the recent ruling on disability rights to workers involuntarily and/or unknowingly exposed to workplace toxins?

The industrial hygienist's job is to provide a safe and healthy environment, considering what is known, at all times for all employees.

SUGGESTED LEARNING EXPERIENCES

1. If you were going to make changes in Meig's course outline, what would they be and why?

2. What do you think is the most important job of the industrial hygienist?

3. Develop a research project involving workplace variables. How did you go about deciding which variables you were going to manipulate?

4. What visuals might you develop demonstrating heat/cold environments in a plant orientation program?

5. Debate: Industrial hygiene is more important than safety engineering in protecting workers.

6. What is the most important contribution of ergonomics to the worker? To industry?

7. Review the state-of-the-art literature on the relationship of benzene to cancer.

8. If you were a member of your state workers' compensation board, how would you evaluate the delayed appearance of symptoms related to occupational exposure of asbestos, cotton dust, or vinyl chloride?

9. If you were a female worker, perhaps interested in having children at some time, what would you like to know about your workplace environment? Would you want any different information if you were a male?

10. Review thoroughly 29 CFR 1910.94. Then make a checklist from it to evaluate the ventilation in a factory.

REFERENCES

1. Baetzer, A. The early days of industrial hygiene—their contribution to current problems. *American Industrial Hygiene Journal,* 1980, **41,** 773–777.

2. *Criteria for a Recommended Standard. Occupational Exposure to Hot Environments* (HSM-72-102-69). Washington, D.C.: GPO, 1972.

3. Edmiston, S., and Szekely, J. What we must know about health hazards in the workplace. *Redbook,* 1980, **33,** 171–174.

4. Federal Standard 313A. *Material Safety Data Sheets, Preparation and the Submissions of. June 1976.*

5. Gloss, D. S., and Wardle, M. G. Development of safe work loads for women employed in physical work. *New England Engineering Journal,* 1977, **5,** 4. (Feb.)

6. Goldfield, J. Contaminant concentration reduction: General ventilation versus local exhaust ventilation. *American Industrial Hygiene Association Journal,* 1980, **41,** 812–818.

7. Kroemer, H. Back injuries can be avoided. *National Safety News,* 1980, **121,** 37–43, (Feb.)

8. Lehmann, G. E., Muller, A., and Spitzer, H. Der Kaloriendbedarf bei gewerblicker arbeit. *Arbeitphysiologia,* 1950, 166–170.

9. Meigs, J. W. *Factors Relating to Occupational Injuries and Diseases: Teaching Outline.* New Haven, Conn.: Department of Public Health; Yale University, 1952.

10. Mutchler, J. E. Principles of ventilation. In NIOSH, *The Industrial Environment—Its Evaluation and Control*. Washington, D.C.: GPO, 1973.

11. NIOSH. *Industrial Hygiene Engineering and Control*. Cincinnati, Ohio, 1978.

12. Nemec, M. Warning: This job may be dangerous to your offspring. *Occupational Hazards*, 1979, **41**, 37–40. (Apr.)

13. U.S. Department of Defense. Military Standard 1472C, Human Engineering Design Criteria for Military Systems, Equipment and Facilities. May, 1981.

14. Occupational Safety and Health Administration. *Code of Federal Regulations*. General Industry Standards (29 CFR 1910). U.S. Department of Labor, Washington, D.C.; June 1981.

10 PERSONAL PROTECTIVE EQUIPMENT

In the workplace, employees are exposed to many substances and conditions which can cause occupational disease or injury. In the early days of the industrial revolution workers assumed certain risks to health and body when they accepted a job. No longer is "assumption of risk" a condition of employment. Protective equipment can be used to safeguard workers against inadequate ventilation, excessive noise, extreme temperatures, and toxic substances.

In many cases engineering and/or environmental controls cannot reduce a hazard sufficiently, so that workers may still be vulnerable to excessive exposure. In these situations, equipment must be provided that will still protect the worker from harm. Personal protective equipment is designed to protect workers who must be exposed to hazards. It can be as simple as work gloves or as complex as body suits with built-in breathing apparatus.

One of the most routine, but critical, tasks of the safety department is to ensure that workers are properly supplied with personal protective equipment. The safety engineer educates the workers about personal protective equipment and monitors equipment quality, ensuring that this equipment is used properly and is adequately maintained.

EXPOSURE

Exposure to any workplace hazard is a function of the duration of exposure and the amount of contaminant.

Assessment of Exposure

If hazards cannot be eliminated or reduced, an orderly procedure should be followed to assess the possible exposure. This includes the following:

1. Identify the substance or substances from which protection is necessary.
2. Obtain full information on the risks of each substance and its significant properties.
3. Determine which, if any, personal characteristics and capabilities are essential to the safe use of the protective devices and procedures required.
4. Determine what equipment is needed for maintenance.
5. From these conditions, select the type of personal protective equipment that will provide protection.[10]

Threshold Limit Values

One of the basic principles of occupational health is that, despite the potential health risk associated with known poisonous substances, for each substance there is a definable and measurable level of human contact below which no significant threat to health exists. This acceptable level of contact, expressed in appropriate terms of magnitude and duration of exposure to the offending agent, is variously called the threshold limit value (TLV), the maximum allowable concentration (MAC), or the permissible dose. Since, in most instances, significant contact with toxic substances is by inhalation of airborne dust, fumes, vapors, and gases, the permissible levels are given in terms of atmospheric concentrations, mg/m^3, particles per m^3 (for mineral dusts), or parts per million of gas. Although there are certain differences of detail in the terms given above, all the terms have the same primary purpose: to identify and locate a point on the scale of dose of the offending agent, above which there is increasing probability of injury, overt illness, and even death, but below which the risk is so limited as to impose no serious threat to health, however long the exposure is continued.[16]

Toxicologists seeking stringent criteria on which to base permissible levels of exposure (TLVs or MACs) have approached the problem from two directions. One approach is to start from the higher levels of demonstrable ill effects and work downward. Another approach is to start from a known safe level in a healthy animal and work upward. The permissible limit is established just under the lowest level of exposure needed to induce any statistically significant deviation from that normal state of the organism. Two basically different concepts and criteria of health are involved. In the first, no serious threat to health is considered to exist as long as the level of exposure does not induce in the organism a demonstrable disturbance of a kind predictive of potential ill health. In the second, a potential for ill health is said to exist, and then an attempt is made to find the dose response level where that potential no longer exists.

In view of the differences in approach and the methods of determining change, it is not surprising that the permissible levels of exposure (TLVs or MACs) in various parts of the world sometimes differ by a factor of 10 or even more (see Table 10.1).

The Joint ILO/WHO Committee on Occupational Health proposed a classification of biological effects of occupational exposure to airborne toxic substances, as follows:

> *Category A (safe exposure zones):* Exposures that do not, as far as is known, induce any detectable change in the health and fitness of exposed persons during their lifetime
>
> *Category B:* Exposures that may induce rapidly reversible effects on health or fitness but that do not cause a definite state of disease
>
> *Category C:* Exposures that may induce a reversible disease
>
> *Category D:* Exposures that may induce irreversible disease or death

Difficulties may be expected in deciding how to classify certain substances encountered in industry in terms of the suggested categories. This is certainly the case with carcinogenic and mutagenic substances where dose–response relationships are not clear. The TLVs set by the American Conference of Governmental Industrial Hygienists (ACGIH) and the MACs prescribed by the health legislation of the USSR are the permissible limits most widely accepted in other countries. In the United States, the list of threshold limit values for airborne contaminants and physical agents contains permissible limits for approximately 600 chemical agents and thresholds in exposure to particulate matter and to physical agents including noise, nonionizing radiation, and heat stress.[16]

ENGINEERING SOLUTIONS

The engineering solutions to these workplace hazards are very promising and many adequate protective devices have been developed. A number of points must be considered including the following:

1. Use of protective equipment should never be considered preferable to removing the hazard in the first place.
2. Protective equipment must be used properly, kept clean, and used for the task and protection for which it was developed.

TABLE 10.1
Maximum Allowable Concentration of Some Harmful Substances in Various Countries, in mg/m³

Substance	U.S.S.R. (1970–1971)	U.S.A. (1981)	Czechoslovakia (1970–1971)	Poland (1970)	Federal Republic of Germany (1971–1972)	United Kingdom (1972)
Acetone	200	2,400	800	200	2,400	2,400
Xylene (Xylol)	50	435	200	100	870	435
Xylidine	3	25	5	–	25	25
Methanol	5	260	100	50	260	260
Lead	0.01	0.05	0.05	0.2	0.15	–
Styrene	5	200	200	50	420	420
Toluene	50	300	200	100	750	375
Trichloroethylene	10	200	250	50	260	535
Carbon dioxide	–	9,000	9,000	–	9,000	9,000
Carbon monoxide	20	55	30	30	55	55
Vinyl chloride	30	770	–	300	260	770

SOURCE: Compiled from International Labor Organization and Code of Federal Regulations 29 CFR 1910.

248

Personal Protective Equipment

Protective Clothing.   Protective clothing is required in a number of industrial environments including those where salt, acids, oils, or radiation are present. Some crucial points to consider when matching protective clothing to job hazards are:

1. What are the specific job hazards?
2. Is a flame-retardant garment needed?
3. Will the exposure be casual or direct?
4. Will the wear be short-term or prolonged?
5. Is temperature a factor?
6. Is color important?
7. What about washability?[9]

The answers to these questions help to determine the best material to use. The next step is to match the right protective clothing with the job hazard. Considered here will be fabrics, coatings, and construction.

The base fabric determines the general characteristics of the clothing: stretch, weight, and tensile strength. Materials are often blended to achieve a mix of properties, but the basic fabrics used are cotton, cotton blends, polyester, and nylon.

Cotton or cotton blends are generally among the least expensive fabrics and are very versatile. They come in light, medium, and heavy weight and knit or square weaves. Cotton is comfortable, helps dissipate body moisture inside the clothing, has good wet strength, and is often blended with rayon or polyester to improve its tensile strength.

Polyester is lighter and tougher than cotton and generally costs less. Polyester knit fabric stretches, is comfortable to wear, and retains its shape well. Polyester can be woven or knitted for a durable two-way stretch, as well as blended to achieve special properties.

Nylon is the lightest, most durable base fabric and generally the most expensive. When tightly woven and coated, it is very tear and puncture resistant.

Coatings are bonded to the base fabrics to seal out the contaminants workers are exposed to: liquids, gases, solids, and flames. Some coatings also improve the tensile strength of the fabric. Others form a tough shell that protects the fabric from abrasion damage.

Most protective clothing is coated only on the outside. However, some is coated on both sides, providing maximum protection from caustic materials or continuous exposure to other work hazards.

Four basic coatings are used on protective clothing, depending on the type of protection that is needed:

1.   SBR is a rubber compound that can be specially compounded with other ingredients to provide superior qualities at relatively low cost. It is flexible, waterproof, and highly abrasion resistant, although long-term exposure to oil will soften SBR. However, for occupations such as highway construction work, SBR can do the job economically. SBR also provides adequate protection against most organic acids.[9]

2.   Neoprene coatings can stand heavy exposure to oil and be relatively unaffected. They also stand up to alcohols, most ketones and aldehydes, organic esters, most salts, alkalies, and many inorganic acids.[9]

3.   Polyvinyl chloride (PVC) is a very versatile, coating that resists a wide range of alcohols, acids, salts, alkalies, and other substances. It makes a remarkably comfortable suit because it is soft and body heat keeps it pliant even in the cold. PVC can be used alone—without a base fabric—for rainwear; it may also be compounded to resist special hazards. For example, Uniroyal has developed a PVC compound, called "Paracril/PVC," which is especially resistant to oils and acids. Another new PVC compound withstands a 72-hour soaking in concentrated industrial acids. When this compound, called "Special PVC," is bonded to both sides of a knitted polyester base fabric, the result is a lightweight, flame-retardant, stretchy material that resists the toughest industrial chemicals.

4.   Polyurethane is the toughest coating. It forms a hard, nonporous shell that is impervious to most industrial hazards except some acids. Its features include high abrasion resistance, high resistance to hydrocarbons, and extreme lightness. Polyurethane is applied in thin coats, so it adds very little weight to the clothing. A polyurethane-coated jacket weighs less than half as much as the same jacket coated with neoprene.[9]

The same considerations should apply to the selection of all personal protective equipment. Although the best equipment may cost more, it provides better protection for a longer period of time.

Cost and Benefits.   There is no question that the costs for effective personal protective equipment are less than the higher insurance rates resulting from increased accident rates. In fact, the cost of personal protective equipment cannot even be compared with the direct cost of an injury. For example, the cost of safety glasses is betwen $5 and $15 versus the cost of losing an eye, valued at somewhere from $6,000 to $115,645 in scheduled workers' compensation benefits plus the medical expenses which could add many thousands of dollars to those amounts. If one were to consider the psychological and indirect costs, then the loss would be much greater.

The psychological effects of wearing adequate personal protective equipment are immense. If workers know they are protected, then they will feel safe and productivity will probably rise. Likewise, working in an unsafe environment can have a demoralizing effect and productivity will probably decrease.

Assessing the Quality of Protective Clothing

There are several ways to test the quality of protective clothing. One is to ask the salesperson; another is to try the clothing on yourself, paying attention to such factors as the size and whether you can bend, sit, kneel, and reach comfortably.

Seams are the most vulnerable part of protective clothing, so seam construction tells a lot about waterproofness. Seams used to be made with thread, which acts like a wick, drawing moisture into the seam. Even modern seam-cementing coatings can eventually crack, exposing threads to moisture.

In assessing the quality of personal protective equipment, what is most important is that the equipment perform the job it is intended to do. For example, goggles protect eyes from particles. No matter how attractive or comfortable they are, they are effective only if they keep foreign matter out of the eyes. Other performance requirements to assess are insulation against electric shock, impact resistance, penetration resistance, nonflammability, and water repellence.[3]

EXPOSURES NECESSITATING PROTECTION

Exposures which necessitate protection include cuts, bruises, abrasions, dirt, heat, cold, toxic chemicals, and radiation. Cuts, bruises, and abrasions require protection because they can be a route of entry for contaminants. The degree of dirt in the environment must be known because either alone or in combination with other pollutants, it can raise the acceptable threshold limit above permissible levels, requiring protective respiratory equipment.

Dermatitis

Dermatitis can be avoided with protective creams and safety gloves. The material used will depend upon the substances handled. Rubber protects against acids and alkalies but provides no protection from carbon disulfide, and aliphatic and aromatic amines.[1]

Noise

Noise is best handled at the source; if this is not feasible or possible, protective devices may be required. Five categories of hearing protection are commonly used in industrial application. Insert earplugs are positioned within the external ear canal. They are available as universal-fit units or premolded in various sizes as custom-fitted units. Earmuffs consist of two ear cups which completely enclose the external ears. Earmuffs are also mounted on safety "hard hats," welding masks, helmets, and similar items

Hearing protection	Expected range of hearing protection (on average) of good protective devices				
	Frequency range in hertz				
	1−20	20−100	100−800	800−8000	78000
Earplugs	5−10	5−20	20−35		30−40
Semi−insert earplugs	5−10	5−20	15−20	25−40	30−40
Earmuffs	0−2	2−15	15−35	30−45	35−45
Earplugs and earmuffs	10−15	15−25	25−45	30−60	40−60
Communication headsets	0−2	2−10	10−30	25−40	30−40
Helmets	0−2	2−2	7−20	20−55	30−55
Space helmet (total head enclosure)	3−8	5−10	10−25	30−60	30−60

FIGURE 10.1.   Approximate maximum noise reduction values in decibels for various types of hearing protection devices (*from National Safety Council*).

of personal equipment. Helmets cover most of the head and may provide hearing protection; however, few helmets are designed with hearing protection as their primary function.[10] Figure 10.1 lists characteristics of the various types of hearing protection devices. The selection of specific types of equipment depends on performance, comfort, application, appearance, and cost. Table 10.2 compares insert- and muff-type protectors. The cost of

TABLE 10.2
Comparison of Insert- and Muff-Type Hearing Protection Devices

Insert-Type Protectors	Muff-Type Protectors
Advantages	
1. They are small and are easily carried.	The protection provided by a good muff-type protector is generally greater and less variable between wearers than that of good earplugs.
2. They can be worn conveniently and effectively with other personally worn items such as glasses, headgear, or hairstyles.	A single size of earmuffs fits a large percentage of heads.
3. They are relatively comfortable to wear in hot environments.	The relatively large muff size can be seen readily at a distance; thus the wearing of these protectors is easily monitored.
4. They are convenient to wear where the head must be maneuvered in close quarters.	Muffs are usually accepted more readily at the beginning of a hearing conservation program than earplugs.
5. The cost of sized earplugs is significantly less than the cost of muffs; however, hand-formed and personally molded protectors may cost as much or more than muffs.	Muffs can be worn even with many minor ear infections.
	Muffs are not misplaced or lost as easily as are earplugs.
Disadvantages	
1. Sized and molded protectors require more time and effort for fitting than do muffs.	Muffs are uncomfortable in hot environments.
2. The amount of protection provided by a good earplug is generally less, and more variable between uses, than that provided by a good muff protector.	Muffs are not as easily carried or stored as are earplugs.
3. Dirt may be transmitted into the ear canal if an earplug is removed and reinserted with dirty hands.	Muffs are not as compatible with other personally worn items such as glasses and headgear as are earplugs.
4. Earplugs are difficult to see in the ear from a distance; hence, it is difficult to monitor groups wearing these devices.	Muff suspension forces may be reduced by usage, or by deliberate bending, so that the protection may be substantially less than expected.
5. Earplugs can be worn only in healthy ear canals and, even in some healthy canals, a period of time is necessary for acceptance.	The relatively large muff size may not be acceptable when the head must be maneuvered in close quarters.
	Muffs are more expensive than most insert-type protectors.

SOURCE: Kroes, D., Fleming, R., and Lempert, B. *List of Personal Hearing Protectors and Attenuation Data*. Washington, D.C.: NIOSH, 1973. (Sept.)

ear protectors varies greatly. Since individual ear canals vary in size, proper fitting is essential.

Ultraviolet and Infrared Radiation

To protect against ultraviolet and infrared radiation as well as against visible glare in inspection operations, protective lenses should be installed in a hand shield or welder's helmet. The shield should be made of a nonflammable material which is opaque to dangerous radiation and a poor conductor of heat. A metal shield is not desirable because it heats under infrared radiation.

Some tinted lenses used in special work afford no protection from infrared and ultraviolet radiation. For instance, most melters' blue glass used in open-hearth furnaces and the lenses used in Bessemer converters afford no protection against either type of harmful radiation. Probably no harm will come from continued use of these lenses if exposures are short. However, new workers learning these flame-reading skills should be provided with lenses that protect against two portions of the spectrum. The chemical composition of the lens rather than its color provides the filtering effect; this factor must be considered when selecting a filtering lense.[8]

OTHER PROTECTIVE EQUIPMENT

This section describes various kinds of protective equipment, including that needed for prevention of external absorption; head, eye, hand, foot, abdominal, and chest injuries; and internal absorption.

Emergencies often require the use of special equipment. Putting out aircraft fires or carrying out repair work in corrosive atmospheres require specially designed personal protective equipment. Thus asbestos and radiant-energy reflecting suits may be required for aircraft fire fighters; rubber suits (or other impervious clothing) may be necessary for use in emergencies in chemical plants. Since the wearer's head must also be enveloped by the protective material, a self-contained oxygen apparatus must be incorporated into the protecting hood.

In the selection of such suits, it is necessary to choose a reliable manufacturer because there are no standards for the design of this equipment. When protective clothing is made of material impervious to moisture, it will hold the moisture given off by the wearer, creating high humidity within the suit and causing discomfort. Such suits should have a supplied-air apparatus, the hose of which can be connected to the facepiece, as well as to the suit, to provide ventilation.

Prevention of Head Injuries

Helmets protect against head injury by cushioning impacts to the head and preventing shock.

Most head injuries occur because the worker was not wearing some form of head protection. The National Safety Council reports that in only 16% of injuries to the head was protection used.[12]

Class A helmets reduce the force of impact of falling objects and the danger from contact with exposed low-voltage conductors. Class B helmets reduce the force of impact of falling objects and the danger from contact with exposed high-voltage conductors. Class C helmets reduce the force of impact of falling objects. This class offers no electrical protection.[1]

Prevention of Eye Injuries

In general, protective eye equipment should fit comfortably and offer a maximum amount of protection to the eyes. For example, protection against flying particles would require that eye equipment fit closely around the eye socket and that ventilation holes be provided in appropriate places to keep the goggles from "steaming." The method of arranging for the ventilation of the goggles will vary; in one case a wire mesh covering is used in place of a glass lens because of its increased capacity for ventilation. However, if the exposure were to irritating vapors, it would be necessary for the protective device not only to fit snugly around the eye socket but also to be so completely enclosed that the vapors in the air could not make contact with the eyes.

Materials used in the construction of protective eye equipment should be noncorrosive, easily cleaned, and in many cases nonflammable; the transparent portion should give the widest possible field of vision without any appreciable distortion or prism effect.

When it is necessary for a worker to wear corrective glasses, it is advisable for the eye protection equipment to be supplied with a corrective lens, ground to the prescription of the wearer. Eye protection goggles that fit over the usual (street-wear type) corrective glasses can be obtained, but it is necessary in such instances for the goggles to be equipped with deep cuts so as to fit comfortably over the prescription glasses.

In general, plastic protective eye equipment has proved successful. However, there frequently is a wide variability in the impact strength of plastic products. In impact tests, the resistance of a plastic is a function of the thickness of the cross section and the area of the sample, as well as the composition of the material. Plastics are not resistant to abrasion, quickly becoming scratched through normal usage. Some plastic eye protective equipment, therefore, is coated with an abrasive-resistant resin which offers better wearing qualities. In cases where plastics are used to protect the eyes from chemicals, the plastic materials may show a surface reaction to certain chemicals but invariably can be expected to stop splashes satisfactorily and give the eyes the desired protection. In operating situations where the eyes are exposed to splashes from hot materials, plastics can be expected to be somewhat superior to glass. Plastic materials can be manufactured which will transmit light as satisfactorily as do glass lenses.

TABLE 10.3
Preventing Eye Injuries While Jump-Starting Batteries[a]

Before Attaching the Cables

Put out all cigarettes and flames. A spark can ignite hydrogen gas from the battery fluid.

Make sure the cars do not touch each other. Set parking brakes and automatic shifts of both cars to "park" (manual transmissions to "neutral") and turn the ignition off.

Take off the battery caps, if removable, and add water if it is needed. Check for ice in the battery fluid. Never jump-start a frozen battery! Replace the caps and cover with a damp cloth.

Do not jump-start unless both batteries are the same voltage. American cars have either 12-volt or 6-volt batteries. (Check owner's manual.) Owners of foreign cars should check their operating manuals for emergency starting directions.

Attaching the Cables

Clamp one jumper cable to the positive (+) pole of the dead battery. Then clamp the other end of the same cable to the positive pole of the good battery.

At the booster battery, connect the second cable to the negative (−) pole. Then clamp the other end of that cable to the stalled car's engine block on the side away from the battery.

Start the car with the good battery and then the disabled car.

Remove the cables, first from the engine block and the booster car's negative terminal and then from the positive poles.

SOURCE: Hirschfelder, D. Better industrial vision testing and safety. *Occupational Health and Safety*, 1980, **49**, 30–34, 56–57. (Nov.–Dec.) Reprinted by permission of *Occupational Health and Safety*.

[a] Thousands of battery explosions occur every year, and nearly two-thirds of the resulting injuries involve the eyes. The victims are often drivers and service station personnel attempting to revive dead cars with jumper cables without following proper safety procedures. To prevent eye-threatening injuries, the National Society to Prevent Blindness suggests these easy-to-follow instructions for jump-starting a car with a dead battery.

For maximum eye safety, the society advises everyone working with car batteries or standing nearby to wear protective goggles to keep flying battery fragments and chemicals out of the eyes. Should battery acid get into the eyes, immediately flush them with water continuously for 15 minutes. Then see a doctor.

However, curved-sheet plastic eye protective equipment may introduce prismatic effects that are undesirable.

Many eye injuries result from jump-starting automobiles. A battery from an operating vehicle is connected to a disabled vehicle by jumper cables. When contact is made, the battery explodes and battery acid is sprayed everywhere. The acid is sulfuric acid (H_2SO_4). Eye injuries may be prevented by following the instructions in Table 10.3.

Prevention of Hand Injuries

More than one-third of the annual occupation-related disabling injuries in the nation involve the fingers, hands, and arms and account for more than 20% of the total compensation paid in the United States. Because of the obvious vulnerability of the fingers, hands, and arms, protective apparel is often required. Manufacturers offer a wide variety of such equipment for many specialized occupations, but the most common is the glove or some adaptation of it.

In general, it should be remembered that gloves are not recommended for operators working around rotating machinery because a glove may get caught in the spinning parts and pull the worker's hand into the machine. Gloves should be carefully selected for specific operations, and protection of the product should not be at the expense of employees' safety.

If gloves are used to protect workers' hands from chemical solutions, the gloves should be long enough to extend well above the wrist. They should not have a flaring cuff, which would allow the chemicals to splash into the glove.

Gloves, hand pads, and mittens reinforced with metal staples along the palm provide greater protective resistance to sharp objects and enable a better grasp on the material being handled. They should not be used around operations involving electrical apparatus because the wearer may receive an electric shock.

Rubber gloves are a practical means of protecting the hands from liquids. However, compounds derived from petroleum products have a deteriorating effect on natural rubber, and it is therefore necessary to choose gloves manufactured of a synthetic material like neoprene.

Linemen and others working around energized high-voltage electrical equipment require specially made and tested rubber gloves that are worn under heavy leather outer gloves to safeguard the rubber gloves from cuts, abrasions, and wire punctures which would reduce their protective value. A regular testing and inspection program for linemen's rubber gloves is absolutely essential, and gloves failing to meet the original specifications should be discarded promptly.

Prevention of Foot Injuries

Foot injuries are another major source of disablement. In 1980 there were 150,000 disabling foot and toe injuries, which amounted to 7% of all disabling injuries. Safety shoes have been used for several generations. However, in a study by the Bureau of Labor Statistics, only 23% of persons with disabling foot injuries were wearing safety shoes or safety boots.

The commonest protective footwear is the metal toe box "safety" shoe. ANSI Specification Z41.1 has recommended a shoe of sturdy, solid construction with a steel toe box flanged at the bottom and resting on the sole of the shoe. Class 75 must be able to support a static load of 2,500 pounds and withstand an impact equivalent to a 75-pound weight dropped a distance of one foot. The inside of the toe box must not come closer than one-half inch to the upper surface of the sole under each of these tests. Classes 50 and 30 must be able to withstand 50 and 30 pounds impact, respectively.

Shoes with metal toe boxes are made for a wide variety of work situations. For example, a department supervisor may select a "dressy" type of safety shoe which would be difficult to differentiate from any other well-made dress shoe, while foundryworkers can obtain shoes which fit snugly around the upper ankle but may be slipped off quickly in order to protect the wearer from hot metal entering the shoe. Other designs are shoes to reduce the possibility of static electricity building on the body of the wearer (conductive shoes); shoes without metallic parts, to reduce the possibility of a spark's being struck when the wearer walks on abrasive surfaces (nonsparking shoes); and shoes which provide insulation to the wearer against electrical conductions through the ground (nonconducting shoes).

Bootguards are required in occupations where objects heavier than 50 pounds may fall and strike the toe or where relatively heavy materials may drop on the instep. These are flanged, heavy gauged metal and corrugated sheet metal covers that protect the foot from toe to ankle. The guard must be able to withstand an impact of at least 300 pounds dropped a distance of one foot. If the flange of the guard is resting on a firm, flat surface, the foot would not be injured.

Prevention of Internal Absorption

Internal absorption must be prevented to protect against unhealthy levels of gases, vapors, dusts, fumes, smokes, mist, and fogs. The first line of attack, as always, is attempting to reduce the hazard at the source. Except in emergencies, respiratory equipment should not be used routinely as the only hazard protection. For all of these hazards, safety glasses and goggles will provide some protection, but some form of respiratory protection is also needed. Figure 10.2 shows a way to decide which form of respiratory device to use. The National Safety Council lists the following factors to consider in choosing respiratory equipment:

1. The nature of the hazardous operation or process.
2. The type of air contaminant, including its physical and chemical properties, physiological effects on the body, and its concentration.
3. The period of time for which respiratory protection must be provided.
4. The location of the hazard area with respect to a source of uncontaminated respirable air.

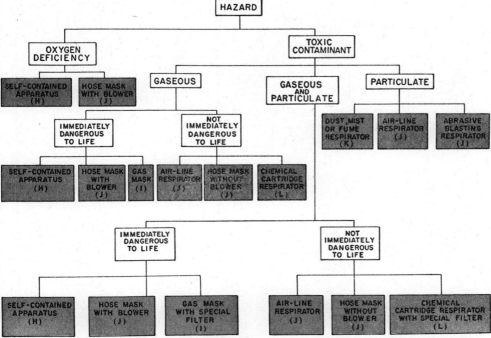

FIGURE 10.2.   Technique for selecting respiratory devices. Letters in parentheses (in shaded boxes) refer to Subparts Title 30 CFR, Chapter I, Part II, which discuss the items (*from Accident Prevention Manual for Industrial Operations, copyright by National Safety Council, Chicago IL 60611; used with permission*).

5. The state of health of personnel involved.
6. The functional and physical characteristics of respiratory protective devices. (Figure 10.3 illustrates an airline respirator.)

The following is recommended procedure for deciding what type of respiratory protective equipment is required: (See Fig 10.4).

1. What is the name of the contaminant to be guarded against?
2. What are its chemical, physical, and toxicological properties?
3. Is it immediately dangerous to life (emergency situations) or is it injurious only after prolonged and continued exposure (non-emergency situations)?
4. What are the limiting factors of the jobs being performed by personnel? (That is, must workers be free in their movements; how long would the device have to be worn per day?)
5. After consideration of all factors, select the type or types of respirators that would satisfy.

Applications
This full facepiece airline respirator is designed to provide respiratory and face protection against irritating gases, vapors, dusts and mists.

1 **Molded Blue Neoprene Rubber**
Resilient, pliable neoprene rubber facepiece provides soft, leak-tight fit, resists aging due to chemicals and petroleum-based vapors.

2 **Large Curved Lens**
Large symmetrical acrylic lens provides wide, distortion-free field of vision. Lens may be replaced quickly and easily by removing two lens retention screws on either side of lens holder. Meets lens requirements of Federal Specification GGG-M-125d.

3 **Anti-Fogging Design**
Inlet ducts direct incoming air over the lens surface.

CAUTION: The Commander Full Facepiece Airline Respirator should not be used in atmospheres containing less than 19.5% oxygen, or in atmospheres containing concentrations of contaminants that are immediately dangerous to life or health.

FIGURE 10.3. Airline respirator (*American Optical Corporation, Southbridge, MA 01550*).

④ Replaceable Lens Cover

Low cost, abrasion-resistant lens cover (.020″ acetate) protects the primary lens from scratching and pitting.

⑤ Adjustable Five-Point Neoprene Head Harness

Provides a snug, comfortable fit.

⑥ Leak-Tight Flap

Special inside flap provides comfortable leak-tight protection for most face sizes.

⑦ Recessed Buckle Rivetheads

Eliminate the possibility of uncomfortable pressure points on head.

⑧ Exhalation Valve Design

Provides low exhalation resistance.

⑨ Non-Kinking Hose

2½ feet of flexible, non-kinking rubber breathing tube with air regulating valve.

⑩ Quick-Connect Coupling

Designed to allow easy hook-up while preventing accidental disconnection of air supply line during use.

PROTECTIVE GLASSES MAY BE WORN

F5500 COMMANDER Frame with special Short Butt Temples

Protective prescription glasses may be worn, by means of special temple bars designed to fit spectacle holder inside the facepiece. Will accommodate most metal frames such as the AO F5500 Series Metal Frames. See Ordering Information for details.

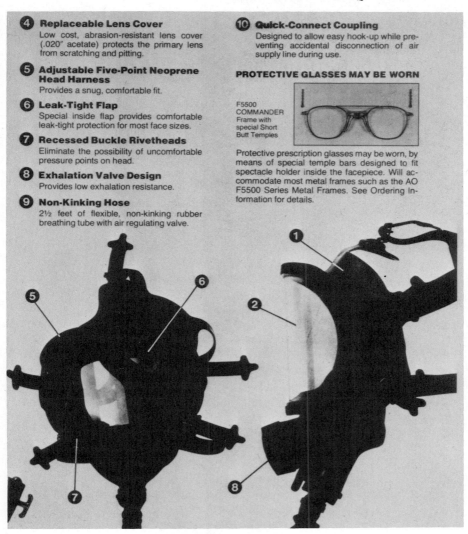

FIGURE 10.3. (*Continued*)

In choosing a respiratory device, remember that chemical cartridge respirators are suitable only in atmospheres which are not immediately dangerous to life and which contain a percentage of contaminants not exceeding 0.1% by volume. These respirators should not be used for contaminants:

1. Which are extremely poisonous in very small concentrations (hydrogen cyanide)
2. Which are not effectively absorbed chemically (carbon monoxide)

Figure 10.4.   Respiratory protective equipment.

3.   Which are not easily identified by odor (methyl chloride, hydrogen sulfide)
4.   Which are irritating to the eyes (sulfur dioxide)

Chemical Cartridge Respirators.   Chemical cartridge respirators may be considered low-capacity gas masks. They consist of a facepiece that fits over the nose and mouth of the wearer and to which is directly attached a small filter-chemical cartridge which is replaceable. These respirators are intended for use in nonemergencies only and in atmospheres which can be breathed without protection but could well cause discomfort or chronic poisoning if breathed for prolonged or repeated periods (at least seven hours per day). They should not be used in atmospheres containing more than 0.1% percent of the contaminant by volume.

This type of respiratory protective equipment has usually been used where there is exposure to the vapors of solvents (in occupations such as spray coating, degreasing, and dry cleaning) and where there may be a low concentration of acid gases (as in smelting sulfide ores).

Gas Masks.   Gas masks consist of a faceplate that fits over the eyes, nose, and mouth and is connected by a flexible tube to a canister containing the chemical absorbent which protects against a specific vapor or gas or groups of vapors or gases. The device is used mainly in emergencies.

Canister-type gas masks should be limited to use in atmospheres that are not deficient in oxygen or where the toxic contaminant does not exceed 1% in concentration (3% for ammonia) by volume.

Gas masks are commonly used where there is handling of volatile or gaseous chemical products when making emergency repairs (in operations such as repairing refrigerating systems, fire fighting, fumigating, and working near containers of poisonous products. It must be remembered that gas masks should not be used in tanks, sewers, or other places where a lack of oxygen may exist. They should not be used for more than a total of two hours, and it should be remembered that gas masks offer no protection in atmospheres deficient in oxygen.

Canisters should be replaced when the wearer first detects a tract of the contaminant coming through the breathing tube. Once a canister has been opened, even though it has been used for as little as two hours, it should be replaced at the end of 12 months. The chemical change that occurs in the canister during that time reduces the efficiency of the chemicals. However, gas-mask canisters with their top and bottom seals intact can be stored for a number of years. A cool place should be chosen for storage.

Mechanical-Filter Respirators.   Mechanical-filter respirators are non-emergency devices used to protect the wearer by filtering a portion of the contaminant out of the air as it is breathed. The device usually consists of a facepiece that covers the mouth and nose and to which is attached a filter element in the form of bags, cylinders, or disks, arranged so that the air as it is inhaled must pass through the filter substance. These respirators provide no protection against gases and vapors. They remove airborne contaminants which are in the the form of particulate matter by physical "trapping" and electrostatic attraction as the air is drawn through the filter by the breather. The efficacy of the filter is determined by the amount of contaminant that is able to get through it. Mechanical-filter respirators (often called dispersoid respirators) are classified by the Bureau of Mines according to the contaminant against which they are intended to offer protection, as follows:

1. Fume respirators, which protect against fumes that are not significantly more toxic than lead
2. Mist respirators, which are used in situations where there are pneumoconiosis-producing mist and nuisance mists

Dust Respirators.   Dust respirators protect against pneumoconiosis-causing dusts such as asbestos, free silica, and nuisance dusts such as coal, flour, gypsum, wood, aluminum, limestone, and cement; and toxic dusts that are not significantly more toxic than lead, such as manganese, arsenic, chromium, selenium, cadmium, vanadium, and their compounds. Each respirator should carry the OSHA approval label or certificate designating the materials for which the respirator offers approved protection.

Self-contained breathing apparatus enables the wearer to be independent of any outside source of air, inasmuch as the device provides oxygen either from compressed-air or oxygen cylinders or by chemical action in the canister attached to the apparatus. This equipment is generally used for emergency situations and may be operable for periods between one-half and two hours. It will provide protection in situations where there are gases, vapors, dusts, fumes, smoke, and mists in concentrations that are tolerated by the skin and in atmospheres deficient in oxygen.

MAINTENANCE

A maintenance program for personal protective equipment is important to have as each piece of this personal protective equipment must be operating at maximum efficiency. This is not a place for errors.

There must be a systematic routine checking of all equipment.

A maintenance program for keeping the personal protective equipment in good and efficient repair assists in promoting the use of the equipment and assures that it is at optimum efficiency. In addition, it has been found that it is less expensive to repair worn devices where possible than it would be to buy new equipment continually. The maintenance of personal protective equipment should be done as a regular assignment, on a full or part-time basis as required, by personnel responsible to the safety specialist in charge. Items such as respirators, which require fitting and some technical knowledge on the part of employees in order to be kept in good condition, should be regularly cleaned, inspected, and maintained according to a schedule. The employees doing this work should be thoroughly responsible and have a complete knowledge of the equipment as well as the duties they are to perform.

Goggles must fit properly if they are to be comfortable and worn. A safety goggle program, therefore, requires that at least one employee should be trained in the fitting and adjusting of the safety glasses. This employee should also be assigned the responsibility of carrying out the maintenance program for safety goggles. The degree to which this phase of the program is conducted by various industrial concerns varies considerably. In some cases, a maintenance shop or unit is established, and personnel are encouraged to bring their safety goggles to this service center for repairs and adjustments whenever desired. In other cases, a goggle service cart is pushed through the departments of the plant or installation. It is equipped to sterilize and clean goggles and is stocked with replacement parts, so the person in charge of the cart is able to make one-the-spot repairs and adjustments.

Respiratory protective equipment requires vigilant attention. Such devices as respirators and airline masks and hoods, which are worn frequently or continuously, should be cleaned daily and inspected for defects. When a considerable number of respiratory protective devices are used at an installation, it is advisable to establish a separate unit or room where the inspections, cleanings, and repairs can be made. This room may be a "safety service room" where all of the personal safety equipment devices that are to be maintained and repaired can be taken care of. Each employee who is required to use a respirator should have his or her own, permanently marked with the employee's name or clock identification number. The respirator should be turned in to the service room at the close of each shift so that it can be cleaned and repaired if necessary.

When inspecting respiratory protective equipment, an examination of the exhalation valve seats should be made to see if they are smooth and clean: the rubber parts should be pulled slightly in order to detect the presence of fine cracks, indicating deterioration of the rubber.

Under OSHA standards, the employer is responsible for making personal protective equipment available, although the employee may have to purchase it. It's the employee's body that's injured if the equipment is lacking, poorly fitted, or not used. However, it is the employer who ultimately pays workers' compensation claims. In the long run, enlightened management knows that both parties suffer when an injury occurs.

SUGGESTED LEARNING EXPERIENCES

1. Examine the hazards faced by an autobody mechanic. Discuss and illustrate the appropriate personal protective equipment for this worker.

2. Should workers be able to choose whether or not to wear the recommended personal protective equipment? Explain.

3. Obtain a catalog of a manufacturer of personal protective equipment and review the contents. Look for the various types of equipment available. Is approval of equipment listed? What limitations are noted by the manufacturer? Are all potential areas of protection covered, or is the equipment limited to only a few types of protection?

4. Your boss told you that in a particular operation in which the workers handle hazardous materials, the workers should be supplied with respirators. You were told to buy a hose-mask type from a local safety supply firm since this type seemed to be the least expensive. What would you do? Develop an outline of the steps you would take to handle the problem.

5. Obtain a canister, hose-mask, airline, or self-contained respirator and try it out. Notice the difficulties that are experienced when breathing. What problems do you have when moving about? How does the mask affect your vision? What is the effect when performing physical tasks?

6. Visit a workplace of your choice. Beforehand, list the kinds of protective equipment you would expect this industry to use. Were you correct? Was there any equipment you thought they should have had but they didn't?

7. How would you as a safety engineer go about instructing the workers in personal protective equipment?

8. How would you as a safety engineer go about monitoring the use of personal protective equipment by workers?

9. What program would you as a safety engineer develop to ensure the proper purchase, use, and maintenance of personal protective equipment in your industry?

10. Discuss temperature extremes in the workplace and the use of personal protective equipment.

REFERENCES

1. American National Standards Institute. *Air Purifying Respirator Cannisters and Cartridges* (ANSI K13.1-1973). New York, 1973.

2. American National Standards Institute. *Men's and Women's Safety Toe Footwear* (ANSI Z41.1-1981). New York, 1981.

3. American National Standards Institute. *Protective Headwear for Industrial Workers* (ANSI Z89.1-1981). New York, 1981.

4. American Optical Brochure. *Commander Facepiece Airline Full Respirator for Comfort, Safety, Convenience.* Southbridge, Mass., n.d.

5. Chamber of Commerce of the United States. *Analysis of Workers' Compensation Laws.* Washington, DC: Chamber of Commerce, 1981.

6. Cluff, G. L. Limitations of ear protection of hearing conservation programs. *Sound and Vibration*, 1980, **14**, 19–24. (Sept.)

7. Engel, P. G. Top-flight programs highlight monitoring respiratory protection. *Occupational Hazards*, 1981, **43**, 31–34. (June).

8. Hirschfelder, D. Better Industrial vision testing and safety. *Occupational Health and Safety*, 1980, **49**, 30–34, 56–57. (Nov.–Dec.)

9. Lynch, P. Matching protective clothing to job hazards. *Occupational Health and Safety*, 1982, **51**, 30–34. (Jan.)

10. Moline, C. H. Protective equipment: Head, eyes, face, body, extremities. In NIOSH, *Industrial Hygiene Engineering and Control.* Cincinnati, Ohio, 1978.

11. NIOSH. *Industrial Hygiene Engineering and Control.* Cincinnati, Ohio, 1978.

12. National Safety Council. *Accident Prevention Manual for Industrial Operations.* 8th ed Chicago, 1981.

13. Nixon, C. W. Industrial hearing conservation. *Audiology and Hearing Education*, 1980, 33–41. (Apr.–May)

14. Schmidek, M. E., Layne, M. S., Lempert, B. L., and Fleming, R. M. In NIOSH, *Survey of Hearing Conservation Programs in Industry* Technical Report (75-178). Cincinnati, Ohio, 1975.

15. Tips on selection and use of respiratory equipment. *Occupational Hazards*, 1979, **41**, 45–47. (Sept.)

16. World Health Organization. *Health Hazards of the Human Environment.* Geneva, 1972.

PART 3 SPECIFIC HAZARDS

11 FIRE CONDITIONS

Fires can and do occur everywhere. They are a concern because of their potential destruction of property and life in residential and nonresidential properties. Fire causes massive loss of life along with massive destruction. In the decade from 1970 to 1980 several thousand persons lost their lives in fires each year. Many persons were killed in large fires. (See Table 11.1.)

The twentieth-century record of major loss of life from building fires in America in five-year periods from 1900 to 1979 is shown in Figure 11.1. The record is limited to building fires that killed at least 25 persons because such fires generally occur in properties covered by legally binding fire codes and because early data are likely to be incomplete for fires that claimed fewer victims. The statistics shown in Figure 11.1 are deaths per 10 million people in the national population, in order to account for the growing population of the United States.

Despite the appalling costs in both lives and property, little is expended to study fire and its prevention except for small amounts spent on collecting statistics and on education. Not even the great Chicago fire started by Mrs. O'Leary's cow or the Cocoanut Grove night club have been substantial enough to facilitate fire research.[6] The Chicago fire burned for more than three days and destroyed over 17,000 buildings. Over 500 persons died as a result of the Cocoanut Grove Fire.

COMBUSTION

Combustion is any chemical process that involves oxidation sufficient to produce heat and light. Combustion is an exothermic, self-sustaining reaction involving a condense-phase fuel, a gas-phase fuel, or both. The process is usually associated with the oxidation of a fuel by atmospheric oxygen with the emission of light. The condensed phase of combustion is

TABLE 11.1
Major Fires Causing Large Loss of Life (10 or More Deaths), 1975 to 1980

Data	Location and Description	No. of Deaths
	Night Clubs	
Oct. 24, 1976	Social club, Bronx, N.Y. (fire blocked escape routes)	25
May 28, 1977	Supper club, Southgate, Ky. (delayed alarm, rapid fire spread, overcrowding, lack of automatic sprinkler protection)	165
	Institutions	
Jan. 30, 1976	Nursing home, Chicago, Ill. (smoke spread through open door to chapel where most residents were attending services)	24
June 26, 1977	County jail, Columbia, Tenn. (incendiary origin)	42
Dec. 9, 1978	Institution for mentally retarded, Ellisville, Miss. (lack of automatic detection and suppression, lack of smoke-stop partitions)	15
Dec. 27, 1979	County jail, Lancaster, S.C. (rapid fire and smoke spread, locked doors)	11
	Residences	
July 7, 1975	Hotel, Portland, Oreg. (home for transients, first floor occupied by businesses, lack of automatic detection and suppression)	11
Dec. 12, 1975	Apartment building, San Francisco, Calif. (incendiary fire, rapid spread through open stairways)	14
Jan. 10, 1976	Hotel, Fremont, Nebr. (natural gas explosion)	20
Feb. 4, 1976	Tenement house, New York, N.Y. (careless smoking, fire spread through pipe recesses)	10
Dec. 20, 1976	Apartment house, Los Angeles, Calif. (delay in alarm transmittal, open stairways)	10
Dec. 23, 1976	Apartment house, Chicago, Ill. (careless handling of liquid fuel)	12
Jan. 28, 1977	Hotel Breckenridge, Minn. (lack of automatic detection and suppression, open stairways)	17
Dec. 13, 1977	Dormitory, Providence, R.I. (rapid spread among Christmas decorations, dead-end corridor)	10
Jan. 28, 1978	Hotel, Kansas City, Mo. (lack of automatic detection and suppression, unprotected vertical openings, inadequate means of egress)	20
Nov. 5, 1978	Hotel, Honesdale, Pa. (incendiary origin)	12
Nov. 26, 1978	Hotel, Greece, N.Y. (incendiary origin, unprotected openings in stairways, inadequate notification of occupants)	10
Jan. 20, 1979	Tenement, Hoboken, N.J. (open stairway, lack of approved egress)	21

TABLE 11.1 (*Continued*)

Data	Location and Description	No. of Deaths
	Residences	
Apr. 1, 1979	Boarding home, Connellsville, Pa. (open stairways, lack of automatic detection and suppression in structure originally built as single-family dwelling)	10
Apr. 2, 1979	Boarding home, Farmington, Mo. (lack of complete detection system, undivided attic space, lack of training of staff and residents in fire evacuation)	25
July 31, 1979	Hotel, Cambridge, Ohio (open stairways, inadequate notification of occupants, combustible interior finish)	10
Nov. 11, 1979	Boarding home, Pioneer, Ohio (combustible interior finish, lack of second means of egress)	14
July 26, 1980	Boarding home, Bradley Beach, N.J. (delayed alarm, lack of early-warning automatic detection, unprotected vertical openings, lack of second means of egress from second and third floors)	24
Nov. 21, 1980	Hotel, Las Vegas, Nev. (improper electric wiring, rapid fire and smoke development, failure to extinguish fire in incipient stage, improperly protected vertical openings, improperly enclosed exit stairs, distribution of smoke by air-handling equipment)	85
Dec. 4, 1980	Hotel, Harrison, N.Y. (failure to extinguish fire in incipient stage, rapid fire development, lack of early-warning detection and alarm)	26
	Stores and Offices	
Dec. 22, 1976	Department store, Brooklyn, N.Y. (structural collapse, dense smoke)	12
Nov. 5, 1978	Department store, Des Moines, Iowa (lack of automatic suppression)	10
	Coal Mines	
Mar. 9 and 11, 1976	Coal mine explosion, Oven Fork, Ky. (two blasts)	26
	Manufacturing Sites	
Aug. 12, 1976	Oil refinery explosion, Chaimette, La.	12
	Storage Facilities	
Dec. 22, 1977	Grain elevator explosion, Westwego, La.	36
Dec. 27, 1977	Grain elevator explosion, Galveston, Tex.	18

TABLE 11.1 (*Continued*)

Data	Location and Description	No. of Deaths
	Transportation	
Jan. 31, 1975	Ship Collision, Marcus Hook, Pa. (chemical cargo ship and oil tanker collision, several explosions, fire spread to dock area)	29
Apr. 30, 1975	LP-gas transport truck explosion, Eagle Pass, Tex. (truck overturned, ruptured, and a section propelled into several mobile homes)	19
June 24, 1975	B-727 airplane crash and fire, New York, N.Y. (crashed on landing during severe storm, spilled fuel ignited)	113
Apr. 27, 1976	B-727 airplane crash and fire, St. Thomas, Virgin Islands	35
Apr. 4, 1977	DC-9 airplane crash and fire, New Hope, Ga.	28
Feb. 24, 1978	Train derailment, Waverly, Tenn. (LP-gas tank car exploded)	16
Nov. 5, 1979	Ship collision, Galveston, Tex. (oil tanker and freighter collided)	32

SOURCE: Reprinted from the *Fire Protection Handbook* (15th ed.). Copyright © 1981 National Fire Protection Association, Quincy, Massachusetts. Reprinted with permission.

usually a glowing combustion or smoldering. The gas-phase combustion is the visible flame. If the combustion process is confined to a space which results in a rapid rise in pressure, it is called an explosion.

Combustion is a reaction which is a continuous combination of a fuel (reducing agent) with certain elements, prominent among which is oxygen in a free or combined form (oxidizing agent). The quality which all these reactions have in common is that they are exothermic (converting chemical energy trapped in the original molecules into the form of thermal energy).

Composition of Combustion

Combustion is composed of flame, heat, fire gases, and smoke.

Flame.   The burning of most materials is an exothermic chemical oxidation process. Energy from the process is evolved as heat, which passes convective (hot gases) and radiative components. The radiative component represents energy released in the visible and infrared portions of the spectrum and is seen as flame or luminosity. Thermal burns can be caused by convective and radiative heat, the contribution of each depending upon the size and intensity of the fire.[6]

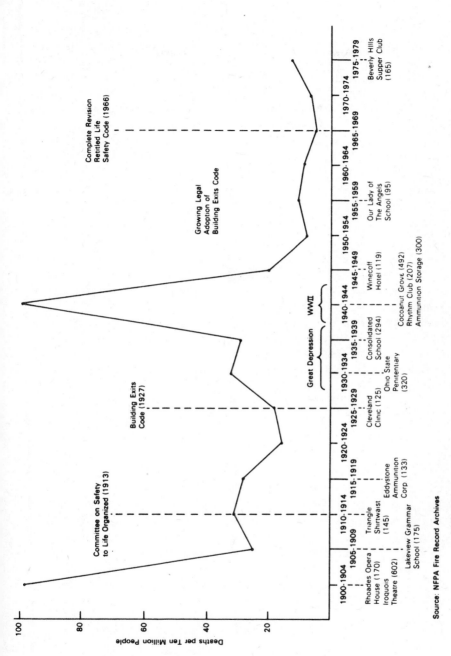

FIGURE 11.1. Major loss of life in twentieth century U.S. building fires taking at least 25 lives (*reprinted from the Fire Protection Handbook, 15th ed., copyright © 1981 National Fire Protection Association, Quincy, Massachusetts, reprinted by permission*).

In consideration of life safety, observation of flame does confirm the presence of fire. However, heat and highly toxic combustion products are often formed without the presence of flame. This is the case with smoldering fires, which generate products of incomplete combustion that are quite hazardous. A significant role of flame may also be the creation of panic conditions leading to subsequent injury from thermal burning and inhalation of smoke.

Heat.   Heat is the combustion product most responsible for the spread of fire in buildings. Heat poses a physical danger to human beings through exposure to hot gases and to radiation. If the total heat energy reacting with the body surpasses the capability of physiological defense processes to compensate, the result may be injury or death.

The effects of exposure to heated air are greatly augmented by the presence of moisture in the fire atmosphere. With higher moisture content, transfer of heat energy is more efficient and the body is less able to rid itself of the heat burden. Moisture can be present in a fire environment as the result of natural humidity, moisture produced from the combustion, and the application of water for extinguishment.

If heat is conducted rapidly to the lungs, a serious decline in blood pressure and capillary blood vessel collapse leading to circulatory failure may result. Severe heat may also cause fluid buildup in the lungs; 300°F has been taken as the maximum air temperature at which one can breathe. A temperature this high can be endured only for a short period and not at all in the presence of moisture.[3] See Table 9.1 for effects of heat stress.

Fire Gases.   Fire gases are the gases which remain when products of combustion are cooled to normal temperatures. Most combustible materials contain carbon which burns to form carbon dioxide when the oxygen supply is ample, but forms carbon monoxide when the oxygen supply is poor. Unless the fuel and air are premixed, the oxygen supply in the combustion zone is usually inadequate for complete combustion. When materials burn, other gases may be formed, including sulfur dioxide, ammonia, hydrogen cyanide, nitrogen oxides, halogen acids, isocyanates, acrolein, and volatile hydrocarbons. Gases formed in a fire depend on many variables, the principal ones being the chemical composition of the burning material, the amount of oxygen available for combustion, and the atmosphere.[6]

Smoke.   Smoke consists of finely divided particulate matter and suspended liquid droplets known as aerosols. The particulate matter is formed from the burning of most materials as a result of incomplete combustion. Under these conditions, incompletely burned organic materials are also evolved in dispersed aerosol form and contribute to visible smoke. Since the average size of particulates and aerosols is about the same as the wavelength of visible light, scattering of the light occurs and vision through smoke is obscured.

Particulates and aerosols cause physical and physiological effects. Since smoke often obscures the passage of light, vision to exits and exit signs is blocked. The development of quantities of smoke sufficient to make exit-ways unusable can be very rapid. Smoke frequently provides the early warnings of fire but at the same time contributes to panic conditions by the very nature of its blinding and irritating effects.

Smoke particulates and aerosols can be harmful when inhaled, and long exposure may cause damage to the respiratory system. Particulates and irritants in the eyes induce tears which impair vision. When present in the nose and throat, they can cause sneezing and coughing at times when persons so affected need their normal faculties. Smoke particulates are often of a size sufficiently small enough to enable them to be inhaled deep into the lungs where absorbed toxicants may produce profound respiratory system damage.

FLAMMABLE/EXPLOSIVE COMPONENTS OF COMBUSTION

Ignition Temperatures

The ignition temperature reported for a flammable liquid is generally the temperature to which a closed or nearly closed container must be heated in order for the liquid in question, when introduced into the container, to ignite spontaneously and burn. Within a given hydrocarbon series, temperature decreases as molecular weight or carbon chain length increases, other factors being equal. Thus methane (CH_4) has a higher ignition temperature, 1004°F, than does hectane [$CH_3(CH_2)_4CH_3$], which is 437°F.

Minimum Ignition Temperatures

The term "lower flammable limit" (LFL) describes the minimum concentration of vapor to air below which propagation of a flame will not occur in the presence of an ignition source. The "upper flammable limit" (UFL) is the maximum vapor-to-air concentration above which propagation of flame will not occur. If a vapor-to-air mixture is below the lower flammable limit, it is described as being "too lean" to burn; if it is above the upper flammable limit, it is "too rich" to burn.

When the vapor-to-air ratio is somewhere between the lower flammable limit and the upper flammable limit, fires and explosions can occur. The mixture is then said to be within its flammable, or explosive, range. When the mixture happens to be in the intermediate range between the LFL and UFL (synonymous with LEL—lower explosive limit—and UEL—upper explosive limit), the ignition is more intense and violent than if the mixture were closer to either the upper or the lower limit.

Flash Point

A vapor mixed with air in proportions below the lower limit of flammability may burn at the source of ignition, that is, in the zone immediately surrounding the source of ignition, without propagating (spreading) flame away from the source of ignition. The flash point of a liquid corresponds roughly to the lowest temperature at which the vapor pressure of the liquid is just sufficient to produce a flammable mixture, or the lower limit of flammability.[10]

Fire Point

The lowest temperature of a liquid in an open container at which vapors evolve fast enough to support continuous combustion is called the fire point. The fire point is usually a few degrees above the lower flash point.

It should be emphasized that fires can spread over liquids whose temperatures are considerably below their lower flash points, provided a source of ignition has been established. In such situations the ignition source or fire itself heats the liquid surface locally so that its temperature rises above the fire point.

Flammable Limits

The limits of flammability are the extreme concentration limits of a combustible in an oxidant which once initiated will continue to propagate at the specified temperature and pressure. For example, hydrogen–air mixtures will propagate flames for concentrations between 4 and 74% by volume of hydrogen at 760 mm of mercury pressure (14.7 psi). The smaller value is the lower (lean) limit, and the larger value is the upper (rich) limit of flammability. When the mixture temperature is increased, the flammability range widens; when the temperature is decreased, the range narrows (see Figure 11.2). A decrease in temperature can cause a previously flammable mixture to become nonflammable by placing it either above or below the limits of flammability for the specific environmental conditions.

Note in Figure 11.2 that for liquid fuels in equilibrium with their vapors in air, a minimum temperature exists for each fuel above which there is sufficient vapor to form a flammable vapor–air mixture. There is also a maximum temperature above which the fuel–vapor concentration is too high to propagate flame. These minimum and maximum temperatures are referred to respectively as the lower and upper flash points in air. For temperatures below the lower flash point there is insufficient fuel vapor in

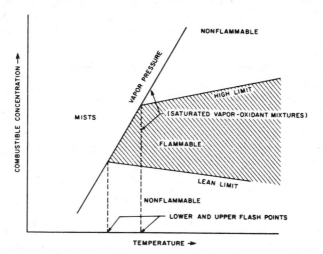

FIGURE 11.2. Saturated vapor-oxidant mixtures should be associated with the sloping vapor pressure line rather than just the upper–flash point vertical line (*reprinted from the Fire Protection Handbook, 15th ed., copyright © 1981 National Fire Protection Association, Quincy, Massachusetts, reprinted by permission*).

the gas phase to sustain a homogeneous ignition. The flash point temperatures for combustible liquids increase with environmental pressure.[6]

Spontaneous Combustion

Oxidation reactions involved in fires are exothermic; that is, one of the products of the reaction is heat. They are often complex, and many are not completely understood.

In order for an oxidation reaction to take place, a combustible material (fuel) and an oxidizing agent must be present. Fuels include innumerable materials not already in their most highly oxidized state. Whether or not a particular material can be further oxidized depends on the chemistry of the material: for practical purposes, it can be claimed that any material consisting primarily of carbon and hydrogen can be oxidized.[5]

The most common oxidizing material is the oxygen in the air. Air is composed of approximately one-fifth oxygen and four-fifths nitrogen.

Flashover

Flashover is part of the combustion process. As the temperature of a fire passes 300°F, the furnishings of the room start outgassing. This includes wood, fabrics, and especially plastic substances. When the gas concentration reaches the lower explosive limit, any flame or spark will set off an explosion-flashover. Instantaneously, every flammable object ignites, and the area becomes an inferno from which no escape is possible.[8]

Dust Explosions

Dust displaced from a surface on which it rests may develop a considerable charge. The ultimate charge depends on the inherent properties of the substance, size of particle, amount of surface contact, surface conductivity, gaseous breakdown, external field, and leakage resistance in a system. Greater charges develop from smooth surfaces than from rough ones, probably because of greater initial surface contact. Electrification develops during the first phase of separation. Subsequent impact of airborne particles on obstructions may affect their charge slightly, but if the impact surface becomes coated with the dust, this effect is slight.[10]

The smaller the size of the dust particle, the easier it is to ignite the dust cloud insofar as the exposed surface area of a unit weight of material increases as the particle size decreases. It is also true that particle size has an effect on the rate of pressure rise. For a given weight concentration of dust, a coarse dust will show a lower rate of pressure rise than a fine dust. The lower explosive limit concentration, the ignition temperature, and the energy necessary for ignition will decrease as dust particle size decreases.

Classifications of Fires

Fires have been classified according to the flammable or combustible substance causing the fire:

1. *Class A Fire.* A fire involving ordinary combustible materials, such as paper, wood, and cloth.
2. *Class B Fire.* A fire involving flammable or combustible liquids, flammable gases, greases, and similar material, and some rubber and plastic materials.
3. *Class C Fire.* A fire involving energized electrical equipment where safety to the employee requires the use of electrically nonconductive extinguishing media.
4. *Class D Fire.* A fire involving combustible metals, such as lithium, magnesium, sodium, potassium, titanium, and zirconium.

CHARACTERISTICS OF FLAMMABLE REACTIONS

Flammability Principles

1. An oxidizing agent, a combustible material, and an ignition source are essential for combustion.

2. The combustible material must be heated to its piloted-ignition temperature before it will ignite or support flame spread.

3. Its subsequent burning is governed by the heat feedback from the flames to the pyrolyzing (chemical change caused by heat) or vaporizing combustible.

4. The burning will continue until one of the following occurs:

    a. The combustible material is consumed

    b. The oxidizing agent concentration is lowered to below the concentration necessary to support combustion

    c. Sufficient heat is removed or prevented from reaching the combustible material to prevent further fuel pyrolysis

    d. The flames are chemically inhibited or sufficiently cooled to prevent further reaction

Ignition

Ignition is the process of initiating self-sustained combustion. If the ignition is caused by the introduction of some small external flame, spark, or glowing object (ember), it is called piloted ignition. If it occurs without the assistance of an external pilot source, it is called autoignition.

The ignition temperature of a substance is the minimum temperature to which it must be heated for it to ignite. Once ignition has occurred, it will continue until all the available fuel or oxygen has been consumed or until the flame is extinguished by cooling.

Rate of Burning

The rate of burning is a function of how fast the chemical reaction of oxidation occurs, as well as the speed at which the vaporized fuel and the oxygen are delivered to the combustion zone. In premixed flames (those in which mixing has occurred before the chemical reaction is initiated), the burning rate is controlled only by the rate at which the substances combine. This is usually very fast. It is for this reason that the mixture of air and combustible vapors is so dangerous; the process, once ignited, is virtually impossible to interrupt.

Since gases mix with one another readily, the burning of a gaseous fuel, such as hydrogen (H_2) or methane (CH_4), is a rapid process. However, the burning of a liquid or solid fuel requires first that the fuel be converted to the gaseous state (volatilization). This process requires the input of considerable heat and is usually slow when compared to the rate of oxidation. The rate of volatilization of a material strongly affects its rate of burning.

Because solid materials are more likely to be found in fire situations where life safety is the primary concern, a great deal of attention has been recently focused on the measurement and reduction of the toxicity of combustion products produced by such materials. In recent years, there has been a great deal of interest in the identity and effects of combustion products.

FIRE HAZARDS

The National Fire Protection Association has developed an identification system designating the hazards of a material. It is divided into four diamonds: one for health, one for flammability, one for reactivity, and one for other variables.

Health

In general, the health hazard in fire fighting is that of a single exposure which may vary from a few seconds up to an hour. The physical exertion demanded in fire fighting or other emergency conditions may be expected to intensify the effects of the exposure. Only hazards arising out of an inherent property of the material are considered. The following explanation is based upon protective equipment normally used by fire fighters.

4.   Materials too dangerous to health to expose fire fighters. A few whiffs of the vapor could cause death or the vapor or liquid could be fatal on penetrating the fire fighter's normal full protective clothing. The normal full protective clothing and breathing apparatus available to the average fire department will not provide adequate protection against inhalation or skin contact with these materials.

3.   Materials extremely hazardous to health, but areas may be entered with extreme care. Full protective clothing, including self-contained breathing apparatus, coat, pants, gloves, boots, and bands around legs, arms, and waist should be provided. No skin surface should be exposed.

2.   Materials hazardous to health, but areas may be entered freely with full face-mask self-contained breathing apparatus which provides eye protection.

1.   Materials only slightly hazardous to health. It may be desirable to wear self-contained breathing apparatus.

0.   Materials which on exposure under fire conditions would offer no hazard beyond that of ordinary combustible material.[13]

Flammability

Susceptibility to burning is the basis for assigning degrees within this category. The method of attacking the fire is influenced by this susceptibility factor.

4. Very flammable gases or very volatile flammable liquids. Shut off flow and keep cooling streams of water on exposed tanks or containers.

3. Materials which can be ignited under almost all normal temperature conditions. Water may be ineffective because of the low flash point.

2. Materials which must be moderately heated before ignition will occur. Water spray may be used to extinguish the fire because the material can be cooled below its flash point.

1. Materials that must be preheated before ignition can occur. Water may cause frothing if it gets below the surface of the liquid and turns to steam. However, water fog gently applied to the surface will cause a frothing which will extinguish the fire.

0. Materials that will not burn.[17]

Reactivity (Stability)

The assignment of degrees in the reactivity category is based upon the susceptibility of materials to release energy either by themselves or in combination with water. Fire exposure is one of the factors considered along with conditions of shock and pressure.

4. Materials which (in themselves) are readily capable of detonation or of explosive decomposition or explosive reaction at normal temperatures and pressures. This includes materials which are sensitive to mechanical or localized thermal shock. If a chemical with this hazard rating is in an advanced or massive fire, the area should be evacuated.

3. Materials which (in themselves) are capable of detonation, explosive decomposition, or explosive reaction but which require a strong initiating source or which must be heated under pressure before initiation. This includes materials which are sensitive to either mechanical or thermal shock. It also includes substances like sodium (Na) or potassium (K) which react violently with water at ordinary temperatures and normal pressure (14.7 psi).

2. Materials which are normally unstable and readily undergo violent chemical change but do not detonate. This includes materials which can undergo chemical reaction that is exothermic at normal temperatures and pressures or which can undergo violent chemical change at elevated temperatures and pressures. It also includes those materials which may react violently with water or which may form potentially explosive mixtures with water. In advance or massive fires, fire fighting should be done from a safe distance or from a protected location.

1. Materials which are normally stable but which may become unstable at elevated temperatures and pressures or which may react with water with some release of energy but not violently. Caution must be used in approaching the fire and applying water.

0.   Materials which (in themselves) are normally stable even under fire exposure conditions and which are not reactive with water. Normal fire-fighting procedures may be used.[16]

The hazard ratings for hazardous or toxic substances are presented as a series of the numbers just described. A totally safe substance would have a rating of 0-0-0. The first 0 indicates that there is no health hazard associated with the material. The second 0 indicates that the substance would not burn. The third 0 indicates that the substance is normally stable. The hazard ratings for some select chemicals are as follows:

Acetaldehyde, CH_3CHO	2-4-2
Carbon tetrachloride, CCL_4	3-0-0
Chromyl chloride, CrO_2H_5	2-4-1
Ethylenimine, $NHCH_2CH_2$	3-3-2
Fluorine, F_2	4-0-3
Potassium, K	3-1-2
Picric Acid, $(O_2N)_3C_6H_2OH$	2-4-4

Materials Capable of Self-Detonation

Sodium metal (Na) ignites spontaneously. It has a ready oxidation and spontaneous ignition in air. It is common practice to ship and store sodium metal under kerosene, usually in a hermetically sealed metal container. Sodium should not be permitted to contact the skin. Fumes from burning sodium are very irritating to skin, eyes, and respiratory system.

Materials Capable of Detonation

Charcoal may react with air at a sufficient rate to cause the charcoal to heat spontaneously and ignite. The principal causes of spontaneous heating of charcoal appear to be:

1. Lack of sufficient cooling and airing before shipping
2. Charcoal becoming wet
3. Friction in grinding of cinder sizes, particularly of material insufficiently aired before grinding
4. Carbonizing of wood at too low a temperature, leaving the charcoal in a chemically unstable condition

Oxides.   Oxides of metals and nonmetals react with water to form alkalies and acids. This reaction takes place violently with sodium oxide (Na_2O). Calcium oxide (CaO), more commonly known as quicklime, reacts vigorously with water with the evolution of enough heat to ignite paper, wood, or other combustible material under some conditions.

Radioactive Substances. Radioactive elements and compounds have fire and explosion hazards identical to those of the same material when not radioactive. An additional hazard is introduced by the various types of radiation emitted, all of which are capable of causing damage to living tissue. Under fire conditions, vapors and dusts (smoke) may be formed that could contaminate not only the building of origin but neighboring buildings and outdoor areas. The fire protection engineer's main concern is to prevent an out-of-control release of these materials under fire conditions or during extinguishment.[9] Radiation will be discussed in Chapter 13.

AIR CONTAMINANTS GENERALLY FOUND IN FIRES

Fire fatalities from the inhalation of fire gases and hot air are far more common than are fire deaths from all other causes combined. Studies of large-loss-of-life fire disasters reveal that in practically all instances the primary cause of death is inhalation of heated, toxic, and oxygen-deficient fire gases.[7] Studies of working fires showed that carbon monoxide was the most serious air contaminant.

Whether the gaseous products of combustion will have a toxic effect on an individual is determined by several variables, including the concentration of the gases, the time of exposure, and the physical condition of the individual. Because the rate of respiration is increased by exertion, heat, and an excess of carbon dioxide, toxic effects on persons inhaling fire gases may be accelerated. Under such conditions, gas concentrations which are ordinarily considered harmless may become dangerous.

Extensive research in the cause of human fire fatalities with respect to exposure to toxic substances has shown carbon monoxide to be the primary toxicant. In approximately 50% of the fire fatality cases studied, blood carboxyhemoglobin levels resulting from inhalation of carbon monoxide were found sufficient to be lethal. In an additional 30% of the victims, combinations of carbon monoxide with preexisting heart disease and/or alcohol intoxication were considered to be the cause of death. Data showed that of the fire fatalities in which alcohol was a factor, 8% of the victims had enough alcohol in the blood to be classified as legally intoxicated.[7]

Persons with some form of preexisting functional impairment, such as heart disease, emphysema, or lung cancer, are particularly susceptible to death from smoke inhalation in a fire. These include the very young, the elderly, the physically disabled, those under the influence of alcohol, drugs, or medication, and those with heart disease. See Table 11.2 for the toxic effects of fire gases. The air contaminants encountered in working

TABLE 11.2
Toxic Effects of Fire Gases

Toxicant	Sources	Toxicological Effects	Dangerous Concentrations (ppm)		
			Short-Term (10-minute) Lethal Concentration	Limit Immediate Danger to Life and Health	Threshold Limited Value
Hydrogen cyanide (HCN)	From combustion of wool, silk, polyacrylonitrite, nylon, polyurethane, and paper	A rapidly fatal asphyxiant poison	350	50	5
Nitrogen dioxide (NO$_2$) and other oxides of nitrogen	Produced in small quantities from cellulose nitrate and celluloid	Strong pulmonary irritant; capable of causing death as well as delayed injury	200	50	5
Ammonia (NH$_3$)	Produced in combustion of wool, silk, nylon, and melamine; concentrations generally low in ordinary building fires	Pungent, unbearable odor; irritant to eyes and nose; capable of causing immediate death as well as delayed injury	1,000	200	50
Hydrogen chloride (HCL)	From combustion of polyvinyl chloride (PVC) and some fire-retardant-treated materials	Respiratory irritant; potential toxicity of HCL coated on particulate may be greater than that for an equivalent amount of gaseous HCL	500[a]	100	5[b]

Combustion product	Source	Physiological effect			
Other halogen acid gases hydrogen bromide (HF) and hydrogen floride (HBr)	From combustion of fluorinated resins or films and of some fire-retardant materials containing bromine	Respiratory irritants	HF 400 HBr 500	50 100	3 3
Sulfur dioxide (SO_2)	From materials containing sulfur	A strong irritant; intolerable well below lethal concentrations	500	100	5
Isocyanates	From urethane polymers; pyrolysis products, such as toluene 2,4-disocyanate (TDI), reported in small-scale laboratory studies; undefined significance in actual fires	Potent respiratory irritants; believed the major irritants in smoke of isocyanate-based urethanes	100 (TDI)	10	0.02
Carbon monoxide	Incomplete oxidation	Hypoxia	400	1,500	5,000
Carbon dioxide	Complete oxidation	Difficult breathing	5,000	50,000	100,000
Acrolein	From pyrolysis of polyolefins and cellulosics at lower temperatures (400°C)	Potent respiratory irritant	30 to 100	5	0.3
Benzene			20,000	2,000	1

SOURCE: Adapted from the *Fire Protection Handbook* (15th ed.). Copyright © 1981 National Fire Protection Association, Quincy, Massachusetts. Reprinted by permission.

[a] If particulate material is absent. If particulate material is present, the short-term lethal concentration is greater than 500 ppm.
[b] Ceiling value.

fires include carbon monoxide (CO), carbon dioxide (CO_2), nitrogen dioxide (NO_2), hydrogen chloride (HCl), hydrogen cyanide (HCn), benzene (C_6H_6), and acrolein (CH_2CHCHO).[4,20]

Carbon Monoxide

Carbon monoxide (CO) is a colorless, odorless, tasteless gas. Carbon monoxide is the major threat in most fire atmospheres. Under controlled burning conditions, the carbon of most organic materials can be oxidized completely to carbon dioxide by supplying an excess of oxygen. In the uncontrolled burning of an accidental fire, the availability of oxygen is never ideal and some of the carbon is incompletely oxidized to carbon monoxide. In a confined smoldering fire, the ratio of carbon monoxide to carbon dioxide is usually greater than in a well-ventilated, freely burning fire.

Carbon monoxide combines with hemoglobin, the oxygen-carrying component of the blood and inhibits gas exchange functions. This causes hypoxia. Carbon monoxide is dangerous at relatively low concentrations.[3] The threshold limit value is only 50 ppm.

Carbon Dioxide

Carbon dioxide (CO_2) is a colorless, odorless gas which results from the complete oxidation of a carbon. Carbon dioxide is usually evolved in large quantities from fires. While not particularly toxic at observed levels, moderate concentrations of carbon dioxide do stimulate the rate of breathing. This condition contributes to the overall hazard of a fire gas environment by causing accelerated inhalation of toxicants and irritants. The rate and depth of breathing are increased 50% by 2% carbon dioxide and doubled by 3% carbon dioxide in air. At 5%, breathing becomes labored and difficult for some individuals, although this concentration of carbon dioxide has been inhaled for up to one hour without serious aftereffects. The threshold limit value of carbon dioxide is 5000 ppm.

Nitrogen Dioxide

Nitrogen dioxide (NO_2) is a red gas which is a respiratory irritant. It can cause pulmonary edema. Brief exposures to it can cause shortness of breath. Nitrogen dioxide may be caused by the oxidation of nitrogen-containing materials on the fixation of atmospheric nitrogen. Most fires show only low concentrations of nitrogen dioxide.

Insufficient Oxygen

Oxygen is consumed from the atmosphere during combustion. When oxygen drops from its usual level of 21% in air to about 17%, a person's motor coordination is impaired; when it drops into the range of 14 to 10%, a person is still conscious but may exercise faulty judgment and will be quickly fatigued; in the range of 10 to 6% a person loses consciousness and must be revived with fresh air or oxygen within a few minutes to prevent death. During periods of exertion, increased oxygen demands may result in oxygen deficiency symptoms at higher pecentages.

IGNITION SOURCES

To prevent fire, it is important to know how fires start. An analysis of many fires show the following are the major causes of industrial fires.

Electrical Causes of Fire

If properly designed, installed, and maintained, electric systems are both convenient and safe; otherwise, they may be a source for both fire and personal injury. Electricity may cause a fire if it arcs or overheats electric equipment; it may cause injury or death through shocks and burns.

When an electric circuit carrying a current is interrupted, either intentionally, as by a switch, or unintentionally, as when a contact at a terminal becomes loosened, arcing or heating from the high-resistance connection is produced. The intensity of the arc and degree of heat depend largely on the current and voltage of the circuit. The temperature may easily be high enough to ignite any combustible material in the vicinity. An electric arc may not only ignite combustible material in its vicinity, such as the insulation and covering of the conductor, but may also fuse the metal of the conductor. Hot sparks from burning combustible material and hot metal may be thrown about or fall and set fire to other combustible material.

When an electric conductor carries a current, heat is produced in direct proportion to the resistance (ohms) of the conductor and to the square of the current (amperage). The resistance of conductors used to convey current to the location in which it is used, or to convey it through the windings of a piece of apparatus (except resistance devices and heaters), should be as low as practical. Metals such as copper, silver, and aluminum are used for this purpose.

As a fire hazard, the heating of electric conductors is negligible under ordinary circumstances. When a conductor is overloaded, the generation of heat becomes a hazard in two ways: (1) through deterioration of the insulation of the conductor and (2) through the excess heat generated. Apparatus or appliances which use electric conductors as heating elements or which use the electric arc to generate heat are likely to be fire hazards unless properly installed and used.

One common method of reducing the degree of hazards is to provide enough air circulation to prevent unsafe temperatures and premature breakdown.

Worn-out-Equipment Fires.   This phrase indicates equipment actually worn out in service or wires which, after serving a full and useful purpose for a period of years, have deteriorated from service, thereby resulting in a fire. "Tired" equipment is responsible for the largest percentage of the electrical fires of known cause. The leading item within this category is the electric motor. Also included in this category are blazes caused by worn-out electronic appliances, lamp and other appliance cords, fixtures, and heated appliances. Aging of electric equipment results in the deterioration of insulation and, in some cases, corrosion or fatigue of the wires themselves.

Improper Use of Approved Equipment.   Although the equipment itself complies with safety standards of recognized testing laboratories, fires have occurred because such equipment is used under conditions not covered by its listing—for example, use of lamp cord extension for an electric heater.

Three of the principle types of misused equipment which cause fires are heating appliances, electric motors, and extension cords.

Accidental Occurrence.   Some electrical fires result from accidental misuse or oversight by the equipment operator. These fires are caused by a variety of actions, including clothes left in contact with lamps, materials accidentally dropped into electric equipment, and heating appliances left on unintentionally.

Defective Installations.   Many electrical fires result from defective installation. Defective installations are those done in a manner contrary to the National Electrical Code—for example, an automatically started fractional horsepower motor which does not have overload running protection as recommended by the code.

Smoking

Smoking is a documented ignition source of many fires. Wherever smoking is permitted, fires may result. Where smoking is allowed, fires are not really preventable. Therefore the safety engineer must attempt to reduce the risk. First emphatic warning signs must be conspicuous in areas where smoking is not allowed. In close proximity to the sign there must be a container to discard smoking material. In addition, information as to where a person can smoke should be provided. Second, in places where smoking is allowed, conspicuous placement of ashtrays and other containers for ash disposal should be made. The risk factor increases when smoking is mixed with alcohol and when smoking is done in bed.

Friction

The mechanical energy used in overcoming the resistance to motion when two solids are rubbed together is known as friction. Friction generates heat. The danger depends upon the available mechanical energy, the rate at which the heat is generated, and its rate of dissipation. Some examples of frictional heat are the heat caused by the friction of a slipping belt against a pulley and the hot metal particles (sparks) thrown off when a piece of foreign metal enters a grinding mill.

Friction Sparks. Friction sparks include the sparks which result from the impact of two hard surfaces, at least one of which is metal. Some examples of friction sparks which caused fires include sparks from a steel ball dropping on a concrete floor, falling tools striking machinery, and shoe nails scraping on concrete floors.

Open Flames

Open flames include the flame from a gas stove, the cutting flame from a torch, and the flames from matches and candles.

Static Electricity

Static electricity is the electrification of materials through physical contact and separation and the effects of the positive and negative charges so formed, particularly where sparks may result which constitute a fire or explosion hazard.

The development of electric charges may not be in itself a potential fire or explosion hazard. There must be a discharge or sudden recombination of separated positive and negative charges. In order for static electricity to be a source of ignition, four conditions must be fulfilled:[1]

1. There must be an effective means of static generation.
2. There must be a means of accumulating the separate charges and maintaining a suitable difference of electrical potential.
3. There must be a spark discharge of adequate energy.
4. The spark must occur in an ignitable mixture.

Static electricity may appear as the result of motions that involve changes in relative positions of contacting surfaces, generally of dissimilar substances (either liquid or solid), one or both of which usually must be a poor conductor of electricity. Examples of such motion that are commonly found in industry are:

1. Flow of fluid through pipes and the subsequent accumulation of a charge on the surface of a nonconducting liquid
2. Breaking up into drops of a stream of liquid and the subsequent impact of such drops onto a solid or liquid surface
3. Steam, air, or gas flowing from any opening in a pipe or hose when the steam is wet or the air or gas stream contains particulate matter
4. Pulverized materials passing through chutes or pneumatic conveyors
5. Nonconductive power or conveyor belts in motion
6. Moving vehicles

Electricity that is present on the surface of a nonconductive body where it is trapped or prevented from escaping is termed static electricity. Electricity on a conducting body which is in contact only with nonconductors is also prevented from escaping and is therefore nonmobile or "static." In either case, the body on which this electricity is evident is said to be "charged."

The generation of static electricity cannot be absolutely prevented because its intrinsic origins are present at every interface. The object of most static-corrective measures is to provide a means whereby charges separated by whatever cause may recombine harmlessly before sparking potentials are attained. If hazardous static conditions cannot be avoided in certain operations, means must be taken to assure that there are no ignitable mixtures at points where sparks may occur.[4] Other important ignition sources include the following: overheated materials, hot surfaces, combustion sparks, spontaneous combustion, cutting and welding, arson, exposure, mechanical sparks, molten substances, and chemical action.[11]

SUGGESTED LEARNING EXPERIENCES

1. Discuss the extent of the problem of fires. What implications does this have for the safety engineer?
2. Discuss the costs of fires from worker and managerial viewpoints. What were the underlying themes? Did you find there were distinctive sides, or was there a more common theme? Explain.
3. Illustrate the principles of combustion.
4. Discuss spontaneous combustion in terms of a workplace of your choice. Did you arrive at any conclusions that you had not had before you did this exercise? Explain.
5. Discuss the classification of fires and why it is important for the safety engineer to know them.

6. Take five substances and discuss their flammability properties.

7. Discuss the fire-fighting occupation in terms of health and safety. Given that the fire fighters' philosophy is to save lives, how does the safety engineer effectively consult and collaborate that there are times when this is not possible? Discuss.

8. Take five substances and discuss their stability.

9. Discuss the air contaminants generally found in fires. What environmental implications abound in this information?

10. Discuss the concept of static electricity, including how this applies to your workplace.

REFERENCES

1. American Industrial Hygiene Association. *Hygenic Guide Series.* Akron, Ohio, 1978.

2. de Ris, J. Chemistry and physics of fire. In G. P. McKinnon, ed., *Fire Protection Handbook* (15th ed.). Quincy, Mass.: National Fire Protection Association, 1981, pp. 3-2–3-14.

3. Dunlap, W. Death by cigarette fire stopping. It stirs debate. *New York Times*, April 19, 1982, pp. B1–B2.

4. Gold, A., Burgess, W. A., and Clougherty, D. Exposure of firefighters to toxic air contaminants. *American Industrial Hygiene Journal*, 1978, **39**, 534–539.

5. Grumman Aerospace Corporation. *A firefighters integrated life protection system. Phase 1. Design and performance.* Requirements report, May 1973.

6. Haessler, W. Theory of fire and explosion control. In *Fire Protection Handbook* (15th ed.). Quincy, Mass.: National Fire Protection Association 1981, pp. 3-24–3-29.

7. Hartzell, G. W., and Switzer, W. G. Combustion products and their effect on life safety. In *Fire Protection Handbook* (15th ed.). Quincy, Mass.: National Fire Projection Association, 1981.

8. Henahan, J. F. Fire. *Science* 1980, **80**, 29–38. (Jan.–Feb.)

9. Henry, M. F. Identification of the hazards of materials. In *Fire Protection Handbook* (15th ed.). Quincy, Mass.: National Fire Protection Association, 1981, pp. 4-108–4-117.

10. Henry, M. F. Flammable and combustible liquids. In *Fire Protection Handbook* (15th ed.). Quincy, Mass.: National Fire Protection Association, 1981, pp. 4-26–4-34.

11. Henry, M. F. Control of electrostatic ignition sources. In *Fire Protection Handbook* (15th ed.). Quincy, Mass.: National Fire Protection Association, 1981.

12. Hom, S., and Karter, M. J. Fire casualties. In *Fire Protection Handbook* (15th ed.). Quincy, Mass.: National Fire Protection Association, 1981, pp. 2-12–2-27.

13. National Fire Protection Association. *Fire Protection Guide on Hazardous Materials,* 7th ed. Quincy, Mass.: National Fire Protection Association, 1978.

14. Occupational Safety and Health Administration. *Code of Federal Regulations.* General Industry Standards (29 CFR 1910). U.S. Department of Labor, Washington, D.C., June 1981.

15. Occupational Safety and Health Administration. *Code of Federal Regulations.* Construction Industry Standards (29 CFR 1926). U.S. Department of Labor, Washington, D.C., October 1979.

16. Ross, J. W. Electrical systems and appliances. In *Fire Protection Handbook* (15th ed.). Quincy, Mass.: National Fire Protection Association, 1981, pp. 7-6–7-34.

17. Sax, N. I. Industrial fire protection. In N. I. Sax (Ed.), *Dangerous Properties of Industrial Materials* (5th ed.). New York: Van Nostrand-Reinhold, 1979, pp. 234–258.

18. Schwab, R. F. Dusts. In *Fire Protection Handbook* (15th ed.). Quincy, Mass.: National Fire Protection Association, 1981, pp. 4-84–4-97.

19. Terrill, J. B., Montgomery, R. R., and Reinhardt, C. F. Toxic gases from fires. *Science*, 1978, **200**, 1343-1347.

20. Treitman, R. D., Burgess, W. A., and Gold, A. Air contaminants encountered by fire fighters. *American Industrial Hygiene Journal*, 1980, **41**, 796–802.

12 FIRE PROTECTION AND CONTROL

Fire protection is usually considered to include fire prevention as well. However, these terms have different meanings. A fire protection program has the purpose of controlling the effects of fire. It is a method of loss control. Fire prevention's purpose is to reduce the incidence of fire.

FIRE PREVENTION

Handling and Storing Flammable and Combustible Liquids

A flammable liquid is any liquid having a flashpoint below 100°F (37.8°C) (except a mixture with components that have flashpoints of 100°F (37.8°C) or higher and make up 99% or more of the total volume of the mixture). Flammable liquids are considered to be class I liquids. Table 12.1 shows the characteristics of flammable and combustible liquids.

A combustible liquid is any liquid having a flashpoint at or above 100°F (37.8°C). Combustible liquids are considered to be either class II or class III liquids. The flashpoint is the minimum temperature at which a liquid gives off a vapor within a test container in sufficient concentration to form an ignitable mixture with air near the surface of the container. (See Table 12.1.)

Whenever flammable and combustible liquids are stored or handled, the liquid is usually exposed to the air at some stage in the operation, except where the storage is confined to sealed containers. Ventilation make-up air, which was discussed in Chapter 5, must be uncontaminated by flammable vapors. There is always the probability of leaks which permit the liquid to escape. Therefore, ventilation is of primary importance to prevent

TABLE 12.1
Classification of Flammable and Combustible Liquids

Type	Flash Point	Boiling Point	Examples
IA	Less than 73°F (22.8°C)	Less than 100°F (37.8°C)	Gasoline, Dimethyl-sulfide [$(CH_3)_2S$]
IB	Less than 73°F (22.8°C)	More than 100°F (37.8°C)	Acetone (CH_3COCH_2) Ethyl Alcohol (C_2H_5OH)
IC	At or above 73°F (22.8°C) but below 100°F (37.8°C)		Nonylene (C_9H_{18}) a-Picone ($C_{10}H_{16}$)
II	At or above 100°F (37.8°C) but below 140°F (60.0°C)		Kerosene, #2 fuel oil, pine tar
IIIA	At or above 140°F (60.0°C) but below 200°F (93.3°C)		Pine oil, poly-vinyl alcohol
IIIB	At or above 200°F (93.3°C)		Cottonseed oil Coconut oil

SOURCE: Occupational Safety and Health Administration, *Code of Federal Regulations*, General Industry Standards (29 CFR 1910). U.S. Department of Labor, Washington, D.C., June 1981.

the accumulation of flammable vapors. It is also good practice to eliminate sources of ignition in the vicinity where flammable liquids with low flash points are stored, handled, or used, even though no vapor may ordinarily be present.[9] Maintenance and/or operating procedures should encourage the control of leakage to prevent the accidental escape of flammable or combustible liquids. Spills should be cleaned up immediately.[20]

Elimination of Ignition Sources

Whenever processes involve flammable liquids in equipment such as arc welding equipment, generators, and hydraulic fluid systems, the liquids should be located in the open. This will lessen the fire potential created by the escape and accumulation of flammable vapors. Gasoline and almost all other flammable liquids produce heavier-than-air vapors which tend to settle on the floor or in pits or depressions. Gasoline vapors may flow along the floor or ground for long distances, be ignited at some remote point, and flash back. The removal of vapors at ground level (including pits) is usually the proper method of ventilation. Convection currents of heated air or normal vapor diffusion may carry even heavy vapors upward, and in such instances, ceiling ventilation may also be desirable. Ventilation to eliminate flammable vapors may be either natural or artificial. Although natural ventilation, where it can be used, has the advantage of not being dependent on electric starting power supplies, it depends upon temperature and wind conditions and is not as controllable as is mechanical ventilation.[8]

Grounding and Bonding

A grounded system is a system of conductors in which at least one conductor (usually the middle wire) is connected to an electrode embedded in the earth either directly or through a current limiting device. Bonding is the electrical connection from one conductive element to another in order to minimize the potential differences to provide suitable conductivity for fault currents. Grounding and bonding must be done to prevent fire and/or explosions. The metal enclosures of conductors (metal armor of cables; metal raceways, boxes, cabinets, and fittings) must be grounded. OSHA requires grounding of exposed noncurrent-carrying metal parts of cord and of plug-connected equipment likely to become energized.[20] OSHA requires grounding of most equipment, especially if the system exceeds 50 volts. Some of these situations include the following:

1.   All three wire DC systems shall have their neutral conduction grounded.
2.   AC circuits of less than 50 volts if installed as overhead conductors outside of building.
3.   AC systems of 50 volts to 1000 volts.

The National Electric Code (NFPA 70-1981) also specifies grounding by use of a ground wire for plug-connected appliances in residential occupancies. This includes refrigerators, freezers, air conditioners, clothes washers and clothes dryers, dishwashers, sump pumps, electric aquarium equipment, and portable and hand-held motor-operated tools and appliances such as drills, hedge clippers, lawn mowers, snowblowers, wet scrubbers, sanders, and saws. An exception is that double insulation may be recognized as providing equivalent protection.[8] Double insulated appliances are specially insulated so that the motor is fully insulated and the case is made of plastic and manufactured in accordance with Underwriter's Laboratory specifications.

FIRE DETECTION

It is always better to prevent a fire from occurring than to detect it once it has occurred; therefore, minimizing fire hazards around the office, home, factory, and school is the first key element. This means, for example, removing all unessential flammable liquids, such as kerosene and naphthalene, and other highly combustible items such as trash; properly storing essential combustibles away from ignition sources; and not overloading electric circuits.[13]

Smoke Detectors

Only about 15% of American homes have smoke detectors, but most fire experts hold that they are the best fire protection one can buy. The authors strongly recommend that every reader of this book have adequate smoke detectors in their living quarters. See the August 1980 *Consumer Reports* for help in selecting one.

Smoke detectors used in combination with escape lanes provide effective protection against death resulting from fires with a wide variety of ignitions.[4] If a fire develops rapidly enough, smoke detectors may not provide enough time for the occupants to escape. Smoke detectors will not always protect you. However, there are rarely any fire deaths in residences protected by properly installed, operating smoke detectors.

A smoke detector will detect most fires much more rapidly than a heat detector. Smoke detectors are identified by their method of operation, which is either by ionization or photoelectricity. Each type has its strengths and weaknesses. As a class, smoke detectors operating on the photoelectric principle respond faster to the smoke generated by low energy (smoldering) fires, as these fires generally produce more of the larger smoke particles. As a class, smoke detectors using the ionization principle provide somewhat faster response to high-energy (open-flaming) fires, since these fires produce large numbers of the smaller smoke particles.[3]

What is the risk of cancer from having a smoke detector in your home? The answer is, very small, for the protection obtained. The risk of contracting cancer as a result of having a smoke detector is one single new case of cancer in the whole United States every seven years. What is the risk of dying in a fire next year?

Ionization.   Smoke detectors utilizing the ionization principle are usually of the spot type. An ionization smoke detector has a small amount of radioactive material (tritium or americium) which ionizes the air in the sensing chamber, thus rendering it conductive and permitting a current flow through the air between two charged electrodes. This gives the sensing chamber an effective electric conductance. When smoke particles enter the ionization area, they decrease the conductance of the air by attaching themselves to the ions, causing a reduction in mobility. When the conductance is less than a predetermined level, the detector responds.

Photoelectric.   For the photoelectric smoke detector, the presence of suspended smoke particles generated during the combustion process affects the propagation of a light beam passing through the air. The effect can be utilized to detect the presence of a fire in two ways: (1) obscuration of light intensity over the beam path or (2) scattering of the light beam.

Light intensity.   Smoke detectors that operate on the principle of light obscuration consist of a light source, a light-beam focusing system, and a photosensitive device. When smoke particles enter the light beam, the light reaching the photosensitive device is reduced, initiating the alarm. The light source is usually a light-emitting diode (LED). The LED is a reliable long-life source of illumination having a low current requirement. When smoke particles enter the light beam, the light reaching the photosensitive device is reduced, initiating the alarm. In practice, most light-obscuring smoke detectors are the beam type and are used to protect large, open areas. They are installed with the light source at one end of the area to be protected and the receiver (the photocell and relay assembly) at the other end.

Scattered-light detectors.   In scattered-light detectors, when smoke particles enter a light path, scattering results. They contain a light source and a photosensitive device so arranged that the light rays do not normally fall onto the photosensitive device. When smoke particles enter the light path, light strikes the particles and is scattered onto the photosensitive device, causing the detector to respond.

Heat Detectors

A heat detector is best-suited for fire detection in either a small, confined space where rapidly changing conditions would not allow the use of other fire-detection devices or where speed of detection is not the prime consideration. Heat detectors respond to the convected energy of a fire and are generally located on or near the ceiling. They respond either when the detecting element reaches a predetermined fixed temperature or to a specified rate of temperature change.[3]

Fixed-Temperature Detectors.   Fixed-temperature detectors are designed to alarm when the temperature of the operating element reaches a specified point.

Rate-of-Rise Detectors.   One effect that a flaming fire has on the surrounding area is to rapidly increase air temperature in the space above the fire. Fixed-temperature heat detectors will not initiate an alarm until the air temperature near the ceiling exceeds the design operating point. The rate-of-rise detector will function when the rate of temperature increase exceeds a predetermined value, typically around 12 to 15 degrees per minute. Rate-of-rise detectors are designed to compensate for the normal changes in ambient temperature which are expected under nonfire conditions.

Pneumatic Detectors. In a pneumatic fire detector, air heated in a tube or chamber will expand, increasing the pressure. This exerts a mechanical force on a diaphragm that will close the alarm contacts.

Flame Detectors

A flame detector responds to the appearance of radiant energy visible to the human eye (approximately 4,000 to 7,700 angstroms) and to radiant energy outside the range of human vision. These detectors are sensitive to glowing embers, coals, or actual flames which radiate energy of sufficient intensity and spectral quality to initiate response of the detector. Due to their fast detection capabilities, flame detectors are generally used only in high-hazard areas such as fuel loading platforms, industrial process areas, hyperbaric chambers, high ceiling areas, and atmospheres in which explosions or very rapid fires may occur. Because flame detectors must be able to "see" the fire, they can be blocked by objects placed in front of them, although the infrared type of flame detector has some capability for detecting radiation reflected from walls.

Infrared Flame Detector. These are made of a filter and lens system used to screen out unwanted wavelengths and focus the incoming energy on a photovoltaic or photoresistive cell sensitive to infrared energy. A major problem is the possible interference from solar radiation in the infrared region. When detectors are located in places shielded from the sun, such as in vaults, filtering or shielding the sun's rays is unnecessary.

Ultraviolet Flame Detector. These generally use a solid-state device with silicon carbide or aluminum nitride as the sensitizing element. They are insensitive to both sunlight and artificial light.

FIRE SUPPRESSION

Fixed Automatic Sprinklers

Automatic sprinklers are devices for automatically distributing water upon a fire in sufficient quantity either to extinguish it entirely or to prevent its spread in the event that the initial fire is out of range or is of a type that cannot be completely extinguished by water discharged by sprinklers.

A properly designed, installed, and maintained sprinkler system is by far the most efficient fire protection system available. Industry would be severely hampered without automatic sprinklers. In the past, 50 automatic sprinklers have extinguished or held in check 96% of the fires in sprinkler protected areas.

Sprinkler System.   The water is fed to the sprinkler through a system of piping, ordinarily suspended from the ceiling, with the sprinklers placed at intervals along the pipes. The orifice of the fusible-link automatic sprinkler is normally closed by a disk or cap held in place by a temperature-sensitive releasing element.

Automatic sprinklers are particularly effective for life safety because they give warning of the existence of fire and at the same time apply water to the burning area. With sprinklers there are seldom problems of access to the seat of the fire or of interference due to smoke, with visibility for fire fighting.

Sprinkler Design.   A sprinkler design usually signifies a combination of water discharge devices (sprinklers), one or more sources of water under pressure, water-flow controlling devices (valves), distribution piping to supply the water to the discharge devices, and auxiliary equipment such as alarms and supervisory devices. Outdoor hydrants, indoor hose stand-pipes, and hand hose connections are also frequently a part of the system that provides protection.

Water supply.   The water supply must be enough for fire protection or supplemental sources must be provided.

The National Fire Protection Association has developed a guide to water supply requirements in sprinkler systems (see Table 12.2). Light hazards include apartment buildings, dormitories, office buildings, seating areas of restaurants, and hospitals. Group 1 ordinary hazards include products of low combustibility, such as those in garages, bakeries, laundries, and canneries. Group 2 ordinary hazards include clothing factories, pharmaceutical manufacturing plants, and shoe factories. Group 3 ordinary hazards include certain woodworking and flour and feed mills, paper mills, piers and wharves, and tire storage facilities.

Extra hazard occupancies consist of properties where flash fires, causing all the sprinklers in a fire area to open, are probable. Group 1 extra hazards include severe fire substances where few or no flammable or combustible liquids are present. These are found in die casting, plywood and particleboard manufacturing, printing and rubber manufacturing operations, saw mills, textile operations, and plastic foam manufacturing.[7,9] Group 2 extra hazards may produce severe fires and contain moderate to substantial amounts of flammable or combustible liquids. These liquids are found in asphalt saturating, flammable-liquid spraying, flow coating, open oil quenching, solvent cleaning, and varnish and paint dipping.

Wet System.   This type of system is generally used wherever there is no danger of the water in the pipes freezing and wherever there are no special conditions requiring one of the other types of systems.

TABLE 12.2
Guide to Water Supply Requirements for Pipe Schedule Sprinkler Systems

Occupancy Classification	Residual Pressure Required[a]	Acceptable Flow at Base of Rise[b]	Duration in Minutes[c]
Light hazard	15 psi	500–750 gpm[d]	30–60
Ordinary hazard			
(Group 1)	15 psi or higher	700–1,000 gpm	60–90
(Group 2)	15 psi or higher	850–1,500 gpm	60–90
(Group 3)	Pressure and flow requirements for sprinklers and hose streams to be determined by authority having jurisdiction.		60–120
Warehouses	Pressure and flow requirements for sprinklers and hose streams to be determined by authority having jurisdiction.		
High-rise buildings	Pressure and flow requirements for sprinklers and hose streams to be determined by authority having jurisdiction.		
Extra hazard	Pressure and flow requirements for sprinklers and hose streams to be determined by authority having jurisdiction.		

SOURCE: Hodnett, R. M. Automatic sprinkler systems. Reprinted from the *Fire Protection Handbook*, 15th ed. Copyright © 1981 National Fire Protection Association, Quincy, Massachusetts. Reprinted by permission.

[a]The pressure required at the base of the sprinkler rise(s) is defined as the residual pressure required at the elevation of the highest sprinkler plus the pressure required to reach this elevation.

[b]The lower figure is the minimum flow including hose streams ordinarily acceptable for pipe schedule sprinkler systems. The higher flow should normally suffice for all cases under each group.

[c]The lower duration figure is ordinarily acceptable where remote station water-flow alarm service or equivalent is provided. The higher duration figure should normally suffice for all cases under each group.

[d]The requirement may be reduced to 250 gpm if building area is limited by size or compartmentation or if building (including roof) is of noncombustible construction.

Dry System. A dry system is one in which the water does not remain in the sprinkler pipe. It is used if an inadvertent leak would cause major damage or if there is danger of freezing.

Fixed Manual application. Standpipe and hose systems provide a means for manual application of water to fires in buildings.

Sprinkler Alarm. An approved device installed so that any water flow from a sprinkler system will result in an audible alarm signal on the premises.[17]

Standpipes. Standpipe systems are designed for fire department use to provide quick and convenient means for obtaining effective fire streams on the upper stories of high buildings or in large low buildings. A Class I standpipe is a 2½ inch (6.3 cm) hose connection for use by a fire department and/or those trained in handling heavy fire streams.

Portable Fire Extinguishers

Virtually all fires are small at first and can be easily extinguished if the proper type and amount of extinguishing agent is promptly applied. The successful use of these agents depends upon the following conditions:

1. The extinguisher must be properly located and in good working order.
2. The extinguisher must be the proper type for the fire which occurs.
3. The fire must be discovered while still small enough for the extinguisher to be effective.
4. The fire must be discovered by a person ready, willing, and able to use the extinguisher.[10]

Fire extinguishers are the first line of defense against fires and should be installed regardless of other fire control measures.

Extinguisher Location. No matter how carefully extinguishers are chosen to match the hazards of an area and the people who will use them, they will not be effective unless they are readily available. Sometimes extinguishers are kept at hand, as in welding operations, but more often one has to travel from the fire to the extinguisher and back before beginning to put the fire out. In such cases, travel distance to the nearest extinguisher is very important. Travel distance is the actual distance (around partitions, through doorways and aisles) that someone must walk to reach the extinguisher.

When placing extinguishers, select locations that will:

1. Provide uniform distribution
2. Provide easy access and be relatively free from temporary blockage
3. Be near normal paths of travel
4. Be near exits and entrances
5. Be free from the potential of physical damage
6. Be readily available[7]

Fire extinguishers must be placed so that the distance to the extinguisher is 75 feet or less for class A and class D materials, 50 feet or less for class B combustible materials.[20]

Extinguisher Inspection.   Once a fire extinguisher has been purchased, it becomes the responsibility of the purchaser or an assigned agent to maintain the device. Adequate maintenance consists of (1) periodically inspecting each extinguisher, (2) recharging each extinguisher following discharge, and (3) performing hydrostatic tests as needed.

An inspection is a quick check that visually determines whether the fire extinguisher is properly placed and will operate. Its purpose is to give reasonable assurance that the extinguisher is fully charged and will function effectively if needed. An inspection should determine that the extinguisher:

1. Is in its designated place
2. Is conspicuous
3. Is not blocked in any way
4. Has not been activated and partially or completely emptied
5. Has not been tampered with
6. Has not sustained any obvious physical damage or been subjected to an environment (such as corrosive fumes) which could interfere with its operation
7. If equipped with a pressure gauge and/or tamper indicators, has each showing conditions to be satisfactory

In addition, the maintenance tag should be checked to determine the date of the last thorough maintenance check.

In order to be effective, inspections must be frequent, regular, and thorough.

Hydrostatic Testing of Extinguishers.   The purpose of hydrostatic testing of portable fire extinguishers that are subject to internal pressures is to protect against unexpected, in-service failure due to:

1. Undetected internal corrosion caused by moisture in the extinguisher
2. External corrosion caused by atmospheric humidity or corrosive vapors
3. Damage caused by rough handling (which may or may not be obvious by external inspection)
4. Repeated pressurizations
5. Manufacturing flaws in the construction of the extinguisher
6. Improper assembly of valves or safety relief disks
7. Exposure of the extinguisher to abnormal heat, as after exposure in a fire[10]

See Table 12.3 for a list of hydrostatic test intervals for extinguishers.

TABLE 12-3
Hydrostatic Test Interval in Years

Extinguisher Type	Test Interval
Soda-acid (stainless steel case)[a]	5
Cartridge operated water and/or antifreeze	5
Stored pressure water and/or antifreeze	5
Wetting agent	5
Foam (stainless steel case)[a]	5
Aqueous film forming foam (AFFF)	5
Loaded stream	5
Dry chemical (stainless steel shell)[a]	5
Carbon dioxide	5
Dry chemical, stored pressure, with mild steel, brazed brass, or aluminum shells	12
Halon 1301 Bromotrifluoromethane	12
Dry powder, cartridge or cylinder operated, with mild steel shells)	12

SOURCE: Occupation, Safety, and Health Administration, *Code of Federal Regulations*. General Industry Standards (29 CFR 1910.157). U.S. Department of Labor, Washington, D.C., June 1981.

[a] Extinguisher having shells constructed of copper or brass joined by rivets or solder shall not be hydrostatically tested and must be removed from service.

Types of Fire Extinguishers

Water. This has been discussed in the section "Fixed Automatic Sprinklers."

Carbon Dioxide. Carbon dioxide will effectively suppress fire in most combustible materials except for a few active metals, such as calcium (Ca) or sodium (Na), and metal hydride materials, such as pyroxylin, a nitrocellulose compound $[C_{12}H_{17}(ONO_2)_3O_7]$, that contain available oxygen.

Halon. Halon is one of the halogenated compounds. These compounds produce dramatic results for class A fires, but their toxic fumes must be taken into consideration. They are used in most computer facilities because they do not disturb or damage printed circuits or electronics, as other methods of extinguishing do.

Dry Powder (Class D Fires). This denotes those chemicals, such as lithium (Li) and magnesium (Mg), that have been designated as useful in extinguishing metal fires. The extinguishing medium usually contains GI powder (pyrene)—graphitized foundry coke plus an organic phosphate; met-L-X-powder—sodium chloride (NaCl) base with additives; and Na-X-powder—sodium carbon base with various additives.

Foam (Class A and B Fires). This material is used to coat the burning surface. For liquid spill fires the foam can be spread over the burning surface by bouncing it off the floor just ahead of the flame front or by standing back and aiming the stream upward so that it will fall lightly on the burning area. On flammable liquid fires or appreciable depth, the discharge should be played against the inside of the back wall of the vat or tank, just above the burning surface, to permit the natural spread of the foam over the burning liquid.[20]

Municipal Water Supply

Hydrants should not be placed further than 800 feet apart. In closely built areas, they should not be more than 500 feet apart. Hydrants should be located as close to a street intersection as possible, with intermedite hydrants along the street to meet the area requirements. Where hydrants are located on a private water system and hose lines are intended to be used directly from the hydrants, they should be so located as to keep hose lines short, preferably not more than 250 feet. At a minimum, there should be a sufficient number of hydrants to make two streams available to every part of the interior of each building not covered by standpipe protection. They should also provide hose stream protection for exterior parts of each building using only the lengths of hose normally attached to the hydrants. It is desirable to have sufficient hydrants to concentrate the required fire flow about any important building with no hose line exceeding 500 feet in length.[1]

Fire Protection in Plants and Factories

Fire Walls.   These are primarily self-supporting and should be designed so as to maintain structural integrity even in cases of complete collapse of the structure on either side of the fire wall. To withstand heat expansion effects, they are commonly made thicker than would be required by normal fire resistance ratings. They must extend through and above combustible roofs to prevent the spread of fire in these roofs when they extend over the tops of the building walls.

Fire Doors.   Doors should swing with exit travel except in small rooms. Vertical or rolling doors are not recognized for use in exits. Fire doors must be kept closed to serve their function of stopping the spread of smoke; if open, they must be closed immediately in case of fire. Ordinary fusible-link-operated devices to close doors in case of fire are designed to close in time to stop the spread of fire but do not operate soon enough to stop the spread of smoke.

Means of Egress.   The access to an exit is that portion of a means of egress which leads to an entrance or to an exit. The access to an exit may be a corridor, an aisle, a balcony, a gallery, a porch, or a roof. Its length establishes travel distance to an exit—an extremely important feature of a means of egress, since an occupant might be exposed to fire during the time it takes to reach an exit. The average recommended maximum distance is 100 feet, but this varies with the occupancy, depending on the fire hazard and the physical ability and alertness of the occupants. The travel distance may be measured from the door of a room to an exit or from the most remote point in a room or floor area to an exit. In those occupancies where there are large numbers of people in an open floor area or where the nature of business conducted makes an open floor area desirable, the travel distance is measured from the most remote point in the area to the exits. The width of an access to an exit should be at least sufficient for the number of persons it must accommodate. Level egresses must be at least 44 inches wide, while inclined egresses (ramps) must be at least 30 inches wide.[20] A fundamental principle of exit access is provision of a free and unobstructed way to exit.[14] A checklist to evaluate means of egress is shown in Figure 12.1.

Fire Alarms.   Municipal fire alarm systems are electrically operated networks divisible into three basic categories: code, voice, and code voice. There should be a sufficient number of boxes located to protect all built-up

Section	Requirement	Compliance			Comments
		Yes	No	N/A	
1910.36	MEANS OF EGRESS				
1910.36a	Application.				
1910.36a	Nothing in paragraph 1910.36 shall be constructed to prohibit a better type of building construction, more exits, or otherwise safer conditions than the minimum requirements specified in this subpart.				
1910.36a	Exits from vehicles, vessels, or other mobile structures are not covered by this requirement.				
1910.36b	Fundamental requirements.				
1910.36b(1)	Every building or structure, new or old, designed for human occupancy shall be provided with exits sufficient to permit the prompt escape of occupants in case of fire or other emergency.				
1910.36b(1)	The design of exits and other safeguards shall be such that reliance for safety to life in case of fire or other emergency will not depend solely on any single safeguard.				
1910.36b(1)	Additional safeguards shall be provided for life safety in case any single safeguard is ineffective due to some human or mechanical failure.				
1910.36b(2)	Every building or structure shall be so constructed, arranged, equipped, maintained, and operated so as to avoid undue danger to the lives and safety of its occupants from fire, smoke, fumes, or resulting panic during the period of time reasonably necessary for escape from the building or structure in case of fire or other emergency.				
1910.36b(3)	Every building or structure shall be provided with exits of kinds, numbers, location, and capacity appropriate to the individual building or structure, with due regard to the character of the occupancy, the number of persons exposed and fire protection available, and the height and type of construction of the building				

FIGURE 12.1.   Safety Checklist: General Industry Occupational Safety and Health Standards (29 CFR 1910). Subpart E: Means of Egress.

Section	Requirement	Compliance			Comments
		Yes	No	N/A	
1910.36	MEANS OF EGRESS				
	or structure, to afford all occupants convenient facilities for escape.				
1910.36b(4)	In every building or structure, exits shall be so arranged and maintained as to provide free and unobstructed egress from all parts of the building or structure at all times when it is occupied.				
1910.36b(4)	No lock or fastening to prevent free escape from the inside of any building shall be installed except in mental, penal, or corrective institutions where supervisory personnel are continually on duty and effective provisions are made to remove occupants in case of fire or other emergency.				
1910.36b(5)	Every exit shall be clearly visible or the route to reach it shall be conspicuously indicated in such a manner that every occupant of every building or structure who is physically and mentally capable will readily know the direction of escape from any point, and each path of escape, in its entirety, shall be so arranged or marked that the way to a place of safety outside is unmistakable.				
1910.36b(5)	Any doorway or passageway not constituting an exit or way to reach an exit, but of such a character as to be subject to being mistaken for an exit, shall be so arranged or marked as to minimize its possible confusion with an exit and the resultant danger of persons endeavoring to escape from fire finding themselves trapped in a dead-end space, such as a cellar or storeroom, from which there is no other way out.				
1910.36b(6)	In every building or structure equipped for artificial illumination, adequate and reliable illumination shall be provided for all exit facilities.				
1910.36b(7)	In every building or structure of such size, arrangement, or occupancy that a fire may not itself pro-				

FIGURE 12.1. (*Continued*)

Section	Requirement	Compliance			Comments
		Yes	No	N/A	
1910.36	**MEANS OF EGRESS**				
	vide adequate warning to occupants, fire alarm facilities shall be provided where necessary to warn occupants of the existence of fire so that they may escape, or to facilitate the orderly conduct of fire exit drills.				
1910.36b(8)	Every building or structure, section, or area thereof of such size, occupancy, and arrangement that the reasonable safety of numbers of occupants may be endangered by the blocking of any single means of egress due to fire or smoke, shall have at least two means of egress remote from each other, so arranged to minimize any possibility that both may be blocked by any one fire or other emergency conditions.				
1910.36b(9)	Compliance with paragraph 1910.36b shall not be interpreted as eliminating or reducing the necessity for other provisions for safety of persons using a structure under normal occupancy conditions, nor shall any provision of the subpart be construed as requiring or permitting any condition that may be hazardous under normal occupancy conditions.				
1910.36c	Protection of employees exposed by construction and repair operations.				
1910.36c(1)	No building or structure under construction shall be occupied in whole or in part until all exit facilities required for the part occupied are completed and ready for use.				
1910.36c(2)	No existing building shall be occupied during repairs or alterations unless all existing exits and any existing fire protection are continuously maintained, or in lieu thereof other measures are taken which provide equivalent safety.				
1910.36c(3)	No flammable or explosive substances or equipment for repairs or alterations shall be introduced in a building of normally low or ordinary hazard classification while the building is occupied unless the condition				

FIGURE 12.1. (*Continued*)

Section	Requirement	Compliance			Comments
		Yes	No	N/A	
1910.36	MEANS OF EGRESS				
	of use and safeguards provided are such as not to create any additional danger or handicap to egress beyond the normally permissible conditions in the building.				
1910.36d	Maintenance.				
1910.36d(1)	Every required exit, way of approach thereto, and way of travel from the exit into the street or open space shall be continuously maintained free of all obstructions or impediments to full instant use in the case of fire or other emergency.				
1910.36d(2)	Every automatic sprinkler system, fire detection and alarm system, exit lighting, fire door, and other items of equipment, where provided, shall be continuously in proper operating condition.				
1910.37	MEANS OF EGRESS, GENERAL				
1910.37a	Permissible exit components.				
1910.37a	An exit shall consist only of the approved components.				
1910.37a	Exit components shall be constructed as an integral part of the building or shall be permanently affixed thereto.				
1910.37b	Protective enclosure or exits.				
1910.37b(1)	When an exit is protected by separation from other parts of the building, the separation shall have at least a 1-hour fire resistance rating when the exit connects three stories or less. This applies whether the stories connected are above or below the story at which exit discharge begins.				
1910.37b(2)	When an exit is protected by separation from other parts of the building, the separation shall have at least a 2-hour fire resistance rating when the exit connects four or more stories, whether above or below the floor of discharge.				
1910.37b(2)	When an exit is protected by separation from other parts of the building, the separation shall be constructed				

FIGURE 12.1.   (*Continued*)

Section	Requirement	Compliance			Comments
		Yes	No	N/A	
1910.37	MEANS OF EGRESS, GENERAL				
	of noncombustible materials, and shall be supported by construction having at least a 2-hour fire resistance rating.				
1910.37b(3)	When an exit is protected by separation from other parts of the building, any opening therein shall be protected by an approved self-closing fire door.				
1910.37b(4)	When an exit is protected by separation from other parts of the building, openings in exit enclosures shall be confined to those necessary for access to the enclosure from normally occupied spaces and for egress from the enclosure.				
1910.37c	Width and capacity of means of egress.				
1910.37c(1)	The capacity in number of persons per unit of exit width for approved components of means of egress shall be as follows: (i) Level of egress components (including Class A ramps): 100 persons. (ii) Inclined egress components (including Class B ramps): 60 persons. (iii) A ramp shall be designated as Class A or Class B in accordance with Table 37-1, CFR 29, Para. 1910.37.				

TABLE 37.1
Width and Slope of Ramps

	Class A
Width	44 inches or greater
Slope	1 to $1\frac{3}{16}$ inches in 12 inches
Maximum height between landings	No limit
	Class B
Width	30 to 44 inches
Slope	$1\frac{3}{16}$ to 2 inches in 12 inches
Maximum height between landings	12 feet

FIGURE 12.1. (*Continued*)

Section	Requirement	Compliance			Comments
		Yes	No	N/A	
1910.37	MEANS OF EGRESS, GENERAL				
1910.37c(2)	Means of egress shall be measured in units of exit width of 22 inches. Fractions of a unit shall not be counted, except that 12 inches added to one or more full units shall be counted as one-half of exit width.				
1910.37c(3)	Units of exit width shall be measured in the clear at the narrowest point of the means of egress.				
1910.37c(3)	Handrail may project inside the measured width on each side not more than 5 inches and a stringer may project inside the measured width not more than $1\frac{1}{2}$ inches.				
1910.37c(3)	An exit or exit access door swinging into an aisle or passageway shall not restrict the effective width thereof at any point during its swing to less than the minimum width specified.				
1910.37d(1)	The capacity of means of egress for any floor, balcony, tier, or other occupied space shall be sufficient for the occupant load thereof. The occupant load shall be the maximum number of persons that may be in the space at any time.				
1910.37d(2)	Where exits serve more than one floor, only the occupant load of each floor considered individually need be used in computing the capacity of the exits at that floor, provided that exit capacity shall not be decreased in the direction of exit travel.				
1910.37e	Arrangement of exits.				
1910.37e	When more than one exit is required from a story, at least two of the exits shall be remote from each other and so arranged as to minimize any possibility that both may be blocked by any one fire or other emergency condition.				
1910.37f	Access to exits.				
1910.37f(1)	Exits shall be so located and exit access shall be so arranged that exits are readily accessible at all times.				
1910.37f(1)	Where exits are not immediately accessible from an open floor area, safe				

FIGURE 12.1. (*Continued*)

Section	Requirement	Compliance			Comments
		Yes	No	N/A	
1910.37	MEANS OF EGRESS, GENERAL				
	and continuous passageways, aisles, or corridors leading directly to every exit and so arranged as to provide convenient access for each occupant to at least two exits by separate ways of travel, except as a single exit or limited dead ends are permitted by other provisions of this subpart, shall be maintained.				
1910.37f(2)	A door from a room to an exit or to a way of exit access shall be of the side-hinged, swinging type. It shall swing with exit travel when the room is occupied by more than 50 persons or used for a high hazard occupancy.				
1910.37f(3)	In no case shall access to an exit be through a bathroom, or other room subject to locking, except where the exit is required to serve only the room subject to locking.				
1910.37f(4)	Ways of exit access and the doors to exits to which they lead shall be so designed and arranged as to be clearly recognizable as such.				
1910.37f(4)	Hangings or draperies shall not be placed over exit doors or otherwise so located as to conceal or obscure any exit.				
1910.37f(4)	Mirrors shall not be placed on exit doors.				
1910.37f(4)	Mirrors shall not be placed in or adjacent to any exit in such a manner as to confuse the direction of exit.				
1910.37f(5)	Exit access shall be so arranged that it will not be necessary to travel toward any area of high hazard occupancy in order to reach the nearest exit unless the path of travel is effectively shielded from the high hazard location by suitable partitions or other physical barriers.				
1910.37f(6)	The minimum width of any way of exit access shall in no case be less than 28 inches.				

FIGURE 12.1. (Continued)

Section	Requirement	Compliance			Comments
		Yes	No	N/A	
1910.37	MEANS OF EGRESS, GENERAL				
1910.37f(6)	Where a single way of exit access leads to an exit, its capacity in terms of width shall be at least equal to the required capacity of the exit to which it leads.				
1910.37g	Exterior ways of exit access.				
1910.37g(2)	Exterior ways of exit access shall have smooth, solid floors, substantially level, and shall have guards on the unenclosed sides.				
1910.37g(3)	Where accumulation of snow or ice is likely because of the climate, the exterior way of exit access shall be protected by a roof, unless it serves as the sole normal means of access to the rooms or spaces served, in which case it may be assumed that snow and ice will be regularly removed in the course of normal occupancy.				
1910.37g(4)	A permanent, reasonably straight path of travel shall be maintained over the required exterior way of exit access.				
1910.37g(4)	There shall be no obstruction of exterior ways of exit by railings, barriers, or gates that divide the open space into sections appurtenant to individual rooms, apartments or other uses.				
1910.37g(5)	An exterior-way-of-exit access shall be so arranged that there are no dead ends in excess of 20 feet.				
1910.37g(5)	Any unenclosed exit served by an exterior-way-of-exit access shall be so located that no part of the exit extends past a vertical plane 20 feet and one-half the required width of the exit from the end of and at right angles to the way of exit access.				
1910.37h	Discharge from exits.				
1910.37h(1)	All exits shall discharge directly to the street, or to a yard, court, or other open space that gives safe access to a public way.				
1910.37h(1)	The streets to which the exits discharge shall be of width adequate to				

FIGURE 12.1. (Continued)

Section	Requirement	Compliance			Comments
		Yes	No	N/A	
1910.37	MEANS OF EGRESS, GENERAL				
	accommodate all persons leaving the building.				
1910.37h(1)	Yards, courts, or other open spaces to which exits discharge shall also be of adequate width and size to provide all persons leaving the building with ready access to the street.				
1910.37h(2)	Stairs and other exits shall be so arranged as to make clear the direction of egress to the street.				
1910.37i	Headroom.				
1910.37i(1)	Means of egress shall be so designed and maintained as to provide adequate headroom, but in no case shall the ceiling height be less than 7 feet 6 inches nor any projection from the ceiling height be less than 6 feet 8 inches from the floor.				
1910.37j	Changes in elevation.				
1910.37j(1)	Where a means of egress is not substantially level, such differences in elevation shall be negotiated by stairs or ramps.				
1910.37k	Maintenance and workmanship.				
1910.37k(1)	Doors, stairs, ramps, passages, signs, and all other components or means of egress shall be of substantial, reliable construction and shall be built or installed in a workmanlike manner.				
1910.37k(2)	Means of egress shall be continuously maintained free of all obstructions or impediments to full instant use in the case of fire or other emergency.				
1910.37k(3)	Any device or alarm installed to restrict the improper use of an exit shall be so designed and installed that it cannot, even in cases of failure, impede or prevent emergency use of such exit.				
1910.37l	Furnishing and decorations.				
1910.37l(1)	No furnishings, decorations, or other objects shall be so placed as to obstruct exits, access thereto, egress therefrom, or visibility thereof.				

FIGURE 12.1. (*Continued*)

Section	Requirement	Compliance			Comments
		Yes	No	N/A	
1910.37	MEANS OF EGRESS, GENERAL				
1910.37l(2)	No furnishings or decorations which are explosive or highly flammable shall be used.				
1910.37m	Automatic sprinkler systems.				
1910.37m	All automatic sprinkler systems shall be continuously maintained in reliable operating condition at all times, and such periodic inspections and tests shall be made as are necessary to assure proper maintenance.				
1910.37n	Fire alarm signaling systems.				
1910.37o	Fire retardant paints.				
1910.37o	Fire retardant paints or solutions shall be renewed at such intervals as necessary to maintain the necessary flame retardant properties.				
1910.37q	Exit marking.				
1910.37q(1)	Exits shall be marked by a readily visible sign.				
1910.37q(1)	Access to exits shall be marked by readily visible signs in all cases where the exit or way to reach it is not immediately visible to the occupants.				
1910.37q(2)	Any door, passage, or stairway which is neither an exit nor a way of exit access, and which is so located or arranged as to be likely to be mistaken for an exit, shall be identified by a sign reading "Not an Exit" or similar designation, or shall be identified by a sign indicating its actual character, such as "To Basement," "Storeroom," "Linen Closet," or the like.				
1910.37q(3)	Every required sign designating an exit or way of exit access shall be so located and of such size, color, and design as to be readily visible.				
1910.37q(3)	No decorations, furnishings, or equipment which impair visibility of an exit sign shall be permitted.				
1910.37q(3)	No brightly illuminated sign (for other than exit purposes), display, or object shall be in or near the line of vision to the required exit sign of				

FIGURE 12.1. (*Continued*)

Section	Requirement	Compliance			Comments
		Yes	No	N/A	
1910.37	MEANS OF EGRESS, GENERAL				
	such a character as to detract attention from the exit sign that it may not be noticed.				
1910.37q(4)	Every exit sign shall be distinctive in color and shall provide contrast with decorations, interior finish, or other signs.				
1910.37q(5)	A sign reading "Exit," or similar designation, with an arrow indicating the directions, shall be placed in every location where the direction of travel to reach the nearest exit is not immediately apparent.				
1910.37q(6)	Every exit shall be suitably illuminated by a reliable light source giving a value of not less than 5 footcandles on the illuminated surface.				
1910.37q(6)	Artificial lights giving illumination to exit signs other than the internally illuminated types shall have screens, discs, or lenses of not less than 25 square inches and made of translucent material to show red or other specified designating color on the side of the approach.				
1910.37q(7)	Each internally illuminated exit sign shall be provided in all occupancies where reduction of normal illumination is permitted.				
1910.37q(8)	Every exit sign shall have the word "Exit" in plainly legible letters not less than 6 inches high, with the principal strokes of letters not less than $\frac{3}{4}$ inch wide.				

FIGURE 12.1. (Continued)

areas of a municipality. It is an undesirable practice to install fire alarm boxes inside buildings because of the lack of accessibility to the general public and the potential hazard of mechanical damage to the circuits extended into the building. To provide proper distribution and protection to the community, a fire alarm box should be visible from the main entrance of any building in congested districts. Another guideline is that in mercantile or manufacturing districts, the travel distance to reach a fire alarm box should not be greater than one block, or 500 feet. In residential areas, the travel distance should not be greater than two blocks, or 800 feet. There should be a fire alarm box at or near the entrance to every school, hospital, nursing home, and place of public assembly. A fire alarm box should be installed near the entrance to all fire stations since many persons are more aware of the location of the nearest fire station than of the nearest fire alarm box.[16] When fire alarms are transmitted by telephone, there is always the possibility that the lines may be down or the exchange may be busy.[7]

Fire Protection in Buildings

Fire escapes should be stairs, not ladders. Fire escapes are at best a poor substitute for standard interior or outside stairs. Their principal use is to correct exit deficiencies of existing buildings where additional stairs cannot be provided. The Life Safety Code (NFPA 101-1981) gives the following criteria for fire escape stair design:

Fire escape stairs ideally extend to the street or ground level. When sidewalks would be obstructed by permanent stairs, swinging stair sections (designed to swing down with the weight of a person) may be used for the lowest flight of the fire escape stair. The area below the swinging section must be kept unobstructed so that the section can reach the ground. A counterweight should be provided for swinging stairs of the type balancing above a pivot; cables should not be used. Fire escapes terminating on balconies above the ground level, with no way to reach the ground except by portable ladders or jumping, are unsafe.[2]

Smokeproof Towers.   Smokeproof towers are the safest form of stair enclosure recommended by the Life Safety Code. Access to the stair tower is only by balconies open to the outside air, vented shafts, or pressurized vestibules, so that smoke and fire will not readily spread into the tower even if the doors are accidentally left open.[15]

Emergency Lighting.   The Life Safety Code requires emergency power for illumination of the means of egress based upon occupancy criteria. Emergency power is required within places of assembly, certain types of educational buildings, health-care facilities, residential buildings with more than 25 rooms or living units, business buildings subject to occupancy by more than 1,000 people, parking garages, and underground and windowless structures subject to occupancy by more than 100 persons.

Well-designed emergency lighting, using a source of power independent from the normal building service, provides the necessary illumination automatically in the event of interruption of power to normal lighting as a result of any failure. Emergency lighting is arranged so that the necessary exit floor illumination will be automatically maintained in the event of failure of the normal lighting of the building, with no appreciable interruption of illumination.

Hotels and Motels. These must have exits wide enough to handle occupants from upper floors as well as occupants from the various public sections of the buiding, which are usually near the main entrance. Two exits are the minimum required for hotel occupancies, and they must be accessible from every floor, including those below the floor of exit discharge which are occupied for public purposes. From every point in sizable open areas within a hotel, and from the entrance door to individual rooms, exit access must be provided so that exits can be reached from at least two different directions.[2,14] Some of the same concerns may be raised for residents of highrise apartment houses and condominiums.[12]

Surviving a hotel fire depends on you. It takes advance planning and clear thinking.

Advance planning:

1. Always pack a roll of duct tape in your suitcase. It can be used to seal your hotel room.
2. At the hotel, before you unpack or become comfortable, do the following:
   a. Find the fire exit.
   b. Check that the fire exit provides a clear route to safety.
   c. Count doorways between your room and the fire exit so that you can find your way in the dark.
   d. Put your key near your bed.
   e. Try to open your window.

In case of fire:

1. Pick up key and put it in your pocket.
2. Crawl to door.
3. Feel door. If it is cold, do the following:
   a. Open door slowly.
   b. If smoke is present, crawl or stay low.
   c. Do not use elevators.
   d. Locate fire exit or alternate.
   e. Go to ground floor if possible; otherwise go to roof.
4. If door feels hot, do the following:
   a. Stay in room.
   b. Telephone for help.
   c. Vent smoke from room, if necessary.

   d. Fill bathtub with water for fighting fire.

   e. Turn on bathroom exhaust fan.

   f. Use duct tape to seal door and other smoke sources.

   g. If duct tape is not available, use wet towels.

   h. Get fresh air from windows if necessary.

   i. Remain calm.

Towers.   If there is only one means of egress, caution must be taken to ensure that occupants cannot be trapped by the construction, occupancy characteristics, or design of a tower. No more than 25 persons may be on any one floor of a tower with a single exit, and all occupants should be of good health. The tower should not be used for sleeping or living purposes. Fire alarm systems arranged so that they sound an alarm in a continuously staffed location are required in towers, except where ladders are provided as a means of egress.

Windowless and Underground Structures.   These present special life safety risks because of the difficulty of venting combustion products. They must have complete automatic sprinkler protection and automatic smoke-venting systems. Outside access panels are required for windowless buildings.

TRAINING AND EDUCATION

Fire Drills

Merely providing well-marked exits does not ensure life safety during a fire; exit drills are needed so that occupants will know how to make an efficient and orderly escape. Exit drills are required in schools and are common in industries with many hazards. Personnel should be assigned to check exits to see that they are available, to search for stragglers, to count occupants once they are outside the fire area, and to control reentry into the building before it is safe. Probably the most important decision is determining when and what to evacuate. In case of doubt, the entire building should be evacuated.

    Responsibility for planning exit drills belongs to the fire loss prevention and control management staff. Plans should be discussed with both middle and line management to ensure their understanding and cooperation.

Fire Brigades

The function of a fire brigade is to provide a force of fire fighters, specifically for the needs of the property, whose members are better acquainted with the property and its problems than are the members of any outside fire department. The availability of fire-fighting assistance from a public or a private fire department may affect the nature of the private fire brigade organization. Fire brigades should be on duty on each working shift and at periods when the plant is shut down or is idle. Members of fire brigades should be physically fit (29 CFR 1910.156).[20]

The fire loss prevention and control manager or a member of his or her staff should (1) provide equipment and supplies for the fire brigade, (2) establish the size and organizational structure of the fire brigade, (3) see that the brigade is suitably staffed and trained, and (4) select the first brigade chiefs.

Training.   A schedule of training should be established for members of the brigade. Members should be required to complete a specified program of instruction as a condition to membership in the brigade. Training sessions should be held at least monthly.

ARSON

Arson is an act of violence.[5] Arson is committed for a variety of reasons, from insurance fraud to vengeance. It is done for profit, for revenge, to conceal other crimes (such as murder and vandalism) and as a manifestation of a psychiatric problem (pyromania).

One of the first signs of arson for spite or revenge is typically a small, purposeless fire that amounts to little or no damage. Fires of unknown or suspicious origin are frequently arson. In 1978 arson resulted in 169,000 fires, 764 deaths, and over 2,800 injuries. The estimated cost of arson in 1978 was over $1.1 billion. Arson statistics are substantially underestimated.

Control measures for arson have not been identified. Instead, this act requires a joint effort of the police and fire department in determining the cause and fighting the fire.[22]

CODES AND STANDARDS

The National Fire Protection Association Life Safety Code (NFPA 101-1981) has already been discussed in this chapter. Municipalities, counties, and states have codes regarding building, zoning, and occupancy rates. OSHA has standards to protect workers. There are two subparts of the general industry standard that are specifically related to fire protection. These are Subpart E—Means of Egress and Subpart L—Fire Protection. Practically every other section of this standard deals with fire protection in some way.[20]

SUGGESTED LEARNING EXPERIENCES

1. Discuss the handling and storing of toxic materials, paying special attention to the fire hazards.

2. What type or types of smoke detectors did the last organization you worked for utilize? Were they effective? Are there any changes you would suggest?

3. Compare and contrast the various types of fire detectors. Which type would be most appropriate for a manufacturing plant, tire recapping facility, welding shop, hospital, plumbing goods distributor, and paint store?

4. If you were confronted with a lithium hazard, what fire controls would you employ?

5. What are the advantages and disadvantages of a wet-type sprinkler system, a dry type, and a halon fire-suppression system?

6. Design an effective egress system for the building in which this class is being held.

7. Compare the recommendations for fire brigades in this chapter with those specified in 29 CFR 1910.

8. A well-known fire protection professional has stated that "when packing your suitcase, you should include a roll of duct tape." What do you make of this?

9. Perform a study of the fire protection materials in one of your school buildings or in your place of business. How well do these comply with the Life Safety Code?

10. As a safety engineer, how would you motivate workers to take fire prevention training seriously?

REFERENCES

1. Anderson, J. Water distribution systems. In *Fire Protection Handbook* (15th ed.). Quincy, Mass.: National Fire Protection Association, 1981, pp. 16-32–16-50.

2. Bryan, J. C. Concepts of egress design. In *Fire Protection Handbook* (15th ed.). Quincy, Mass.: National Fire Protection Association, 1981, pp. 6-8–6-23.

3. Burkowski, R. W., and Zimmerman, C. E. Household warning devices. In *Fire Protection Handbook* (15th ed.). Quincy, Mass.: National Fire Protection Association, 1981, pp. 15-19–15-24.

4. Burkowski, R. W., and Zimmerman, C. E. Automatic fire detectors. In *Fire Protection Handbook* (15th ed.). Quincy, Mass.: National Fire Protection Association, 1981, pp. 15-25–15-33.

5. Carter, R. E. Fire and arson investigation. In *Fire Protection Handbook* (15th ed.). Quincy, Mass.: National Fire Protection Association, 1981, pp. 14-46–14-55.

6. Cirigliano, V. Waterflow alarms and sprinkler system supervision. In *Fire Protection Handbook* (15th ed.). Quincy, Mass.: National Fire Protection Association, 1981, pp. 17-41–17-48.

7. Cooper, . F. *Electrical Safety Engineering.* London: Newnes-Butterworth, 1979.

8. Ford, C. L. Halogenated agents and systems. In *Fire Protection Handbook* (15th ed.). Quincy, Mass.: National Fire Protection Association, 1981, pp. 18-11–18-22.

9. Henahan, J. F. Fire. *Science*, 1980, **80**, 29–38. (Jan.-Feb.)

10. Henry, M. F. Flammable and combustible liquids. In *Fire Protection Handbook* (15th ed.). Quincy, Mass.: National Fire Protection Association, 1981, pp. 1-26–4-34.

11. Hodnett, R. M. Automatic sprinkler systems. In *Fire Protection Handbook* (15th ed.). Quincy, Mass.: National Fire Protection Association, 1981, pp. 17-2–17-23.

12. Mendes, R. E. Risk management in high density housing. *Professional Safety*, 1980, **25**, 11–13. (Feb.)

13. National Safety Council. *Accident Prevention Manual for Industrial Operations* (5th ed.). Chicago, 1974.

14. Peterson, M. E. Inspection and maintenance of fire extinguishers. In *Fire Protection Handbook* (15th ed.). Quincy, Mass.: National Fire Protection Association, 1981, pp 19-22–19-26.

15. Peterson, M. E. The role of extinguishers in fire protection. In *Fire Protection Handbook* (15th ed.). Quincy, Mass.: National Fire Protection Association, 1981, pp. 19-2–19-6.

16. Sax, N. I. Industrial fire protection. In N. I. Sax (Ed.), *Dangerous Properties or Industrial Materials* (5th ed.). New York: Van Nostrand-Reinhold, 1979.

17. Schulman, M. R. Fire department emergency communication systems. In *Fire Protection Handbook* (15th ed.). Quincy, Mass.: National Fire Protection Association, 1981, pp. 15-2–15-8.

18. Slitka, M. J. Residential occupancies. In *Fire Protection Handbook* (15th ed.). Quincy, Mass.: National Fire Protection Association, 1981, pp. 6-24–6-31.

19. Slye, D. M. Unusual occupancies. In *Fire Protection Handbook* (15th ed.). Quincy, Mass.: National Fire Protection Association, 1981, pp. 6-68–6-69.

20. U.S. Department of Labor, Occupational Safety and Health Administration. General Industry Standard, 29 CFR 1910. Washington, D.C. G.P.O., June, 1981.

21. Williamson, H. V. Carbon dioxide and application systems. In *Fire Protection Handbook* (15th ed.). Quincy, Mass.: National Fire Protection Association, 1981, pp. 18-2–18-10.

22. Combating the arson menace. *Occupational Hazards*, 1981, 43, 59–62. (May)

13 RADIATION

Radioactivity has been with us for a long time. It has been with us since the beginning. The "big bang," is the hypothesized beginning of the universe, a gigantic explosion which occurred 100 billion years ago. It sent nuclear debris to the far corners of the universe; some of this debris became our solar system. Radiation has been a companion of the human race since the first homo sapien. There is radiation in the air we breathe, the food we eat, the water we drink. We live in a world in which we are exposed to much radioactivity. A first step might be to examine how much radiation we are exposed to ourselves. By using Table 13.1, you may determine your own annual dose of radiation.

Some radioactive substances were already affecting humankind before the phenomenon was even known. Since the Middle Ages, miners have extracted ore from mines in central and eastern Europe and many of them developed a fatal lung disease. This disease was finally determined in the twentieth century to be lung cancer. The miners inhaled air containing a radioactive gas (radon). More recently, lung cancer has been reported among uranium miners.

Although radiation has existed since the hypothesized big bang, it was identified less than 100 years ago. In 1895 Wilhelm Roentgen discovered that images created by the bones of the hands could be seen when the hand was placed between a Crookes tube and a screen covered with a fluorescent chemical which re-emitted the energy as light. These images were called x-rays.

In the same time period, Henri Becquerel discovered natural radiation. Becquerel placed photographic plates in darkness near a uranium sample; he found when he developed them that the plates had somehow been exposed. This study convinced him that uranium was like x-rays but was a natural form of radiation.

Pierre and Marie Curie discovered radium and polonium while search-

323

TABLE 13.1
Annual Exposure to Ionizing Radiation, in Millirems of Radiation

Source of Radiation	Annual Dose	My Dose
Cosmic radiation at sea level	30	
Altitude at which you live:		
Add 1 mrem/100 feet (30m) of altitude		
Air, water, and food	25	
Radioactive substances in your house construction		
Wood	30	
Concrete	34	
Brick	34	
Stone	36	
Television viewing		
Black and white:	1 mrem/hour	
Color:	2 mrem/hour	
Jet aircraft travel:	2 mrem/trip	
X-rays		
Chest:	150 mrem/x-ray	
Dental:	20 mrem/x-ray	
Total		—

ing for other sources of radiation. In 1901 Becquerel and Curie reported an extract of radium which had 500 times as much activity as metallic uranium.

In 1905 Albert Einstein published his famous equation $E = mc^2$ where m is mass and c is the speed of light, 3×10^{10} cm /sec. When mass is converted to energy, a tremendous amount of energy is released. This work led to the development of nuclear energy.

The first radiation injuries were permanent skin burns on the hands of experimenters working with x-rays. By 1902 there were already 147 cases of radiation injuries reported in the literature. However, the impact of radiation on health was not fully realized until the years following the nuclear detonations of Hiroshima and Nagasaki.

The first occupational safety and health problem associated with radioactive materials occurred during and after World War I from the use of radium in luminous paints. At that time luminous watch dials came into mass production. About 40 workers, mostly women, who painted numbers on the dials of watches are known to have died. They "tipped" their brushes with their lips and then dipped their brushes into the paint. The radium that they ingested caused what was originally called "radium jaw" but was really bone cancer (osteosarcoma). In addition, many suffered from a cancer called aplastic anemia.

With the increasing use of x-rays and radium in medicine, the harmful effects of radiation became better known. This led to public awareness of the problems and the present comprehensive radiation protection system.

SIZE OF THE PROBLEM

Every person on earth is continuously exposed to a variety of forms of radiation. The major risk is posed by excessive doses of ultraviolet and infrared radiation (excessive sunlight, heat, and cold). Ionizing radiation is the type that most people associate with health risks. Table 13.2 shows no fatal accidents attributed to ionizing radiation between 1975 and 1978. Several million workers are exposed to ionizing radiation. Work-related illnesses resulting from radiation are shown in Table 13.3.

The percentage of the work force affected by radiation-related illness is quite small, because radiation illnesses include those caused by ionizing, ultraviolet, infrared, and microwave radiation.[5] The incidence rate of radiation illnesses is difficult to determine for several reasons:

1. There is usually a long latency between exposure and onset of disease.
2. Radiation illness is a mixed group of syndromes ranging from sunburn to radiation sickness.
3. Only incidents resulting in workers' compensation claims or death are included.
4. Most diseases have multiple causes.

The incidence of radiation illness claims for workers' compensation in Colorado was 45 cases in 1978 and 59 cases in 1979.[2,3] This amounted to 2.2% and 2.0% of occupational workers' compensation claims in those years. However, there were 37,029 and 42,415 workers' compensation claims in Colorado in 1978 and 1979, respectively. Radiation claims were 0.1% and ionizing radiation claims were less than 0.01% in a state where uranium is mined and milled.

TABLE 13.2
Accidental Deaths Resulting from Radiation

Type of Accident or Manner of Injury	1978	1977	1976	1975
All accidental deaths	105,561	103,202	100,761	103,030
Excessive heat	267	308	100	190
Excessive cold	652	634	434	359
Ionizing radiation	0	0	0	0

SOURCE: National Safety Council. *Accident Facts*. Chicago, 1981. Reprinted with permission.

TABLE 13.3
Work-Related Radiation Illness in Colorado, by Industry

	Agriculture	Mining	Construction	Manufacturing	Transportation and Public Utilities
All illnesses					
1978	39	52	289	521	208
1979	54	72	451	761	304
Radiation					
1978	0	3	12	15	2
1979	0	2	22	19	5

SOURCE: Colorado Department of Labor, with permission.

FUNDAMENTALS OF IONIZING RADIATION

Radiation comes from several sources: the emission of particles by spontaneous degradation of a naturally occurring radioactive substance; nuclear fission (splitting the atom); and nuclear fusion (fusing two atoms together to make a single atom).

Kinds of Radioactivity

An element which is naturally radioactive emits either alpha particles or beta particles, which may also be accompanied by gamma radiation.

Alpha Particles.   When the nucleus of an atom emits an alpha particle, the atom is transformed into a new atom; it is decreased by four atomic mass units, and its atomic number is decreased by two. For instance, the isotope (which will be discussed later in this chapter) uranium 235 emits alpha particles by spontaneous disintegration and changes to thorium 231. Alpha particles are fast-moving ones with a velocity of 109 cm/sec. An alpha particle has the same form as the nucleus of a helium atom and consists of two protons and two neutrons. Most solid objects, including skin, will effectively stop an alpha particle. However, the alpha particle is extremely dangerous inside the body.

Beta Particles.   The beta particle is a charged particle emitted from a nucleus with a mass and a charge equal to an electron. An electron is usually associated with an orbit around the atom and is very small in comparison to an atom. It is approximately $\frac{1}{7500}$ of an alpha particle. Many radioactive isotopes emit beta particles, including antimony 122, arsenic 74, and cadmium 115.

Like the alpha particle, the emission of a beta particle changes the organization of the nucleus. It does not change the weight significantly, but it does change a neutron to a proton. This changes an isotope of an atom to an isotope of the next higher atom. Silver could theoretically be changed into cadmium and gold into mercury, as the alchemist once sought to turn lead into gold.

Gamma Radiation.   Gamma rays are the short-wavelength electromagnetic radiation emitted during nuclear disintegration. Gamma rays are often found to accompany the emission of alpha and beta particles. It has been shown that gamma rays have the same characteristics as x-rays. The distinction is made on the basis of their origin. Gamma rays have rather high energy levels which range from 0.15 million to several million electronvolts. From a safety viewpoint, gamma radiation is important because it has deep penetration power and can cause serious health problems. One measure of its penetration power is the half-value layer. This is the thickness of a substance which will reduce the value of a specified radiation quantity by one-half. The half-value layer for 1 million electronvolts of gamma radiation is equal to 0.5 inches of steel plate.

EXPOSURE

Exposure is the measure of ionization produced in air by gamma rays or x-rays. It is the sum of all ions of one sign produced in the air. The basic unit of exposure is the roentgen (r). One roentgen equals 2.58×10^{-4} coulombs per kilogram of air. One roentgen will produce 2×10^9 ion pairs in the air.

There are two types of radiation exposure—acute and chronic. Acute exposure is radiation exposure of a short time duration. Chronic exposure is radiation exposure of a long duration.

Units of Measure—Energy

In radiation safety the units of measure are the same ones used in physics. The basic unit is the newton (N). This is a unit of force which when applied to a one-kilogram mass will give it an acceleration of one meter per second per second ($1 N = 1 kg \times 1 m/sec^2$). Another unit is the joule (J), which is equivalent to one newton expended along the distance of one meter ($J = N \times m$). An erg is equal to 1×10^{-7} joules.

An electronvolt (eV) is the standard unit of atomic and nuclear activity. It is actually the energy gained by an electron in passing through a potential difference of one volt. KeV is a thousand electronvolts, while MeV is a million electronvolts.

$$1 \text{ eV} = 1.6 \times 10^{-12} \text{ ergs}$$

Dose.   Dose is the quantity of ionizing radiation absorbed, per unit of mass, by the body or by any portion of the body.

Rad.   Rad is a measure of the dose of any ionizing radiation to body tissue in terms of the energy absorbed per unit of mass of the tissue. One rad is the dose corresponding to the absorption of 100 ergs per gram of tissue. [One millirad (mrad) = 0.001 rad. An international unit called the gray (Gy) will replace the rad; 1 gray = 100 rads.] An acute dose of 100 rads to the whole body is likely to cause radiation sickness.

Rem.   Rem is a measure of the dose of any ionizing radiation to body tissue in terms of its estimated biological effect relative to a dose of one roentgen (r) of x-rays [1 millirem (mrem) = 0.001 rem]. The relation of the rem to other dose units depends upon the biological effect under consideration and upon the conditions for irradiation. Each of the following is considered to be equivalent to a dose of 1 rem:

1. A dose of 1 r due to x- or gamma radiation.
2. A dose of 1 rad due to x-, gamma, or beta radiation.
3. A dose of 0.1 rad due to neutrons or high-energy protons.
4. A dose of 0.5 rad due to particles heavier than protons and with sufficient energy to reach the lens of the eye.
5. If it is more convenient to measure the neutron flux, or equivalent, than to determine the neutron dose in rads, 1 rem of neutron radiation may be assumed to be equivalent to 14 million neutrons per square centimeter incident upon the body. If there is sufficient infor-

TABLE 13.4
Neutron Flux Dose Equivalents

Neutron Energy (MeV)	Neutrons/cm^2 equivalent to a does of 1 rem	Average Flux to Deliver 100 Millirems in 40 Hours (Neutrons/cm^2)
Thermal	970×10^6	670
0.001	720×10^6	500
0.005	820×10^6	570
0.02	400×10^6	280
0.1	120×10^6	80
0.5	43×10^6	30
1.0	26×10^6	18
2.5	29×10^6	20
5.0	26×10^6	18
7.5	24×10^6	17
10	24×10^6	17
10–30	14×10^6	10

SOURCE: Occupational Safety and Health Administration *Code of Federal Regulations*. General Industry Standards (29 CFR 1910). U.S. Department of Labor, Washington, D.C., June 1981.

mation to estimate with reasonable accuracy the approximate distribution in energy of the neutrons, the incident number of neutrons per square meter equivalent to 1 rem may be estimated from Table 13.4.

Units of Activity

The basic unit of radiation activity is the spontaneous disintegration of an atom of a natural radioactive substance. The becquerel is the term used to designate this single disintegration. A becquerel is the activity of a radionuclide having one spontaneous transition per second. A curie (Ci) equals 3.7×10^{10} nuclear transformations per second. A millicurie (mCi) is a thousandth of a curie and equals 3.7×10^7 becquerels. A microcurie (μCi) is a millionth of a curie and equals 3.7×10^4 becquerels. A picocurie equals 2.22 becquerels. The units of activity can be related to potential exposure.

TYPES OF NUCLEAR REACTIONS

There are several types of nuclear reactions. Some of these occur naturally. Natural radioactivity is the process of radioactive decay in which new elements are formed when a nuclide releases a product. Another type of nuclear reaction is called fission and is commonly known as splitting the atom. The atomic nucleus releases a large amount of energy when it is broken apart. A third type of nuclear reaction is called fusion, in which atoms are combined to make a new atom. This is the process by which heat is produced by the sun. Two atoms of hydrogen on the sun are combined into one atom of helium.

Radioactive Decay

Radioactive decay is the disintegration of the nucleus of an unstable nuclide by spontaneous emission of charged particles and/or protons. A nuclide is a species of atom characterized by the constitution of its nucleus. The nuclear constitution is specified by the number of protons (P), number of neutrons (N), and energy content. To be regarded as a distinct nuclide, the atom must be capable of existing for a measurable period of time. Nuclides having the same number of protons in their nucleus and hence the same atomic number but a different number of neutrons are isotopes. A stable isotope is one which is not radioactive. An isotope is an atom which has the same atomic number (Z) of the element but a different mass. Oxygen, for example, has eight isotopes ranging in mass from 13 to 20. A second example of an isotope is the element hydrogen. Hydrogen has two heavy isotopes. Hydrogen's normal state has one atomic mass unit (AMU).

TABLE 13.5
Half-Lifes of Selected Radionuclides

Radionuclide	Half-life
Hydrogen 3	12.3
Carbon 14	5,730 years
Nitrogen 16	7.2 seconds
Sodium 22	2.60 years
Sodium 24	15.0 hours
Sulfur 35	88 days
Argon 41	1.83 hours
Cobalt 57	5.26 years
Krypton 87	76 minutes
Strontium 90	28.1 years
Iodine 125	60 days
Iodine 131	8.05 days
Thorium 232	1.41×10^{10} years
Uranium 235	4.51×10^{9} years

One isotope which has two AMUs is called deuterium. Another isotope which has three AMUs is called tritium.

The measure of radioactive decay is called the half-life. For a single radioactive decay process, the time required for the activity to decrease to half its value is called the half-life. Some substances decay rapidly, some slowly. The half-lifes of a number of isotopes are shown in Table 13.5. This time is very important for the safety engineer faced with radioactive contamination. If the substance decays rapidly, one can simply wait for the radiation to be greatly reduced. Take, for example, an isotope with a half-life of one day. After six days only 1.5% of the original radiation would remain (see Figure 13.1).

Isotopes release energy in an effort to find a more stable level. Depending upon the type of reaction, the element may change. The isotope decays into other radionuclides or forms daughter products such as iodine 39 forms iodine 38 when it disintegrates. The general effects of nuclear decay are as follows:

Emission	Outcome
Alpha particle	New element
Beta particle	New element
Neutron	Same element, new isotope
Gamma radiation	Same element

Nuclear Fission

Nuclear fission is the division of a heavy nucleus into two (or rarely, more) parts with masses of equal size, usually accompanied by emission of neu-

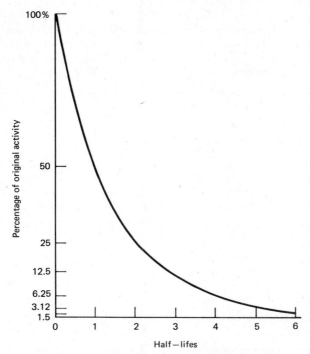

FIGURE 13.1.   Half-lifes for a hypothetical nuclide.

trons, gamma radiation, and rarely, small charged fragments. Nuclear fission is thought of as splitting the atom. It can occur spontaneously without the addition of particles or energy to the nucleus, or it can occur by bombarding a fissile material (a nuclide capable of undergoing fission by interaction with a slow neutron). One sample of nuclear fission of uranium is the following. Uranium, Z = 92, atomic weight (AW) = 236, produces barium, Z = 56, AW = 141, Krypton, Z = 36, AW = 92, and releases 3 neutrons. The following equation depicts this reaction:

$$_0n^1 + {}_{92}U^{235} \rightarrow ({}_{92}U^{236}) \rightarrow {}_{56}Ba^{141} + {}_{36}Kr^{92} + {}_3n^1 + 200 \text{ MeV}$$

Nuclear Fusion

Nuclear fusion is the opposite of nuclear fission. In fission an atom is literally split apart, while in nuclear fusion a reaction between two light nuclei results in the production of at least one nuclear species heavier than either of the initial nuclei, together with excess energy. The simplest fusion reaction is that of one deuterium (heavy hydrogen) atom with another deuterium atom to produce helium.

$$H^2 + H^2 \rightarrow He^3 + n + 3.25 \text{ MeV}$$

This reaction produces a large amount of energy. One gram of deuterium could produce 5 kilowatt hours of energy. Although uses of fusion on earth are limited to the hydrogen bomb, studies have been conducted into the feasibility of using nuclear fusion reaction for power production. The supply of deuterium is without limit. However, the safety implications are almost beyond imagination.

RADIATION ACCIDENTS

In this chapter many aspects of radiation have been discussed. What is the nature of radiation? How much exposure can a person be allowed to receive? How can persons be best-protected from radiation? What happens if the protective measures fail and a person or persons are exposed to excess ionizing radiation.

There are four types of radiation accidents.

1. *Radiation Exposure.*   The individual receives excess whole or partial body external radiation.
2. *Internal Contamination.*   The individual ingests or inhales radioactive materials.
3. *External Contamination.*   The individual's body surface and/or clothing is contaminated with radioactive liquids and/or particles.
4. *Contaminated Wounds.*   The individual is contaminated externally and also has a wound.

Immediate Steps to Take in Radiation Accidents

The following steps are to be taken in case of an accident involving ionizing radiation:

1. Conduct evaluation of situation in regard to (1) levels of external radiation exposure and (2) contamination by radionuclides.
2. If external radiation levels are high, evacuate exposed personnel from accident area. (If possibility of contamination exists, assure that evacuees are confined until monitored).
3. Confine contamination in the accident area to prevent further spread. If liquid, use absorbent material to keep from spreading. If possible, close off air circulation and seal doors and windows. Prevent further personnel access to radiation area. Remove contaminated clothing and shoes before going to a clean area.
4. Locate and monitor all persons who may be contaminated. Perform simple decontamination, if necessary, and remonitor. Give first aid if needed.

5. Obtain medical and health physics' assistance promptly.
6. Obtain a complete history of accident.

Acute Radiation Syndrome

Overexposure of a worker to either an internal or external radiation source is a medical emergency. An acute radiation incident involves a whole body exposure of 100 roentgens over a very short period. The initial symptoms of acute radiation syndrome are nausea, vomiting, diarrhea, weakness, and profound depression of vital processes (shock). In about two days the person becomes feverish and has an indefinite feeling of sickness or illness.

PERMISSIBLE EXPOSURE

The permissible dose for an individual is that dose accumulated over a long period of time or resulting from a single exposure. Any severe somatic injuries, for example, leukemia, that might result from exposure of individuals to the permissible dose would be limited to an exceedingly small fraction of the exposed group; effects such as shortening of life-span, which might be expected to occur more frequently, would be very slight and would likely be hidden by normal biological variation. The permissible doses can, therefore, be expected to produce effects that could be detectable only by statistical methods applied to large groups.

There are risks to workers associated with ionizing radiation. However, these risks are very small; when compared to risks involved in activities of daily living, they do not significantly endanger workers or the general public.

The Nuclear Regulatory Commission has set up standards of permissible doses, levels, and concentrations of radioactive material a worker may come in contact with. These regulations and standards are published in 29 CFR 1910, paragraph 96, and 10 CFR 20, paragraph 101.

A worker may be exposed to radiation which includes alpha rays, beta rays, gamma rays, x-rays, neutrons, high-speed protons, and other atomic particles. The exposure may take place in either a restricted area or an unrestricted area. A restricted area includes any area access which is controlled for the purposes of protecting workers from ionizing radiation. An unrestricted area includes any area access not controlled for the purposes of protecting the worker from ionizing radiation and any areas which are used for workers' residences. The total permissible exposure for workers in restricted areas for a calendar quarter is 3 rems. A calendar quarter generally means a three-month period (12 to 14 weeks) which begins in January, April, July, and October. This permissible dose is shown in Table 13.6. The whole body dose may under certain circumstances be as much as 3 rems per quarter. Persons under 18 years of age may be exposed to only 10% of the amounts shown in Table 13.6.

TABLE 13.6
Permissible Limits for Ionized Radiation

Part of Body	rems per Calendar Quarter
Whole body: Head and Trunk, active blood-forming organs, lens of eyes, or gonads	$1\frac{1}{4}$
Hands and forearms, feet and ankles	$18\frac{3}{4}$
Skin of whole body	$7\frac{1}{2}$

SOURCE: Occupational Safety and Health Administration, *Code of Federal Regulations*. General Industry Standards (29 CFR 1910). U.S. Department of Labor, Washington, D.C., June 1981.

Warning Signs

The American National Standard N.2.1 specifies that a radiation symbol (see Figure 13.2) shall be used "to signify the actual or potential presence of ionizing radiation and to identify objects, devices, materials or combination of materials, which emit ionizing radiation."[1] This symbol is purple on a yellow background. However, some international signs are black on a contrasting color. International Standards organization does not specify a color in recommendation R361-1963, "Basic Ionizing Radiation Symbol."

Any area in a plant or laboratory must be conspicuously marked with a warning sign in addition to the standard radiation symbol: "Caution Radiation Area."

Any area of a plant or laboratory in which a person might receive a dose of 100 millirems per hour must be marked with a warning sign: "Caution High-Radiation Area."

In addition, when a person enters such an area, an alarm is set off, enabling both the person and his or her supervisor to be immediately aware of such entry.

Any area of a plant or laboratory in which there is airborne radioactive material in excess of the amounts specified in the Code of Federal Regulations (Table 1 of Appendix B to 10 CFR 20) must have a warning sign. If the radioactive isotope in question exceeds the limit, then the area must be marked with the radiation symbol together with a warning sign: "Caution Airborne Radioactivity Area."

SHIELDING

Shielding is the process of placing a barrier between a radiation source and people. A shield is material used to prevent or reduce the passage of particles or radiation. The material may absorb the radiation either partially or totally. The effectiveness of the shield is a function of its own physical

Form NRC-3
(1-80)
10 CFR 19
10 CFR 20

UNITED STATES NUCLEAR REGULATORY COMMISSION
Washington, D.C. 20555

NOTICE TO EMPLOYEES

STANDARDS FOR PROTECTION AGAINST RADIATION (PART 20); NOTICES, INSTRUCTIONS AND REPORTS TO WORKERS; INSPECTIONS (PART 19)

In Part 20 of its Rules and Regulations, the Nuclear Regulatory Commission has established standards for your protection against radiation hazards from radioactive material under license issued by the Nuclear Regulatory Commission. In Part 19 of its Rules and Regulations, the Nuclear Regulatory Commission has established certain provisions for the options of workers engaged in NRC-licensed activities.

YOUR EMPLOYER'S RESPONSIBILITY

Your employer is required to—

1. Apply these NRC regulations and the conditions of his NRC license to all work under the license.

2. Post or otherwise make available to you a copy of the NRC regulations, licenses, and operating procedures which apply to work you are engaged in, and explain their provisions to you.

3. Post Notices of Violation involving radiological working conditions, proposed imposition of civil penalties and orders.

YOUR RESPONSIBILITY AS A WORKER

You should familiarize yourself with those provisions of the NRC regulations, and the operating procedures which apply to the work you are engaged in. You should observe their provisions for your own protection and protection of your co workers.

WHAT IS COVERED BY THESE NRC REGULATIONS

1. Limits on exposure to radiation and radioactive material in restricted and unrestricted areas;

2. Measures to be taken after accidental exposure;

3. Personnel monitoring, surveys and equipment;

4. Caution signs, labels, and safety interlock equipment;

5. Exposure records and reports;

6. Options for workers regarding NRC inspections; and

7. Related matters.

REPORTS ON YOUR RADIATION EXPOSURE HISTORY

1. The NRC regulations require that your employer give you a written report if you receive an

exposure in excess of any applicable limit as set forth in the regulations or in the license. The basic limits for exposure to employees are set forth in Sections 20.101, 20.103, and 20.104 of the Part 20 regulations. These Sections specify limits on exposure to radiation and exposure to concentrations of radioactive material in air.

2. If you work where personnel monitoring is required pursuant to Section 20.202;

(a) your employer must give you a written report of your radiation exposures upon the termination of your employment, if you request it, and

(b) your employer must advise you annually of your exposure to radiation, if you request it.

INSPECTIONS

All activities under the license are subject to inspection by representatives of the NRC. In addition, any worker or representative of workers who believes that there is a violation of the Atomic Energy Act of 1954, the regulations issued thereunder, or the terms of the employer's license with regard to radiological working conditions in which the worker is engaged, may request an inspection by sending a notice of the alleged violation to the appropriate United States Nuclear Regulatory Commission Inspection and Enforcement Regional Office (shown on map at right). The request must set forth the specific grounds for the notice, and must be signed by the worker or the representative of the workers. During inspections, NRC inspectors may confer privately with workers, and any worker may bring to the attention of the inspectors any past or present condition which he believes contributed to or caused any violation as described above.

POSTING REQUIREMENTS

Copies of this notice must be posted in a sufficient number of places in every establishment where activities licensed by the NRC are conducted, to permit employees working in or frequenting any portion of a restricted area to observe a copy on the way to or from their place of employment.

UNITED STATES NUCLEAR REGULATORY COMMISSION

A representative of the Nuclear Regulatory Commission can be contacted at (1 at) the following addresses and telephone numbers. The Regional Office will accept collect telephone calls from employees who wish to register complaints or concerns about radiological working conditions or other matters regarding compliance with Commission rules and regulations.

Regional Offices

REGION	ADDRESS	TELEPHONE	
		DAYTIME	NIGHTS AND HOLIDAYS
I	Region I, Office of Inspection and Enforcement, USNRC 631 Park Avenue King of Prussia, Pennsylvania 19406	215 337-5000	215 337-5000
II	Region II, Office of Inspection and Enforcement, USNRC 101 Marietta St., N.W., Suite 3100 Atlanta, Georgia 30303	404 221-4683	404 221-4683
III	Region III, Office of Inspection and Enforcement, USNRC 799 Roosevelt Road Glen Ellyn, Illinois 60137	312 932-2500	312 932-2500
IV	Region IV, Office of Inspection and Enforcement, USNRC 611 Ryan Plaza Drive, Suite 1000 Arlington, Texas 76012	817 334-2841	817 334-2841
V	Region V, Office of Inspection and Enforcement, USNRC 1990 N. California Boulevard, Suite 202, Walnut Creek Plaza Walnut Creek, California 94596	415 943-3700	415 943-3700

Figure 1-1 NRC Form 3

PUERTO RICO — REGION II

VIRGIN IS — REGION II

CANAL ZONE — REGION II

ALASKA — REGION V

HAWAII — REGION V

FIGURE 13.2. Notice to employees (from United States Nuclear Regulatory Commission, Washington, D.C. 20555).

335

properties and the characteristics of the radiation. Therefore, each type of radiation is stopped most efficiently by a particular type of shield. When several types of radiation coexist, a shield may have to be made of a combination of materials.

Shielding from Alpha Particles

Alpha particles can be blocked by almost any material. A sheet of paper or ordinary clothing is an effective shield for alpha particles. Because of the short range of these particles, even the skin is a good barrier. However, alpha particles can enter the eyes. Alpha particles are dangerous only when the source is ingested.

Shielding from Beta Particles

Beta particles penetrate much further than alpha particles. Beta particles will not be stopped by clothes, skin, or a sheet of paper. Therefore, a shield must be used with beta radiation. However, beta particles can produce x-radiation which must also be shielded against. Because of this, a light metallic shield is necessary. Aluminum is a preferable shield against beta radiation. A 1 cm sheet will absorb beta particles up to 5 MeV, and a sandwich of aluminum and lead will absorb both the beta rays and the protons generated.

Shielding from Neutrons

Neutrons penetrate most materials, including the body. Neutrons do not produce ionizing radiation themselves. However, they cause it to be produced when they collide with other nuclei. Some substances which may be used for shielding from neutrons are lithium, boron, wax, and water.

Shielding from Gamma Radiation

Gamma radiation is very energetic and can penetrate very deeply. The "half-value layer" is used to measure penetration; this is the thickness of a substance which reduces the amount of radiation by one-half. (Compare this term with half-life.) The most effective materials for gamma shields are those made up of elements having high atomic numbers and high densities. Such elements are uranium, thorium, lead, gold, and tungsten. The weight and cost of these metals limit their use as shielding materials. Therefore, less costly medium-weight metals like iron, lead, aluminum, nickel, and chromium are used.

A 1 MeV gamma radiation source would need a half-value layer of 0.125 inch of lead to reduce its strength by 50%. One inch of lead will reduce its strength to 0.039%. Can you determine the thickness needed to reduce its

TABLE 13.7
Thickness of Lead Required to Reduce Beam to 5%

Potential	Half-Value Layer (mm)	Required Lead Thickness (mm)
60 kVp	1.2 Al[a]	0.10
100 kVp	1.0 Al	0.16
100 kVp	2.0 Al	0.25
100 kVp	3.0 Al	0.35
140 kVp	0.5 Cu[b]	0.7
200 kVp	1.0 Cu	1.0
250 kVp	3.0 Cu	1.7
400 kVp	4.0 Cu	2.3
1,000 kVp	3.2 Pb[c]	20.5
2,000 kVcp	6.0 Pb	43.0
2,000 kVcp	14.5 Pb	63.0
3,000 kVcp	16.2 Pb	70.0
6,000 kVcp	17.0 Pb	74.0
8,000 kVcp	15.5 Pb	67.0

SOURCE: Bureau of Radiological Health, DHEW, *Radiological Health Handbook* (Rev. ed.), Washington, D.C.: GPO, 1970.
[a] Al = aluminum.
[b] Cu = copper.
[c] Pb = lead.

strength to 3.13%? Another approach is to use Table 13.7, which specifies the thicknesses of lead required to reduce the radiation to 5%.

Distance

Another way to protect people from radiation is to keep them away from it. Distance is effective for protection in all cases. Alpha emissions can travel only a few centimeters in air. Beta particles may travel up to 20 meters. As one can see with alpha and beta particles, distance is an inexpensive and effective means of shielding. Even with gamma radiation, the square of the distance in meters may be used to reduce the exposure. A person who is 10 meters from a source will receive only 1% of the radiation that a person who is 1 meter away will receive.

Time

All radioactive substances decay with time. Some substances decay very rapidly, while others decay very slowly. Half-lifes range from a few seconds to over a billion years.

In cases of radiation contamination the safety engineer must know the source of the radiation. If the half-life is relatively short, then it might be convenient to wait until the radiation is safer to handle.

By the combined use of shielding, distance, and time, the safety engineer can effectively protect personnel from excess radiation. There are three basic rules in protecting personnel from radiation.

1. Keep personnel as far away as possible from a radioactive source.
2. Work with radiation for as short a time as possible.
3. Use shielding.

INDUSTRIES THAT USE RADIOACTIVE MATERIALS

Radioactive materials are used in a wide variety of industrial processes and equipment, ranging from nondestructive testing in quality assurance to low-temperature sterilization of drugs and foods. It has been suggested that milk be sterilized in this manner. Nuclear explosions, such as those that occurred at Hiroshima and Nagasaki, have changed the whole character of nuclear energy. There is an entire industry involved in the design, manufacture, and storage of nuclear devices.

The nuclear fuel reprocessing industry treats the used or spent reactor elements to recover reusable, fissionable radioactive material. Spent fuel from reactors usually contains substantial amounts of fissionable materials, but the fuel is no longer satisfactory for producing heat which can be used to drive turbines. These materials can be recovered and reprocessed into a form suitable for reuse in nuclear reactors.

GENERATING ELECTRICITY BY NUCLEAR POWER

Nuclear power is generally conceived of as electricity derived from nuclear fission. Actually, nuclear fission produces steam to drive a turbine which produces electricity. The fission which produces steam cannot be used for a nuclear explosion. A rough comparison of energy output between coal and uranium 235 shows why nuclear power seems popular. One ton of U^{235} will produce as much energy as 2.5 million tons of coal. One atom of U^{235} will produce 50 million times as much energy as one atom of coal. Can you explain the difference?

> On Wednesday, March 28, 1979, 36 seconds after the hour of 4:00 A.M., several water pumps stopped working in the Unit 2 nuclear power plant on Three Mile Island (TMI), 10 miles southeast of Harrisburg, Pennsylvania. Thus began the accident at Three Mile Island. In the minutes, hours, and days that followed a series of events—compounded by equipment failures,

inappropriate procedures, and human errors and ignorance—escalated into the worst crisis yet experienced by the nation's nuclear power industry.[4]

The Three Mile Island accident has intensified the concern for radiation health. "The accident has resulted in redirection of reactor safety priorities."[6] The accident was started by a minor failure. However, misoperation and human error, compounded by mismanagement, created a near panic. The main issues in nuclear safety involve the following: What is the likelihood of a meltdown (China syndrome)? A meltdown is a serious nuclear incident in which the reactor loses its coolant and literally melts down (possibly all the way to China), releasing a catastrophic amount of radiation into the environment. The dose to an average individual within 50 miles of TMI was 1.7 millirems. In a population of 2,000,000, approximately 325,000 would normally contract a fatal cancer. Because of the TMI accident, now 325,000.6 will die from cancer. Significant controversy continues as to whether low-level radiation is more dangerous than people believed. However, none of the recent studies have produced any dose–response relationships.

The Lessons Learned

Human factors play a major role in safety. TMI woke up industry to the fact that operators do interact with machines and the person–machine interface is critical for system operation. The former director of safety for IBM has continually stated that if a person could take only one course in the area of safety, it should be one in human factors. Human reliability has become a major area of study. The Institute for Nuclear Power Operations was established to improve the quality and training of nuclear power plant operators.

The TMI accident showed that work rules placed safety in very low priority. It might even be stated that safety rules were ignored. As a result, the operating public utilities realized that they must take the lead in safety, and they formed the Nuclear Safety Analysis Center.

In 1981, the Nuclear Regulatory Commission (NRC) suspended the operating license of the Pacific Gas and Electric Company's new Diablo Canyon nuclear plant due to company error in quality assurance. The plant is also in an earthquake zone—it is located only a few miles from a fault. New York Consolidated Edison Company (Con Ed) closed one of its nuclear power plants because of cooling system malfunction. The NRC has come down heavily on nuclear power plants because of shoddy workmanship. It has been reported that the walls are crumbling at some of the reactor plants. Many utilities did not understand the inherent dangers and technical problems of building such plants. Some did not even have competent managers to handle large projects as sophisticated as nuclear plants.[6] Thus the day of a safe nuclear plant does not seem to be at hand.

SUGGESTED LEARNING EXPERIENCES

1. In functioning in your role as a safety and health professional, what information would you want to have made available to you regarding the radiation possibilities in your place of employment? How would you go about obtaining this information?

2. Discuss protective equipment in relation to radiation.

3. Develop a position paper discussing your philosophy regarding the increased use of radiation in industry and in life in general.

4. In what ways has knowledge about radiation affected your functioning as a professional?

5. As a safety professional, what role might you play in your local community as a consultant on an issue regarding radioactivity? Role play or in some other way dramatize the situation.

6. Determine the daily dose of radiation to which you are exposed. Were you surprised at the amount? Do you foresee any long-term effects? Is there a way by which you plan to decrease your radiation exposure?

7. Develop a lesson plan for teaching the main components of radiation to a group of new employees. After developing your lesson plan, including teaching strategies you will use, reassess it to be sure you are covering the areas you think are most important and will impart the information you think is necessary.

8. After a radiation contamination accident at a work site, what kind of support and preparation might you anticipate workers would need before they would go back to work and function safely? How would you suggest they be provided?

9. What ideas do you have for more safe regulations regarding use of radiation in industry? What would you like to see accomplished in this area? What role would you like to take?

10. Depict graphically a radiation accident, including the steps leading up to it. Illustrate the correct steps to take during both the accident and its aftermath.

REFERENCES

1. American National Standards Institute. *Radiation Symbols* (ANSI N. 2.1-1963). New York, 1963.

2. Colorado Division of Labor. *Annual Report: Work Related Accidents and Illnesses in Colorado—1978.* July 1979.

3. Colorado Division of Labor. *Annual Report: Work Related Injuries and Illness in Colorado—1979*. October 1980.

4. Kemeny, J. G., Babbitt, B., Haggerty, P. E., Lewis, C., Marks, P. A., Marrett, C. B., McBride, L., McPherson, H. C., Peterson, R. W., Pigford, T. H., Taylor, T. B., and Trunk, A. D., *Report of the President's Commission on the Accident at Three Mile Island*. Washington, D.C.: GPO, 1979.

5. Moss, C. E., Conover, D. E., Murray, W. E., and Kulre, A. N. Estimated number of U.S. workers potentially exposed to electromagnetic radiation. *Annual American Industrial Hygiene Conference, 1977*, p. 431, 1977.

6. Mynatt, F. R. Nuclear reactor safety since Three Mile Island. *Science*, 1982, **216**, 131–135.

7. National Safety Council. *Accident Facts*. Chicago, 1981.

8. Robertson, J. C. *A Guide to Radiation Protection*. New York: Wiley, 1976.

14 NOISE

Hearing loss is an affliction which has existed throughout the history of the human race. It occurs in all walks of life and is due to many causes. Until the development of the audiometer in recent years, there was no means of measuring the degree of hearing loss with appreciable accuracy. Now, partial losses are easily measurable by use of commercially available audiometers.

There is growing interest on the part of management in sensorineural hearing loss which is due to noise. This has been stimulated by (1) the trend toward coverage for partial loss of hearing under state workers' compensation laws and (2) the planned hearing conservation program regulations of OSHA. Total hearing loss may result in a large compensation claim—up to $145,000 or more.[1] Nebraska considers permanent total loss of hearing to be compensated as permanent total disability. These developments have had considerable impact upon many companies.[3]

That workers in noisy occupations develop hearing loss has been long known. Prior to 1940, hearing was not regarded as a significant factor in compensation. Some claims occurred, but these were primarily due to traumatic injuries from such causes as blasts, concussions, blows to the head, and foreign objects or infections in the ears. Since 1948, all states except Wisconsin and Kentucky have passed hearing loss legislation or revised existing laws to include noise-induced hearing loss as a basis for workers' compensation benefits. Most states have provided compensation in varying degrees for partial hearing loss due to industrial noise exposure.[1]

HOW NOISE AFFECTS OUR LIVES

Noise affects our lives in a multiplicity of spheres. Much of what workers do affects or is affected by noise. Noise (depending upon duration and

TABLE 14.1
Decibel Values for Some Typical Sounds

Decibels	Example
20	Studio for sound pictures
30	Soft whisper (5 feet)
40	Quiet office
50	Average residence
60	Conversation speech (3 feet)
70	Freight train (100 feet)
74	Average automobile (30 feet)
80	Very noisy restaurant
80	Average factory
90	Subway
90	Printing press plant
100	Looms in textile mill
100	Electric furnace area
110	Woodworking
110	Casting shakeout area
120	Hydraulic press
120	50 hp siren (100 feet)
140	Jet plane
180	Rocket launching pad

intensity) can improve or degrade work performance. Noise can also have a psychologically positive or negative effect, depending upon how the hearer perceives the sound—whether it is wanted or unwanted. Unwanted sound can often have a negative effect.

When is it quiet? This depends upon the recipient of the sound. To some people, for a work environment to be called quiet, there would have to be no audible sound; other workers are able to "tune out" unwanted sounds—unless the sounds are distracting—and the workplace becomes quiet for them. (Table 14.1 lists some typical sounds and their decibel values.)

PROPERTIES OF SOUND

The properties of sound include frequency, intensity, and duration.

The frequency of a sound wave is defined as the number of pressure changes per second (hertz, or Hz). There are several instruments used to record this characteristic. Middle C has a frequency of 1,024 Hz, while noise often has no single predominant frequency. High-frequency (high-pitch) noises are more damaging to hearing than are low-frequency noises.

Intensity is measured in such absolute units as the watt and ergs per second. It is proportional to the square of the pressure. The decibel (db) is the unit of measurement that conveys the intensity of a sound. The louder the noise, the higher its intensity.

The longer the duration of a noise, the greater the potential nuisance. Long-duration exposure to high-frequency, high-intensity sound is likely to produce sensorineural deafness.

ANATOMY AND PHYSIOLOGY OF HEARING

The ear is composed of the outer, middle, and inner ear and is shown in Figure 14.1.

The Outer Ear

Sound waves travel down the external auditory meatus, the "hole" in the ear, and strike the tympanic membrane (eardrum) at its end. This tube is lined with stiff hairs and glands which secrete a bitter wax; this serves to prevent the entrance of foreign bodies. The tympanic membrane is a tight band of skin which separates the outer and middle ear. It is suspended between two bodies of air at roughly equal pressures. This membrane vibrates freely. It transmits the frequency and amplitude of sound waves, unless the intensity is too great and its vibrations are protectively damped (reduced).

The Middle Ear

Firmly attached to one side of the inner surface of the tympanic membrane is the handle of the malleus (hammer), one of a series of three small bones, which transmit vibrations from the tympanic membrane to the oval window of the inner ear. Back-and-forth movements of the malleus cause a corresponding movement of the incus (anvil), into whose socketlike end fits the head of the stapes (stirrups). The footplate of this third bone fits securely over the oval window. The stapes moves in a rocking motion rather than in and out as one might expect. Attached to the malleus is one muscle, the tensor tympani, and to the stapes, another, the stapedius. When both muscles contract reflexively, the tympanic membrane is made more taut and the footplate of the stapes is pushed inward. Both of these effects play a protective role: as the intensity of stimulation increases beyond certain limits, these muscles contract and the amplitude of vibration of the membranes is effectively damped. This is most pronounced with tones of low frequency. If an intense stimulus is repeated too rapidly—faster than the 14 to 16 msec reaction time for these muscles—they fail to serve their protective function, and the eardrum is often ruptured and

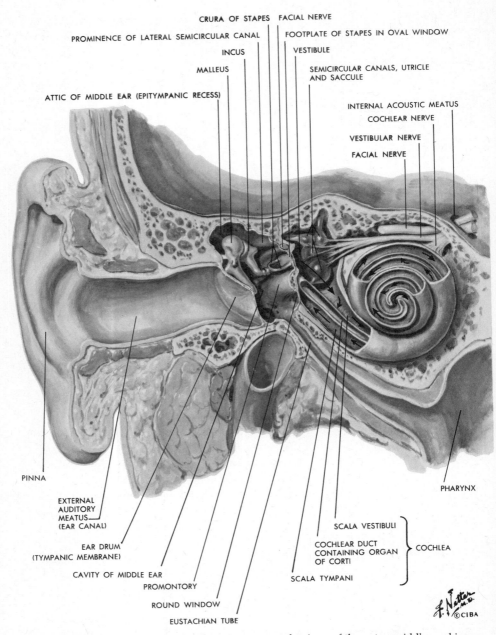

CRURA OF STAPES   FACIAL NERVE

PROMINENCE OF LATERAL SEMICIRCULAR CANAL   FOOTPLATE OF STAPES IN OVAL WINDOW

INCUS   VESTIBULE

MALLEUS   SEMICIRCULAR CANALS, UTRICLE AND SACCULE

ATTIC OF MIDDLE EAR (EPITYMPANIC RECESS)

INTERNAL ACOUSTIC MEATUS
COCHLEAR NERVE

VESTIBULAR NERVE

FACIAL NERVE

PINNA

PHARYNX

EXTERNAL AUDITORY MEATUS (EAR CANAL)

EAR DRUM (TYMPANIC MEMBRANE)

CAVITY OF MIDDLE EAR

PROMONTORY

ROUND WINDOW

EUSTACHIAN TUBE

SCALA VESTIBULI

COCHLEAR DUCT CONTAINING ORGAN OF CORTI   } COCHLEA

SCALA TYMPANI

FIGURE 14.1.   Cross section showing the gross mechanisms of the outer, middle, and inner ear (from National Safety Council).

346

other damages are incurred within the inner ear. This type of damage may be suffered by unprotected ears in battle (explosion deafness). Equalization of air pressure between the middle ear and the outside is accomplished by opening the Eustachian tube during swallowing or yawning. The lower end of this tube opens into the nasopharynx. The "cracking" in our ears, experienced when driving up or down steep gradients, is produced by sudden changes in pressure within the middle ear.

Transmission over the three bones of the middle ear is an extremely efficient process. Although fairly extensive punctures in the tympanic membrane cause only slight loss in hearing, disarticulating the three bones by removing a small piece of the incus results in an elevation of threshold by 60 db on the average. Distortions in the middle ear, however, can arise.

The Inner Ear

The inner ear serves to translate the mechanical inputs from the middle ear into a neural message which may be transmitted to the brain. This transmission is accomplished by a cranial nerve, the acoustic nerve. The inner ear consists of the cochlea, which is shaped like a snail shell, broad at the base and tapering towards its apex.

The auditory message is transmitted in a pattern of impulses to the auditory cortex. Loudness is thought to depend somehow on total frequency, but theorists have not agreed upon how pitch is determined. Because of this controversy, there are several theories of hearing which have been advanced.

THEORIES OF HEARING

Helmholtz proposed in 1863 that pitch discrimination depended on the particular fibers activated by a given tone. He specified the rods of Corti as resonators in the inner ear. He later named the basilar membrane as the resonator. This membrane is known to vibrate under stimulation, and it is in intimate association with the hair cells on which terminate the auditory nerve fibers—it is much as if there were a tiny harp in each ear which, when "sung into" by the stimulus, responds with sympathetic vibrations to match its frequency characteristics.

Nonresonance–Place Theory

This view purports that the cochlea is a hydraulic system contained in a vessel with elastic walls. This analysis demonstrates that the locus of maximal displacement along the basilar membrane will vary with the frequency of the stimulating tone. This theory holds that different auditory fibers are selectively excited by tones of different frequencies.

Frequency Theories

Theorists ascribe to the position that the pitch characteristics of tones below certain frequencies are determined by the frequency of impulses set up in the auditory nerve as a whole. Successive pressure waves, excite all available fibers; if the frequency of these energy changes is slow enough, all auditory fibers are excited on each peak. The total number of fibers firing at each peak will increase as the intensity of stimulation is increased. Thus loudness becomes a function of the total number of impulses per unit time.

The essential notion underlying the frequency theories is that the pitch quality one experiences is based upon the frequency of impulses being delivered to the auditory centers.

Resonance–Volley Theory

Perhaps the most widely accepted theory of hearing is a compromise between place and volley. This theory advances the dual concepts of pitch being represented on the basilar membrane and in impulse frequency. Frequency serves for the low tones, place the high tones, and both perform in the broader ground between.

AUDITION (HEARING)

Thresholds

The normal young adult can hear sounds over a wide range of frequencies, from about 20 to 15,000 Hz (cycles per second). The ability to hear middle- and high-frequency sounds usually decreases with age. Those persons exposed to high noise levels for extended periods experience hearing losses over and above that due to advancing age.

The human ear responds differently to different frequencies. Because different sound frequencies may have differing psychological and physiological effects, it is important that knowledge of the frequency components of a noise be obtained. In detailed industry hygiene work, it is customary to separate the total noise into eight frequency ranges called octave bands. The octave bands most commonly used in noise measurement start with the 37.5 to 75 Hz band and continue through the 4,800 to 9,600 band. In each octave the highest frequency is twice that of the lowest.

Loudness

In certain abnormalities of hearing, the relation of loudness to frequency is distorted. Of chief concern are cases of selective hearing loss. Some individuals have narrow regions of lowered sensitivity bounded by regions of normal hearing. The progressive impairment in hearing high tones that accompanies aging is an example of this. Progressive atrophy of the auditory nerve supplying the basal turn of the cochlea has been correlated with these age changes. Most forms of transmission deafness appear as selective loss of acuity for low tones, and they exaggerate the normal effects of tension on those muscles that damp the action of the ossicles. Selective losses in hearing for specific frequencies are more compatible with a place theory than with any form of frequency theory.

The Effect of Noise on Hearing

If the ear is subjected to high levels of noise for a sufficient period of time, some loss of hearing may occur. A number of factors can influence the effect of the noise exposure. Among these are:

1. Variations in individual susceptibility
2. The total energy of the sound
3. The frequency distribution of the sound
4. Other characteristics of the noise exposure, such as whether it is continuous, intermittent, or made up of a series of impacts
5. The total daily time of exposure
6. The length of employment in the noise environment[3]

Because of the complex relationships of noise and exposure time to threshold shift (reduction in hearing level) and the many possible contributory causes, OSHA established a criterion for protecting workers against hearing loss which is spelled out in 29 CFR 1910.95.[4] Protection against the effects of noise exposure shall be provided when the A-weighted sound level exceeds the levels shown in Table 14.2. When the daily noise exposure is composed of two or more periods of noise exposure of different levels, their combined effect should be considered. Use the following formula to determine time-weighted noise exposure:

$$C_1/T_1 + C_2/T_2 + C_3/T_3 = \text{exposure}$$

where C is the total time of exposure at a particular noise level and T is the total exposure time allowed at a particular noise level taken from Table 14.2.

If the sum exceeds 1, then the employee is considered to have been exposed to excess noise. Exposure to impulse or impact noise should not exceed 140 db peak sound level. Whenever employee noise exposure equals or exceeds an eight-hour time-weighted-average (TWA) sound level of 85 db or, equivalently, a dose of 50 percent, the employer shall administer a continuing effective hearing conservation program.[4]

TABLE 14.2
Permissible Noise Exposures

Duration per Day, Hours	Sound Level, dbA
8	90
6	92
4	95
3	97
2	100
$1\frac{1}{2}$	102
1	105
$\frac{3}{4}$	107
$\frac{1}{2}$	110
$\frac{1}{4}$	115

SOURCE: Occupational Safety and Health Administration, *Code of Federal Regulations*. General Industry Standards (29 CFR 1910) U.S. Departments of Labor, Washington, D.C., June 1981.

Etiology of Hearing Impairment. Loss of hearing may be defined as any reduction in the ability to hear. Temporary hearing loss from exposure to loud noises for a few hours, with normal hearing usually returning after a rest period, is called a temporary threshold shift. The recovery period may be minutes, hours, days, or even longer. Permanent hearing loss may occur as a result of the aging process, diseases, injury, or exposure to loud noises for extended periods of time. The hearing loss associated with exposure to industrial noise is commonly referred to as "acoustic trauma." This type of hearing loss is the result of nerve- or hair-cell destruction in the hearing organ and it is not reversible. However, such hearing losses usually are only partial. Total hearing loss is most frequently associated with disease or traumatic injury.

In general, hearing losses are of three major types: conductive, sensorineural, and acoustic trauma. The conductive type is not caused by sustained noise exposure. The sensorineural type may be caused by noise exposure as well as by other causes, including presbycusis (the decrease in hearing due to the aging process). If a hearing loss is produced by a noise, such as an accidental explosion, it is called acoustic trauma, and is considered an occupational accident.

Occupational Hearing Loss. The most common levels of industrial noise exposures are well below the pain threshold. There is a wide range of noise levels and frequencies to which long-time exposure may cause a slowly developing impairment of hearing. The part of the inner ear that may be damaged depends on the frequency components of the noise field that are present at the level of exposure. Individual susceptibility of the exposed worker may also be a factor. Noise-induced permanent hearing loss is first evident in a reduction in the ability to hear high-frequency sounds. As the exposure continues, the reduction progresses to the lower frequency sounds in the speech range. Exposure to noise that will produce this slow damage may sometimes be accompanied by other signs, such as a sensation of tingling or ringing in the ear when one moves out of the noise field.

Environmental noise damages the Organ of Corti in the cochlea of the inner ear. Sound-induced motion of the fluid in the cochlea induces shearing and bending movements in the Organ of Corti which, in turn, result in electrical stimuli to the auditory nerve. Prolonged and excessive noise injures and finally destroys the hair cells, and thus disrupts the sound transmission mechanism. The damage generally occurs first to hair cells associated with the perception of frequencies higher than 2,000 Hz and below 8,000 Hz. As exposure continues over many years, the damage spreads to higher and lower frequencies (2,000, 1,000, and 500 Hz), and loss of the hearing of speech becomes apparent.

In general, workers' compensation laws permit payment only for hearing loss in the speech frequencies (500, 1,000, and 2,000 Hz). The fact that initial noise-induced hearing loss generally occurs at frequencies of about 2,000 Hz provides a powerful tool for the detection of the condition and for preventive action before the loss spreads to the speech–hearing frequencies.

The word "deafness," commonly used to refer to industrial or occupational hearing loss, may be somewhat misleading when applied to the early phases of the condition. Although the word implies obvious difficulty in the hearing of speech, the difficulty during the early phases lies not so much in hearing speech as in understanding it. Hence, unless one specifically looks for this lack of consonant discrimination, the diagnosis of industrial deafness is likely to be missed. Managers may be unaware that deafness exists in their plants, since they can walk repeatedly through their noisy shops and converse satisfactorily with their employees. Too frequently, management fails to recognize that conversing in a noisy environment requires one to speak above the level of the noise and that the vocabulary used in these brief conversations is such that any employee may be able to carry on a kind of limited and stereotyped conversation even though he or she has a substantial deafness. It is not until the hearing loss extends to the lower frequencies and becomes more profound in the higher ones that poor hearing, as well as understanding, of speech becomes a prominent symptom. By then the damage to hearing is severe, irreversible, and often seriously handicapping. The major handicaps in the severe cases are in communication. Often these individuals cannot hear warning signals, such as fire alarms and telephone bells.

MEASUREMENT OF NOISE

Sound Pressure Level

Sound pressure or intensity follows the inverse-square law; that is, as the distance from the source increases, the sound level decreases.

Decibel.   To avoid working with unwieldy, large numbers in evaluating sound intensity, a logarithmic scale is used with the decibel as the unit of measure. Because decibels are logarithmic units, they cannot be added or subtracted arithmetically. If the intensity of a sound is doubled, there will be a corresponding increase of only 3 db, not double the number. If one machine caused an exposure of 90 db, a second identical machine placed adjacent to the first would result in a noise exposure of 93 db, not 180 db. On the decibel scale, zero decibels is the threshold of hearing, and 120 decibels is the threshold of pain.

Decibel Addition.   The calculation involved in decibel addition is fundamental to safety. If one knows the sound levels of two separate sources, then what is the total noise when they are operating simultaneously? The following can be used:[2]

1.   When two decibel levels are equal or within 1 db of each other, their sum is 3 db higher than the higher individual level. For example, 89 dbA + 89 dbA = 92 db; 72 dbA + 73 dbA = 76 db.

2.   When two decibel levels are 2 or 3 db apart, their sum is 2 db higher than the higher individual level. For example, 87 dbA + 89 dbA = 91 db; 76 dbA + 79 dbA = 81 db.

3.   When two decibel levels are 4 to 9 db apart, their sum is 1 db higher than the higher individual level. For example, 82 dbA + 86 dbA = 87 db; 32 dbA + 40 dbA = 41 db.

4.   When two decibel levels are 10 or more db apart, their sum is the same as the higher individual level.

When adding several decibel levels, begin with the two lower levels to find their combined level, and add their sum to the next highest level. Continue until all levels are incorporated. Figure 14.2 gives an example of how several levels can be added to find their decibel total.

Sound-Level Measurement

The pure-tone air conduction audiogram provides an accurate map of hearing acuity for a range of selected frequencies. The frequencies monitored by audiometry should cover the range of 500 to 8,000 Hz. Early noise-

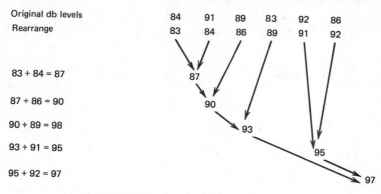

FIGURE 14.2.   Example of decibel addition.

induced hearing loss is usually centered about 1,000 Hz and thus is detected as a 4,000 Hz dip, or notch. A deepening and spreading of this notch, as exposure continues, may be predictive of eventual loss in the speech-hearing frequencies and should trigger the initiation of a hearing conservation program.

SUGGESTED LEARNING EXPERIENCES

1.   Research several workplace environments of your choice, determining their noise levels. Were the results what you expected? Explain.

2.   What do you see ahead for work-related hearing loss over the next 20 years? Explain.

3.   Prepare a training program on noise. What turned out to be your emphasis?

4.   As a safety professional being used as a consultant to a group of industrial workers, how would you account for hearing loss due to age and that due to workplace exposure?

5.   What, as a worker, do you want to know about threshold shifts?

6.   Discuss noise and the environment at large. What do you envision occurring in this sphere?

7.   Discuss in detail preventive measures in relation to workplace noise exposure.

8.   Illustrate noise measurement.

9.   Debate exposure to noise from the side of management and the side of workers.

10. Where do you place workplace noise in your list of workplace hazards? Explain.

REFERENCES

1. Chamber of Commerce of the United States. *Analysis of Workers' Compensation Laws* (1981 ed.). 1982.

2. Jensen, P., Jokel, C. R., and Miles, L. H. *Industrial Noise Control Manual* (Rev. ed.). Cincinnati, Ohio: NIOSH, 1978.

3. National Safety Council. *Accident Prevention Manual for Industrial Operations* (7th ed.). Chicago, 1977.

4. Occupational Safety and Health Administration. *Code of Federal Regulations.* General Industry Standards (29 CFR 1910). U.S. Department of Labor, Washington, D.C., June 1981.

PART 4 COMPONENTS OF SAFETY ADMINISTRATION

15 PLANT SAFETY MANAGEMENT

Plant safety managers are responsible for the day-to-day operation of the safety department, which is usually a one-or-two-person operation reporting to the chief executive operating officer (CEO) at the plant. Besides hands-on safety engineering, which includes safety inspections, accident investigations, and safety training, they must administer the safety department.

Safety management consists of applying general management principles, such as span of control (in which the manager receives authority from top management and delegates it to as many subordinates as his or her management can effectively span, usually about seven), for the protection of people, material, and equipment.

PRINCIPAL DUTIES OF THE SAFETY DEPARTMENT

The principal duties of the safety department depend on the company's business, but in general terms they consist of providing safety training, conducting safety inspections and surveys, investigating accidents, overseeing workers' compensation claims, coordinating safety activities with the insurance carrier, supplying and supervising the correct use of personal protective equipment, assuring that OSHA requirements are met, and assuring that company safety policies and safety procedures are followed.

Safety Effectiveness

Safety engineering has not always been as effective as it might have been, mainly because management has been primarily concerned with productivity—sometimes at the expense of safety. It is important that the safety manager report directly to top management. This gives safety corporate visibility and the "clout" that such official support confers. In most organizations employees know whom they should report to. See Figure 1.1 in Chapter 1 for an ideal approach. In Figure 1.1 the safety manager reports directly to the CEO.

Relationships among Groups that Handle Safety Problems

Several other groups are also concerned with safety problems and can take action if plant safety is not deemed effective enough. These are quality assurance, quality circles, personnel departments, legal departments, and security departments.

Quality assurance inspects the output of the plant to ensure it meets the specifications. These may give an indication of potential hazards and product safety problems.

Quality circles are rather new in the United States but have been used extensively in Japan to analyze problems and to improve workplace conditions. Their solutions can enhance safety.

The personnel department is directly concerned with safety problems because the burden of hiring replacements for workers on leaves of absence due to work-related injuries and/or illness falls upon it.

The legal department plays a significant role in the handling of safety problems. It represents management in safety-related litigation and investigation of claims. It may become involved in workers' compensation claims, public liability, and product assurance.

The security department also has an important role in handling safety problems. It inspects the premises so that intruders will be kept out. In its routine security patrols it can often observe safety hazards and potentially dangerous situations. This aids the safety department by providing additional observers who may be in locations that safety may not inspect.

The First-Line Supervisor

First-line supervisors represent the goals of both management and the workers and, therefore, must have a clear understanding of each. Supervisors are in an ideal position for safety training and direct safety supervision. They are also in a focal position in relation to the safety engineer, because they are responsible for safety policies and for reinforcing their use by the employees. It is very important for the plant safety professional to be allied with first-line supervisors, in the interests of plant and safety.

The Safety Committee

Safety committees are composed of management and employee representatives. They are a means of involving employees in the safety decisions that affect them.

Committees should be small enough (no more than eight members) for effective communication, including discussions, voicing of differences of opinion, and resolution of conflicts; at the same time they should be large enough to accomplish the committee's task. If the safety committee is composed of shop stewards, it often becomes a place for union grievances to be aired.

The safety committee's duties depend on what management has designated as its functions. In most companies committees make recommendations to management; they are not decision-making bodies. They assess employee complaints and management concerns and attempt to examine issues objectively and make recommendations. It is critical that the safety committee's duties be clear in relation to management and employees, so that it serves both constituencies responsibly and effectively.

The effectiveness of the safety committee varies, depending on the authority conferred on it by management and the level of commitment of the individual members to developing and implementing a safety program. The safety committee can perform activities from routine inspections, to recommendations for personnel protective equipment, to establishing penalties for noncompliance with safety policies.

The safety committee can be a very important, respected, and influential means for employees to voice their concerns and assure that something is done about them, or it can be another meaningless committee that keeps minutes and wastes time.

Liaison with Company Insurance Carrier

One very important function of the plant safety engineer is the role as liaison with the company insurer(s) and with workers' compensation boards to work out differences and keep communication channels open. Serving as liaison can be difficult if the safety professional has strong biases and beliefs which are contrary to those of the insurance carrier and/or workers. The plant safety engineer must always remember that his or her job is to present facts and must not let personal values interfere so as to affect the worker.

Relationship with the Community

Safety becomes an important community concern if the plant's products or effluents in any way affect the local environment. If an industrial concern does not have the support of the community in which it is operating, unnecessary ill will can be generated. It is much easier and more efficient to involve the community from the beginning, as much as practical, in safety concerns which may have an impact on its citizens and their environment. It makes more sense to educate the members of the community about possible hazards than for them to find out after the fact and respond by trying to close the plant down or, at least temporarily, halt operations.

Although community members need not be informed of day-to-day plant operations or safety decisions, it is important to keep them abreast of key decisions that might affect their environment and to keep communication channels open. Being a good neighbor is good business.

Professional Safety Societies

It is important for safety engineers to belong to professional safety societies such as The American Society of Safety Engineers. They can benefit from contacts with other safety professionals and from exchanges of information. The professional contacts can serve as a boost to morale and as a way to learn strategies for effectively handling current problems.

It is also helpful in the implementation of safety at the plant level for managers to belong to some of these professional societies and to attend their meetings. Group affiliation with safety professionals helps the implementation of safety principles. See Chapter 5 for further information about safety organizations. It is also important that safety engineers obtain their Certified Safety Professional credentials.

Selling Safety to Management

How do we convince management of the long-range cost effectiveness of safety?

Management is interested in profits and the return on investments. What will be the return on investment per dollar spent for safety? If management can be convinced that safety measures are cost effective, that is, that they will reduce operating expenses and increase profit, it is more likely to agree to support a plant safety program. To convince management, the safety manager must marshal an array of facts to support the contention that safety is in the company's best interests as well as the employee's. Safety factors which can be quantified include:

1.   Dollar losses—uninsured, such as deductibles and uncovered losses
2.   Indirect costs
3.   Insurance premium
4.   Insurance experience modifications
5.   Insurance loss ratio—the ratio of insured loss to total loss

Direct losses refer to time lost from work by an injured employee, time lost by fellow employees, and money that the company's insurance carriers pay to the employee. Sometimes these costs are difficult to determine exactly because the long-term effects and, therefore, costs of an accident can't be assessed when it occurs.

Indirect Costs.   Indirect costs are even more difficult to assess than direct costs. They include cost of training new employees, losses through failure to complete contracts, overhead cost while work is disrupted, loss in earning power, and economic loss to employee's family. These were discussed in Chapter 4.

Insurance Premiums.   Insurance premiums are paid directly by a company to the insurance carriers for protection. The rates are prorated and are based upon likelihood of an accident and a company's past record.

Insurance Experience Modification.   "Experience rates" are workers' compensation rates determined for each company based upon its accident record. The accident record is based upon the risks inherent in the business, weighting such factors as the kind of business; the types of hazards; hazards beyond the control of the company, such as electric cable explosions or severe climatic conditions; and the company's program for accident and injury prevention. It is this last factor over which workers have the most control and can effect the most change.

Safety Requirements

Several factors must be taken into account in developing safety requirements for a company. These are:

1. Plant safety policy and administrative procedures. What policy has been established by the company to deal with safety results peculiar to its business, and what company procedures have been instituted to accomplish these goals?
2. State and local regulations concerning the industry in general and the company in particular. What specific regulations must the company comply with in order to stay within legal bounds?
3. Occupational Safety and Health Act (OSHA) limitations and specifications to which the company must adhere.
4. Other federal regulations published for the company's particular type of operation.
5. Regulations of the Nuclear Regulatory Commission (NRC) that apply to the company's operation.

6. Environmental Protection Agency (EPA) regulations that the company must take into consideration.

7. Applicable Equal Employment Opportunities Commission (EEOC) regulations.

Every company must have someone who is knowledgeable about these issues so they can be sure that they are operating within the limits prescribed by these agencies and can plan and function accordingly.

Accountability

In considering the implementation of safety at the plant level, responsibility can be determined by work activities, by the results of work activities, or by a combination of both. Who is responsible for safety? Is management to be held responsible if it does not assign a qualified person to implement safe working conditions? Is the safety professional accountable for every accident that occurs? Is safety the responsibility of the first-line supervisor? Should an employee's accident go on the supervisor's record? Is each employee responsible for his or her own accidents? Is it not the employee's responsibility to be aware of risks involved in his or her work and to question any procedure or task that is not safe?

Who will make these decisions? The important fact here is that every employee, no matter what his or her position, must be very clear about who is accountable for what and be sure that it seems fair and, if not, that it is negotiated.

Companies regard accountability in terms of safety results, number of accident claims, and amount of insurance premiums, all of which are based upon accident and injury rates.

Safety results are based upon a company's safety record. How many incidents have occurred that are in some way related to safety factors?

Do some supervisors have better employee safety records than others? What factors are involved? One way to document safety results is by sampling safety procedure. This is a systematic observation of workers on the job to determine if, and how often, they are working unsafely. The first step is to list work procedures and identify each as either safe or unsafe. Then during an unannounced walk-through inspection of the plant, employees are observed as they perform the listed procedures. The number of unsafe and safe actions that were performed is noted. One then determines whether the unsafe acts are above the average for that procedure. As with any sample, one must be careful not to generalize results, since only a portion of the population was used. However, trends can be identified, pinpointing increases or decreases in accident rates.

The last step is to prepare a report for management, showing each employee's rating (expressed as a percentage of safe to unsafe actions). The report will also include each supervisor's rating based on the number of the employees supervised.

It is one thing to develop an interest in safety, but is is even more important to sustain interest. Management understanding and agreement must be obtained and maintained before any safety incentive program is established.

Many programs or campaigns can be developed to maintain an interest in safety.

There are many other ways that workers can be reinforced for their participation in a working safety program. These incentives can be very diverse and can depend in large measure on what is important to the employees. What are the possibilities in their work situations, and what will make them feel good about the work they have put in and feel good about themselves and want to continue? To one employee a formal award might be important, to another employee a special luncheon would be important, to another employee a free hour would be important, while to another employee being placed on a prestigious committee might be important.

This area of sustained interest in safety is an important one to develop in order for plant safety not only to become an industrial priority but also to stay a priority.

Safety Inspections and Surveys

The purpose of safety inspections and surveys is to identify hazards, to determine possible sources of hazards, to find pollutants in the work environment, and to assess safety precautions in the workplace.

Inspections should be a routine part of the safety professional's functions, and safety surveys should be done as often as necessary.

Accident Investigation

There should be thorough investigation of every industrial accident, describing details of the accident, the sequence of events leading to the accident, possible causes, and how similar accidents might be prevented. The investigation must be conducted impartially and objectively, citing all facts regardless of who is implicated. To prevent accidents, we must know cause–effect relationships.

Overseeing of Workers' Compensation

This is a priority function because, more than any other single factor, it reflects the cost effectiveness of the safety program in the form of lower insurance premiums resulting from fewer accidents and a safer working environment. The safety professional will make sure that the company is receiving the lowest rating to which it is entitled. Insurance ratings are determined in one of two ways: (1) by the manual rating, which is based on the industry classification assigned by the National Council on Compensation Insurance and determined by the risks inherent in a particular industry, or (2) by the experience rating, which is a combination of the manual rating and the individual company's accident rate. The latter method could mean an increase or decrease in insurance premium, depending on accident rate.

Supply and Use of Personal Protective Equipment

It is one thing to distribute protective gear such as earplugs, hard hats, or heavy shoes. It is another to insist that these items be worn. Despite education about their necessity and the obvious intent to minimize risk of injury, many workers seem reluctant to use them. This resistance, which is encountered mainly in men, may arise from the "macho" myth—the image of the bold, brave male scornfully refusing "sissy" protective equipment. Others may complain that it hampers or impedes them in some way (if so, it could be poorly designed). Regardless of the reasons, the safety engineer must insist that safety gear be worn and impose penalties for nonuse.

Fulfilling OSHA Requirements

The safety professional must see that the appropriate documentation is maintained to satisfy the Occupational Safety and Health Code. Some of the reporting can be tedious, but portions can be assigned to others to avoid a frantic rush when an OSHA inspection is announced. Thorough records should be kept and routinely used to assess and pinpoint how to prevent accidents or eliminate hazards in specific places.

Safety Training

To implement an effective safety program, all employees and supervisory personnel must receive safety training. By means of demonstrations, all employees must be shown the safest way of doing their jobs, especially those requiring the use of hazardous materials or machinery. They should be informed about company policy for specific procedures and be kept up to date on new rules or technologies which may affect their work.

Special programs, such as a back-injury prevention program, can be helpful in industries where improper lifting or turning can result in back injury. This is especially important because so few people know the correct way to lift heavy objects.

In terms of staff development and education, it is important to provide information on the most recent developments in hazard analysis of substances that are handled by the staff. At information meetings, existing hazard analysis procedures should be compared with the new suggested methods and plans should be developed.

SUGGESTED LEARNING EXPERIENCES

1. What pointers would you provide safety management for improved plant safety?

2. If you were a safety manager, what committee would you want to have in your plant to help in the implementation of safety and health principles? Why?

3. Debate: The worker and the union representative should decide plant safety issues that relate to employees, the union, the safety committee, top management, and the safety professional.

4. Make a visual presentation, such as an organization chart, on employee interrelationships in a plant setting and how they enhance or detract from productivity.

5. Besides the interfaces shown in Learning Experience 4, what aspects of quality assurance affect plant safety?

6. If you were a safety manager, what would be some strategies you would use to improve plant safety?

7. What would you like the safety committee in your industry to be responsible for?

8. What do you consider your most important role as a safety manager at the plant level? Discuss.

9. Conduct a survey of employees, asking what three factors about their job they would like to change. Did they include anything regarding health and safety?

10. Review the literature on safety management. Select a general management principle other than span of contol and apply it.

REFERENCES

1. Dionne, E. Motivating workers with incentives. *National Safety News*, 1980, **99,** 75–79. (Jan.)
2. Nertney, R. J. *Management Factors in Accident and Incident Prevention.* U.S. Department of Energy, Washington, D.C., 1978.

16 CORPORATE SAFETY

The corporate safety department writes safety policies and procedures and publishes a corporate safety manual. It also has some supervisory and administrative control over the plant safety programs. While the plant safety engineer relates to a number of groups at the plant level, the corporate safety department has line and staff relationships with other corporate departments. The plant safety manager reports to the plant manager, while the corporate safety director reports to the chief executive officer (CEO). Where safety is placed in the organization reflects its relative importance in terms of corporate priorities.

DIRECTOR OF SAFETY

The director of safety will decide how safety is to be handled within the corporate structure. The relative degree of visibility and importance of the department is reflected by the number of safety personnel assigned to it, the cooperation given by line employees, and its relationship with other departments on the same corporate level. The safety director performs most activities through coordination, collaboration, and consultation, which are defined in Table 16.1.

Safety must be handled from the corporate level with all departments involved, and it must have the approval and support of management.

The organizational structure in a large corporation and a small company may be similar, but in the large one the relationships are more complex. In the large corporation the safety department may report to the director of environmental health and safety, while in a small company it reports directly to the CEO.

Safety must have the support of management. If it doesn't, it does not really matter to whom the director of safety reports.

TABLE 16.1
Coordination, Collaboration, and Consultation Defined

Coordination: Process whereby different segments are brought together.
Example: The director of safety will often function in the coordination role in bringing together two departments who have differing views on the same subject.

Collaboration: Process whereby two or more persons discuss the same concept, project, and proposal and together arrive at ways to handle it.
Example: The director of safety, the industrial hygienist, and the environmental toxicologist assess a workplace spill and present a proposal to management on the next steps to take.

Consultation: Process whereby one person asks another for advice and perceives him or her as an expert in that particular situation. It is important to remember that the consultant is in an advisory role and the consultee (person seeking the advisee) has the choice of whether to use this advice or not.
Example: The safety director provides consultation to the epidemiologist on examining workplace data concerning what appears to be a significant increase of accidents during a particular industrial operation.

NOTE: One of the differences between consultation and supervision is that consultation does not employ responsibility whereas supervision does.

Petersen thinks that the placement of safety depends on the organization and the personalities of people.[2] But many managers and safety professionals take exception to this view and believe that the organization should be built around the task at hand, using tested and proven structures.

There are several criteria to determine how safety is placed in an organization. These are:

1.   *Report to a boss with influence.*   In part this is a personal evaluation; in part it is an evaluation of the structure. If the boss is line, does his or her line authority encompass the hazards to be controlled? If staff, can his or her voice reach an executive whose command will cause necessary action?

2.   *Report to a boss who wants safety.*   Problems arise when a chief executive wants results, but the needs of the safety director are unmet by an immediate boss who is devoted to other problems.

3.   *Have a channel to the top.*   Management properly sets the priorities between production results and safety results, between sales expansion and elimination of unsafe driver-salespersons, between security of confidential research and the prying eyes of the safety director. This is not to say that safety must be placed in the upper echelons; but it does assert that channels to the upper echelons must exist so that all parts of the organization can be heard. Too often the only channel is a bypass with all its frictions.

4.   *In the organizational structure, locate the safety function under the executive in charge of the major activity.*   The safety function in this case eliminates

the channel to the top. However, if the acute need is in the shop, it is better that the safety personnel work with the shop executives directly. If the acute need involves truck operation, it works better if the transportation executive handles the problem with the safety director as staff assistant.[2]

Every corporation should be able to look at this criteria and decide what is best for its own needs as to the maximum placement of safety. This safety placement will take into account the goals and objectives of management, as well as those of safety and health, and governmental and public demands and regulations.

But most of all, this placement of safety will be determined by management goals and objectives and by negotiations with the correct safety person(s) that will give both management and the safety person what they want.

BUILDING A SAFETY DEPARTMENT THROUGH ORGANIZATION DEVELOPMENT

One of the ways that has proved to be effective in the development and building of work relationships involving personnel from different sectors of the work population is organization development (OD).

Organization development is a prescription for a process of planned organizational change. French and Bell state some characteristics which differentiate organization development from traditional interventions such as small groups, T-groups, and separate-division-level meetings. These stated characteristics can also be seen as reasons why organization development interventions continue to be used by corporations that have found that organization development outcomes are measurable and show results. However, for a corporation to support organization development interventions, the corporate structure must be committed to change if that is considered necessary for positive growth.

The characteristics stated by French and Bell are[1]:

1. An emphasis, although not exclusively so, on group and organization processes in contrast to substantive content
2. An emphasis on the work team as the key unit for learning more effective modes of organizational behavior
3. An emphasis on the management of the culture of the total system and total system ramifications
4. An emphasis on the collaborative management of work team culture
5. The use of the action model
6. The use of a behavioral scientist change agent or catalyst
7. A view of the change effort as an ongoing process

One of the very helpful ways that organization development has aided the corporate safety department is in its relationships with all employees, those above them, those on line with them, and those under them.

From being exposed to organization techniques, the corporate safety professional has assessed his or her impact upon people, relationships with others, and strengths and weaknesses; has learned ways to impart information in a more receptive way; and has learned strategies for being assertive without being aggressive if aggression has the possible capabilities of acting as a disclaimer or an obstruction to obtaining the following:

1. The information (data) the corporate safety professional needs to obtain
2. The information regarding work safety that is to be imparted and heard
3. The necessary cooperation from all employees in order to have a safe working environment

TABLE 16.2
Sequence of Events as They Might Occur in an Organization Development Program Focusing on Improvement of Safety

---

1. Middle or top management of an organization becomes interested in OD and feels that the organization has problems which can be met through training.
2. Management invites an outside OD consultant to visit the organization.
3. After the consultant's entry and contact with a variety of organization roles and groups, the organization works out a contract with the consultant specifying the nature of the projected relationship and its goals and general procedures.
4. The consultant, working with insiders, collects data about the organization via interviews, questionnaires, and observations.
5. These data form the basis of a joint diagnosis of the points of difficulty in the organization and, if appropriate, between the organization and its environment. Goals for change are explicitly identified.
6. A first "intervention" (usually an intensive meeting involving several key persons, a group, or more than one group) is planned. The data collected earlier are often fed back and discussed. Exercises for training in communication skills or group functioning are often used as constructive vehicles for discussing the data.
7. The intervention is evaluated following a new collection of data. Often the future success of the effort depends on the degree to which key figures have been "freed up" to be more open, concerned, and creative about improvement. (In this case, it would be improvement of safety.)
8. Subsequent steps in intervention are planned on the basis of these data, and the process continues.
9. More and more people are involved and become supportive of the desired outcomes.

SOURCE: Reprinted from: Miles, Matthew, B., & Schmuck, Richard A., *Organization Development in Schools*, San Diego, CA: University Associates, Inc., 1971. Used with permission.

TABLE 16.3
Modes of Intervention Used in Organization Development

1. *Training or education.*   Procedures involving direct teaching or experience-based learning. Such methods as lectures, exercise, simulations, and T-groups are examples.
2. *Process consultation.*   Watching and aiding ongoing processes and coaching to improve them. (Videotapes might be used here.)
3. *Confrontation.*   Bringing together units of the organization (persons, the roles that the persons play in the work situation, or groups), which have previously been in poor communication. Usually accompanied by supporting data.
4. *Data feedback.*   Systematic collection of information, which is then reported back to appropriate organizational units as a base for diagnosis, problem solving, and planning.
5. *Problem solving.*   Meetings essentially focusing on problem identification, diagnosis, and solution intervention and implementation.
6. *Plan making.*   Activity focused primarily on planning and goal setting to re-plot the organizations's future.
7. *OD task force establishment.*   Setting up ad hoc problem-solving groups or internal teams of specialists to ensure that the organization solves problems and carries out plans continuously.
8. *Refocusing activity.*   Action which has as its prime focus the alteration of the organization's structure, work flow, and means of accomplishing tasks.

SOURCE: Reprinted from: Miles, Matthew, B., & Schmuck, Richard A., *Organization Development in Schools,* San Diego, CA: University Associates, Inc., 1971. Used with permission.

The description of an organization development program sequence shown in Table 16.2 is just an example of how an organizational program could be implemented. The modes of intervention used in organization development are shown in Table 16.3.

The purpose of organization development is to improve a corporation's functioning by enabling individual members to examine their effect on the organization and their communicating patterns. This is done so that all members of an organization can accomplish tasks without getting in each other's way but instead complementing each other. Many organizational concepts sound like common sense, and an employee may wonder why a communication expert has to make him or her aware of such communication patterns. An example would be the safety director's inability to communicate to the CEO because the director's authoritative manner makes the CEO want to rebel against anything the director says and do just the opposite.

Safety's relationship with other departments is crucial because of changing priorities depending on corporate objectives, private and governmental funding, public needs, consumer demands, employee motivation and satisfaction, and accident rates of the corporation. In the competition for corporate funds, the safety director must make sure his or her department does not get cut back to the point where personnel safety is endangered. Instead, the director must use tact, diplomacy, and perseverance at all levels to effectively gain employee cooperation in having a safe working environment.

CORPORATE SAFETY POLICY

The corporate safety department is responsible for carrying out the company's safety policy. The underlying principle of such a policy is to publish safety regulations and procedures and see that every employee knows and abides by the rules. Because of the importance of the safety manual and the potential for trouble if it isn't followed, it is very important that it be distributed and employees required to read it and understand their roles and responsibilities and stated policies. Table 16.4 lists the minimum content of a corporate safety policy.[3] The corporate safety policy has several components, including management's commitment to safety, the directives issued by management, the management policy manual, and the safety manual.

The importance of top management's commitment to safety cannot be overstressed. This commitment must include not only the official company policy but the actions to support it. Failure to enforce safety rules subtly betrays a lack of real commitment on the part of management.

TABLE 16.4
Format for an Occupational Safety and Health Manual

1. *Introduction.* A brief statement of management's objectives.
2. *Scope of safety activities.* Is the manual for corporate wide use or does it apply to some narrower group? What safety and health activities are covered: on-the-job safety, off-the-job safety, industrial hygiene, public safety, laboratory safety, product safety, fleet safety, hazardous material, hazardous waste products?
3. *Authority.* Who prepares policies? Who approves policies? Who is responsible for implementation of policy? Who will enforce the policy? How much authority does each of the above have?
4. The basic elements of the safety program.
5. Documentation, correspondence, inspection results, industrial hygiene survey, safety programs, accident investigations, safety committee meetings.
6. *Safety policies and standards.* Some examples are personal protective equipment, flammable and combustible liquid handling, dispensing, and storage, chemical substance control, hearing conservation program, radioactive materials, welding in closed containers, testing of high pressure systems, visual display units.

TABLE 16.5
Safety Activities for Top Management

1. Safety meetings
2. Safety committee meetings
3. Safety inspections
4. Safety problem diagnosis and solutions
5. Accident/incident investigations
6. Safety training classes and/or programs
7. Uses personal protective equipment when appropriate
8. Supports employee participation in safety programs
9. Attends professional safety meetings with safety personnel

SOURCE: Nertney, R. J. *Management Factors in Accident and Incident Prevention.* U.S. Department of Energy, Washington, D.C., 1978.

Employees pick up such nonverbal cues and sense the contradiction between official policy and an unofficial laissez-faire attitude. There are ways to evaluate management's support for the corporate safety program, of which Nertney lists eight.[2] These are shown in Table 16.5.

It is up to management to communicate its concerns and plans in clearly stated policies and directives. This is especially important in the implementation of safety policies of the organization. Safety policies must be congruent with directives. They must be thoroughly assessed and evaluated; otherwise, employees will handle discrepancies by doing nothing, by attempting to resolve the differences, or by choosing a particular rule to follow. These options cost time and energy which could be spent more productively.

The corporate policies and procedures manual includes managerial decisions, directives, and procedures with which all employees should be familiar. The management policy manual must address all managerial issues that employees have a need to know.

Corporate Safety Manual

The company safety manual comprises every policy and procedure required to assure a safe working environment and promote safe work habits. The safety manual is for all workers and their responsibility for carrying it out must be emphasized. Safe working conditions can be assured only with the full cooperation of everyone. Examples of what may be included in a safety manual are shown in Figures 16.1 through 16.3. It is, of course, the responsibility of each company to decide what safety policies are applicable to their business and, therefore, what to include in the safety manual.

SAFETY POLICY FOR RADIOLOGICAL CONTROL

Safety Management Manual: *Policy Statement*
Prepared by Systems and Procedures: *Harry Emmanuel*
Approved by: *General Manager R. K. Elizabeth*
Effective date: *8/8/82*
Subject: *Radiological Control*

Policy

The Division will implement and maintain a Radiological Control Program in accordance with contract and statutory requirements to permit the accomplishment of radiological work associated with nuclear research. The program shall be strictly enforced to ensure that radiological operations are not hazardous to the health and welfare of employees, the environment, or the community. All division employees shall be aware of and comply with the appropriate radiological practices and controls necessary to perform their work.

Policy Implementation

A. The Radiological Control Department is responsible for ensuring that the Division meets all contractual and statutory requirements for radiological control. Specific responsibilities include:

1. Establish, direct, maintain, and be responsible for the overall Division Radiological Control Program to ensure all contractual and statutory requirements are met.
2. Identify radiological control support requirements to the appropriate Division managers or directors. Identify unfulfilled requirements and interprogram priority conflicts to the Deputy General Manager or General Manager for resolution.
3. Participate in evaluating the performance of Division management and key supervision on whom satisfactory performance of the Division Radiological Control Program depends.
4. Monitor the amount of radiation exposure received by personnel and emitted to the environment, and assure that limits are not exceeded.
5. Monitor radiological operations, conduct area and environmental surveys, and control radioactive material to preclude hazard to personnel and the environment.
6. Maintain an effective radiological accident plan to control radiological emergencies. Initiate emergency action when necessary to safeguard personnel and the environment from the effects of ionizing radiation.
7. Evaluate radiological problems, deficiencies, and incidents for basic causes, and initiate corrective action.
8. Maintain accountability for all radioactive material in the laboratory. Minimize the amount in the laboratory by (a) developing means for rapid handling of radioactive material which is ultimately to be shipped out, (b) establishing adequate control of all radioactive material in the laboratory, (c) assisting in reducing the amount of solid waste products in radiation work, and (d) reviewing requirements for radioactive material with procurement. Engineering and Nuclear Engineering shall minimize the amount of radioactive material coming into the laboratory.
9. Document radiological activities, personnel exposure to radiation, radiation effects on the environment, radiological qualifications, radioactive material discharges and shipments, and other records and reports as required.
10. Provide technical assistance with radiological control facilities and processes, including those relating to purchase specifications and material handling.
11. Provide radiological control training to include (a) indoctrination in basic radiological control safety requirements and (b) special radiological control training programs. This includes review of training provided by other departments to assure

FIGURE 16.1. Safety policy for radiological control.

that radiological control input is sufficient to assure personnel and environmental safety to meet the needs of the program.

12. Develop and conduct radiological control monitor training and retraining programs to maintain an adequate force of qualified monitors.

13. Determine radiological conditions of arriving equipment, maintain updated radiological conditions of equipment during repairs.

14. Provide technical direction and services in the field of health physics.

15. Provide and maintain instrumentation and equipment to support monitoring of radiation, contamination, and personnel radiation exposure.

B. The Operations Department is responsible for the following:

1. Direct and perform laboratory radiological operations, ensure compliance with procedures, and ensure work is performed in a manner which minimizes radiation exposure to personnel.

2. Train or obtain assistance to train personnel in correct radiological control work practices.

C. The Nuclear Construction Engineering Department is responsible for the following:

1. Direct and perform radiological operations under their cognizance, ensure compliance with approved procedures, and ensure work is performed in a manner which minimizes radiation exposure to personnel.

2. Provide specific radiological control directions in trade work paper in accordance with shipyard procedures and contractual and statutory requirements.

3. Control the list of personnel eligible for work in radiation areas.

4. Conduct independent audits of radiological operations and recommend programs to improve performance.

5. Conduct a continuing program to reduce personnel radiation exposure.

6. Plan research work in a manner which minimizes radiation exposure to personnel.

7. Train or obtain assistance to train personnel in satisfactory performance of work associated with radiological controls.

D. Each Department Manager is responsible to implement radiological control practices in compliance with approved procedures, minimize the number of personnel assigned to work in radiation areas, and accept responsibility for and take action on department's portion of the Radiological Control Program.

E. Each Division employee is responsible for the following:

1. Comply with specific radiological control instructions, either written in work paper or given verbally by Radiological Control.

2. Minimize radiation exposure to himself or herself and to others.

3. Contact Radiological Control in any radiological situation where there is any doubt as to the proper action.

FIGURE 16.1. *(Continued)*

Some safety policies state that under certain situations the safety specialist must be consulted and must approve or temporarily step in and assume command (stop the operation). When it is necessary to do this, the safety professional must remember that effective carrying out of safe procedures is done by influencing line management and not by directing it. Anytime that safety must step in and assume direction of the line, a failure has occurred in that the line had not taken care of the situation first.[3]

ACCIDENT PREVENTION PROGRAM SAFETY POLICY.

Safety Manual
Approved: *M. E. Brownfield*
General Manager: *R. K. Elizabeth*
Subject: *Accident Prevention Program*
Effective Date: *5/1/82*

Policy

The strength of our company lies in our people. Therefore, I am personally dedicated to the principle that the protection of our employees is an essential prerequisite to the attainment of our objectives. In simplest terms, I have given the Safety Program my fullest support and expect that each employee will do the same.

The elimination of accidents and health hazards must be a continuing concern and objective of every one of us. It is my policy that we provide and maintain a work environment free from hazards that could cause injury to any employee and that we comply with all State, Federal, and Division regulations established to safeguard the life and health of our work force.

I trust that each of you will join me in pledging your leadership ability to gain personal commitments to total accident prevention and loss control as a way of life. To attain the maximum results from our program, each of us must be dedicated to the idea that every accident and unnecessary loss can be prevented.

Policy Implementation

To make this program effective, every member of management shall ensure that:

1. Work is not assigned which is hazardous or is located in a hazardous area until all steps have been taken to provide for the safety of the employee.
2. All employees have received proper job instruction and are familiar with pertinent safety and health laws, regulations, and instructions, including using personal protective equipment.
3. Work areas are frequently examined to ascertain that the work environment is safe and that employees are working in a safe manner.
4. All safety and health deficiencies are corrected immediately and not repeated.
5. A monthly review is conducted of the accident experience in his or her area and an action plan is initiated to improve that experience where necessary.
6. Supervisors set a good example by complying with all safety rules.

Our goal is zero accidents. You must carry out your part of the program as you would any other responsibility if we are to achieve this goal.

FIGURE 16.2.   Accident prevention program safety policy.

Such decisions are based on what is most effective for the place of operation. The important point is that complete safety information must be supplied in written form to all employees. Details depend on the industry. For example, in an industry doing piecework, where all heavy lifting is automated, the safety policy would not include instructions for safe lifting of loads; on the other hand, a business like the postal service must define safe lifting loads and correct lifting procedures in the safety manual. An

Key Safety Responsibility	Yes, Sometimes, or No?	What Should Be Done to Improve?

1. Have I established group safety goals and objectives to back up company safety policy and company safety goals and objectives?

2. Have I provided adequate guidance to my people in determining:
   a. Acceptable risks?
   b. Risks that they can accept on their own?
   c. Required approvals for accepting risks?

3. Have I established rules regarding procedures and methods to create them? When required? What kind? Content? Sign-off? Required approvals?

4. Have I established rules regarding safety analysis? Job safety analysis? Safe work permits?

5. Have I established rules relating to compliance with codes, standards, and regulations?

6. Do I require a safety information search when new materials, hardware, processes (or changes in processes) are used?

7. Do I require full life-cycle safety analysis work—from concept to closeout, decommission, and disposal? Do I require preliminary hazard analyses for planning purposes?

8. Have I established availability of technical information systems so that I know what risks and hazards exist in work being done in my group or organization?
   a. Do my people have the technical information to do their work safely?
   b. Do my managers know what risks and hazards are involved in work that my group is doing?

9. Have I established work-monitoring systems so that I really know what my people are doing and what their safety problems are?

10. Is the proper technical information for work being done in my group being collected and analyzed?

11. When problems arise, do troubleshooting and analysis lead to:
    a. Fixing what went wrong?
    b. Fixing defects in the system that allowed things to go wrong?

12. Is work done in my group or organization properly tied into the company's independent audit and appraisal system (especially Q/A and safety review)?

FIGURE 16.3. Checklist for the line manager's role in goal setting, establishment of safety concepts and requirements, and technical information flow. Source: Nertney, R. J. *Management Factors in Accident and Incident Prevention.* U.S. Department of Energy, Washington, D.C., 1978.)

industry requiring precision work would specify safety levels of illumination, recommendations for breaks to prevent eye strain, and general information on small, repetitive tasks, while coal miners would not need this type of information in their work.

What to include in a safety policy manual is an important research project for the safety professional, especially if no manual yet exists. In the areas of hazard prevention and health promotion, what goes into the safety manual and how the safety procedures are carried out will directly affect corporate safety and health of each organization.

The way in which management delegates responsibility has a bearing on how effectively it will be implemented. Nertney has developed a checklist (see Figure 16.3) which lists the safety manager's role in safety.[2] This is an excellent tool for evaluating what the safety manager is doing regarding safety and can do in order to become more effective in development, maintenance, and establishment of safety policies within an organization.

SUGGESTED LEARNING EXPERIENCES

1. Develop an overall corporate safety program for a multiplant corporation which produces either cosmetics or plumbing equipment.

2. Compare the corporate strategies for safety in manufacturing plants, construction sites, oil rigs, and hospitals.

3. Write a job description for a corporate safety director.

4. One of your plants has been accused of "midnight dumping." How would you handle this problem at a corporate level? How would you go about developing a corporate policy and/or procedure to control hazardous waste?

5. The bottom line in many corporations is return on investment (ROI). How would corporate safety coexist in this sort of management?

6. Discuss the pros and cons of the following: No safety manager should accept a position unless it includes direct access to the chief operating officer.

7. You have been invited to give a presentation to a group of safety managers. Develop an outline and your visuals on the topic of corporate safety. Remember, these people also manage safety programs in corporations.

8. Discuss the differences between corporate safety and safety management at a plant level.

9. What should be the educational and training requirements for a corporate safety director?

10. What would your company's policy be regarding "walk around pay" for its employees?

REFERENCES

1. French, W. L., Bell, C. H., and Zawacki, R. I. *Organization Development: Theory, Practice and Research*. Plano, Tex.: Business Publications, 1978.

2. Nertney, R. J. *Management Factors in Accident and Incident Prevention*. U.S. Department of Energy, Washington, D.C., 1978.

3. Petersen, D. *Techniques of Safety Management*. New York: McGraw-Hill, 1971.

4. Sell, R. G., and Shipley, P. (Eds.). *Satisfactions in Work Designs: Ergonomics and Other Approaches*. London: Taylor & Francis, 1979.

5. Shaber, E., Byrom, J., Chandler, D., and Eicher, R. *ERDA Guide to the Classification of Occupational Injuries and Illness* (ERDA 76-45/7). Idaho: EG&G, Inc., 1976.

17 STARTING A SAFETY TRAINING PROGRAM

Safety training is one way to gain maximum return on money invested in a safety program. A well-trained work force is less apt to have accidents and more likely to inform the safety department of hazardous conditions. Safety training should include every worker from the newest person most recently hired to the person nearing retirement and from the cleaning person to the president of the corporation.

Safety training is always required but is needed most urgently when accident rates have increased or are abnormally high, when new equipment is installed or new processes are used, or when there is high personnel turnover. Most safety training focuses on accident prevention, which should include instruction in new equipment or processes.

This chapter describes how to start a plant-level training program. It tells how to perform an analysis of the safety issues within a plant; how to perform a job safety analysis; and how to do task and skill analyses, which are translated into safety training. It discusses forms of learning and reinforcement and methods of presentation in terms of maximum functioning of personnel. Safety training is one of the most important jobs a safety engineer performs.

A high or increasing accident rate definitely calls for a safety improvement program. Safety training with employer and employee cooperation and safer operating methods may be sufficient to improve safety performance.

Safety training should result in more accurate reporting of accidents and injuries, although willingness to report depends on several factors, such as the reward for reporting versus the punishment for reporting and the reporter's assessment of what action will be taken. Often when safety is discussed what is meant is frequency of injuries. If greater emphasis is placed on plant safety, people will be more likely to report minor accidents which may lead to major accidents or injuries in the future.

Effective safety training should help improve reporting of errors, since human error rates are one of the most important measures that safety engineers are interested in.

High labor turnover means an influx of new employees who must be indoctrinated in company safety procedure. If they are not trained or not trained adequately, they themselves constitute a hazard and increase the risk of accidents to themselves and others. It is important to be sure that other areas of orientation do not crowd out safety training.

All employees should be given safety training at the start of their employment and on a regular basis; thereafter, thorough refresher courses should be given for everyone—new employees, old employees, transferees from other plants, supervisors, and managers. Many variables affect accident rates and must be considered in developing accident prevention programs. These include defective vision, lack of perceptual motor skills, the element of risk taking, lapses in attention, the hypnotic effect of monotonous or repetitive tasks, and lack of physical fitness. No doubt you can add several more of your own. Every one of these factors could easily be overlooked in veteran employees, who must be informed if methods or procedures they have been using for a long time, even years, are now considered inappropriate. Age can also alter reflexes and eyesight; this deterioration can be hazardous.

Lapses in attention are common, but if they happen when an employee is engaged in work demanding close concentration, they can result in accidents. If there is the possibility of accidents, preventive measures such as providing more frequent work breaks, playing background music, or providing a bright, stimulating work environment should be taken.

Monotonous or repetitive tasks can cause a dangerous trancelike state; these should be carefully analyzed and methods of breaking the monotony instituted. Poor physical fitness affects both physical and mental conditioning, with accompanying sluggishness of mind and body, dulling alertness and efficiency. For jobs requiring heavy labor, good physical fitness will be more apparent but not necessarily more important. It is often employees with sedentary jobs who are in the poorest physical shape and who should be encouraged to get in shape and stay in shape, perhaps through company-sponsored exercise programs and facilities. In many communist countries exercise is conducted at the workplace. The difference between physically active and sedentary workers was shown in an examination of health records of postal workers. Postal clerks had significantly more physical ailments than mail carriers.

BASIC PRINCIPLES OF LEARNING

In designing a safety training program, some basic principles of learning will help the trainer decide how to get the employees to absorb more safety training and how to retain what they have learned. Some of the questions a safety engineer should answer are: What are appropriate learning approaches for the subject? Should the learner be reinforced (given a reward) for learning the material? If so, what should be used for the reinforcer? How frequently should the material be presented?

Positive Reinforcement

A basic learning principle is that of reinforcement. Reinforcement refers to the principle that when an activity is rewarded it tends to reoccur. Safety behavior which is rewarded tends to be repeated. Therefore, if one wants a safe workplace, one will use the principle of reinforcement to keep safe behavior at a high level. Optimal reinforcement is given immediately after the positive behavior occurs. A reinforcement might be a small material item or verbal praise.

Negative Reinforcement

Punishments such as being docked pay for a safety violation or not getting a pay raise for a safety violation are not an effective means of promoting safe behavior. Punishment serves to inhibit the undesired behavior, but the worker does not unlearn the unsafe act. When punishment is removed, the worker will revert to the undesired behavior.

Transfer of Learning

In terms of a safety training program, transfer of learning is important. Transfer of learning is being able to utilize the skills learned in a formal-teaching work setting or to carry over the safety behavior learned in one situation into a different situation. Thus safety training about machine guarding for a press will be used when the operator uses a shear.

Retention of Safety Information

Based on experimental studies, the amount of information retained from oral or visual safety presentations decreases over time. Minimal retention means a high probability of accident or accidents.[12] Figure 17.1 demonstrates this theory. Retention drops off rapidly following the initial orientation but is increased sharply when reinforced, thus resulting in a wavelike series of highs and lows. In periods when retention is low, the potential for accidents can be dangerously high. To maintain the retention of safety information above hazard potential, it is essential to provide periodic reinforcement in the form of review or refresher courses. The amount of information retained initially depends on the instructor's capabilities and the employee's receptivity at the time. In addition, the decay rate will vary

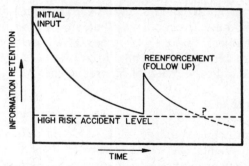

FIGURE 17.1. Typical individual's safety information retention pattern (*from Vandergrift, E. F., How to measure safety information retention, in Professional Safety, Sept. 1979*).

among individuals. The information retained at any time is unknown as is the level at which the retention factor drops below the hazard zone. What is needed is a means of evaluating each employee's retention, as well as statistical data to establish how frequently training must be repeated.

Establishing a Base Line. Figure 17.2 presents a safety quiz developed at Westinghouse Electric Corporation, Energy Systems Operation, which could provide the safety professional with information about how effective a training program is, how well the supervisors have reinforced that training, what specific points of safety awareness have not been retained, which employees bear closer surveillance, and the optimal frequency for safety training. It is suggested that the questions be geared to as many specific items as possible, such as:

1. Type of protective equipment to wear for a particular job or in a particular work area
2. Priorities to establish in case of serious plant incident
3. Immediate responses to hypothesized situations
4. Proper use of specific safety equipment
5. Location of standby emergency equipment
6. Preliminary signs to look for which might indicate the onset of equipment or process failure
7. Safety rules which are to be followed at all times[9]

Motivation

In planning a safety training program, it will be helpful for the safety professional to know some basic motivation principles. These are based on the premise that all persons have needs which, when identified, can be a

This is a brief quiz to determine your comprehension and awareness of the safety program regarding operations at the Waltz Mill site of Westinghouse Energy Systems Division. In answering the questions, first rely on your recall of the safety information which you have been given to read, and failing in that, answer the questions to the best of your ability relying on your past experience and best judgment. The results of the quiz will be discussed with you in order to optimize your personal safety and that of your co-workers on the job. You will be given 30 minutes to complete this quiz.

1) In your own words, list 5 of the 36 General Safety Rules which apply at the WESO site.

_____

2) Indicate the order of importance of the following criteria which will govern emergency actions in case of a serious accident.

_____ Minimize equipment loss

_____ Minimize personnel hazards

_____ Carry out test program goals

3) If you are involved in operation of the Process Development Unit (PDU), what would be your immediate action should you be the first person to observe:

3a) A fire caused by leaking gas?

_____

3b) A coal, char, or waste fire outside the bins?

_____

3c) An explosion?

_____

4) When performing work on the structure, when are you required to wear a safety life belt?

_____

5) If you are asked to work in a pit which is several feet deep, what safety precautions should you take?

_____

_____

_____

6) If you are working on the structure and start to become light-headed and/or sick in the stomach, what should you do?

_____

What could be happening to make you feel that way?

_____

7) When and where should you use the following personal safety equipment?

Hard Hat: _____

Safety Glasses: _____

Leather Gloves: _____

Respirator: _____

8) If you are monitoring the air on the PDU structure during operations with a Sentox 2 instrument, what would you do if the Lower Explosion Limit (LEL) meter registers:

10%? _____

20%? _____

50%? _____

100%? _____

9) You know that loud noises can be injurious to your hearing ability. In an unmarked area, what is a simple way to tell if the noise level is beyond a reasonably safe level?

_____

10) If you are working on the 3rd level of the structure, where would you find the nearest fire extinguisher?

_____

11) Name the 3 hazardous conditions for which the air is monitored on the PDU.

_____

12) Where are the emergency evacuation areas for the PDU or lower site?

_____

the upper site?

_____

13) To report an emergency to the ARD security force, what number would you call and what general information would you give?

_____

14) In the event of a fire in Test Cell B, what are you expected to do?

_____

15) If you are working alone and sustain a minor injury such as a minor cut or a pinched finger, what should you do?

_____

16) If in the event you are the first person to discover a seriously injured co-worker, what should you do? If working daylight?

_____

If working afternoon or midnight shift?

_____

17) What is the proper way to transport a compressed gas cylinder?

_____

18) If you have been using a partial face respirator with a nuisance dust filter, what should you do with it when the job is completed?

_____

19) Why is it important to observe good housekeeping rules?

_____

20) If you observe a co-worker working unsafely, what should you do? Why?

_____

*21) How important do you consider a safety program in relationship to your job? (check one)

_____ Important

_____ Some help

_____ Might come in handy

_____ Not necessary for me

*22) In your opinion, is there any information which you have not as yet received which would add to the safe performance of your job? List.

_____

NAME: _____ DATE: _____

*Not graded.

FIGURE 17.2. WESO site safety quiz (*from Vandergrift, E. F., How to measure safety information retention, in Professional Safety, Sept. 1979*).

source for high levels of motivation. If these needs are sufficiently reinforced, the individual can be motivated to do almost anything.

Psychological research has shown that for most employees, motivators are achievement, recognition, quality of work, responsibility, and
advancement. "Dissatisfiers" are company policies, supervision, working conditions, salary, work relationships, and status. A safety training
program must minimize dissatisfiers and reinforce the satisfiers, or
motivators.

Research has also shown that very anxious and very bored workers are
very difficult to motivate. The very anxious person is one who freezes
when his or her car stalls on a railroad crossing because his or her anxiety is
so high. There is little the safety engineer can do to arouse this type of
person for safety. The person who is constantly saying "I'm so bored" or
acts like any assignment is too much is also difficult to motivate.[6]

Figure 17.3 illustrates the interrelationships between task content, the
work situation, job satisfaction, and motivation to work. This figure also
shows the relationship between worker satisfaction and absenteeism and
labor turnover. Studies on work and motivation suggest that motivation to
work depends on the employees' anticipated satisfaction with the results
of their effort and performance or the degree to which they believe their
results can be attained.[3] For example, workers can choose between such
alternatives as how many breaks they take, when they will take them, how
fast they work, how "hard" they work, whether they will get further
schooling, and what they expect of themselves and others. According to
motivational concepts and learning theory, employees will choose the alternative or alternatives that have optimum results. The choices may vary
among employees, depending upon their individual needs.

How strong of a motivator is money? To what extent can money be a
way of obtaining feelings of achievement, competence, and power? What
needs are currently met by money? What needs might be met by money?

In studies of money as a motivator it was concluded that more precise
research is needed before any generalizations for industry can be made.
The authors of the studies raise some interesting questions regarding
money and suggest areas of research that might be fruitful.[8]

The effect of a pay raise or the length of time before that raise acts as a
reinforcer is not known. Also not known is the optimal reinforcement
schedule to be used in giving salary increases for obtaining desired
changes in job behavior. Monitoring work outputs on jobs where production is under direct employee control and where it is easily assessed may
provide valuable insights. Such information would have important implications for how often and in what amounts incentive raises should be built
into the compensation package.[5] Safety should be one of the primary incentives for pay raises.

We also need to investigate the relation between amount of money and
the amount of behavior money motivates.[5] Is there some point beyond

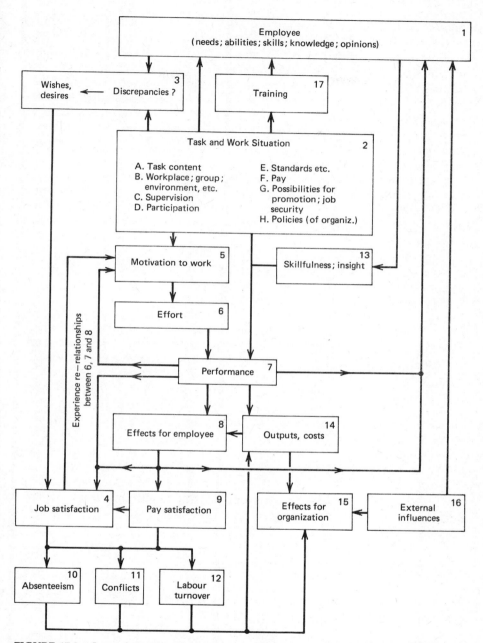

FIGURE 17.3. Interrelationships between task content, work situation, the employee, job satisfaction and motivation to work, and other factors (*from de Jong, in Sell, R. G., and Shipley, P., eds., Satisfactions in Work Design, Taylor & Francis, 1979, p. 46*).

which increases in compensation are no longer related to improvement in the relevant behavior? Do humans show the same sharp decline between amount of reward and strength or number of responses that lower organisms display? Does money reinforcement promote greater amounts of safe behavior? These are only a few areas—there are many more; such as, Does a person who has a higher-status position need a larger raise than a lower-status employee for the raise to be an incentive? What is the effect of a fixed pay schedule on incentive? (A fixed interval is reinforcement following a fixed period of time after the last reinforced response.) This is how most industrial employees get paid, yet one would assume from learning and motivation theory that the resulting work performance would be low. What would happen if the paycheck were divided into several parts, such as so much for tenure, so much for minimum service rendered, so much for maximum service rendered, so much for excellent performance, and so much for outstanding safety record?

The effects of motivation on safety are far-reaching. A worker with no safety motivation is more likely not only to be injured on the job but to cause injury to others. No matter what programs are developed, what work facilities are provided, or what incentives are built into the work system, they will be ineffective unless workers are motivated to do their work as safely as possible and see that others in supervisory jobs do likewise. In planning, developing, and implementing the safety program, what is known about motivation can be used to gain maximum worker cooperation.

The largest concern for research in safety motivation is that data have not been collected. It is necessary to base studies on actual worker performance to evaluate factors such as using money as an incentive and a satisfier.

Job satisfaction is related to motivation, rewards, job needs, individual needs, and job possibilities. Figure 17.4 suggests a way to assess job satisfaction by looking at the relationships of these various factors.[1] Figure 17.5 illustrates an outline schema for work design, identifying those aspects that would motivate the worker to perform better.[2]

THE SAFETY TRAINING PROGRAM

The occupational training committee of the National Safety Council cites six steps for developing a training program. These are:

1. Identifying training needs
2. Formulating training objectives
3. Gathering materials and developing course outlines
4. Selecting training methods and techniques

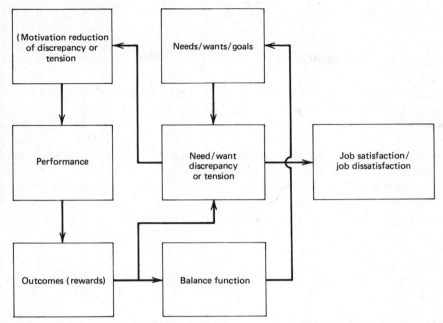

FIGURE 17.4. Proposed model of the production of job satisfaction (*from Cope, D., in Sell, R., and Shipley, P., eds., Satisfactions in Work Design, Taylor & Francis, 1979, p. 61*).

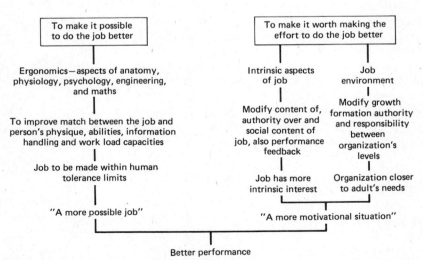

FIGURE 17.5. An outline schema for work design (*from Corlett, E. N., in Sell, R., and Shipley, P., eds., Satisfactions in Work Design, Taylor & Francis, 1979, p. 32*).

5. Conducting training programs
6. Evaluating training programs

Identifying Training Needs

Identifying training needs requires examining and correlating data on accident rate and turnover, wildcat strikes, excessive scrap, customer complaints, sudden decrease in sales, production decreases, major company purchases, or company expansion. Any or all of these could signal either that something is wrong (as when strikes occur) or that training could prevent accidents (as when a company suddenly expands).

Once the need has been established based on this information or from employee concerns, supervisor concerns, exit interviews, evaluations, or new employee observations, it is time to choose a program title that accurately conveys what the program is. It should sound interesting but not so intimidating that employees feel they won't be able to handle the material. Sample titles are "Emergency First Aid Treatment," "Basic Techniques of Cardiopulmonary Resuscitation (CPR)," and "Emergency Fire Equipment Handling."

Formulating Objectives

Objectives quantify what it is that workers will gain by completing the training program. These objectives are also used to determine if workers have attained the desired level of proficiency. It is important not only for workers to see what they have accomplished as a result of the training program but also for the safety professional to have this information so management can measure the cost effectiveness of the program.

Course Description

A course description lists the several areas to be covered, such as safety applications, human and environmental factors, safeguarding processes and machines, safety in product design, occupational health and hygiene, toxicology, organization of a safety program, training and education, and motor fleet safety.

A course syllabus (see Table 17.1) contains course title, course objectives, course outline, evaluation, teaching methodology, and references. It should be developed before the training program begins and handed out the first day of class. It conveys to the learner the expectation, the knowledge to be acquired, the work to be accomplished, and evaluation criteria. Most of all, it communicates the importance of the program.

After the title has been selected and objectives listed, review the relevant literature and choose textbooks and other written materials that will best meet the course objectives. Finally, prepare an outline of the course.

The safety instructor should have packets of materials to hand out to employees. Manuals, films, or programmed courses may also be used.

Courses developed by the National Safety Council or by NIOSH can also be helpful and may even meet all or most of the safety training objectives.[7,11]

Training Methods

Training methods selected depend on the material (some materials are more conducive to lectures than others), the size of the audience (more than 25 is too big for a seminar), and the teacher's style.[10] Some teachers are excellent lecturers and can hold an audience's attention, while others do much better with smaller groups and prefer using a discussion–seminar method.

Conducting Training Programs

Demonstrations and other visual methods are important teaching tools. In deciding what is to be covered in each section, teachers should also decide what teaching methods to use, approximately how much time it will take, what visuals or demonstrations to use, and what materials must be ready to hand out to employees.

In most companies, except for very small ones, safety professionals will not be the only ones giving classes but will in most cases be responsible for coordinating the classes, preparing a schedule, and making sure responsibilities are clear. They may also be responsible for selecting adequate classroom space, providing equipment and supplies, conducting evaluations, maintaining attendance records, and certifying that a course has been taken.

Evaluating the Training Program

Evaluating the training program is necessary to measure the effectiveness of the training. Pretesting determines the need for the program; posttesting quantifies what has been learned. Written examinations and demonstrations can be used.

It is also important to assess whether the previously identified lack of safety knowledge has been corrected by the training and to determine if more training is needed. Ideally the new skills will be reinforced at appropriate intervals.

Part of the evaluation is to obtain study feedback on the program. Was it helpful? Did the students get what they wanted? Was there anything they would have liked done differently? What are their goals for themselves in future programs? Although it is advantageous for the safety professional to have a long-range program, training programs will also be required. These

TABLE 17.1
Syllabus for a College Course in Occupational Safety and Health

1.0 *Course*
    1.1   A one-semester course meeting three hours per week, three credits.
2.0 *Course Description*
    2.1   An introduction to industrial safety and health as applicable to general industry. Study of federal law, Occupational Safety and Health Act, and workers' compensation laws. General history of occupational safety. Factors comprising safety technology, organization, and development of safety programs, safety training, and record keeping. Accident prevention, hazard recognition and analysis, personal protective equipment and industrial hygiene. Specific occupational hazards such as noise, radiation, and fire protection will be discussed.
3.0 *Course Objectives*
    3.1   The course is designed to provide the student with a basic understanding of industrial safety and health, along with laws and general safety practices applied to industry.
    3.2   Specific Objectives.
        3.2.1   To enable students to learn the foundations and history of occupational safety.
        3.2.2   To acquaint the student with the Occupational Safety and Health Act of 1970 and its effect on industry.
        3.2.3   To make students aware of the federal standards and the *Federal Register* covering all phases of occupational safety.
        3.2.4   To have students develop a knowledge of the basic elements of occupational safety organization, management, employer and employee responsibilities, and the maintenance of a safe working environment.
        3.2.5   To provide students with some insight to in-plant safety training, supervision, accident record keeping and analysis, risk management, and emergency planning.
        3.2.6   To develop student awareness of machine hazards, noise problems, materials handling problems, storage problems, chemical hazards, and electrical hazards.
        3.2.7   To familiarize students with personal protective equipment.
4.0 *Course Outline*
    4.1   History of occupational safety.
        4.1.1   From the Middle Ages to the present day.
        4.1.2   Safety progress in the United States over the last 50 years.
    4.2   The Occupational Safety and Health Act of 1970 (OSHA).
        4.2.1   Provisions of the law.
        4.2.2   Implications of the law and its impact on industry.
        4.2.3   Compliance inspections, violations, enforcement of the law.
    4.3   Standards under OSHA.
        4.3.1   *Federal Register*.
        4.3.2   Codes and other references.

TABLE 17.1. (*Continued*)

4.0 *Course Outline*
  4.4 Basic elements of occupational safety technology and organization.
    4.4.1 Management's responsibilities.
    4.4.2 Employees' responsibilities.
    4.4.3 Assignment of responsibilities.
    4.4.4 Safety training at the plant level
  4.5 Accident and injury record-keeping systems as required by law.
  4.6 Medical and first aid programs.
  4.7 Machine and equipment hazards.
  4.8 Noise problems.
  4.9 Materials handling and storage problems.
    4.9.1 Chemical hazards.
    4.9.2 Electrical hazards.
    4.9.3 Personal protective equipment.
5.0 *Evaluation*
  5.1 Course evaluation will be based on results of midterm and final examination, plus a research paper on a specific area of occupational safety and health selected by the student.
  5.2 The midterm examination will be 30% of the course grade.
  5.3 The final examination will be 35% of the course grade.
  5.4 The research paper will be 35% of the course grade.
6.0 *Texts and Supplementary Materials*
  6.1 The text: Gloss, D. S., and Wardle, M. G. *Introduction to Safety Engineering*. New York: Wiley, 1984.
  6.2 References: National Safety Council, *Accident Prevention Manual for Industrial Operations*. 8th ed Chicago: National Safety Council, 1980.
    6.2.1 NSC's *Supervisors Safety Manual*.
    6.2.2 The *Federal Register*.
    6.2.3 Codes published by National Fire Protection Association and American National Standards Association.
  6.3 Films and visual aids.
    6.3.1 *Sound of Sound*, American Optical Company.
    6.3.2 "Safety Management" film series, NSC.
    6.3.3 "Human Factors in Safety" series, NSC.
    6.3.4 Other visuals obtainable from the National Safety Council and industry-produced films.
7.0 *Teaching Methodology*
  7.1 The course will be conducted by lectures, discussions, visual aids, and field trips to observe functioning industrial programs. Guest lecturers from federal offices and safety leaders with expertise in special safety areas will be invited to speak.

may include first aid, specific handling of a substance, and individual instruction in using equipment because of a specific deficit, such as poor depth perception.

HEALTH AND SAFETY TRAINING FOR WOMEN

Health and safety training for women is no different than the training for men, except that it requires more careful monitoring of certain work to be sure the woman does not hurt herself. However, it has been found that women who are employed in occupations that were traditionally male-dominated have injury rates that are the same as those of their male counterparts. Wardle and Gloss in their review on women and work performance, question women's abilities in gripping (hand strength), lifting (upper arm strength), and heavy moving.[13] In our culture, few women lift heavy objects daily; therefore, they may require weight training to do it safely. It is imperative that workers not overstep their limits. The wise acknowledgment of one's limitations is certainly not seen as a weakness. All workers must remember that they aren't helping anyone if they add to the number of workplace injuries.

SUPERVISOR TRAINING

Supervisor training is crucial to the success of a safety training program. The following suggestions will help supervisors deal with the changes that a safety program implies:

1. Guidance from an external or internal change agent with experience in the technology of implementation of change
2. Modification of reward systems, performance appraisal schemes, and job evaluation procedures compatible with the principles and objectives of participation
3. Provision of effective communication networks and mechanisms to promote efficient communication
4. Allowance of extensive periods of time for learning and changes to occur before evaluating participation[4]

The goals of supervisor training are to describe company policy, to present accident prevention fundamentals, and to suggest methods of applying them to the safety training program. Supervisors must be comfortable with these skills before they can be expected to train others.

JOB SAFETY ANALYSIS

A job safety analysis (JSA) is a written procedure for reviewing job methods, uncovering hazards, and recommending safe job procedures.

Tasks which require a number of sequential steps are especially suitable for a JSA. They include materials handling (heavy drums), work on high-energy electric systems, acid/caustic cleaning, crane repair, crane operation, trenching and excavating, erection and use of scaffolding, beryllium and zirconium machining, repairing caustic/acid tanks, car/bus service on hoists, and operation of x-ray and low-power laser units.

A JSA consists of selecting the job to be analyzed, breaking it down into steps, identifying the hazards, determining how to reduce or control the hazards, applying the controls, and evaluating the controls.

After determining which jobs are appropriate for a JSA, the next step is to establish priorities. The tasks are ranked in descending order from highest to lowest accident potential. Highest risks should be analyzed first. Ranking risks requires frequency and severity studies. (See Chapter 27 for further discussion of these requirements.)

Accident Frequency

Any job that repeatedly produces accidents is a candidate for a JSA. The more accidents associated with the job, the higher its priority should be.

Every job that has produced a disabling injury should be analyzed, because the injury suggests previous preventive measures were not successful. Some jobs may have no record of accidents but may nevertheless have the potential for causing a severe injury.

Judgment and Experience

Many jobs, such as handling volatile substances or exceptionally heavy equipment, qualify for immediate safety analysis because of potential hazards. Such work can result in fatalities.

Routine Jobs

An employee engaged in routine or repetitive tasks with inherent hazards is continuously exposed to hazards. For example, exposures to high levels of noise (over 85 db A) over a period of time will affect hearing.

Job Changes

Changes in processes, equipment, or materials can introduce new hazards which may not be immediately apparent. Changes can also increase the work load of employees and overtax their ability. Job safety analysis should be done to prevent accidents in these cases.

In identifying potential hazards, the safety professional should consider the following steps:

1. Is there a danger of striking against, being struck by, or otherwise making injurious contact with an object (for example, being struck by a suspended drill casing or piping as it is moved into place)?
2. Can an employee be caught in, on, or between objects (for example, an unguarded V-belt, gears, or reciprocating machinery)?
3. Can the employee slip, trip, or fall on some level or to another (for example, slipping in an oil-changing area of a garage, tripping on loose or worn carpet, or falling from a scaffold)?
4. Can an employee strain himself or herself by pushing, pulling, or lifting (for example, pushing a file cabinet into place or pulling it away from a wall in a confined area)? Back injuries are common in every type of industrial operation; therefore, do not overlook the lifting of heavy or awkward objects.
5. Is the environment hazardous and/or toxic; that is, does it contain gas, vapor, mist, fumes, or dust, heat, ionizing, or nonionizing radiation (for example, the production of toxic fumes and nonionizing radiation from arc welding or galvanized sheet metal)?

Potential for Human Error

The potential for human error must be taken into account in a job safety analysis. The following factors should be considered:

1. Equipment can be utilized correctly. What effects could there be if used incorrectly?
2. Personnel tend to take shortcuts to avoid arduous, lengthy, uncomfortable, or unintelligible procedures.
3. Equipment that is difficult to maintain will suffer from lack of maintenance.
4. Requirements for special employee training should be kept to a minimum.
5. Written procedures which are difficult to understand without further explanation may result in employee confusion.
6. Stress in the work environment and in private life could result in accidents.
7. The meaning to employees of their jobs affects their performance.

Compilation of this information should result in specific recommendations for job procedures, which may vary depending upon the specific task, the workers doing the task, and the environment in which it is done. After controls are implemented, they should be assessed for effectiveness in terms of frequency of workplace accidents or injuries.

SUGGESTED LEARNING EXPERIENCES

1. Design a physical fitness program for employees in a sedentary occupation.

2. Illustrate three workplace adaptations of general learning theory.

3. Using Vandergrift's quiz for retention of safety information, what questions would you add or delete for an employee group of your choice? Are there other ways you could use the data collected than those the text states?

4. What would be the best motivation to you in your job situation? Explain.

5. Develop a job safety analysis for a hazard you have identified in your workplace.

6. Prepare a safety training program dealing with a subject of your choice.

7. What are ways that you as a system safety professional would attempt to control human error in your workplace?

8. Utilizing learning theory principles, how would you develop a training module for orientation of new employees?

9. If you were a member of management staff and had been employed in your industry for 18 years, what would be the most useful way for a young recent college graduate to relate to you? What principles are you utilizing to arrive at this decision?

10. If you were going to assess your safety training needs, what would you say they are?

REFERENCES

1. Cope, D. Understanding the concept of job satisfaction. In R. Sell and P. Shipley (Eds.), *Satisfactions in Work Design: Ergonomics and Other Approaches.* London: Taylor & Francis, 1979, pp. 57–64.

2. Corlett, E. N. Isolation and curiosity as sources of work attitudes. In R. Sell and P. Shipley (Eds.), *Satisfactions in Work Design: Ergonomics and Other Approaches.* London: Taylor & Francis, 1979, pp. 51–56.

3. Damodaron, L. Managerial learning needs for effective employee participation. In R. Sell and P. Shipley (Eds.), *Satisfactions in Work Design: Ergonomics and Other Approaches*. London: Taylor & Francis, 1979, pp. 161–168.

4. De Jong, J. Roles and instruments and work design and work re-design. In R. Sell and P. Shipley (Eds.), *Satisfactions in Work Design: Ergonomics and Other Approaches*. London: Taylor & Francis, 1979, pp. 41–50.

5. Dionne, E. Motivating workers with incentives. *National Safety News*, 1980, **121,** 75–79. (Jan.)

6. Hebb, D. O. *The Organization of Behavior*. New York: Holt, 1960.

7. National Safety Council. *1981–82 College and University Safety Courses*. Chicago, 1981.

8. Opsahl, R., and Dunnette, M. The role of financial compensation in industrial motivation. *Psychological Bulletin*, 1966, **66,** 94–118.

9. Pavlov, P. *Job Safety Analysis*. Technical Report. Idaho Operation Office, Department of Energy, November, 1979.

10. Tucker, M. E. *Audiovisual Resources*. National Institute of Occupational Safety and Health, Washington, D.C.: GPO, 1981.

11. U.S. Department of Health and Human Services. *NIOSH Schedule of Courses, 1981–82* (DHHS 81–138), Washington, D.C., 1981.

12. Vandergrift, E. F. How to measure safety information retention. *Professional Safety*, 1979, **24,** 48–51. (Sept.)

13. Wardle, M. G., and Gloss, D. S. Women's capacities to perform strenuous work. *Women and Health*, 1980, **5,** 5–15.

18 SAFETY PROGRAM EVALUATION

The safety program, like any other corporate program, must be evaluated for effectiveness. The evaluation should be formal. While safety engineers are on the job, they should be aware that their behavior and their department's performance will be evaluated. The basic principles of program evaluation are the same for safety as for any other program.

Safety program evaluation is *a systematic method of measuring the effectiveness of a planned group of activities aimed at improving workplace safety*. The program is evaluated by examining improvement in safety, costs, and any further improvements that can be made. The results of the evaluation will determine whether the safety program is recommended to be continued, expanded, cut back, changed, or abandoned.

One way to conduct a safety program evaluation is to compile frequency and severity statistics for industrial accidents and workers' compensation claims. These indices may be compared with industry rates or with the previous year's rates for the company. Another method is to evaluate the records of the safety department. Are accidents thoroughly investigated and documented, and are suitable actions taken? How much safety information is retained by the workers from safety training? Finally, how many unsafe actions are performed which might lead to accidents?

PROCESS OF EVALUATION

There are two kinds of evaluations. The aim of one is to improve a program, whereas the aim of the other is to assess a program. Improvement and effectiveness evaluations are always distinguished from one another by the way information is used rather than by the kinds of information

399

collected or the stage at which it is gathered. In an improvement evaluation, information is used to modify the program. In an effectiveness evaluation, information is used to assess the program's quality and results.

The purpose of an improvement evaluation is to identify a program's strengths and weaknesses, determine causes of the latter, and recommend ways to eliminate them. Evaluations conducted for improvement focus on finding out what goes on within a program. Because of this, comparisons with other programs are not as useful as information about how well the program itself has been implemented and how well it is achieving its goals.

Effectiveness evaluation appraises a program's overall impact and determines the consistency with which it produces certain outcomes. Several types of information are collected in an effectiveness study. One type involves identifying how well a program's goals have been met. Another type determines the impact of a program for special groups. Still other types of information needed for an effectiveness evaluation focus on the comparative value of a program and examine its side effects and costs.

Short- or Long-Term Goals

The evaluation should assess both short- and long-term goals, weighting the relative importance of the safety data obtained and assessing any unanticipated results. Often these unexpected results can be most meaningful and provide information that probably would not have been obtained if the evaluation had not been done.

USES OF RESULTS

Different people will be hoping to have different questions answered. People who may be interested in evaluation results include top managers, safety program directors, safety supervisors, employees, the public, and consumers.

Top managers need information that will help them address such issues as changes that must be implemented. The effectiveness of the program and the source of most accidents and injuries will need to be identified. Suggestions for different allocations of funds will also be examined.

Safety directors desire information concerning the effectiveness of the safety program. They also desire information about program strategies.

The safety supervisory staff needs information about technique. How effective is safety training? Are safe habits being learned? What training methods have workers found most helpful in learning safety material?

Employees carrying out day-to-day workplace activities want answers to questions on the technical aspects of daily operations. Which ones are unsafe? Does this concur with their own perceptions? What changes may occur in their workday as a result of suggested changes?

Members of the public will be concerned with the outcome of a safety program evaluation to the extent that industrial activities impinge upon their environment and affect the quality of their lives. Their primary interest is in whether the corporation is observing maximum safety precautions, for example, in the disposal of toxic substances. Industries must respond to these legitimate public concerns. The public has become very powerful in terms of supporting legislation and supporting or not supporting corporations. It has engaged in effective protests against industrial growth which has been proved unsafe. Companies can no longer function as if in a vacuum; they must recognize their direct responsibility to the surrounding community.

Consumers have grown in number and awareness and have demanded that their safety be taken into consideration. Product liability has become an area of major concern to industry. Litigation and adverse publicity negatively affect any industry.

A thorough safety program evaluation obtains data on the overall effectiveness of the program and on the effectiveness of specific components. Compiling both kinds of data can produce more complete information on the program.

EVALUATION RESEARCH

Purpose of Evaluation Research

The purpose of evaluation research is to measure the effects of a program against the goals it set out to accomplish as a means of contributing to subsequent decision making about the program and improving future programming.

Evaluation research is thought to result in rational decision making by having objective measurements on the effectiveness of a safety program before decisions are made regarding alterations in the program.

Evaluation research is a way to increase effective policymaking. With objective information on the outcomes of programs, intelligent decisions can be made on budget allocations and program planning. Programs that yield good results will be expanded; for those that make poor showings, suggestions for improvement will be made.

Evaluators are asked to supervise or conduct a number of activities. The following are suggested categories:

1. Formulating credible evaluation questions
2. Developing data-collecting methodology
3. Planning data collection
4. Collecting data

5. Planning and conducting data analyses
6. Reporting evaluation information
7. Managing an evaluation

To be effective, an evaluation must produce timely information that is useful in improving or establishing the effectiveness of a program. It is helpful that the evaluator formulate questions that will give clients the information they need. First, the evaluator must know the program's goals and activities and the kinds of information that will be acceptable as evidence of program effectiveness. Has the program been implemented as planned? Is it making progress toward its goals? How is it applicable to special situations or special groups?

TYPES OF SAFETY PROGRAMS WHICH SHOULD BE EVALUATED

Training programs are the largest safety programs that need to be evaluated because they are available for almost every workplace hazard that has been identified. One example of such an evaluation was a review of a hearing conservation program developed by a company which produces electric appliances. It featured an orientation session for all persons who might be exposed to noise and advised employees of correct ear protection against high noise levels. Features of this program included a safety conference for devising the program; noise measurement, analysis, and control; hearing tests for all employees exposed to noise levels of 90 or more decibels; education and indoctrination programs for employees; and selection and fitting of ear protectors for employees.

Another safety training program involved more than training. A noise control and hearing conservation program was developed for a tobacco company, involving 13,000 employees. The program involved company-wide noise surveys; measurement and evaluation of noise hazards; installation of sound muffling enclosures, materials, and systems; installation of warning signs for hazardous noise areas; provision of mobile audiometric testing of employees; and provision of personal protective equipment, with mandatory-use enforcement.

Accident Investigation and Reporting

Accident investigation and reporting must be evaluated, since many of the statistics compiled in the workplace are based upon this information. Part of a program evaluation will determine whether or not there is adequate accident investigating and reporting.

Some factors examined include the following:

1. Methods of identifying hazards
2. Detailed studies of hazards of planned and proposed facilities, operations, and services
3. Hazard analysis of existing facilities, operations, and products
4. Evaluation of the loss-producing potential
5. Analysis of the personnel and material losses resulting from accidents and occupational illnesses
6. Identification of errors or omissions involving incomplete decision making, faulty judgment, administrative miscalculation, and poor practices
7. Identification of potential weaknesses found in existing policies, directives, objectives, or practices

Noise Control and Abatement

Noise control and abatement is another component that must be evaluated. The noise must first be assessed, using appropriate equipment. For simple noise measurement, an OSHA-approved sound level meter which evaluates decibels can be used. The noise levels would then be measured at appropriate locations.

Safety Committees

The safety committees of an industry are another factor that must be assessed in a safety program evaluation. The functions can vary from investigating incidences of lead poisoning to developing safety programs, setting periodical medical examinations, and studying worker safety orientation. The safety committee's functions will vary depending on the power and duties it has been given by each industry. These committees will need to be evaluated to ascertain if they are accomplishing what they have been set up to do.

The Safety Inspection Policies

The safety inspection policies are another aspect to evaluate in a safety program evaluation. This is an area that is at the crux of a safety program. Without this aspect it would be difficult to meaningfully evaluate a safety program.

WHO SHOULD PERFORM THE SAFETY PROGRAM EVALUATION

There are some advantages to having the evaluation done by company evaluators and other advantages to having it done by outside evaluators.

Several factors affecting evaluators that are "inside" are confidence, objectivity, understanding of the program, autonomy, and the use of collected data.

Objectivity is imperative in a safety program evaluation. This includes the correct use of statistical tests no matter what the results look like. Evaluators need to know "real issues" and the events that are occurring in the workplace if the evaluation is to be useful. For the evaluation to be meaningful, evaluators must not only collect data but interpret results. Even though the data collected may be thorough and complete, if it is not adequately interpreted, results will be difficult to apply.

This question of who will perform the evaluation becomes an important one. Other issues are, Who will have access to program evaluation results and to what use will the results be put? In general, outside evaluators are presumed to be more objective than are inside ones, based on the assumption that they have no vested interest. Fink and Kosecoff state the following in regard to this issue:

> Regular staff members who take responsibility for evaluation activities are called internal evaluators, while outside consultants are called external evaluators. Internal evaluators have the advantage of being closer to the program and its staff are consequently less obtrusive. They are less likely to be objective than an outside evaluator, however, because they are personally involved in the program and because their job depends upon it. External evaluators, on the other hand, have the advantage of being objective and of having a fresh perspective. Their disadvantages are their image as outsiders and critics and sometimes their physical isolation from the program. In improvement evaluations, it is possible to use either an internal or external evaluator since the evaluation is likely to require close contact with the project and staff. Even an external evaluator can be expected to become somewhat involved in a project that grows because of his or her recommendations. In effectiveness however, the evaluator should be an outsider who is professionally independent of the effect of any recommendations or the good will of the program staff.[1]

It is generally better to use an external evaluator than an internal evaluator, regardless of the evaluation's context. External evaluators are more likely to be objective and to have an independent perspective, and as a result, the evaluation's findings are more likely to be accepted.[1]

The data obtained from the evaluation research will only be as accurate as the methodology used to gather the information. Every effort should be made to assure validity by choosing professionals to do the evaluation.

It is important to understand the difference between evaluation for a safety program and basic research. Evaluation is intended for use where basic research puts the emphasis on the production of knowledge and leaves its use to the natural processes of dissemination and application.

Evaluation starts out with use in mind. In its ideal form, safety program evauation is conducted for an organization that has decisions to make and looks to the evaluation for answers on which to base its decisions. Use is often less direct and immediate than that, but it always provides the rationale for evaluation.

Evaluation research considers decision maker's questions rather than evaluator's questions. Unlike basic researchers who formulate their own hypotheses, evaluators deal with program concerns. They have a lot to say about the scope of the study and approach it from the perspectives of their discipline. They are usually free to bolster it with investigations of particular concern to them. But the core of the study represents matters of administrative and programmatic interest. The underlying evaluation hypothesis is that the program is accomplishig what it set out to do.

Evaluation compares "what is" with "what should be." Although investigators remain unbiased and objective, they are concerned with data that demonstrate whether the program is achieving its intended goals. The question, however formulated, is concerned with measuring how the program meets stated criteria. The element of judgment against criteria is basic to evaluation research and differentiates it from other kinds of research. A statement of program goals is therefore essential to an evaluation.

Interpersonal frictions are not uncommon between evaluators and safety professionals. The safety practitioners' roles tend to make them unresponsive to research requests. From their point of view, service is imperative; evaluation research is not likely to contribute enough to improving service to be worth the disruptions and delays it may cause. Often they believe strongly in the worth of the program and see little need for evaluation at all. The judgmental quality of evaluation research means that the merit of their activities is being weighed. If the evaluation results are negative, that is, the program is not accomplishing the purpose for which it was established, then the program and possibly some jobs may be in jeopardy. While basic research is published, the majority of evaluation reports are unpublished. Program administrators and staff often believe that the information contained in reports was generated only to answer their questions and, consequently, are not eager to have any "dirty linen" that may be uncovered revealed in public. Evaluators are sometimes so pressed for time or so discouraged by the compromises necessary in research design that they submit a report to the company and go on to the next study. Yet if we are to understand why one program works and another doesn't, a data base is essential. Only through publication will useful studies accumulate. When results show that a program has had little effect, it is important that other interested parties be informed of this.

There is no best way to conduct a program safety evaluation. All of these factors must be considered and the best safety evaluation done taking into account the constraints of the workplace.

EVALUATION RESEARCH TOOLS AND TECHNIQUES

Evaluation research tools are important in the evaluation procedure. Before these can be chosen, an evaluation design must be developed. However,

> . . . constructing an evaluation design involves deciding how people will be grouped and which variables will be manipulated during an evaluation. As part of the process, the evaluator must identify independent and dependent variables and assess the internal and external validity of the design. Improvement evaluations often use case studies and other noncomparison design strategies, while effectiveness evaluations tend to use more powerful designs that involve comparison groups.[1]

Evaluation Research Tools

Selecting techniques to be used is very important in collecting data for the safety program evaluation. There are many techniques that can be used; the choice will depend on which ones are best-suited to yield the required information. Examples of evaluation tools include:

> Checklist—a form on which the observer records the absence or presence (or frequency of occurrence) of specified events and behaviors
>
> Rating scale—a tool that requires the observer to rate a phenomenon in points along a descriptive continuum
>
> Observational sampling—a method for obtaining representative examples of the observed behavior
>
> Time sampling—a method that involves periods during which the sampling will take place
>
> Event sampling—a method in which a specific event is selected for observation
>
> Questionnaires and interviews

Investigators must be sure these tools are measuring what they want to measure and that workers understand the methods and can supply appropriate information.

Before using these measurement tools, operational definitions of the concepts being studied must be prepared. An operational definition divides the procedure or behavior being observed into measurable components. The more components there are and the more specifically each unit being measured is defined, the easier it will be to know what information should be collected, how to collect it, and how to analyze and interpret it. Operational definitions take time to develop, but completeness and preciseness will produce more reliable results. After determining the components to be studied, the researcher will choose the techniques to measure these variables. These techniques can include self-reports, structured interviews regarding workplace procedures, time samplings of concentrations in the air, checklists recording observations of whether workers are wearing appropriate personal protective clothing and using the appropriate equipment for their work.

The scope of a detailed questionnaire assessing the minute-by-minute activities of workers can be immense. It can encompass not only specific job activities but hourly, daily, and weekly activities as well. In fact, questionnaires can be as detailed and for as long a period of time as deemed necessary. By comparison, the scope of air-sampling techniques is smaller because a more specific measure is being ascertained and because identified variables or clusters of variables are being addressed. These might include the concentration of gases, vapors, mists, and fog in a certain work area.

The scope of the tabulation of workplace injuries and accidents would be great because so many variables could be included, such as:

Date
Time of accident
Events that ocurred before the accident
Other workers' accounts of the accident
The injured employee's account of the accident
First aid measures
Extent of the injury

This information could be compiled along with the number of injuries per employee, places, and times of injuries, and causes. Such data can be interpreted in many ways.

ACCURACY

It is also important to determine the accuracy of the measurement techniques. Two kinds of validity, internal and external, must be considered, as well as the reliability of the techniques.

"Internal validity" means that the evaluation research can distinguish between changes arising from the evaluation itself (for example, improved use of machine guarding because the evaluator is present) and those resulting from other sources. The following are considered to be threats to internal validity.

History.   Historical threats are changes in the environment which occur simultaneously with the program. For example, if a program designed to change workers' attitudes coincided with a major event, it would be difficult to determine whether any observed changes were due to the program or to the event.

Maturation.   Maturation threats are changes within the individuals participating in the program that result from natural biological or psychological development. This would not usually be a concern in the adult working population.

Testing.   Testing threats are the effects of taking a pretest or a posttest. For example, in an evaluation of a three-month safety program for which workers took the same test before and after instruction, it is possible that any observed gains in learning resulted primarily from the effects of taking the pretest rather than from the program. That is, the pretest gave workers practice and made them so familiar with the test that their performance on the posttest would have improved even without the benefit of instruction.

Instrumentation.   Instrumentation threats are due to changes in the calibration of an instrument, the instrument itself, the observers, or the scores.

Statistical regression.   It is a statistical fact that when people are chosen to participate in a program on the basis of extremely high or low scores on some selection measure, the high scores will perform less well and the low scores will perform better if the people take the same or a similar test a second time. Consequently, if workers are selected to participate in a program on the basis of very low scores on a pretest, their average score on a similar posttest will probably increase whether or not the program has any impact.

Selection.   To evaluate a program that rewarded workers for attending health and safety training classes, the attendance records of participating and nonparticipating workers can be compared. If the two groups were not equivalent, it would be difficult to know whether any observed differences were due to the program or to inherent differences (like sex or ethnicity) between the students who did and did not participate. Selection threats result when assignments produce groups with innately different characteristics.

Mortality.   Mortality threats are the result of participants' dropping out of the evaluation. For an evaluation that compared a new dermatitis treatment with the traditional treatment, the 60 patients who volunteered for each group were asked to visit the clinic at the end of their treatment program for a brief physical examination. Fifty-seven patients from the traditinal group and 43 from the treatment group came for the examination, but the evaluation could not be sure if any changes in health status were the result of the treatment or the result of changes in the groups due to their differing drop-out rates.

''External validity'' is the criterion for deciding whether the evaluation findings will hold true for other people. Threats to external validity include:

Reactive effects of testing. Reactive effects of testing are threats due to a pretest sensitizing participants to a program. If a high-protein weight reduction program required dieters to be weighed when they first enrolled in the program and every month thereafter, then it is possible that any weight losses would result primarily from participants' reaction to being weighed rather than to the diet itself.

Interactive effects of selection bias. Even though a program was found to be effective for a certain kind of participant in one part of the country, it does not follow that the program would be effective for a different kind of participant in another part of the country.

Reactive effects of innovation. Participants may perform better simply because they are excited about taking part in an innovative program and/or an evaluation study.

Multiple-program interference. If an evaluation of a new management training program revealed that the students were also in a computer language program, then it is possible that any observed changes in behavior were the product of the two programs in combination and that they could not be reproduced if either was carried out independently.

The evaluation design must be internally valid for findings to be reliable and usable. External validity is important whenever the findings are to be applied to other people or other settings, as in most large programs, or when findings derived from current program participants are used to make decisions affecting future participants.

The evaluator should select a design with built-in controls of the most likely threats to validity. For example, statistical regression is more of a threat to internal validity, say, than instrumentation. A design should be selected that controls for statistical regression but not for instrumentation.

Reliability is the degree of consistency and accuracy with which a technique measures what it was intended to measure. The reliability of any measuring tool depends upon the extent to which internal and external threats have been controlled and the adequacy of the design itself, including the sampling techniques.[7]

Choosing Representative Samples

Evaluators are frequently required to select a representative sample of all participants in the program being evaluated. A representative sample is a limited number of members of the population to which the evaluations' findings can be applied. There are three methods for obtaining representative samples:

1. *Simple random sampling.* The total population is considered as a single group and from it a number of people is randomly selected. For example, a lottery system could be used to select 10 industrial concerns from all industries in a state.

2. *Stratified random sampling.* The total population is divided into subgroups and then from each group a certain number of people is randomly selected so that their number in the sample is proportionate to the subgroup's size in the total population. For example, if there are twice as many men as women, in order to obtain a representative sample of 120 people, 80 men would be randomly selected from all men and 40 women would be selected from all women. Simple and stratified random sampling techniques are the most accurate means of creating the similar or equivalent comparison groups necessary for experimental design strategies.

3. *Purposive sampling.* Individuals or groups are deliberately selected for a particular reason. For example, a joint committee of chemical engineers and chemists might decide that a representative sample of a program's participants should include industrial hygienists and toxicologists to evaluate a health hazard.

No matter which sampling technique is used, it is always necessary to compare the sample to the total population in terms of relevant factors such as ethnic composition, socioeconomic status, and achievement levels.

Independent variables from the structure of a design are manipulated in an evaluation. Dependent variables are observed and measured to determine the results of manipulating the independent variables. For example, if an evaluation question asks about the effects of workers' achievement on accident reduction, then the independent variable is workers' behavior and the dependent variable is the number of accidents.

Collection of Data

Safety program data can be collected in several ways, according to the design used. These data collection methods could include any of the following:

Physiological or biophysical measures (temperature, blood pressure, weight, strength, speed)

Chemical measures (hormone levels, blood toxin levels, radioactive levels)

Microbiological levels (lead counts)

Observational methods (checklists, unstructured observation, rating scales, videotapes, audiotape recording)

Interviews

Questions, scales, and psychological measures (attitude scales), Likert scale (attitude scale), Guttman scale (attitude scale), and semantic differential (attitude scale)

Personality measures

Intelligence, aptitude, and achievement tests

Content Analysis

Content analysis, another method of collecting data, is commonly used to evaluate documents. The researcher decides what is to be coded and examines each document in exactly the same way.

Each of the major data collection methods can be used to evaluate safety programs. Many kinds of industrial data have already been collected and stored in computers, and the safety evaluator needs only to retrieve the data and analyze it.

Program Records.   Program records provide another format for collecting and analyzing workplace data. As with every other kind of data, the more thorough and complete the records are, the more accurate will be the conclusions based on them.

Analyzing evaluation information is summarizing and synthesizing the data to find the answers to the evaluation questions. Information analysis involves preparing the evaluation data for analysis, applying the appropriate analytic methods, and interpreting the results.

In an improvement evaluation, the evaluator usually can be flexible and creative in applying analysis methods to get the most information about the success of the program and its parts. In an effectiveness evaluation, the evaluator should select well-defined analytic methods that will permit inferences regarding the program's value in comparison to other programs and the broad consequences of participation in it. Most important, this material should be cast in an understandable format.

Reporting Data

The evaluator is responsible for reporting evaluation information in a form that is easy to use. In an improvement evaluation, the evaluator usually provides frequent reports to the program staff. These may include formal written reports, memoranda, presentations at meetings, or telephone calls. An effectiveness evaluation requires a formal and detailed report.[7]

SAFETY PROGRAM EVALUATION APPLICATIONS

There are abundant examples of safety evaluation programs in many industries; one example is cited here.

An evaluation of safety programs at Three Mile Island was conducted.[2] The evaluators reviewed operations at the Radiation Protection Program at the Nuclear Reactor Site, Unit 2. They found management and technical deficiencies in the program. They concluded that since the accident in 1979, General Public Utilities/Metropolitan Edison (GPU/ME) had adopted a stronger commitment to personnel and operations safety. The current safety program was capable of supporting limited work activities while maintaining employee exposures to radiation at a low level. The program required continuous upgrading. Recommendations were made to allow GPU/ME to continue limited recovery operations, provided that management maintained its present attitude of safety concern. It was recommended that major recovery operations should be postponed until the safety program was upgraded and evaluated by an independent committee. It was also urged that GPU/ME provide a management plan and time schedule for safety program improvements.

SUGGESTED LEARNING EXPERIENCES

1. How might you as a safety professional present the evaluation process of the safety program to management? to supervisors? to workers?

2. Debate: Evaluation research is or is not vital to a comprehensive and effective safety program.

3. How would you as a safety professional utilize the data from a safety program evaluation?

4. As a safety professional, whom would you want to have conduct the safety program evaluation? Discuss your rationale.

5. Select an industry of your choice and present the evaluation research tools that you would have utilized in gathering the information that you wish to gather.

6. Develop a detailed program evaluation plan for a workplace of your choice.

7. Attend safety audits for two workplaces of your choice. Critique both of them in terms of their purpose, data collected, and results conveyed. If you had been conducting the audit, what might you have done differently?

8. Select a workplace of your choice. From your knowledge thus far, what workplace hazards would you expect to exist? Hypothetically conduct a workplace audit. What factors emerged as hazards?

9. Discuss the pros and cons of a safety audit done by "internal" evaluators as opposed to "external" evaluators.

10. In terms of learning principles, including motivation, how would you want the results of a safety program evaluation to be presented to the involved employees? Include your rationale.

REFERENCES

1. Fink, A., and Kosecoff, J. *An Evaluation Primer*. Washington, D.C.: Capitol Publications, 1978.
2. Meinhold, D., Murphy, T. D., Neely, D. R., Kathren, R. L., Rich, B. L., Stone, G. F., and Casey, W. R. *Three Mile Island, Unit 2, Radiation Protection Program. Report of the Special Panel, Office of Nuclear Reactor Regulation, Radiological Assessment Branch. United States Regulatory Commission*, Washington, D.C.: GPO, 1979.

19 INITIATING A CORPORATE SAFETY PROGRAM

A total safety program refers to a specific approach to reducing accidents and occupational illness from the workplace. Although an accident-free environment probably is not attainable, it is the goal. In order to keep accidents to a minimum, it is necessary first to initiate a total safety program. Then one can identify hazards and other causes of workplace accidents and institute a system of planned change.

Before a safety program can be instituted, a number of decisions must be made by the safety manager in collaboration with his or her manager, who is preferably the chief executive officer (CEO).

The response to the following questions will have a significant effect on the development of a safety program.

1. How is responsibility for safety tasks to be allocated in the organization?
2. Is this allocation based on management awareness of and commitment to safety, or is it merely a grudging concession to rules and regulations imposed by OSHA?
3. How much authority will the safety department have?
4. How much funding will the safety department receive?

If management is only responding to external pressures like OSHA, then the safety engineer will have to be more vigilant in conducting safety audits and monitoring safety performance. However, total safety is always the goal of every safety engineer.

The place to start a safety program after the questions of philosophy have been assessed, principles established, and priorities set is determination of the safety organization.

A total loss control unit has been proposed which is a full service safety organization.[1] It eliminates much overlap (which was discussed in Chapter 5) and reduces role confusion among safety professionals. It can determine in a clear way what the company's safety functions are and can make the necessary alterations to achieve the development and implementation of a total safety program.

Many companies are moving in the direction of appointing a certified safety professional as director of safety. The duties implied by this role will be partly determined by federal and local legislation regarding safety and health, partly by the nature of the business operations, and partly by the expertise of the safety professional.[1]

Table 19.1 lists 10 general functions of the safety director in the role of organizer, stimulator, and leader (guide) of the safety program. It seems imperative for all safety directors to implement these functions, although in smaller companies the director might have more tasks than these and in larger companies, fewer. The safety director's functions may also vary depending upon what is happening in the industrial organization at the time. Is there an oncoming audit? Has there just been a major accident? Has new management just taken over?

These tasks might shift in terms of focus at varying times, but the general gestalt (purpose, doing) would be the same.

A general safety program encompasses everything from monitoring hazardous substances to maintaining health records for all employees.

DETERMINING THE NEED

In any company, no matter what size, a safety program can be developed on paper, but to be effective it must, of course, be implemented and evaluated. The initial step in developing a safety program is to perform a job safety analysis (JSA) and a safety audit.

Job Safety Analysis

A JSA would examine all hazards or potential hazards and the means necessary to control them. This was completely discussed in Chapter 17. Some of the hazards and controls in an auto body shop are shown in Table 19.2. The hazards are then assessed in two ways: by estimating the frequency of potential and by the severity of the hazard. These will be discussed in detail in Chapter 27 as part of a hazard analysis report.

TABLE 19.1
General Functions of a Safety Director

1. The formulation and administration of the safety program.
2. The acquisition of the latest and best hazard control information.
3. The representation of management before the public, employees, insurance companies, and governmental agencies as its safety resource.
4. The advisement on safety-related issues to managers at all levels.
5. The collection and recording of pertinent data on safety-related operational matters, including work injury causes and statistics.
6. The reporting to top management periodically, on a regular basis (that is, monthly, quarterly, or annually), on the state of the organization's safety effort.
7. The advisement to supervisors of safety training programs.
8. The coordination with the organization's medical department (or part-time physician and medical advisor) on the safe placement of new or convalescing employees.
9. The inspection of the facilities for compliance with federal, state, and local regulations as well as with the safety program's established operating procedures and any insurance company recommendations that are offered.
10. The participation in the review of purchase specifications to ascertain whether danger points in inherently hazardous machinery and equipment are guarded correctly and in the design of new facilities, equipment layout, or process arrangements to determine whether all needed considerations for safety have been satisfied.

SOURCE: De Reamer, R. How to establish an effective safety program. *Professional Safety*, 1981, **26**, 28–30. (May)

TABLE 19.2
Example of Possible Hazards and Controls in an Auto Body Shop

Hazard	Control
Electric service	
440 volts AC	Lockout protection
	Barrier to prevent access
Welding machine	
Current	Disconnect switch
Electric arc	Face shield
	Welding gloves
Housekeeping	Readily available bins for dissassembled parts
	Monitor for trash control
	Dumpster for scrap metal
	Absorbant material for oil spills

Safety Audit

A total safety program should be systematically developed, even though few programs in existence have been built in this manner. A safety audit consists of identifying safety and health hazards, weaknesses, in safety program components, and the adequacy of supervisor and manager safety know-how and effort.

A safety audit is conducted which examines the entire spectrum of potential hazards. Safety audits should be conducted by the most qualified people in the safety department, and the audit team should include line managers.

The safety performance of each activity or safety design piece of equipment should be assessed.

The audit usually includes surveying to assess hazardous waste, conducting inspections as necessary, training workers in the proper use of materials and safe handling of equipment, educating about exposures, and auditing personnel exposed to hazardous material. The extent of the auditing is determined by the possible amount of data to be collected. If only a small amount of data is expected, it can probably be computed manually. If, on the other hand, a large amount is to be collected, then a computer will be used to analyze the data. Monitoring, surveillance, inspections, and even training can be easier and more accessible as people become familiar with the computer and more facile in using it. All safety program components can be stored in a computer. It could contain due dates, length of time required to carry out a procedure, and personnel designated to perform it. Spaces could be provided to input data, a program set up for analyzing the data, and a printout obtained of survey results.

Total Safety

Total safety is the bottom line for all places of employment. Will this ever be achieved? If enough training programs are implemented, if enough reinforcements are given, if the importance of safety is stressed enough so that every employee will value it, if management totally supports safety in all ways, then perhaps the realization of total safety for all will be a reality.

Do We Have Any Other Choice?

After the safety audit has been conducted, and its results compiled, the first question that must be asked would be, Is the company a safe and healthful place to work? The second and third questions would probably be, What standards must be met and how can the safety department meet them? The third question is more easily answered than the first. For most companies the safety standard that sets the minimum requirements is the general industry standard of OSHA (see Chapter 3). It has been mentioned earlier in this chapter. Other safety standards are published by the American National Standards Association, the International Standards Organization, the National Fire Protection Association, and others. These standards set the minimum objectives by which a safety engineer can judge how much safety the organization is providing.

The first question can only be judged qualitatively. How well does your organization compare with others in your industry? Accident rates for a number of industries have been shown in various chapters in this book. These are only numbers. They do not measure the human suffering, disrupted lives, or broken families.

The general safety policy of a company reflects the philosophy and biases of top-level executives who must be in full agreement about the direction the company is taking in regard to safety. Safety policy cannot successfully be established without official, explicit management support.

This responsibility for representing a whole unit is going to have to come from management, as individual workers are not in a position to know how management functions.

Does management in other occupational areas use implied coercion for increased productivity, disregarding safety aspects? For the sake of production, foremen at a New England shipyard made it difficult for workers to replace lost or broken safety glasses. When OSHA made a wall-to-wall inspection, the shipyard was fined several hundred thousand dollars.

Putting direct responsibility for safety at the supervisory level is one way to cut through some of the directives (oral or written) from management which can lead to unsafe practices. If supervisors are directly responsible for safety, they will monitor all their employees to make sure they are doing their work safely and in accordance with the company's safety policy. Supervisors will receive directives from management and relay employee concerns to management. They can be involved in both management and employee sides of the operation and are also in a pivotal position to serve as management's liaison with the union regarding safety matters. They may also serve as consultants on safety and health to the union. The latter role will depend upon the organizational chart, which defines relationships and responsibilities. If supervisors don't work directly with the union, whoever does should collaborate with them. This person might very well be someone designated by the safety professional. Once management has become convinced of safety's importance and the safety promoter and supervisors understand management's commitment, a successful safety program may be initiated.

The most important aspect of instituting a safety program is through safety training and safety promotion. Safety training is so critical that an entire chapter is devoted to it. (See Chapter 17.)

"Promotion" has been defined as persuasion through motivation. The goal of a safety promotion program is to enlist the cooperation of every employee for the promotion and maintenance of optimal health and freedom from accident and injury. A successful safety promotion program would result in the acceptance of responsibility by employees for performing their own work safely. In launching a safety promotion program, the safety professional and management must focus on areas meaningful to employees. It is best to avoid generalities and stick to specifics, that is, to aim at situations that employees are familiar with and can relate to directly. In instituting a safety program, the effectiveness of safety incentives and adequate motivators should not be underestimated.

A safety program is evaluated by measuring the number of accidents and accident-related insurance claims that are filed.

SAFETY PROGRAM FOR A SMALL COMPANY

The small company has safety problems similar to those of the large multinational corporation. The auto body shop often has housekeeping hazards. Assembling and reassembling wrecked automobiles cause much clutter. Pieces of ragged metal can cause rather severe wounds. Hydraulic jacks, used for lifting vehicles, stretching and bending sheet metal, and straightening the frame of a vehicle, can fail, crushing a worker. Acetylene and oxygen are used for cutting and welding; acetylene is highly flammable. Gas welding produces fumes. Electric welding produces an arc which is dangerous to the eye. Hazards associated with an auto body shop include electrical hazards, paints and thinners, epoxys, carbon monoxide, and lead fumes.

A large organization would require an industrial nurse, an industrial hygienist, and a safety engineer to deal with this variety of hazards.

In a small company, many of the safety tasks may be handled by one person. Very often the small company will not have medical personnel or some of the safety-related professionals on site. If this is so, it is imperative that there be backup personnel, that every employee knows what the procedure is in case of an accident, that there be assigned first aid personnel, and that the plant manager or the safety professional has an active consultation network that can be contacted as necessary for advice. In Ontario, the Workmen's Compensation Board requires that at least one member per shift have extensive first aid training. This is provided free of charge by St. John's Ambulance. Other examples would be using the services of a local toxicologist in a materials handling spill or seeking the advice of the National Safety Council on questions of safety and industrial concerns. This organization has a fund of information on numerous topics; if it doesn't have the needed information, the NSC can suggest to the inquirer where the information might be obtained. NIOSH has amassed a tremendous amount of information on all facets of industrial operations. It is important for the manager or safety professional of a small company to assess what safety resources of the community in which he or she operates before an accident so he or she will know what is available when it is needed.

It is also important that emergency procedures can be identified (no matter how small a company is) for one does not know when an industrial accident or a sudden illness, such as a heart attack, could occur. It is certainly better to be as prepared as one can be in a situation like this than to be unprepared and have an injury or accident which occurred at the workplace be fatal or needlessly crippling.

In a small company, a few people may be expected to have many skills. Oftentimes in a small company the only professional on site besides the manager may be the occupational nurse. He or she can be trained to handle many of the safety demands of a small company. These tasks could include using first aid for employees as necessary, collecting the data of an on-the-job injury, giving influenza injections, being supportive to the workers after a workplace injury, and collecting air samples to determine concentration levels in a specified area. If this is so, the nurse will be a key figure in supporting, developing, and implementing safety policy and will need to keep abreast of what is happening in occupational health and safety and of new developments that are occurring. Might there be more efficient and safe ways of doing certain procedures? Are the workers, for the type of work they are doing, taking correct breaks?

The nurse's influence can be very important; he or she can be helpful in supporting a physical fitness program for all interested employees, in providing information on diets and high energy foods, and in encouraging active involvement in rehabilitation programs for drugs and alcohol. The roles the occupational health nurse performs are various and will continue to grow as he or she continues to expand boundaries and to discover new ways of functioning.

SUGGESTED LEARNING EXPERIENCES

1. Develop an overall safety program for a workplace.

2. Compare the supervisory process in manufacturing plants, construction sites, oil rigging operations, and hospitals.

3. Make a chart listing the qualities of a good manager.

4. Write a job description listing the criteria of a plant supervisor.

5. Organize a panel discussion of workers' perceptions of safety.

6. Simulate an employees' group discussion which is being held immediately after a coemployee was injured. Have someone critique your style and comment upon what you did. At the conclusion of the session have each participant comment upon how it felt playing his or her role.

7. Discuss your philosophical biases regarding governmental funding and concomitant regulations. How do make decisions when what you

are involved in is against your beliefs and you are in conflict? Is the amount of money involved a factor?

8. Develop a research proposal to compare two similar company's safety programs.

9. As a safety professional, how would you go about identifying the causes of workplace injuries?

10. Illustrate the development of unions, depicting their meaning to employees.

SUGGESTED LEARNING EXPERIENCES

1. Develop an overall safety program for a workplace.

2. Compare the supervisory process in manufacturing plants, construction sites, oil rigging operations, and hospitals.

3. Make a chart listing the qualities of a good manager.

4. Write a job description listing the criteria of a plant supervisor.

5. Organize a panel discussion of workers' perceptions of safety.

6. Simulate an employees' group discussion which is being held immediately after a coemployee was injured. Have someone critique your style and comment upon what you did. At the conclusion of the session have each participant comment upon how it felt playing his or her role.

7. Discuss your philosophical biases regarding governmental funding and concomitant regulations. How do make decisions when what you are involved in is against your believes and you are in conflict? Is the amount of money involved a factor?

8. Develop a research proposal to compare two similar company's safety programs.

9. As a safety professional, how would you go about identifying the causes of workplace injuries?

10. Illustrate the development of unions, depicting their meaning to employees.

REFERENCES

1. De Reamer, R. How to establish an effective safety program. *Professional safety*, 1981, **26**, 28–30.

20 RISK MANAGEMENT

Because risk management applies to a very complex field, insurance, this chapter presents only a limited introduction to the subject.

Risk management is the process by which management decisions are made about controlling and minimizing hazards and accepting residual risks.[1]

Risk analysis is a mathematical calculation of the probability of the annual cost of insurance, omitting profit and overhead, required to cover all losses averaged over a great many years.

COST-BENEFIT ANALYSIS

Cost-benefit analysis is another risk assessment technique. In simple terms cost-benefit analysis is a measurement of the cost of performing a certain action or actions and a measurement of the benefits that occur as a result of these actions.[2] A calculation of the relation betwen these two factors is then made.[10]

When goals have been established, constraints or limits to their achievement should be estimated. In a linear programming problem, you analyze the effects and constraints upon an optimal solution of the problem. These constraints may be categorized as follows:

1. Physical constraints based on the current limitations of human inputs, technologies, and production possibilities
2. Legal constraints consisting of laws, court rulings, administrative rulings, and international conventions
3. Administrative or managerial constraints caused by the limitations of the people and the physical resources available to deal with the objective

4. Distributional constraints consisting of limitations imposed by the geographic regions, income classes, and age distribution of those clients or customers who are recipients of the value provided by the organization

5. Political constraints consisting of limitations imposed upon the organization by the political processes in a system

6. Financial or budget constraints imposed by fixed budgets, costs of capital, foreign exchange rates, or financial policies established by the organization

7. Traditional, social, and religious constraints imposed by the social customs, mores, and values in the system which limit available options and possible actions

After goals have been established and constraints estimated, external effects should be examined. These side effects have been classified as follows:

1. *Production to Production.* An activity by one producer which affects the output of other producers, for example, the water pumped from one mine affecting the nearby mines

2. *Production to Consumption.* An activity by one producer which affects consumers, for example, noise and pollution

3. *Consumption to Consumption.* An activity by one consumer which affects other consumers, for example, radio noise in a public park

4. *Consumption to Production.* An activity by one consumer which affects producers, for example, an individual hunting through a planted field

The first step in a cost-benefit analysis is to calculate the risk inherent in a given activity. For years, insurance companies have concentrated on weighing risks in order to calculate premium rates. The relative riskiness of many activities, such as flying (1 death per 2.5 million miles) can be found in any good almanac.

For other activities, the job is more difficult. For one thing, the consequences of many actions don't closely follow the event. Excessive exposure to x-rays, contact with certain chemicals, or use of dangerous drugs often does not result in any apparent damage until months or years after the initial exposure. For another, it is difficult to assess risks when more than one hazard is involved. Although it has been determined that asbestos can cause cancer, many asbestos workers also smoke; thus when lung cancer develops, the actual cause is uncertain unless the cancer is mesothelioma, the form of cancer only caused by asbestos.

Workers at risk for asbestos exposure are shipyard workers employed during World War II, asbestos products manufacturing employees, chemical and refinery maintenance personnel, power plant workers, stationary engineers, and firefighters in commerical, industrial, and residential buildings. Other groups exposed to asbestos include railway steam engine workers, marine engineers, and U.S. Navy engine room personnel; even their families may have some significant exposure. The land-based construction industry used either directly or incorporated in finished products more than two-thirds of all asbestos mined in or imported into the United States. Among workers exposed were painters who sanded wallboard joint spackle; carpenters who cut cement panels, wallboard, and acoustic tile containing asbestos; floor tile installers, if previously installed surfaces were sanded; and plasterers who sprayed asbestos-containing materials on steel work for fireproofing high-rise office buildings and on room surfaces of auditoriums, night clubs, and schools for acoustic insulation. Indirect exposures, similar to those in shipyard construction, are important for workers on jobs where asbestos materials were sprayed and for children attending school with asbestos insulation applications.[4]

More troubling is the fact that some of the greatest hazards can't be statistically determined because history is not a reliable guide. Just because there has never been a catastrophic accident in a nuclear power plant doesn't mean there can't be one. Scientists have estimated the risk of a nuclear power plant disaster as one in a hundred years. With 75 operating nuclear plants, what is the risk that an accident will happen next year? To calculate risks, economists have been forced to construct elaborate computer models that identify the various things that can go wrong and, from past experience with similar types of situations, determine within a broad range how likely it is that particular failure will occur.

COST TO ELIMINATE RISK

After estimating the risk, it is necessary to calculate the cost of reducing it. This is customarily expressed either as the cost of extending a life by one year or the cost of increasing the average longevity of the average person taking the risk. These calculations are made for different levels of risk reduction and plotted on a graph. This process ignores the issue of how much value we place on life.

Efforts to assign economic weight to human lives actually go back a generation, when organizations such as the U.S. Public Health Service and the military services began trying to quantify the cost to society when an individual becomes sick or disabled. Using age, income, and other data, it has been calculated that when a person is killed on the highway, an average of $300,000 is lost in potential productivity, court time, ambulance services, funeral expenses, and other expenses. This approach fails to consider the emotional impact of the loss of a human being.

How much risk is acceptable depends on the amount of benefits; that is,

Table 20.1
Risk: How People See It[a]

Activity and Deaths per Year (Estimated)	League of Women Voters	College Students	Business and Professional Club Members
1. Smoking (150,000)	4	3	4
2. Alcoholic beverages (100,000)	6	7	5
3. Motor vehicles (50,000)	2	5	3
4. Handguns (17,000)	3	2	1
5. Electric power (14,000)	18	19	19
6. Motorcycles (3,000)	5	6	2
7. Swimming (3,000)	19	30	17
8. Surgery (2,000)	10	11	9
9. X-rays (2,300)	22	17	24
10. Railroads (1,950)	24	23	20
11. General (private) aviation (1,300)	7	15	11
12. Large construction (1,000)	12	14	13
13. Bicycles (1,000)	16	24	14
14. Hunting (800)	13	18	10
15. Home appliances (200)	29	27	27
16. Fire fighting (195)	11	10	6
17. Police work (160)	8	8	7
18. Contraceptives (150)	20	9	22
19. Commercial aviation (130)	17	16	18
20. Nuclear power (100)	1	1	8
21. Mountain climbing (30)	15	22	12
22. Power mowers (24)	27	28	25
23. High school & college football (23)	23	26	21
24. Skiing (18)	21	25	16
25. Vaccinations (10)	30	29	29
26. Food coloring[b]	26	20	30
27. Food preservatives[b]	25	12	28
28. Pesticides[b]	9	4	15
29. Prescription antibiotics[b]	28	21	26
30. Spray cans[b]	14	13	23

SOURCE: Niles, H., and Antilla, S. What price safety? The "zero-risk" debate. *Dun's Review*, 1979, **117**, 48–57. (Sept.) Reprinted with the special permission of Dun's Business Month, (formerly Dun's Review), September 1979, Copyright 1979, Dun & Bradstreet Publications Corporation.

[a] The public's idea of what is most risky usually differs widely from the facts. When three groups were asked to rank 30 products or activities from most to least risky, they came up with the ordering below. The accurate list, based on past experience, is shown on the left.

[b] Not available.

the higher the risk, the more benefits must accrue. The public will accept a higher level of voluntary risk—about 1,000 times higher—than of involuntary risk.[6] At the same time people were asked what they thought were acceptable risks in a variety of situations. Table 20.1 shows that different groups of people see risks differently. Risks are rated more on a psychological basis than on a reality basis. Column 1 of Table 20.1 lists different activities and the number of deaths per year associated with each activity. Smoking, which statistically has the highest risk, is not judged to be as risky as nuclear power plants, which are ranked as the number-one risk by members of the League of Women Voters and by college students.[5]

Table 20.2 presents a list of actions which increase an individual's risk of death. In other words, when you behave in a certain manner, you increase the risk of having an accident or contracting a disease. In this table the increased risk is very small and you may consciously say that you are willing to take this increased risk. But consider the situation, for example, when you are married to a smoker and live in a stone house in Denver, Colorado. At what point do you say that you are not willing to increase your risk any more?

Paradoxically, nuclear power has stirred a great deal of public consternation, while dangers of cigarette smoking leave people relatively unconcerned. Yet benefit figures clearly indicate that more lives can be saved by getting smokers to quit than by investing additional sums on nuclear power plant safety.

Given such choices, where should our priorities lie? Such decisions can only be made by well-informed public officials. But risk–benefit advocates insist that the methods are not in conflict.

RISK ANALYSIS

The preceding section described how risk management can be applied. In this section, the principal functions and methods of risk analysis are discussed. Risk analysis is a process for evaluating potential risks associated with particular activities. It considers two aspects of accidents: frequency and severity. Frequency refers to the likelihood of occurrence. Severity refers to the impact of loss in terms of dollars or degree of physical impairment. Using these factors, a risk analysis develops classes of severity and frequency. These classes are used to rank the relative risk of various events. They are presented in Chapter 27. Each class is defined in terms meaningful to the organization. (For example, "catastrophic" may, to one organization, refer to losses of more than $500,000 and which involve death or dismemberment to more than one person; another organization may have lower or higher standards.) A risk analysis matrix is prepared and risk categories determined combining both variables. (Figure 20.1 displays one such matrix.) Events are ranked in terms of probability and

TABLE 20.2
Actions Increasing Risk of Death by One in a Million

Action	Nature of Risk
Smoking 1.4 cigarettes	Cancer, heart disease
Drinking 0.5 liter of wine	Cirrhosis of the liver
Spending 1 hour in a coal mine	Black lung disease
Spending 3 hours in a coal mine	Accident
Living 2 days in New York or Boston	Air pollution, heart disease
Traveling 6 minutes by canoe	Accident
Traveling 10 miles by bicycle	Accident
Traveling 30 miles by car	Accident
Flying 1,000 miles by jet	Accident
Flying 6,000 miles by jet	Cancer caused by cosmic radiation
Living 2 months in Denver on vacation from New York	Cancer caused by cosmic radiation
Living 2 months in average stone or brick building	Cancer caused by natural radioactivity
One chest x-ray taken in good hospital	Cancer caused by radiation
Living 2 months with a cigarette smoker	Cancer, heart disease
Drinking heavily chlorinated water (e.g., in Miami) for 1 year	Cancer caused by chloroform
Drinking 30 12-oz. cans of diet soda	Cancer caused by saccharin
Living 5 years in the open at site boundary of a typical nuclear power plant	Cancer caused by radiation
Drinking 1,000 24-oz soft drinks from recently banned plastic bottles	Cancer from acrylonitrile monomer
Living 20 years near PVC (polyvinyl chloride) plant	Cancer caused by vinyl chloride
Living 150 years within 20 miles of a nuclear power plant	Cancer caused by radiation
Eating 100 charcoal-broiled steaks	Cancer from benzopyrene
Risk of accident by living within 5 miles of a nuclear reactor for 50 years.	Cancer caused by radiation

SOURCE: Wilson, R. Analyzing the daily risks of life. *Technology Review*, 1979, **81**, 41–46. (Feb.) Reprinted with permission from *Technology Review*, Copyright 1979.

severity, and the ones that fall in the categories of high combined risk (imminent and catastrophic) are given priority. (In this way, potentials are ranked so that the worst ones can be addressed first.)

Another type of risk analysis considers two components of "severity": intensity and extensiveness. (See Figure 20.2.) Intensity refers to the dollar impact of the severity of injury; extensiveness is the number of people

Severity

	I	II	III	IV
A	IA	IIA	IIIA	IVA
B	IB	IIB	IIIB	IVB
C	IC	IIC	IIIC	IVC
D	ID	IID	IIID	IVD

Probability

Imminent danger: IA or IIA
Seriously deficiency: IB, IC, ID, IIB, IIC, or IID
Nonserious deficiency: IIIA, IIIB, IIIC, or IIID
Minimal deficiency: IVA, IVB, IVC, or IVD

FIGURE 20.1.   A risk analysis using two variables (*from NIOSH*).

affected per incident. This method takes human involvement more into consideration in the risk analysis and evaluation, but adding a third variable makes the process more complex. Figure 20.3 illustrates in a three-dimensional analysis the interaction of probability, severity, and extensiveness. If you are in Block IV-D-1, as demonstrated in this figure, you better run for cover as a tidal wave is just about to arrive. Computers

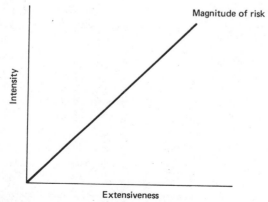

FIGURE 20.2.   A risk analysis considering two components.

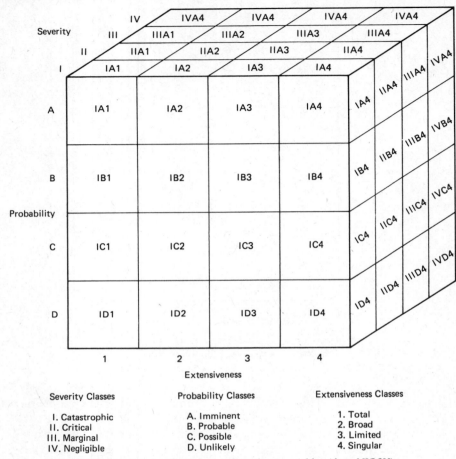

FIGURE 20.3.   A risk analysis using three variables (*from NIOSH*).

Severity Classes	Probability Classes	Extensiveness Classes
I. Catastrophic	A. Imminent	1. Total
II. Critical	B. Probable	2. Broad
III. Marginal	C. Possible	3. Limited
IV. Negligible	D. Unlikely	4. Singular

should eliminate problems posed by such complexity. Variables could be programmed in the computer and the risk analysis quickly completed.

Exposures to Substances

A risk analysis requires defining the conditions of exposure by analyzing exposure in terms of severity, frequency, and adverse reactions; relating it to effects; and estimating the risks. The organization's financial ability to assume risks must be evaluated, controls designed to reduce or eliminate the risks, and funding sought for implementing risk-reduction measures.

Much can be learned by reviewing inadvertent occupational exposure. People are exposed to hazards through accidents, wars, and natural disasters. Occupational exposures provide important evidence. Because workers typically experience long-term exposure at rather high intensity (compared to exposure levels of the general population), investigation of their health may lead to early identification of hazards to which the rest of society is exposed at much lower levels. An added advantage is that it is often possible, because of the close and consistent association between workers and their environment, to convincingly link effects with their causes. No one doubts that coal miners' black lung disease is caused by coal dust. Asbestos fibers are a serious hazard, aggravating the delicate inner lining of the lungs and causing the debilitating scarring known as asbestosis.[5]

Much of what we know about the effects of asbestosis comes from studies of miners, processors, insulation installers, and others who have frequently been exposed to the material for years. Comparing their health with that of people rarely exposed has provided insights available in no other way. In assessing noise exposure, for example, the degree of noise from a jackhammer or a blasting site may be more significant than the noise level in quieter but busier work areas. Likewise, air pollution from a toxic spill may require more immediate intervention than the daily pollution at a work site.

Chemical Risks. Chemicals may be degraded as they pass through the environment by the action of bacteria, light, or other agents; that is, they may be metabolically converted by the body into other substances. Therefore, in assessing toxicity, it is first necessary to determine the extent to which a chemical is degraded or metabolized; the products must then be analyzed to determine their toxicity. Are they stored in the liver? Are they excreted? For instance, when DDT is released into the environment, its molecular architecture is slowly converted into the related chemicals DDE, DDD, and other substances which may be less or more toxic than DDT. The correct question is, What is the effect of DDT and its breakdown products?

The amount of a chemical in a mixure is expressed as parts of that chemical relative to the whole mixture. For example, if there is one ounce of a chemical in a thousand ounces of a total mixture, it is said to be present at one part per thousand. By midcentury it was possible to detect chemicals at the level of one part per million. In 1951, DDT was detected in human milk at one-tenth of one part per million.[7]

Adverse Reactions. The severity of physiological damage caused by chemical poisoning depends on the chemical, the size of the dose, and the dose–response relationship. Irritation and sensitization are common reactions to chemical exposures in the workplace. Skin diseases are the most common of all occupational illnesses.

Less immediately detectable reactions are those associated with carcinogens—chemicals, viruses, and radiation. The FDA states that a carcinogen is a substance that, when administered in appropriate dose, causes an increased incidence of malignant tumors in experimental animals as compared with a controlled series of untreated animals.

Data is accumulating on the link between exposure to a substance and its long-term effects. Among survivors of Hiroshima and Nagasaki, leukemia has been a long-term effect; neck cancer occurs in fishermen exposed to solar radiation for long periods of time; there is increased incidence of miscarriages or difficulties in conceiving among hospital operating-room personnel because of exposure to anesthetic gases. A dose–response relationship is quite apparent in these cases.

Estimating Overall Risk

Estimating overall risk consists of summarizing information gathered about exposures, severity, adverse reactions, and other effects and using it to prepare estimates of the risk potential for individual workers, groups of workers, and consumers.

This section presents an example of how risk assessment may be accomplished. The following assumptions about the risks from radiation were made:

1. Manufactured radiation is not substantively different from radiation emanating from natural resources.
2. Illnesses induced by manufactured radiation are not generally different in kind from diseases caused by natural ionizing radiation.
3. Among the several effects of radiation, the genetic ones demand special attention because of their importance to our descendants.
4. The total amount of genetically significant radiation currently received by the average person from manufactured sources is less than the amount received from natural sources.[3]

Having narrowed the scope of the problem with a set of limiting statements, or constraints, it is now possible to restate the problem so that the risk may be estimated with greater precision. Ionizing radiation has multiple effects on the human body. Excess radiation exposure can cause illness and can change the genetics of the germ cells. The adverse effects of radiation exposure include growth impairment, mental retardation, cataracts, sterilization, and the induction of cancers of the thyroid, bone, skin, breast, lung, and white blood cells. Genetic effects include changes which range from minor to major, trivial to tragic, in the first generation following gene or chromosome alteration; other effects are postponed to succeeding generations. Overall genetic risk to the human population from radiation includes the following:

1. Risk relative to natural background radiation
2. Risk estimates for specific genetic conditions
3. Risk relative to current prevalence of serious disabilities
4. Risk in terms of overall ill health[3]

The analysis defined the conditions of exposure, detailing who is exposed to how much of what kind of radiation and comparing the contributions from various sources; it identified the adverse effects on health, both genetic and somatic; it related exposure to effect; and then, on these bases, it estimated the overall risks.

In weighing potential health risks against potential benefits, it must never be forgotten that even the most far-seeing view may provide erroneous in light of unexpected scientific developments.

In risk management specific terms identifying components of loss are used. "Loss reduction" is any action which reduces the losses incurred. "Loss transfer" means shifting the recovery cost following the loss. (Insurance is a good example of loss transfer.) "Loss retention" refers to circumstances under which the loss is neither removed nor transferred. Loss reduction may also be a decrease of the physical destruction (as by reducing the amount of material burned or the number of persons injured), or a reduction of the operational loss from a given amount of destruction (as by having standby equipment or more effective medical care for the injured). It includes the concepts of loss prevention and control as well as the concept of risk avoidance—the refusal to accept a given risk.

One of the most important aspects of risk management is allocating funding. Although risk management is related to insurance, which certainly implies secondary and tertiary health promotion and prevention (treatment and rehabilitation), risk management's main purpose is primary prevention. This means identifying sources of accidents and injuries before they occur and taking measures to prevent them.

The Occupational Safety and Health Act has had decided impact upon risk management. With OSHA setting permissible limits and acceptable workplace standards, management has been compelled to use risk management concepts not only to make the workplace safer but to determine cost benefits and risk factors.

OSHA has been instrumental in persuading corporate executives to regard risk management as an aid instead of a necessary evil. Because of OSHA's demands for maximum safety irrespective of cost, managers have been required to understand more fully employee needs, consumer needs, and insurers' concerns. At the same time they must keep their fingers on the pulse of productivity and profit. Safety engineers must be wise enough to see that if a company cannot meet its payroll and make a profit, it will not survive. Hence risk management will be used increasingly to enable corporate managers not only to adequately assess the current work environment in terms of risks, costs, benefits, and losses but to provide data on variables like costs, workplace injuries, and market demands. This information is an important prerequisite to assessing what must be done to improve safety. Included are the changes that OSHA has mandated. By using the results of risk management studies, corporate managers may decide to abolish an operation, make alterations, or purchase new equipment which meets OSHA's standards.

Risk management will become increasingly important, not only as legislation makes demands on industry but also as employees demand a safe environment.

SUGGESTED LEARNING EXPERIENCES

1. Describe the benefits you would want to be included in an insurance program at your workplace.

2. Investigate three similar industries and the insurance benefits they provide for their employees. Based on their accident/injury record, have they chosen the best insurance? Justify how you arrived at your decisions.

3. If you were the president of a corporation, how would you use risk management?

4. If you were the safety professional in an organization, how would you use risk management?

5. Prepare a debate focused around the area of a known exposure to a workplace contaminant. What decisions would you make concerning whether you would work in that position.

6. Develop a program titled "Risk Management" for a staff development meeting.

7. Set up a linear programming problem to solve a risk appraisal for either noise or vibration. How many variables and how many constraints would you need?

8. What risk factors in the environment would you like most to see eliminated? How would you suggest doing this?

9. Develop a fault tree and a hazard matrix for three known workplace hazards. Discuss the implications therein.

10. Develop a safety poster illustrating the key concepts of risk management.

REFERENCES

1. Briscoe, G. J. *Risk Management Guide* (ERDA 76-45/11). Idaho: EG&G, Inc., 1977.

2. Mehr, R. I., and Hedges, B. A. *Risk Management: Concepts and Applications.* Homewood, Ill.: Irwin, 1974.

3. National Acadeny of Sciences/National Research Council, Advisory Committee on the Biological Effects of Ionizing Radiation. *The Effects on Populations of Explosure to Low Levels of Ionizing Radiation.* Wahington, D.C.: GPO, 1972.

4. Nicholson, W. J. The role of occupation in the production of cancer. *Risk Analysis,* 1981, **1,** 77–79.

5. Niles, H., and Antilla, S. What price safety? The "zero-risk" debate. *Dun's Review,* 1979, **117,** 48–57. (Sept.)

6. Sheriff, R. E. Loss control comes of age. *Professional Safety,* 1980, **25,** 15–18. (Sept.)

7. U.S. Department of Health, Education, and Welfare. *Report of the Secretary's Commission on Pesticides and their Relationship to Environmental Health,* Washington, D.C.: GPO, 1979.

21 PLANNING FOR EMERGENCIES

Planning for emergencies should be part of every safety program. The first major problem to decide upon is what constitutes an emergency and how it differs from routine operations. Emergencies can arise from natural phenomena, such as meteorological events, or from human causes; regardless of how they arise, the planning process is similar. A contingency plan must be in effect: there may be large numbers of injured workers and property damage may be extensive. There must also be someone who can make the decisions necessary for the organization as a whole. A chain of command must be established, and these persons must be trained to make emergency decisions. Not preparing properly for emergencies may put the organization permanently out of operation and many lives may be adversely affected.

In this chapter the various classifications of disasters will be addressed. These include natural emergencies such as earthquakes, tornadoes, and communicable-disease outbreaks; and human-caused emergencies such as wars, airplane accidents, fires, and hazardous material spills. Also discussed will be the responsibilities of various personnel during an emergency situation. This will include the communications system and personnel accounting. Some of the concomitant factors, such as psychological stress, of a disaster will be included in the section on emergency decision planning. Then development of a contingency plan, along with emergency training, will be discussed.

DISASTERS

Disasters occur all over the world (see Figure 21.1 for an example), and the magnitude is huge. In industry there is always a concern for fires and

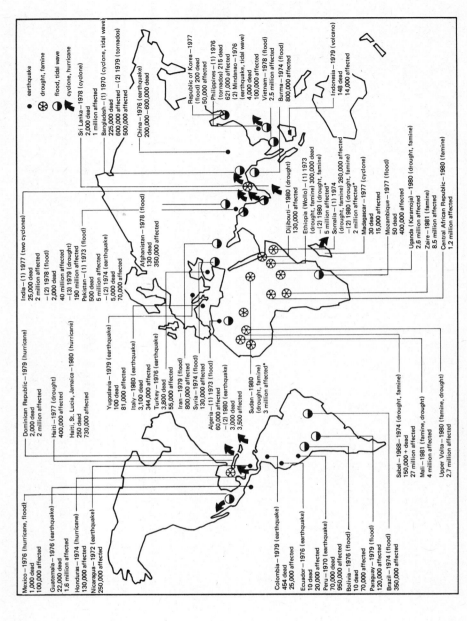

FIGURE 21.1. Victims of natural disasters, 1970–1980 (*from WHO*).

438

TABLE 21.1
Classification of Disasters

A. Natural disasters
  1. Natural phenomena beneath the earth's surface
     a. Earthquakes
     b. Tsunamis
     c. Volcanic eruptions
  2. Natural phenomena at the earth's surface
     a. Landslides
     b. Avalanches
  3. Meteorological/hydrological phenomena
     a. Windstorms (cyclones, typhoons, hurricanes)
     b. Tornadoes
     c. Hailstorms and snowstorms
     d. Floods
     e. Droughts
  4. Biological phenomena
     a. Locust swarms
     b. Epidemics of communicable diseases
B. Human-caused disasters
  1. Caused by warfare
     a. Conventional warfare, including siege and warfare
     b. Nonconventional warfare (nuclear, biological, chemical)
     c. Sabotage
  2. Caused by accidents
     a. Vehicular (planes, trains, ships, cars)
     b. Drowning
     c. Collapse of buildings and other structures
     d. Explosions
     e. Fires
     f. Biological
     g. Chemical, including poisonings by pesticides and pollution

SOURCE: Western, K. A. *The epidemiology of natural and man-made disasters: The present state of the art.* Academic dissertation, London School of Hygiene and Tropical Medicine, London University, 1972.

explosions and materials accidents. It is important in terms of overall safety operation to include planning for and action during a disaster. Western has classified the etiology of disasters (see Table 21.1).[7]

Natural Disasters

Earthquakes. Earthquakes are believed to result from the gradual change in weight distribution of the earth's crust. This sets up enormous strain and stresses on the rock layer, and the rock fractures, causing shock waves which travel outward at several miles per second. Most deaths and injuries occur from collapsing buildings. Resultant fires and water-main problems can cause additional damage. In 1906, an earthquake shook San Francisco. It started fires which produced 20 times as much damage as the quake itself. In 1923, an earthquake shook Tokyo and Yokahama; fires started, and these fires killed an additional 100,000 people. Some other examples of major earthquake damage were Iran in A.D. 893 with 180,999 recorded deaths; China in 1556 with 830,000 deaths; India in 1737 with 300,000 deaths; China in 1920 with 180,000 deaths; Japan in 1923 with 140,000 deaths; and Iran in 1968 with 12,000 deaths.[6]

The Richter scale is a measure of the amount of energy involved in an earthquake, ascertained by means of calibrated seismographs. It is a logarithmic scale in which each higher number of the scale represents approximately 31 times the energy of the preceding number. A 4 on the Richter scale is 31 times as violent as a 3. A magnitude of 2 is about the lowest that is ordinarily reported as being felt by humans. The earthquake which destroyed San Francisco in 1906 measured 8.3 on the Richter scale. The "Good Friday Quake" which killed more than 100 people and changed the face of a wide area of Alaska in March 1964 measured 6.5 on the Richter scale. The worst quakes ever recorded measured 8.9 on the Richter scale— one was off the coast of Colombia and Ecuador on January 31, 1906, and the other was off the east coast of Honshu, Japan, on March 2, 1933.

Tsunamis.   Tsunamis (tidal and seismic waves) are ocean waves caused by an earthquake in the floor of the ocean. Tsunamis travel at speeds of about 400 miles per hour and can move from one end of the ocean to the other. They occur frequently in the Pacific and Indian oceans. Most loss of life is caused by drowning and by being crushed by debris which is carried along by the water at high speeds. If the water level at an ocean beach rapidly lowers (the tide is disturbingly low), it should be assumed that a tsunami will strike in a few minutes, and everyone should move quickly to high ground.[6]

Volcanic Eruptions.   A volcanic eruption is an outpouring of molten rock from an opening in the earth's crust. When a large amount of lava and other material is blown out of a volcano with great force near a populated area, it causes a disaster. In 1966 an eruption of Mount Pelee in Martinique killed 40,000 persons. In the eruption of Mount St. Helens in the state of Washington, 25 persons were killed. A volcano does give warnings, such as the escape of smoke, and evacuation is a must. Most victims of a volcanic eruption are killed by poisonous fumes, superheated gases, or masses of volcanic dust. The inhabitants of Pompeii are believed to have been smothered under a 10- to 14-foot layer of volcanic dust which fell on them.

Avalanches.   An avalanche is a sudden fall of a large mass of snow, ice, rock, or earth from a high elevation to a lower one. Sometimes, terms like "rock slide" or "landslide" are used. Avalanches caused by snow frequently occur on steep mountain slopes throughout the world. In 1982 one occurred in Colorado, killing several people. Also in 1982 there was a mud slide that occurred in California, causing the loss of thousands of homes. Earth which was once firmly held to slopes by the roots of live plants loses its cohesiveness when the plants die or are killed. Subsequent rains turn the earth into a thick mud which suddenly starts to move down a hill or mountain. People living on slopes or in valleys should always regard heavy, prolonged rain as a warning sign. The danger is much greater if there are few trees on the slopes. Most avalanches occur when heavy layers of snow and ice on a mountainside become warmed by the sun and slip from their former anchorage.

Hurricanes.   A hurricane is a large storm, 100 to 1,000 kilometers in diameter and about 15 kilometers high, with a small, clam central area called the eye. The greatest winds in Atlantic hurricanes are about 160 miles per hour, but more common velocities only half as large do great damage. Hurricanes form most frequently in late summer in six tropical ocean areas both north and south of the Equator. The entire Gulf and Atlantic coasts of the United States are subject to hurricanes in late summer, when the sea-surface temperature is high and the deep layer of subtropical easterly winds extends furthest north. Westward-moving tropical disturbances are frequent, and some develop into hurricanes.

Tornadoes.   Tornadoes are the most severe storm phenomenon on this planet. Their winds of about 360 kilometers (225 miles) per hour are 2 times more forceful than the 180 kilometers (115 miles) per hour of moderately severe hurricanes. Few structures can withstand the direct onslaught of a major tornado, which can overwhelm whole communities. Tornadoes occur most frequently in the midwestern and southeastern United States. In the southeast they usually occur in early spring, and their occurrence center drifts north and west as summer advances.

In 1965 tornadoes in the midwest of the United States killed 272 persons. In 1925 in Missouri, Illinois, and Indiana 689 lives were lost. In general, the southwest corner of a basement is the safest part of a home during a tornado; it is safer than being outdoors. If caught out of doors, however, it is usually safe to lie flat in a shallow ditch or depression. In the midwest, special shelters have been built for protection. Remember, in *The Wizard of Oz* Dorothy and her dog did not get to the shelter before the tornado struck. The most important step if a tornado strikes is the prevention of fire. All electric power should be turned off and fires stopped. A usual indication of a tornado is a heavy cloud formation in which the clouds seem to be moving in all directions at once rather than in a single major direction. An absolute sign of a tornado is the sighting of a funnel cloud. When the tornado is very close, it sounds like an express train overhead— and then there are only a few seconds left before it strikes. Death and injury from tornadoes are caused by the blowing force of the tornado during the suction phase, in which the atmospheric pressure drops by as much as 2 pounds per square inch and structures are blown outward and people and furniture are lifted into the air. There is a tornado pattern: high winds, suction phase, then high winds again. People are injured or killed by flying debris, by being smashed against buildings, by falling trees, or by heart attacks.

Hail. Hail occurs throughout the United States, with increased frequency in the south, the midwest, and the elevated areas of Colorado, New Mexico, and Wyoming. Hail forms when a frozen particle is formed and it collects liquid particles which freeze on contact. The particles which form range in size from that of a small pellet to that of a tennis ball. Most hail damage represents crop losses, but occasional extra-large hail carried by extra-strong winds wreck havoc with whole communities.

Snowstorms. Snowstorms claim lives every year in many parts of the world. Snow covers much of the United States every winter. The winters of 1981 and 1982 were severe for the United States, and 1982 was also particularly harsh for Great Britain. Some meteorologists are predicting that we are entering another ice age and can expect severe winters to be on the increase.

Snow can precipitate an emergency if a large amount falls in a short period of time. A snowstorm on December 24, 1982, dropped 2 to 3 feet of snow on Denver, Colorado, paralyzing the city. Persons may be injured or killed. Equipment may be overloaded. Roofs may collapse under the snow load, since snow may be quite heavy.

Floods. A flood is a rising and overflowing of a body of water (lake, river, or ocean) beyond its usual confines and onto normally dry land. Excessive rainfall in a short time period overloads a body of water and causes it to overflow its banks. Floods of disaster magnitude on the Atlantic seacoast areas in the United States are frequently associated with hurricanes. In 1889 in Johnstown, Pennsylvania, a flood resulted from a collapsing dam and caused 2,200 deaths. Drowning causes many deaths from floods. Some injuries and death are caused by floating debris. Another cause of death during a flood can be fires.

Droughts. A drought, the severe lack of rain to an area, can be another cause of disasters. Because of improved methods of irrigation, much farmland is now protected. But long periods without rain cause major failure of plants and animals.

Epidemics. Large-scale communicable-disease outbreaks are commonly referred to as epidemics, which are incidences of disease affecting large numbers of people in a short time and usually in a common location. The influenza epidemic of 1919 caused the deaths of 20 million people (500,000 of these were Americans). In the history of the human race, epidemics have been by far the greatest of all disasters. It was a form of the bubonic plague which raced through Europe in 1347 and 1348. No one knows how many died in the Black Death because medieval statistics were very primitive. A general conception is that from a third to a half of Europe's population was killed. This was the greatest disaster to occur on this planet. The primary cause of epidemics are infectious microorganisms. The Black Death was caused by a bacilli and transmitted to humans by a flea.

Famines.   Famines are severe shortages of food. This may result from crop failure due to drought or too much rain, farming worn-out land, or lack of farming skills. Famine causes everyone to suffer, and the weakest suffer the most and frequently perish.

People-Caused Disasters

War.   The number of people killed on both sides in World War I was about 5 million, of which 116,000 were Americans. A world war today would probably result in billions of deaths, if not complete annihilation of the human race, because of the sophisticated weaponry that would probably be utilized. Lives lost by sabotage must always be considered as part of war. Many of these deaths go unreported, so exact figures are difficult to obtain. Nonconventional warfare, such as a nuclear war, would be a disaster for which there are no precautions because of the unknown long-range effects. Heads of governments must fully realize the possible consequences of such a disaster and make wise decisions with the understanding that in a nuclear war there will be no winners.

Sabotage.   Sabotage is an attempt to cause interruptions or shutdown in a workplace. Fencing and guards can be used to deter this, and a strict visitors' pass system can be instituted. In emergency times it is important that visitors are not permitted on the premises and that specific orders are given on the evacuation procedure. The guard(s) must have an adequate backup system to maintain vigilance.

Accidents.   Accidents cause many disasters. Cars, trucks, and buses cause many accidents with millions of lives lost each year. Driving while intoxicated has become an increasingly documented cause of vehicular accidents. These are discussed in Chapter 24.

Railroads.   Railroads tend to be one of the safest methods of travel. In most cases the number of deaths is small compared to the number caused by other forms of transportation. In 1950 and 1951 commuter trains were involved in accidents which resulted in 163 deaths. Railroad accidents are usually caused by defective equipment, poor roadbeds, poor management, inappropriate signals, or bad judgment on the part of operating personnel.

Aviation.   Airplanes cause disasters by the very fact of the large number of victims usually involved. It is amazing with today's sophisticated equipment and airplane training that airplane accidents occur at such a frequent rate. There have been hundreds of lives lost because of airplane crashes. In September, 1983 an unarmed 747 with 269 passengers aboard was destroyed by a Soviet fighter plane.

Explosions.   An explosion is a sudden release of a large amount of energy, accompanied by shock and pressure waves. Explosions occur worldwide, and their frequency is increasing because of the increased use of explosives by industry and by terrorist organizations. The ignition of methane in mines and the ignition of natural gas also produce large explosions.

An explosion of cyclohexane occurred at Flixborough, England, in June 1974, resulting in the deaths of 28 employees and in injuries to others working at the plant, which itself had resultant structural damage. The number of fatalities was as low as it was because the explosion occurred on a Saturday when only a skeleton crew was on duty. Injuries to the public and damage to property also occurred beyond the plant boundary. Fires on the plant resulting from the explosion burned for several days. The explosion followed the ignition of a cloud of cyclohexane vapor mixed with air which was produced by the failure of one of the pipes between five interconnected oxidation vessels containing cyclohexane.

It was concluded that the blast effect was similar to that produced by 16 ± 2 tons of TNT. It was estimated that about 30 tons of cyclohexane had formed an explosive cloud. This cloud came into contact with a source of ignition and subsequently burned, producing a violent explosion.[4]

Structure Collapse.   The collapse of a structure is another example of a person-caused disaster. Building collapses during construction are the most common. In Brighton, Massachusetts, in 1974 an apartment tower collapsed while under construction, killing several workers. An investigation showed that inferior concrete had been used in the construction materials. In Kansas City, a hotel bridge collapsed in 1980, killing over 100 persons. In industry we are also concerned with the disasters that occur when structural elements collapse, such as a scaffold breaking, killing many workers. (The recent collapse of a concrete form killed hundreds of workers.)

Fires.   Fires if not controlled can become a disaster. They are often person-caused disasters, and many lives are involved. In 1941 in Boston,

Massachusetts, a fire at the Coconut Grove Club claimed over 400 lives. Most fire deaths are caused by inhalation of smoke and toxic gases, not by flames. Fires were discussed in Chapters 11 and 12.

Hazardous Material Spills.   Hazardous materials can claim many lives and cause disasters if they are not handled properly and if adequate control measures are not utilized. In October 1981 the Environmental Protection Agency published a list of 115 waste dumps which may be hazardous to health. It will cost industrial chemical producers $1.6 billion to clean them up. They include New York's Love Canal, Kentucky's Valley of Drums, and New Jersey's Chemical Control Corporation.

Disturbances in the Ecological Balance.   Disturbances in the ecological balance are caused by pollutants. These can be caused by radioactive fallout, organic pesticides, radionuclides in fertilizers, heavy metals, salts from irrigation waters, industrial and household wastes, salts used on roads for the purpose of deicing, lead from fuel combustion, pathogenic microorganisms, growth-regulating chemicals, mulching materials, oil spills, and improper handling of materials which can alter the ecological balance.

There are many graphic examples that can be used to illustrate the effects of ecological imbalance. The Hooker Chemical Company dumped toxic materials into the Niagara Falls Region (Love Canal) with disastrous results. Lake Michigan's drinking water was polluted when the Youngstown Sheet and Tube Company dumped waste oil into the lake, which was later described in a government report as a visual aesthetic nuisance characterized by oil, fish, detergent foam, sludge, debris, and dark water with a septic odor.[6] The pollution of land and waterways and the resultant ecological imbalance is far from over. Daily, more is discovered about failure of industries to adequately dispose of their wastes. Sometimes the devastating results do not show up until years later.

A nuclear disaster occurred in Kyshtym in Russia in 1957–1958, and a large area became contaminated. This contamination supposedly was caused by improper handling of nuclear waste material, specifically plutonium production. Soran and Stillman hypothesize that the radioactive liquid waste was ground-disposed in the Techa River Valley.[5] From the late 1940s there had been waterborne radioactive contamination, and by 1954 radiation sickness was appearing in people and cattle. Waterborne radioactive contamination was diverted down the river from the contaminated reactor cooling water. This nuclear waste disposal provided another mechanism for the spread of radioactive contaminants through the valley. The Soviets successfully created a contaminated area. The Kyshtym disaster is just that—a record of the disastrous, long-lasting effects humans can wreck on their environment if they fail to take adquate steps to protect it.[5] The area may never be fit for human habitation, since the half-life of some of the contaminants is measured in millions of years. (See Chapter 13 for a discussion of radiation).

The Resource Conservation Recovery Act (RCRA) sets forth guidelines for managing hazardous materials from the "cradle to the grave." In other words, this act specifies how hazardous materials are to be handled from their creation by a chemical manufacturer to their final disposal. The purpose of this law is to ensure environmental protection through safe handling, proper storage, safe transportation, and control and treatment of waste products.

The United States Environmental Protection Act lists four characteristics for identifying hazardous materials:

1. *Ignitability.* Material which poses a fire hazard during routine management and which may spread harmful particles over a wide area (gasoline)
2. *Corrosivity.* Material which requires special containers because of its nature to corrode standard materials or which requires segregation from other materials because of its ability to dissolve toxic contaminants (concentrated sulfuric acid)
3. *Reactivity.* Material which during routine use tends to react spontaneously and vigorously with air or water, to be unstable to shock or heat, to generate toxic gases, or to explode (lithium metal)
4. *Toxicity.* Material which when improperly managed may release toxicants in sufficient quantities to pose a substantial health hazard (cyanide)

All hazardous materials must have hazardous material data sheets (Federal Standard 313A) made up and available in case of a spill. These sheets and their uses were discussed in Chapter 5.

EMERGENCY PLANNING

Emergency planning is an important component of an overall safety program plan for any workplace.

Length of Plan

The length of documents is also important. No emergency plan should be more than 10 pages long; otherwise, it won't be read, let alone understood. To be as short as 10 pages, it will have to be carefully drafted and may have appendices. The written words should clearly describe the purpose and limits of the plan.[1]

Emergency Communications

It is imperative in disaster planning to have excellent communication links for the plan to work. Hence one of the first priorities will be the establishment of an emergency control system through which all external messages will be delivered. The internal communications system will be most effective if radio transmissions are utilized as much as possible along with telephones. This implies thorough training of the telephone operators, who must be located in a safe area and who must not leave their posts until relief arrives.

Emergency Training

Training should be divided into four modules:

1. The first training module consists of itemized steps that would have to be taken in case of a spill, an explosion, and so forth. It will lay out complete sets of task analyses. This training would be available to manufacturing, materials, and security.
2. The second level of training will be provided to safety, security, materials, laboratory, and manufacturing management. It would provide training in planning for the future occurrence of a disaster.
3. The third level of training would be for management safety and would assist the professional in identifying and categorizing the status of the situation for which he or she is responsible as to possible breaks in techniques and possible hazards. What are the safest areas? What are the least safe areas? What areas are at high risk for spills?
4. Emergency planning will be done on the highest level.

Responsibilities

Responsibilities during an emergency situation include designations of people to give directions, to provide security, to staff the emergency control center, to be in charge of safety, to be in charge of fire protection, and to provide salvage direction. Each of these tasks is most important in any disaster. The inherent responsibilities must be clearly and concretely defined before a disaster occurs; during a disaster is no time to decide on tasks and responsibilities. The most important factor in emergency planning is to have a clear reason and objective for what is going to be done. The reason for the plan is to set down a specific course of action to pursue in an emergency, when there is no time to be wasted. The plan should first specify the organizational responsibility during an emergency. The safety engineer would probably assume responsibility for the protection and

TABLE 21.2
A List of Management Information Systems Functions in the Event of a Disaster

1. Perform emergency shutoff of all computers and peripheral equipment.
2. When time is limited, evacuate all personnel from the premises according to the fire drill plan.
3. As time permits:
   a. Store tapes, cards, and disks in a vault.
   b. Notify off-shift employees not to report to work.
   c. Notify alternate computer facilities of the impending need to use their services.
   d. Prepare a list of those systems stored on backup tapes, indicating the place of storage.
   e. Update the backup tapes to make them as current as possible.
   f. Protect equipment against damage from elements and/or plant protection equipment.
   g. Prepare a list of computer services needed in the case of disaster.
   h. Provide for removal of confidential material.
   i. Remove personnel inventory tape and latest system tapes from the building.
   j. Notify equipment suppliers and insurance agency.

evacuation of employees before and during the emergency and for cleanup or salvage after the emergency. The safety engineer should have the power to make decisions to obtain needed equipment and machines on his or her own authority. Next, the plan should have evacuation contingencies.

Evacuate:

When a tank car catches fire and releases chlorine or other chemicals

When a chlorine tank car ruptures and releases chlorine gas

Do not evacuate:

When there is a temperature inversion which threatens a repeat of the infamous November 1948 air pollution episode in Donora, Pennsylvania, which killed 40 people

When fire occurs (as at chemical dump of the Chemical Control Corporation in Bayonne, New Jersey) which, with a change of wind direction, could engulf a nearby city with toxic fumes

Give warning by siren (to evacuate):

When there is a nuclear accident in which radiation is being released to the atmosphere

Do not give warning by siren:

When there is a nuclear accident in which radiation is not released outside of the containment building

TABLE 21.3
Sample of Emergency Shutdown Procedure

It is imperative that plant equipment and utilities be shut down as quickly as possible to prevent further damage or personal injury and that equipment and utilities systems be restored as quickly as possible to enable resumption of normal operations.

Emergency situations can occur at any time and for a variety of reasons, natural or of human origin. In any case, the advance warnings can vary greatly, depending upon the nature of the emergency and the potential extent of damage.

Emergency shutdown of equipment and services will depend upon available time. The following describes the general procedure for a "crash" shutdown or for an orderly shutdown for each building.

Whenever it becomes necessary to shut down any system, the proper authorities should be notified prior to shutdown it time permits or after shutdown if it is on a "crash" basis.

Parties to be notified for various conditions are:
    Any shutdown conditions:
        Superintendent
        Plant engineer
        Safety engineer
        Personnel manager
        Utilities supervisor

For the following conditions, notify in addition:
    Power shutdown:
        Electric company
        Fire department
        Maintenance supervisor
    Water shutdown:
        Water department
        Maintenance supervisor
    Gas Shutdown:
        Gas company
    Sprinkler shutdown:
        Fire department
        Fire insurance company
        Maintenance supervisor

In developing a disaster plan, it is important to have a well-developed contingency plan. The elements of the plan must include specific steps that need to be taken for specific disasters. It should be clearly labeled and easily accessible to the necessary personnel. The locations of the buildings that will be used in case of a disaster should be defined first. See Table 21.2 for a sample list of management information systems functions that might be done in the event of an emergency. For sample procedures to follow in the event of an emergency shutdown procedure, see Tables 21.3 to 21.5.

TABLE 21.4
Plant Engineering and Maintenance Disaster Preparations

1. Prepare master plans of all utilities systems for each building showing areas served, main shutoff and isolation valves, switches, breakers.
2. Prepare a list of contractors who would be available on short notice and who have agreed to assist the company to repair damage. Contractors should represent all trades that might be needed under various emergency conditions.
3. Maintain a central file of installation, operation, and maintenance manuals for all equipment. Manuals should include wiring and piping diagrams and a troubleshooting guide.
4. Maintain an inventory of spare parts for all critical equipment, and catalog all items.
5. Make available to key personnel a list of all plant engineering employees, with home address, phone number, area of responsibility, or specialty trade.
6. Color code all utility systems to assist in rapid identification of systems.
7. Tag all critical valves and switches and show corresponding identification on master plans.
8. Periodically conduct mock disaster exercises to instruct plant engineering personnel of their responsibilities.
9. Prepare an equipment and utilities service priority list for each department. This list should indicate the relative priorities of the utilities and equipment within each department so that the most critical service will be restored first.
10. Review critical needs of each department with department managers, and make necessary adjustments to disaster plans.

TABLE 21.5
Disaster Equipment List

Tarpaulins	Pipe threaders
Speedy-Dri	Gasoline-driven pump
Plastic sheeting	Extension cords
Shovels	Portable lights
Picks	Ladders
Wheelbarrows	Portable fans
Water hose	Respirators (air packs)
Chain fall	Monitoring instruments for
Cutting torch with tanks of acetylene	radioactivity
and oxygen	Explosion meters
Arc welder	Fire axes
Portable generator (gasoline)	Crowbars
Pipe cutters	Spare fire extinguishers

TABLE 21.6
Methods of Handling Multiple Casualties[a]

1. Initial effort of treatment must be restricted to the simplest possible procedures on the greatest number of victims likely to survive.
2. Treatment procedures are designed to preserve life over limb and function over appearance.
3. The control of severe bleeding, the maintenance of normal breathing (artificial respiration), the treatment of shock, and the closure of chest wounds which may allow the lungs to collapse are the critical lifesaving measures that must receive first priority.
4. If possible, all major fractures should be splinted on the spot.
5. No stretcher case should be removed from the stretcher until the patient has reached a place of definitive treatment.
6. The use of a tourniquet is restricted to cases of severe bleeding which cannot be controlled by compression (direct pressure applied to the wounded area). Once a tourniquet is applied, a special note must be made on the identification tag, and the doctor or nurse must be advised immediately of its presence.
7. First aid care of eyes involves the use of clean water to flush out dirt and chemicals. If there is evidence of laceration or severe contusion to an eye, both eyes should be patched with gauze and bandages and the patient given priority attention by the nearest physician or nurse practitioner.
8. Unconscious victims should be placed on a stretcher in a prone or semiprone position, with the head turned to one side to prevent aspiration and to allow secretions to drain away.

[a] A list of first aid kit contents can be obtained from any medical supplier. Consult your safety supplier for the appropriate size and contents for your needs.

A civil disturbance plan also needs to be developed and should include specific steps on how to handle the public at the time of a disaster. It should specify the exact time limitations that will be permitted for the public to leave the area(s) of concern and the exact steps that will be taken. The responsibilities and the duties of the security police will need to be clearly enumerated, as a disaster is not the time for added confusion. In addition, a plan for handling multiple casualties (see Table 21.6) will be needed.

In order to have an effective emergency plan, emergency training is necessary. The training of each of the service units is the responsibility of the service chief in charge. However, the director of disaster control should arrange for frequent coordinated practice drills for the entire installation at irregular intervals. In the beginning, it is advisable for practice alerts to be announced. When the director is satisfied that all plant personnel have rehearsed their duties and move to their appointed places in a satisfactory manner, the practice may then be called without warning. The drills serve two purposes. First, they assist in developing correct habits for times of emergency. When an emergency occurs, there generally is not sufficient time to think of what to do. It is necessary to act intuitively, and this is only accomplished through practice. Second, skill in handling of emergency situations and procedures is sharpened by practice drills.

It is recommended that all training and drills be conducted on company time or that employees be compensated for any additional time that may be required for their participation after the close of the workday.

In order to provide some semblance of reality and to acquire an optimum or practical experience, it is advisable that training be carried out under conditions that simulate an actual emergency as closely as possible. All the necessary apparatus and equipment assigned to each unit should be used, and personnel should be asked to play the parts of casualties. As an example, the fire squads could use a suitable empty field to burn waste (such as packing crates, oil residues, and quantities of paper), and the proper types of apparatus should be employed to extinguish the fire. This would afford the squads the opportunity of working with the equipment and acquiring a knowledge of its limitations and capacities. For example, it could quickly be demonstrated that full-sized hoses cannot be handled by untrained personnel.

A part of the training procedure might be the presentation of talks by personnel or outside experts who have had experience in severe catastrophes. The talks should be completely descriptive, including all the gory details. Although this may seem unnecessarily shocking, it must be remembered that members of the service units should be prepared for the most severe conditions they are apt to face at the time of disaster.

There may also be specialized training that will need to be done, such as learning CPR (see Figure 21.2) or some specialized techniques which are specific to certain industries.[3]

Decision Making

When we speak of disasters, it is important to consider the concomitant factors involved in emergency decision making. Effective emergency decisions are best-made when there is:

1. Awareness of serious risks if no protective action is taken
2. Awareness of serious risks if any of the salient protective actions are taken
3. Moderate or high degree of hope that a search for information and advice will lead to a better (that is, less risky) solution
4. Belief that there is sufficient time to search and deliberate before any serious threat will materialize

When one or another of these conditions is not met, a defective coping pattern, such as defensive avoidance or hypervigilance, will be dominant, which generally leads to maladaptive actions.[7]

HEN a person's heart and lungs stop functioning because of a heart attack, shock, drowning or other causes, it is possible to save a life by administering CPR, or cardiopulmonary resuscitation.

CPR provides artificial circulation and breathing for the victim. External cardiac compressions administered manually are alter-

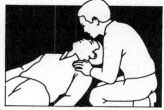

1. Determine if victim is unconscious — Tap or gently shake the victim's shoulder. Shout, "Are you OK?" If no response shout "Help!" (Someone nearby may be able to assist.) Do the Airway step next.

nated with mouth-to-mouth resuscitation in order to stimulate the natural functions of the heart and lungs.

This article contains an overview of CPR training and is not intended as a complete guide. Contact your local chapter of the American Red Cross for further information on how you can learn this life-saving procedure.

Airway step — Place one hand on the forehead and push firmly backward. Place the other hand under the neck near the base of the skull and lift gently. Tip the head until the chin points straight up. This should open the airway. Place your ear near the victim's mouth and nose. Look at the chest for breathing movements, listen for breaths and feel for breathing against your cheek. If no breathing occurs do the Quick step next.

3. Quick step — Give 4 quick full breaths, one on top of the other. To do this keep the head tipped and pinch the nose. Open your mouth wide and take a deep breath, making a good seal. Now, give the 4 breaths without waiting in between. Do the Check step next.

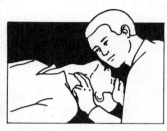

4. Check step — Check the pulse and breathing for at least 5 seconds but no more than 10. To do this, keep the head tipped with the hand on the forehead. Place the fingertips of your other hand on the adam's apple, slide your fingers into the groove at the side of the neck nearest you. If there is a pulse but no breathing give one breath every 5 seconds. If no pulse or breathing is present send someone for emergency assistance (dial 911 or operator) while locating proper hand position. Begin chest compressions.

Hand position for chest compressions
1. With your middle and index fingers find the lower edge of the victim's rib cage on the side nearest you. 2. Trace the edge of the ribs to the notch where the ribs meet the breastbone. 3. Place the middle finger on the notch, the index finger next to it. Put the heel of the other hand on the breastbone next to the fingers. 4. Put your first hand on top of the hand on the breastbone. Keep the fingers off the chest.

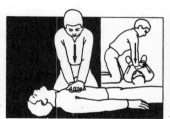

6. Chest compressions — Push straight down without bending your elbows while maintaining proper hand position. Keep knees shoulder width apart. Shoulders should be directly over victim's breastbone. Keep hands along midline of body. Bend from the hip, not the knees. Keep fingers off the chest. Push down about 1½ to 2 inches. Push smoothly. Count, "1 and, 2 and, 3 and, etc."

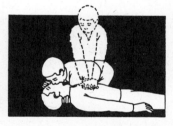

7. Push 15 - Breathe 2 — Give 15 compressions at a rate of 80 per minute. Tip the head so the chin points up and give 2 quick full breaths. Continue to repeat 15 compressions followed by 2 breaths. Check the pulse and breathing after the first minute and every few minutes thereafter. Note: Do not practice chest compressions on people as it could cause internal injuries.

FIGURE 21.2. Learning cardiopulmonary resuscitation (*from Occupational Hazards, September 1982*).

453

Janis and Mann's theory specifies the contrasting conditions that determine whether the stress engendered by decisional conflict will facilitate or interfere with an effective search for and appraisal of alternative courses of emergency action.[2] Janis and Mann have identified five basic patterns of coping behavior that affect the quality of decision making. One of the patterns—vigilance—generally results in thorough information searching, unbiased assimilating of new information, and effective planning.[2] The other four patterns, which usually result in defective decision making when the decision maker is confronted with a genuine emergency, are (1) unconflicted inertia, (2) unconflicted change to a new course of action, (3) defensive avoidance, and (4) hypervigilance.

Withey theorizes that the cognitive appraisal of persons exposed to disaster warnings includes estimates of (1) the probability that the dangerous event will materialize, (2) the severity of personal losses if it does materialize, and (3) the probable advantages and disadvantages of the alternative means available for averting or minimizing the danger.[8] In short, people are capable of making emergency decisions on the basis of essentially the same types of cognitive judgments that they use when they make other consequential decisions.

There are two main ways in which emergency decisions in the face of oncoming disaster differ from executive or professional decisions and the usual decisions of everyday life. One is that there is much more at stake in emergency decisions—often the personal survival of the decision maker and of the people he or she values the most. A second important difference is in the amount of time available to make a choice before crucial options are lost. The time is generally much shorter for emergency decisions, sometimes only a few seconds. Because of the very high emotional arousal evoked by emergency warnings, we can see in extreme form in emergency decisions the ways in which psychological stress affects cognitive functioning.

SUGGESTED LEARNING EXPERIENCES

1. Design an emergency plan for an industry of your choice.

2. In terms of natural disaster, where would you think is the safest place in America to work? Why? In the world? Why?

3. If you were to take over as the safety manager in an industry that had no emergency plans, where would you place your initial energies?

4. Has studying the material in this chapter, and doing some outside reading on the same, made you feel any differently about where you live? Explain.

5. In terms of psychological principles, what do you want to remember in training for major emergencies?

6. Illustrate a communication system in a disaster.

7. Select a disaster of your choice. Illustrate and describe the possible events (causes), interventions, and outcomes.

8. How many persons would you want trained in your workplace to be able to administer first aid? Discuss.

9. Plan and debate both sides of the statement "In a nuclear war, there will be no winners."

10. Hazardous material spills represent one of the greatest threats to our environment. Contact a chemical manufacturer, transporter, or user on the EPA, and learn the steps taken to protect the environment.

REFERENCES

1. Garb, S., and Eng, E. *Disaster Handbook*. New York: Springer, 1979.
2. Janis, I. L., and Mann, L. Emergency decision making: A theoretical analysis of responses to disaster warnings. *Journal of Human Stress*, 1977, **3**, 35–48.
3. Red Cross. *You can learn CPR* (pamphlet).
4. Sabee, C., Samuels, E. E., and O'Brien, T. P. The characteristics of the explosion of cyclohexane at the Nypro (UK) and Flixborough plant on 1st June 1974. *Journal of Occupational Accidents*, 1976, **1**, 203–215.
5. Soran, D. M., and Stillman, D. B. *An Analysis of the Alleged Kyshtym Disaster*. Los Alamos, N.M.: Los Alamos National Laboratory, 1982.
6. Turner, B. A. *Man-made Disasters*. London: Wykeham Publications, 1978.
7. Western, K. A. *The Epidemiology of Natural and Man-made Disasters: The Present State of the Art*. Academic dissertation, London School of Hygiene and Tropical Medicine, London University, 1972.
8. Withey, S. Reaction to uncertain threat. In G. W. Barker and D. Chapman (Eds.), *Man and Society in Disaster* (pp. 93–123). New York: Basic Books, 1962.

PART 5 SAFETY ENGINEERING IN PRACTICE

22 MANUFACTURING

Virtually everything we use in our world is manufactured, from buttons to Boeing 747s; manufacturing is performed in small local plants and in giant multinational corporations composed of many diverse industries. In this chapter we will present a brief overview of the types of hazards encountered in industrial operations. Many manufacturing processes require using hazardous materials which differ from industry to industry. Hazards in a woodworking plant differ from those in a plastics plant. Hazards vary within the different manufacturing processes, from converting raw materials to partially finished products to the final products ready for sale, and hazards differ with the kind of production. In a manufacturing facility, safety engineers must know all aspects of the operations. They should be able to understand appropriate OSHA standards in relation to equipment, facilities, machines, and operations.

Common sources of safety and health problems in manufacturing are the machinery and its operation. Other concerns include manual materials handling, electrical contact, welding and brazing, high temperature, high noise levels, and disposal of hazardous materials. These are only the tip of the iceberg. The safety engineer must be constantly aware of any change in the facility and the manufacturing processes which may affect safety.

In this chapter the epidemiology of injuries and occupational illnesses related to manufacturing will be discussed. Several typical operations will be described. These include forming, cutting, assembling, and finishing. They will be discussed in terms of their inherent hazards and the safety precautions that must be taken to mitigate them. Following this, special hazards in manufacturing will be addressed, including unguarded machines, noise and vibration, electric shock, housekeeping, manual materials handling, the conveyor system, forklift trucks and welding gases. A number of health hazards in manufacturing will then be discussed. These include degreasing agents, plating solutions, plastics, and toxic materials.

459

SIZE OF THE PROBLEM

In 1979, there were almost 20 million workers in the manufacturing indus-try, and there were over 2.2 million injuries/illnesses.[8] This means that over 11% of the workers were injured on the job in a single year. Table 22.1 compares the injury/accident rate of manufacturing with the rates of other industries. The construction industry is the only one which exceeded manufacturing's incidence rate. However, when the more serious lost workday cases are considered, the incidence rate for manufacturing is substantially below the rate for mining and construction and almost identi-cal with the rates for agriculture, forestry, and fishing and transportation and public utilities.

Occupational illnesses are only a small portion of the disabling condi-tions in industry. There were about 0.2 cases per 100 workers. About 60% of the occupational illnesses were in manufacturing, which has only about 20% of the workers.

Manufacturing includes many risky procedures such as welding, using heavy equipment, forging, heavy cutting, handling of hazardous materi-als, and manipulating dangerous substances like cadmium.

TYPICAL MANUFACTURING OPERATIONS AND ACCIDENTS WHICH ARE RELATED TO THEM

Casting

Casting is a method of forming metals. The molten metal, which is often iron or bronze, is poured into a mold made of sand. Pipe fittings and engine blocks are generally made by casting. Casting may cause many possible types of workplace injury or disease (involving eyes, face, respira-tory system, arms, hands, legs, and feet).

After cooling, the casting is finished to its final form. It is often sandblasted to remove burrs. Casting involves a procedure that produces very fine dust particles which permeate the respiratory system. Hence, in sandblasting, the foundry should have dust-tight rooms with doors kept closed; therefore, castings should be thoroughly dusted before they are removed from the room. Protective clothing and equipment should be mandatory for workers exposed to this dust. Castings may be polished using abrasive wheels which also generate excessive dust. Full face protec-tion should be worn, and the dust should be controlled with an exhaust system. The dust can be minimized if castings are precleaned in a barrel, mill, or abrasive chamber.

Forming

Forming operations are operations in which the material is shaped by mold, either using a liquid material or a powder. Some operations which

TABLE 22.1
Deaths and Disabling Injuries, 1980

Industry Group	Workers[a] (000)	Deaths 1980	Deaths Change from 1979	Death Rates[b] 1980	Death Rates[b] 1970	Death Rates[b] % Change	Disabling Injuries[c] 1980	Disabling Injury Rate
All industries	96,800	13,000[d]	−300	13	18	−28%	2,200,000[d]	2.2
Trade	21,800	1,400	0	6	7	−14%	400,000	1.8
Service	24,300	1,800	0	7	12	−42%	390,000	1.6
Manufacturing	19,900	1,600	−200	8	9	−11%	450,000	2.2
Government	15,600	1,700	0	11	14	−21%	310,000	2.0
Transportation and public utilities	5,300	1,500	−100	28	36	−22%	170,000	3.2
Construction	5,600	2,500	−100	45	61	−26%	240,000	4.3
Mining, quarrying	1,000	500	0	50	100	−50%	40,000	4.0
Agriculture[e]	3,300	2,000	+100	61	66	− 8%	200,000	6.1

SOURCE: National Safety Council estimates (rounded) based on data from the National Center for Health Statistics, state departments of health, and state industrial commissions; numbers of workers are based on Bureau of Labor Statistics data. (Reprinted from National Safety Council. *Accident Facts* (p. 23). Chicago, 1981, with permission.

[a] Workers are all persons gainfully employed, including owners, managers, other paid employees, the self-employed and unpaid family workers, but excluding domestic servants.

[b] Deaths per 100,000 workers in each group.

[c] Disabling beyond the day of the accident, see below and page 2

[d] About 4,500 of the deaths and 200,000 of the injuries involved motor vehicles.

[e] See also "Work deaths and death rates in agriculture," page 85, and "Uniqueness of farming," page 86, for comparability.

use forming are drug companies making pills or precision parts manufacturers for jet engines.

Forging

In olden times, forging was hammering red-hot metal against an anvil. The smith was a mighty man with large and dirty hands. Today, forging is similar. A piece of metal is heated in a furnace until it is red hot. It is often taken manually and placed in a die where a mechanical hammer pounds it into shape. Forging is a very common manufacturing operation in which metal is formed by hammers. There are many opportunities for injuries. The injury rate for the forging industry (SIC 3462) was 13.91 cases per 100 workers in 1981.[4] The most frequent causes of injury are being struck; improperly using materials handling equipment such as tong lifts; crushing fingers, hands, or arms between the dies; crushing fingers; and being burned.

Operating a hammer with a worn cylinder sleeve is hazardous. When the sleeve is so worn that the swing of the ram cannot be controlled, the hammer should be shut down and repaired. Operating a hammer with broken piston rings is also dangerous. A piece of broken piston ring passing through the steam ports and lodging in the throttle valve can cause the ram to drop out of control, often catching the operator's tongs or the transfer tool and causing serious injury.[3]

Repair persons, in particular, are exposed to crushing injuries when they remove and install parts around the hammer, because of the possibility of the hammer's dropping. Therefore, the hammer must be lowered prior to maintenance.

The upsetter is a horizontal forging machine which forges hot bar stock, usually round, into a great many forms by a squeezing action instead of impact blows, as in the case of forging hammers. This machine is used in making bolts. Changing the dies is very hazardous. See Table 22.2 for an example of detailed instructions for safe handling of dies.

Unless properly confined, hot scale produced during the forging operation can cause serious burns. Air or steam curtains at the front of the die and directed onto the die facing will prevent scale from coming out the front of the press. Combination scale and smoke exhaust hoods should be installed at the back of the press to confine the scale and exhaust the smoke created by die lubricant. Safety hats, eye protection, leather aprons, safety gloves, and safety shoes should be worn by operators at all forging-press operations. All forging presses should be equipped with pedal guards and nonrepeat devices. A forging press should never be operating on continuous stroke. The operating controls should require depression of the pedal for every cycle of the slide. The press controls should also permit inching of the slide for die setting or other adjustments.[3]

TABLE 22.2
Safe Procedure for Changing Dies

1. Check to make sure that all tools of correct size and other equipment are at hand.
2. Check dies and headers for defects which could develop into hazards.
3. Shut off all power, and lock the master switch in OFF position.
4. Turn off the water or header lubricant, and set the air brake. If the upsetter does not have an air brake, wait until the flywheel has stopped completely.
5. Loosen all setscrews, locknuts, and hold-down bolts on the die clamps.
6. Remove the dies by means of a swivel or an arm crane. (Eyebolt holes 2 in. deep should be drilled into all dies. Eyebolts or swivels should be stocked only in $\frac{1}{4}$-in. graduations to a minimum of $\frac{3}{4}$ in. in diameter.)
7. Remove all die packing and thoroughly clean the die-set seat; inspect for burrs, especially along the die keys.
8. Remove headers and dummies and any shims which may have been used.
9. Measure the new dies to determine the amount of packing needed. At least a $\frac{1}{4}$-in. liner should be used to help protect the die seats.
10. Set in the new stationary die. Check to make sure that it is seated properly and packed correctly.
11. Tighten the hold-downs by hand.
12. Turn on the power, open the safety catch, release the brake and the flywheel. Set the machine on INCH, and slowly inch the header slide forward to bring the toolholders into correct position for the headers.
13. Disconnect the power, and lock out the master switch. Make certain that the flywheel has completely stopped.
14. Assemble the new headers, tool dummies, and toolholders. Be sure that the headers to the die cavity are correctly matched and on correct center. It is good practice to set up according to a die layout which shows all principal dimensions on the equipment.
15. Check for the need to use shims behind the headers. Shims should be the washer type, not the horseshoe type, which fit around the shank of the header and cannot fall out.
16. Make a complete check of the assembly before finally tightening the header and dummy setscrews.
17. Insert the moving die, make sure that it is properly located, and then tighten the hold-down bolts.
18. Turn on the power. Set the machine on INCH, and slowly close the dies.
19. Check for match alignment and proper amount of packing. If too much packing is used, the safety pin should open the dies.
20. Allow for expansion of the dies when they become hot. It may be necessary to remove a shim from in back of the die.
21. If everything checks out, shut off the power, lock out the master switch, and allow the flywheel to come to a complete stop.
22. Tighten hold-down bolts, locknuts, and setscrews.
23. If needed, attach a front gauge or backstop gauge.
24. Turn on the power, and try out one forging. Check dimensionally. If dies or headers need further adjustment, again turn off the power, lock out the master switch, and wait until the flywheel comes to a complete stop.

SOURCE: from National Safety Council, *Accident Prevention Manual for Industrial Operations* (7th ed.). Chicago, 1977, copyright by National Safety Council, Chicago, IL 60611. Used with permission.

Bending

Bending is the operation which takes a piece of metal and changes it from a flat sheet to a piece with an angle such as 90°. The equipment used most frequently is a brake press. The ductwork of a hot-air heating system is usually formed by bending. What must be protected are the face, eyes, and hands. The equipment must be properly used. The danger of bending lies at the point of operation. Where the stock is inserted or withdrawn fingers should be protected. Hands are also at risk for being cut with the sharp metal. Personal protective equipment is needed in bending (sheet or sheet metal industry). The best safety gloves, leather apron, and machine guarding are essential.

Extruding

Extruding is a process of forming metal in which a piece of barstock is fed through a die and its form is changed. The metal is often red hot. This process is used in making "I" beams or manufacturing railroad tracks.

In the aluminum extruded-product industry (SIC 3354) there were 15.40 injuries per 100 workers in 1980.[4]

What needs to be underlined is the prevention of finger injury and the respiratory system, head injuries, and heat exposure.

Cutting

In this operation a piece of material is shaped by cutting. In its most simple sense, making paper dolls is a cutting operation. Cutting is another large area in manufacturing operations. Needless to say, there are many safety precautions which must be considered. Primary among these is the protection of fingers, utilizing all the machine guarding devices available. The danger of cutting action exists at the movable cutting edge of the machine as it approaches contact with the material being cut. Examples of cutting machines are band and circular saws, milling machines, and grinders (see Figure 22.1).

Sawing

Sawing is a very common operation which ranges from sawing a two-by-four piece of lumber to sawing a complex shape from a piece of plate steel. Saws should be selected for the work they must do. For example, for cutting across the grain of wood, use a crosscut saw; for cutting with the grain, a rip saw. The difference between them is the angle of their teeth. It is most important when using saws to keep the hands away from the cutting edge, to protect the eyes from flying particles, to have adequate illumination, and to frequently clean the floor. Electric saws must be

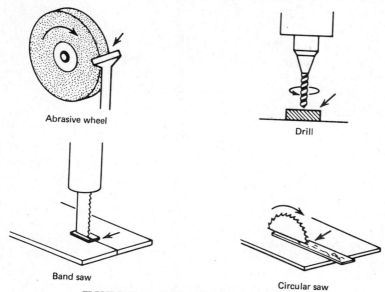

Abrasive wheel

Drill

Band saw

Circular saw

FIGURE 22.1.   Examples of cutting machines.

equipped with guards above and below the faceplate—the lower guard must automatically retract to cover the exposed saw teeth. Employees must be trained to use the guard as intended. OSHA standards require that frames and exposed metal parts of portable saws operating at over 90 volts should be grounded. In addition, circular saws should not be jammed or crowded into the work. The saw should be started and stopped outside the work. At the beginning and end of the stroke, or when the teeth are exposed, the operator must keep his or her body out of the line of cut. The saws must be equipped with "dead-man" controls or a "trigger" switch that shuts off the power when pressure is released.[8]

Drilling

Drilling is an operation in which a bit is rotated at speeds high enough to cut a hole in a material. In drilling, pieces of work should be clamped or anchored to prevent moving. Hands are at risk for puncture wounds and for injury in this operation. Eyes must be protected from flying material.

Stamping

Stamping is a method of cutting identically shaped pieces from a piece of sheet stock. A punch press is often used which ranges in amount of pressure exerted from one to several hundred tons. Stamping is used to make

flat bolt washers or the pieces of automobile frames. In stamping operations, the operator must be sure the hands are out of the way of the
operating presses and that long hair is kept away from the machine. Likewise, as in other cutting operations, the point of operation is the hazardous
area and must be guarded. The flywheel presents a major guarding problem since it can become loosened from its position and place the worker at
risk for severe injury.

Welding

Welding involves melting of a metal by either a flame or an electric arc in
the presence of a flux or shielding gas. It produces particulate flames and
gases from the base metal, the welding rod, the flux, surface coatings or
surface contaminants, or a combination of these sources. Welding, arc
welding, resistance welding, brazing, and other processes are used for
many different manufacturing processes. Each process has specific applications according to the metals being joined, the types of products being
made, the production facilities available, and the site location of the work.
Most require some form of worker protection to ensure that exposure to air
contaminant concentrations does not exceed the maximum safe-exposure
limits. The welder must frequently breathe the gases and fumes produced
by the welding operations since welding fumes rise vertically, due to heating of the gases and air in the area of the arc or flame, and the welder's
head is usually positioned close to the fumes. Fumes and gases flow
around the welding hood and some enter the breathing zone. In some
cases, nearby workers may also be exposed to harmful air contaminants
but not usually to the same extent as is the welder. Control methods used
for general dilution ventilation, local ventilation—either fixed or portable
hoods—"smoke exhaust" welding guns, air-supplied helmets, and complete automation (elimination of the human welder.)[1]

Welding problems can vary with the method of welding employed, the
location of work, the materials being welded, and the control measures
employed. Certain hazards are common to most arc welding processes.
Principally, these are damage to eyes and skin from infrared and ultraviolet
radiation and from the molten metal and electric arc, burns from contact
with hot metal or sparks, adverse health effects from breathing metal
fumes and gases, accidents from materials handling, and electric shock.

Eye exposure to ultraviolet radiation from the arc can result in a condition known as "welder's flash" or "arc eye"—an irritation with a sensation
of sand in the eyes. These flash burns are very painful and repeated exposure may result in permanent eye injury. Overexposure to ultraviolet radiation from the molten metal produces a sensation of burning on the skin
but is usually not of sufficient intensity to cause permanent skin damage.[2]

Fumes and Gases. The hazard potential from the fumes and gases
generated by a welding operation depends upon the chemical composition

of the materials being used, the concentrations of the chemicals in the worker's breathing zone, and the duration of exposure. The composition of the fume itself depends upon the various materials involved in the welding process. Air contaminants may arise from the various components of the welding rods, oxides of the metals, and alloying elements being joined—ozone, carbon monoxide, and the oxides of nitrogen.[1]

Ozone and the oxides of nitrogen are the principal toxic gases produced by the arc welding process. Ozone, an intensely irritating gas, is produced by the action of the electric arc through air. Toxic fumes are generated from welding on metals coated with or containing alloys of lead, zinc, cadmium, or beryllium. Some paints may also produce toxic fumes when heated with the welding torch.[1]

Fumes from many metals, including cadmium, copper, nickel, and zinc, are capable of producing metal-fume fever. The symptoms are similar to those of influenza and usually occur a few hours after exposure. The symptoms include a metallic taste in the mouth, dryness of the nose and throat, weakness, fatigue, muscular and joint pain, fever, chills, and nausea. These symptoms usually last less than 24 hours and a temporary immunity follows. As a result, welders are more susceptible to this condition on Mondays or on workdays following holidays than on other workdays.

Some silver solders contain cadmium. Even a brief overexposure to cadmium fumes can produce a severe lung irritation which may be fatal. The welding of metal alloys, particularly those containing beryllium and its compounds, can also cause serious lung problems.

Electricity.   The avoidance of electric shock is largely within the control of the welder. Electric shocks experienced at welding voltages usually do not cause severe injury; however, under certain conditions they can be lethal. Even mild shocks can produce involuntary muscle contractions which can lead to injurious falls. The severity of the shock is determined largely by the path of the current and the amount of current flowing through the body. This is determined by the voltage and the contact resistance of the area of skin involved. Wearing clothing which is damp from perspiration or working in wet conditions reduces skin contact resistance and thus increases the danger from electric shock.

Arc Welding Hazards.   Arc welding hazards can be controlled by using effective ventilation, by following safety working practices such as heeding warning labels, and by using respirators and personal protective equipment.

Proper application of local mechanical exhaust ventilation or general dilution ventilation will vary with the type of welding being performed. In open-air welding or in large, well-ventilated maintenance shops where welding is not done constantly, the hazards from airborne contaminants may not be significant (except for the more toxic fumes such as cadmium, beryllium, lead, and zinc). For heavy or production welding indoors, local or general dilution ventilation must be provided. Specifically designed local exhaust ventilation and/or approved supplied-air respirators will usually be required for welding on or with materials containing fluorides, lead, cadmium, beryllium, or other highly toxic materials and for welding in confined spaces such as tanks or boilers. Local exhaust ventilation should be located as near to the welding operation as possible for greater capture efficiency.[2]

Welders' helmets and goggles with proper filter plates and filter lenses must be worn during arc welding to protect from harmful rays and from flying sparks and debris. All welders and welders' helpers should have adequate eyesight to avoid positioning themselves too close to the arc—thus increasing exposure to gases and fumes. The helpers or attendants and other workers in the adjacent area should also be provided with proper eye protection. Barrier curtains around the operation can provide protection for workers in adjacent areas during the welding process.

Finishing Procedures

In finishing procedures such as painting, sealing, and plating it is particularly important to have adequate ventilation to protect the skin against unnecessary spills. Cleaning off such spills can remove the outer layer of the skin, thus leaving it exposed to chemicals and other elements without its natural protective barrier. Paint ingredients include pigments, binders, solvents, drying agents, and many other elements. Solvent-based paints and thinners often have a narcotic effect when the solvent, which becomes airborne during spraying and drying operations, is breathed in excessive concentrations. Long-term exposures can cause liver and lung damage. Benzene (C_6H_6) used as a solvent can cause leukemia. Of the many paint pigments in common use, lead, chromium, and cadmium compounds are particularly toxic. For these reasons, as well as for fire protection, adequate ventilation is required in paint spray areas and in spray booths. Although use of water-soluble paints can solve the problem of inhaling organic solvents and eliminate the hazard presented by paint flammability, it should be noted that it does not reduce the respiratory problems posed by toxic pigments, driers, or other materials which become airborne during spraying or even curing.

MACHINE GUARDING

The methods of machine guarding may be grouped under four main classifications: enclosure guards, interlocking guards, automatic guards, and two-handed operating devices.[3]

Fixed Enclosure Guards

Fixed enclosure guards should be used in preference to all other types. They always prevent access to dangerous parts by completely enclosing a hazardous operation, and they can also be effective in controlling dust or chips generated by the operation. Because of limited feed-size openings, enclosure guards admit stock but will not admit an employee's hands into the danger zone. They may be constructed so as to be adjustable to different sets of tools and dies of varying thicknesses of stock, but once adjusted, they must be fixed. As a general rule, power transmission apparatus is best-protected by enclosure guards.

Interlocking Guards

When a fixed enclosure guard is not practical, an interlocking enclosure or barrier should be considered as the first alternative. With the machine turned off, it can be opened to feed in stock and can be adjusted as the particular material and operation requires. These guards utilize an electrical or mechanical interlocking connection to the operating mechanism that prevents the operation of the machine until the guard is returned to a closed position and the operator can no longer reach the point of danger. Workers should never use the interlock as a substitute for the machine's power switch.

Automatic Guards

When neither an enclosure guard nor an interlocking guard is practical, an automatic guard may be used. An automatic guard acts independently of the operator, repeating its cycle as long as the machine operates. This type of guard removes the operator's hands, arms, or body from the danger zone as the machine cycles. It is operated by the machine itself through a system of linkages connected to the operating mechanism. Common types of automatic guards are sweep and push-away devices which create a moving barrier across the danger zone and push the operator's hand away from the area.

Two-Handed Operating Devices

Two-handed operating devices are also designed to protect a machine operator from point-of-operation hazards. Although they are not guards in the technical sense, they accomplish the same effect. These devices may be used to activate the machine cycle. They require simultaneous action of the operator's hands on electric switch buttons, air control valves, and mechanical levers. The actuating controls must be located so as to make it impossible for the operator to move his or her hands from the controls to the danger zone before the machine has completed its closing cycle, because they are too far apart. The two-handed controls must be designed so as to prevent the locking, tying down, or holding down of one control, allowing a hand free access to the danger zone. See Figure 22.2 for examples of guarding hazards.

SPECIAL HAZARDS IN MANUFACTURING

Noise and Vibration

What needs to be stressed here is that there is no reason for noise at any level to be a hazard, since very adequate protective equipment has been developed for this. Individuals find that a loud noise is more annoying than a less-loud noise; noise varying in intensity and frequency is more annoying than continuous, steady-state noise; nondirectional noise is more annoying than directional noise; and noise that appears to be moving is more annoying than noise that appears to be stationary.[3]

Whole-body vibration is thought to affect visual acuity and the ability to perform certain motor activities. Segmental vibration is vibration to a specific part of the body only. Reynaud's phenomena ("dead fingers") may result from vibration. This affects 50% of the exposed workers. The hands become white and have no feeling. It usually disappears when the fingers have been warmed for some time. Complete recovery is rare. The syndrome has been observed in a number of occupations, including those requiring the use of fairly light vibrating tools, such as air hammers, for scraping and chipping metal surfaces, stone cutting, lumbering, and cleaning departments of foundries.[7]

Electric Shock

Another special hazard in the manufacturing industry is that of electric shock, although the involved principles certainly cut across many industries. Electric shock occurs when the body becomes a part of the electric circuit or closed loop. Electricity travels only in a closed loop and from generator to equipment to earth. Electric shock occurs when the body touches the energized wire while it is in contact with the ground. The severity of the shock is determined by the following:

1. Amount (rate of current flow in amperes) and duration (exposure time of the current flow)
2. The part of the body through which the electric current flows

Type of Guarding Method	Action of Guard	Advantages	Limitations	Typical Machines on Which Used
		ENCLOSURES OR BARRIERS		
Complete, simple fixed enclosure	Barrier or enclosure which admits the stock but which will not admit hands into danger zone because of feed opening size, remote location, or unusual shape.	Provides complete enclosure if kept in place. Both hands free. Generally permits increased production. Easy to install. Ideal for blanking on power presses. Can be combined with automatic or semiautomatic feeds.	Limited to specific operations. May require special tools to remove jammed stock. May interfere with visibility.	Bread slicers Embossing presses Meat grinders Metal square shears Nip points of inrunning rubber, paper, textile rolls Paper corner cutters Power presses
Warning enclosures (usually adjustable to stock being fed)	Barrier or enclosure admits the operator's hand but warns him before danger zone is reached.	Makes "hard to guard" machines safer. Generally does not interfere with production. Easy to install. Admits varying sizes of stock.	Hands may enter danger zone—enclosure not complete at all times. Danger of operator not using guard. Often requires frequent adjustment and careful maintenance.	Band saws Circular saws Cloth cutters Dough brakes Ice Crushers Jointers Leather strippers Rock crushers Wood shapers
Barrier with electric contact or mechanical stop activating mechanical or electric brake	Barrier quickly stops machine or prevents application of injurious pressure when any part of operator's body contacts it or approaches danger zone.	Makes "hard to guard" machines safer. Does not interfere with production.	Requires careful adjustment and maintenance. Possibility of minor injury before guard operates. Operator can make guard inoperative.	Calenders Dough brakes Flat roll ironers Paper box corner stayers Paper box enders Power presses Rubber mills
Enclosure with electrical or mechanical interlock	Enclosure or barrier shuts off or disengages power and prevents starting of machine when guard is open; prevents opening of the guard while machine is under power or coasting. (Interlocks should not prevent manual operation or "inching" by remote control.)	Does not interfere with production. Hands are free; operation of guard is automatic. Provides complete and positive enclosure.	Requires careful adjustment and maintenance. Operator may be able to make guard inoperative. Does not protect in event of mechanical repeat.	Dough brakes and mixers Foundry tumblers Laundry extractors, driers, and tumblers Power presses Tanning drums Textile pickers, cards

FIGURE 22.2. Point of operation protection (*from Accident Prevention Manual for Industrial Operations, copyright 1980 by National Safety Council, Chicago, IL 60611; used with permission*).

471

Type of Guarding Method	Action of Guard	Advantages	Limitations	Typical Machines on Which Used
AUTOMATIC OR SEMIAUTOMATIC FEED				
Nonmanual or partly manual loading of feed mechanism, with point of operation enclosed	Stock fed by chutes, hoppers, conveyors, movable dies, dial feed, rolls, etc. Enclosure will not admit any part of body.	Generally increases production Operator cannot place hands in danger zone.	Excessive installation cost for short run. Requires skilled maintenance. Not adaptable to variations in stock.	Baking and candy machines Circular saws Power presses Textile pickers Wood planers Wood shapers
HAND REMOVAL DEVICES				
Hand restraints	A fixed bar and cord or strap with hand attachments which, when worn and adjusted, do not permit an operator to reach into the point of operation.	Operator cannot place hands in danger zone. Permits maximum hand feeding; can be used on higher-speed machines. No obstruction to feeding a variety of stock. Easy to install.	Requires frequent inspection, maintenance, and adjustment to each operator. Limits movement of operator. May obstruct space around operator. Does not permit blanking from hand-fed strip.	Embossing presses Power presses
Hand pullaway device	A cable-operated attachment on slide, connected to the operator's hands or arms to pull the hands back only if they remain in the danger zone; otherwise it does not interfere with normal operation.	Acts even in event of repeat. Permits maximum hand feeding; can be used on higher speed machines. No obstruction to feeding a variety of stock. Easy to install.	Requires unusually good maintenance and adjustment to each operator. Frequent inspection necessary. Limits movement of operator. May obstruct work space around operator. Does not permit blanking from hand-fed strip stock.	Embossing presses Power presses
TWO-HAND TRIP				
Electric	Simultaneous pressure of two hands on switch buttons in series actuates machine.	Can be adapted to multiple operation. Operator's hands away from danger zone. No obstruction to hand feeding. Does not require adjustment. Can be equipped with continuous pressure remote controls to permit "inching." Generally easy to install.	Operator may try to reach into danger zone after tripping machine. Does not protect against mechanical repeat unless blocks or stops are used. Some trips can be rendered unsafe by holding with the arm, blocking or tying down one control, thereby permitting one-hand operation. Not used for some blanking operations.	Dough mixers Embossing presses Paper cutters Pressing machines Power presses Washing tumblers
Mechanical	Simultaneous pressure of two hands on air control valves, mechanical levers, controls interlocked with foot control, or the removal of solid blocks or stops permits normal operation of machine.			

FIGURE 22.2. (*Continued*)

472

Type of Guarding Method	Action of Guard	Advantages	Limitations	Typical Machines on Which Used
MISCELLANEOUS				
Limited slide travel	Slide travel limited to ¼ in. or less; fingers cannot enter between pressure points.	Provides positive protection. Requires no maintenance or adjustment.	Small opening limits size of stock.	Foot power (kick) presses Power presses
Electric eye	Electric eye beam and brake quickly stop machine or prevent its starting if the hands are in the danger zone.	Does not interfere with normal feeding or production. No obstruction on machine or around operator.	Expensive to install. Does not protect against mechanical repeat. Generally limited to use on slow speed machines with friction clutches or other means to stop the machine during the operating cycle. Can be circumvented	Embossing presses Power presses Rubber mills Squaring shears Press brakes
Special tools or handles on dies	Long-handled tongs, vacuum lifters, or hand die holders which avoid need for operator's putting his hand in the danger zone.	Inexpensive and adaptable to different types of stock. Sometimes increases protection of other guards.	Operator must keep his hands out of danger zone. Requires unusually good employee training and close supervision.	Dough brakes Leather die cutters Power presses Forging hammers
Special jigs or feeding devices	Hand-operated feeding devices of metal or wood which keep the operator's hands at a safe distance from the danger zone.	May speed production as well as safeguard machines. Generally economical for long jobs.	Machine itself not guarded; safe operation depends upon correct use of device. Requires good employee training, close supervision. Suitable for limited types of work.	Circular saws Dough brakes Jointers Meat grinders Paper cutters Power presses Drill presses

FIGURE 22.2.   (*Continued*)

3. The resistance of the body (determined in part by whether it is moist or dry)
4. The type of circuit with which contact is made (60 Hz, 120 volts AC is more dangerous than equivalent uninterrupted DC currents)
5. Frequencies (since at higher frequencies the body experiences heat instead of a shock)
6. The place in the heart cycle when the shock occurs
7. Age of the person
8. Size (height and weight) of the person
9. Physical condition of the person

TABLE 22.3
Effects of Electric Current on People

Effect	Current in Milliamperes			
	Direct		60 Hertz	
	Men	Women	Men	Women
Slight sensation on hand	1	0.6	0.4	0.3
Perception threshold	5.2	3.5	1.1	0.7
Shock—not painful, muscular control not lost	9	6	1.8	1.2
Shock—painful, muscular control not lost	62	41	9	6
Shock—painful, let-go threshold	76	51	16	10.5
Shock—painful and severe, muscular contractions, breathing difficult	90	60	23	15
Shock—possible ventricular fibrillation effect from 3-second shocks	500	500	.100	100
Short shocks lasting to seconds			$\dfrac{165}{\sqrt{t}}$	$\dfrac{165}{\sqrt{t}}$
High-voltage surges[a]	50	50	13.6	13.6

SOURCE: Gloss, D. S., and Wardle, M. G. Electrical safety in hospitals. *Professional Safety*, 1980, **25**, 25–30. (June)

[a] Energy in watt-seconds or joules.

See Table 22.3 for the relationship of the degree of shock to the electric current.

Housekeeping

Good housekeeping is necessary in the manufacturing industry. Good housekeeping reduces operating costs, increases production, opens aisles to permit faster traffic, lowers accident rates, reduces fire hazards, and raises employee morale. Workers do not operate with top efficiency in an environment booby-trapped with hazards. OSHA standards specifically require that good housekeeping be an integral part of industry's concern. All places of employment, passageways, storerooms, and service rooms must be kept clean and orderly, and the floor of every workroom must be kept clean and, as far as possible, dry. Where mechanical handling equipment is used, sufficient safety clearances must be allowed for aisles, at loading docks, through doorways, and wherever turns or passage must be made. Aisles and passageways must be kept clear without obstructions. Permanent aisles and passageways must be appropriately marked.

Conveyors

Conveyors are used to move materials from one point to another. Powered conveyors should be adequately guarded with mesh enclosures or railings to prevent workers from being caught on moving parts or being injured by falling materials. When moving parts of driving are within 7 feet of the floor or walkway, they should be guarded or otherwise arranged to prevent possible injury to employees.

Persons should not ride the conveying-medium element of a conveyor, except on conveyors which specifically incorporate platforms and control rooms. No one should ride on any element of a vertical conveyor. Conveyors should have warning devices and controls for emergency stops at numerous and convenient places. Access and sufficient clearance should be provided to permit safe repair work and lubrication. The power shutoff switch must be locked in the OFF position when that work is being done.

Forklift Trucks

Forklift trucks are mechanical materials moving equipment composed of a truck body with two forks in front which can be raised, lowered, or tilted. The following principles are important in regard to forklift trucks:

1. Use an extension if the load is particularly deep.
2. Lift from the broadest side of the load, and set the forks at the greatest width the pallet allows, for maximum stability.
3. Level the top of the forks.
4. Do not lift a load with only one fork.

The stability of the loaded truck increases as the center of gravity of the load is brought closer to the front axle; the steering is easier when the load is as close to the truck as possible. The forklift operator must not pass forks or attachments over anyone, nor may anyone pass under them. This applies whether the truck is loaded or empty. There is danger of striking someone with a fork or attchment. Part of the load may fall off and hit someone, or the load may be lowered onto someone.[2]

The forklift is a heavy piece of workplace equipment and can be dangerous if not used properly. Related injuries can range from damaging the load to death of the operator.

HEALTH HAZARDS

Several health hazards in the manufacturing industry are degreasing agents, plating solutions, plastics, and toxic materials. A major health hazard in the manufacturing industry is contact dermatitis. It pervades a wide variety of manufacturing operations involving degreasing, painting and spraying, and handling of hazardous materials.

Solvents

Solvents have many applications in manufacturing: degreasing, paint stripping, paint thinning, and solvent welding. Exposure (particularly to organic solvents) presents a potential threat to the health of the workers. All organic solvents have some effect on the central nervous system and skin. The principal modes of exposure are inhalation of vapors and skin contact. The chlorinated organic solvents are especially hazardous from a health standpoint. Inhalation of organic solvent vapors may initially cause headache, lack of coordination, and drowsiness. Such inhalation exposure is not thought to have permanent effects on health if exposure is not prolonged or frequent, but it may well increase the risk of accidents.[3]

Some organic solvents may have permanent effects despite low or infrequent exposures and should not be used at all unless stringent engineering controls are applied. This group of solvents includes carbon tetrachloride (CCl_4), benzene (C_6H), carbon disulfide (CS_2), gasoline, and chloroform ($CHCl_3$). Higher organic-solvent exposure levels can result in serious immediate effects (acute effects), including respiratory arrest (for example, in tank cleaning operations). Prolonged overexposure, such as for months or years, may result in serious damage to the blood, lungs, liver, kidneys, or gastrointestinal system. Although many organic solvents are capable of causing such long-term damage, the chlorinated organic solvents are especially toxic, since they tend to accumulate in the body.[3]

Skin contact with solvents may cause dermatitis, ranging in severity from simple irritation (reddening) to significant skin damage (skin ulcers). Even the most inert solvents can dissolve the body's natural outer barrier of fats and oils, leaving the skin unprotected. Once this skin barrier is removed, the skin becomes susceptible to disfiguring and sometimes disabling dermatitis.[3]

Mechanical Exhaust Ventilation

Mechanical exhaust ventilation is frequently needed to control the level of solvent vapors in workplace air. In addition, employees must be trained to avoid skin contact with solvents as well as inhalation of solvent vapors. For example, workers should be advised to position their work so that vapors will be pulled away from their breathing zone, rather than through it, and should wear gloves that are impervious to the solvents they handle.[7]

Acrylic Plastics. Acrylic plastics, such as ethyl acrylate [Ch_2= $CHC(O)OC_2H_5$], are widely used by firms producing plastic fixtures. Inhalation of and skin contact with vapors released during molding, cutting, and similar operations are the primary routes of employee exposure. Overexposure to vapors of the acrylics can lead to liver and kidney damage. The threshold limit value of ethyl acrylate is 25 ppm. Due to the highly combustible nature of the acrylics, fire and explosion precautions are also necessary.[3]

Polyvinyl chloride (PVC) is used in manufacturing primarily for wire insulation, containers, and film packaging. Heat-sealing, plastic packages, and hot-wire cutting of PVC films produces irritating vapors that may be the cause of a condition known as "meat-wrapper's asthma." Polyvinyl chloride is now considered a carcinogen.[3]

Adhesives. Adhesives such as methylene chloride (CH_2Cl_2) (used for solvent welding of acrylics) present an inhalation hazard. The body converts methylene chloride to carbon monoxide, which lowers the ability of the blood to carry oxygen. Symptoms of overexposure include headache, nausea, irritability, confusion, and shortness of breath.[3] The threshold limit value of methylene chloride is 200 ppm.

Caustics. Caustics, such as sodium hydroxide (NaOH) or potassium hydroxide, are often used to clean metal parts. Caustic chemicals are all corrosive. Eye injury is the most severe result of contact with them. Damage to the skin, mucous membranes, and respiratory tract can also result from exposure to caustics. Strict procedures for handling them should be followed, and adequate safeguards, such as emergency showers and eyewashes, should be provided. Caustic mists should be controlled at the source of dispersion by effective methods such as full enclosure, local exhaust ventilation, or other means.[3]

Electroplating. Electroplating is used extensively in manufacturing to provide decorative and tarnish-resistant finishes. The chief hazard associated with electroplating is exposure to toxic chemicals, including strong acids and alkalis. Emergency eyewash and shower facilities are required in areas where corrosive materials are handled or used. Splash goggles, rubber gauntlet gloves, and a rubber apron should be routinely worn, and protective creams should be used on hands and forearms. Employees working with plating chemicals should shower at the end of the shift and change into street clothes. Chrome plating requires the use of chromic acid. Employees who work with chromic acid must have periodic medical examinations of the mouth, nose, and other parts of the body to detect any ulceration in its early stages.

Copper plating baths may be either acid or alkaline. The cyanide compounds in alkaline baths present the greatest hazard in copper plating. These chemicals may become airborne when the tanks are charged. Since cyanide solutions are readily absorbed, skin contact must be avoided. Local exhaust ventilation systems are required to draw off vapors. Respirators may be needed to reduce inhalation exposure. Employees must limit skin contact by wearing impervious gloves. Good personal hygiene practices should be stressed, including frequent washing of exposed skin areas, particularly before eating, smoking, or using the bathroom.

If a cyanide salt is mixed with an acid, deadly hydrogen cyanide (HCN) gas can result. All traces of acid (such as from acid cleaning washes) must be rinsed away from parts before they are immersed in the cyanide vat. Nickel plating solutions can cause an itch rash ("nickel itch"). Workers who are sensitive to nickel compounds should wear full-body protective clothing, gloves, and face masks.[3] Careful handling and use is imperative.

The manufacturing industry cuts across every domain of life. It is far-reaching and there are many persons at risk for injury.

SUGGESTED LEARNING EXPERIENCES

1. If you were the safety engineer in a manufacturing industry, how would you go about setting safety priorities?

2. What kind of safety training in a manufacturing industry would you insist that all employees have?

3. What epidemiological information would you want to have regarding the possible occupational hazards in a manufacturing industry of your choice?

4. What role should management take to assure a safer manufacturing work environment? What role should the employee take? The consumer?

5. Choose a manufacturing industry and make a poster depicting both the hazards and the safety preventive measures needed.

6. Debate the issue: "There is no way in the manufacturing industry—considering the emphasis upon production—that workplace safety can be a priority."

7. In terms of productivity, how would you as a safety professional convince management of its role in the emphasis of workplace safety?

8. Demonstrate several ways you as a safety professional could convince big, heavy, brawny workers in the manufacturing industry of the need for performing all workplace operations in the safest way.

9. Visit a manufacturing industry, paying particular attention to the guarding equipment that is used. Critique these measures.

10. If you as a safety professional and a member of top management of a manufacturing plant were preparing a full-page advertisement in the *National Safety News*, what would it look like?

REFERENCES

1. OH Series No. 002, U.S. Department of Health, Education and Welfare, Welding Safety. Public Health Service, NIOSH HSM 72-10261, 1972.
2. U.S. Department of Health, Education and Welfare. *NIOSH Outline for Training Powered Industrial Truck Operators.* National Institute for Occupational Safety and Health, October, 1978.
3. National Safety Council. *Accident Prevention Manual for Industrial Operations.* (7th ed.). Chicago, 1977.
4. National Safety Council. *Accident Facts.* Chicago, 1981.

23 CONSTRUCTION SAFETY

The construction industry, which employs over a million persons, or about 6% of the private-sector work force, accounted in 1980 for nearly 11% of all work injuries and more than 20% of the fatalities.[3] Only mining and agriculture had higher work accident and fatality rates. Not surprisingly, it has one of the highest premiums for workers' compensation. Besides the direct cost of personal injuries, equipment is often damaged and work ceases while an injured worker is cared for.

Construction sites abound with hazards—beginning with excavation and continuing until the key is turned over to the owner. They include ironworks, welding, lifts, hoists, and scaffolds. At no time during a construction project can safety be dispensed with; yet most construction sites don't have a safety engineer.

Safety equipment is frequently unavailable at a site, because of management's fear that it will be lost or stolen. Construction workers rarely use protective equipment, because it is not available or because management's fear that it will interfere with production. However, hard hats have become an accepted form of protective equipment for many workers in the construction industry. The classic picture is of a construction worker at the site working without a hard hat in front of a sign announcing, "This Is a Hard Hat Job."

Alcoholism is so common with construction workers that one could consider it to be of epidemic proportions. Alcoholism undoubtedly contributes to many accidents.

Clearly, safety engineers have their work cut out for them if they are to reduce construction accident rates. Moreover, this must be accomplished on a short-term basis and not as part of long-range safety planning.

SIZE OF THE PROBLEM

The construction industry trades with very high incidence rates were structural steel erection workers (SIC 1791) with 30.3 cases per 100 full-time employees; ornamental metal, glass, excavation, and demolition workers (SIC 170) with 29.7 cases; residential building construction workers with 27.2 cases; and nonresidential building construction workers with 27.0 cases.[13]

Types of Injuries

Fractures, strains and sprains, and cuts and abrasions were the three leading kinds of injuries. Thirty percent or more of the cases were fractures which occurred in the following industries: structural steelwork, highway (street) construction, and painting and decorating. Strains and sprains clustered heavily (at least 25% but less than 30%) in masonry and stonework, concrete work, plumbing, heating, and air conditioning; electrical and general building contractors also had high injury rates in this category. Cuts and abrasions occurred frequently in carpentry and wood flooring, concrete work, structural steel erection, and general building contracting. The three major parts of the body affected in injuries in the construction industry were the trunk, lower extremities, and hands and fingers.[6]

Principal Causes

Construction accidents were mainly associated with certain causes. The principal agents for accidents were manual materials handling, working surfaces like scaffolds and steel erections, hand tools like saws, and backhoe vehicles. In certain specialty construction, such as roofing, burns resulted from hot substances (tar). In highway and street construction, the motor vehicle (mostly, the truck) is a leading agency of accidents.

Occupational illnesses in masonry, stonework, plastering, and concrete work were mainly skin diseases or disorders. In the 1972–1975 period, skin diseases or disorders ranged from 40 to 54% of all recorded illness cases in masonry, stonework, and plastering and from 41 to 71% of all cases in concrete work. General building contracting was the only construction industry with a relatively high percentage of respiratory conditions due to toxic agents, ranging from 10 to 17% of all illness cases during the 1972–1975 period. The proportion of cases involving dust disease of the lung, poisoning disorders due to physical agents, and disorders associated with repeated trauma was low. Other occupational illnesses were sometimes high, ranging from 35 to 47% of total illness cases in plumbing, heating, and air conditioning, and from 44 to 77% in carpentry and flooring.[5]

HAZARDOUS OPERATIONS

The remainder of this chapter addresses construction industry hazards and the ways that personal protective equipment can help reduce accidents. It also discusses sanitation, medical and first aid, and fire protection.

Excavation, Trenching, Tunneling

Excavation equipment used for digging and moving heavy objects offers many opportunities for accidents. Rules for operating equipment properly and safety must be established and enforced.

The assumption that construction workers are always healthy and fit is unwarranted. They should have thorough medical checkups. Excavations, trenches, tunnels, and foundations are not places for workers who have dizzy spells from high blood pressure, chest pains from exertion, or double vision. Operators of excavation equipment should definitely not be abusing drugs or alcohol. Unfortunately, construction crews often drink heavily. Drug abuse is strongly related to alcoholism. Safety engineers should institute a drug and alcohol education and referral service as part of any program.

Although alcoholism is a culturewide problem, it is particularly dangerous in an industry where the consequences are so potentially devastating. Alcoholism must be confronted and treated, and the treatment must be supported by the work force. (This was discussed more thoroughly in Chapter 6.)

Excavating, trenching, and tunneling involve moving large quantities of earth from a site. When earth is removed, the sides of the excavation or trench are liable to cave in, burying workers under tons of earth. The sides must be shored up to prevent that from happening.

Tunneling. A tunnel is an underground passage. Tunneling is one of the oldest and most difficult and hazardous jobs in the world. In the 1840s, a tunnel aqueduct built by Hadrian in the second century was uncovered. It is 15 miles long and still usable. In the fifteenth century, the Paris sewer system was begun, and many Roman quarries and mines were rediscovered. In 1935, when the Moscow subway was under construction, a secret underground city, presumably built by Ivan the Terrible in the sixteenth century, was discovered. Many tunneling deaths in the early twentieth century were caused by the bends, explosions, cave-ins, floods, and suffocation from fire. The Japanese Tanna Tunnel, begun in 1922, cut across an earthquake fault line, triggering floods which drowned large numbers of workers.

Electricity

Electrical hazards are more prevalent in the construction industry than in almost any other. Given the huge array of power tools and equipment, the necessity for being around live power lines in wiring buildings, and the presence of water on construction sites, this is not surprising.

The National Electrical Code (NFPA 70-1981), prepared by the National Fire Protection Association, gives specific guidelines for each electrical requirement; it has been adopted as a national consensus standard by OSHA. The purpose of the Code is the practical safeguarding of persons and buildings and their contents from hazards arising from the use of electricity. The Code sets basic minimum safety standards. Compliance with the Code and proper maintenance will result in an installation that is generally free from electrical hazards. This freedom from electrical hazards does not insure that the installation is

1. efficient,
2. convenient,
3. adequate for good service,
4. adequate for future expansion.

Hazard may occur because of overloading of the electrical system. This occurs because initial wiring did not provide for increases in the use of electricity.

Contractors often attempt to cut corners to save time and money. Violations of standards for use of electric equipment have been cited frequently in all industries but especially in construction.

Cranes

A crane is usually a large vehicle with a boom. Personnel switching from the older cab-down design to the newer, higher-capacity cab-up model must be trained in the setup instructions provided by the manufacturer of the crane. The increase in the number of instructions and notes on manufacturers' capacity charts provided with new equipment is testimony to the fact that these newer cranes require closer supervision during setup and operation. Due to the maneuverability and greater capacities of these cranes, ground bearing pressures become increasingly important and spotting and locating the equipment becomes more crucial.[8]

Hydraulic cranes of this type appear "easy" to operate. The hydraulic controls and pumps give instant control over the movements of the crane. Yet being able to distinguish the swing control level and the hoist control level is not sufficient. Because of the seeming ease of operation, some companies provide little training or do not give operators enough time to become familiar with the new equipment's capabilites and the operator's manual. See Figure 23.1 for a crane safety checklist.

In some localities, where operators are licensed at different levels, the self-propelled hydraulic crane is in the lowest category, indicating the relative indifference to safety. Also, the setup and capacity determinations are usually made by an operator, often without any consultation with supervisory or other technical personnel. If these determinations are in error, accidents can occur. Lack of awareness of the importance of the setup requirements on newer cranes has proved to be of grave consequence. The objective is to improve the quality of the operator's decision-making capabilities.

It is obvious even to a casual observer that large-capacity cranes with long booms represent a special problem. The amount of pressure exerted on the ground by this large equipment is a critical factor. In the early sixties, a 75-ton crane was considered a giant. But crane technology has increased at a tremendous pace to meet the demands of modern construction. The trends are toward the increasing use of "modular assembly" manufacturing techniques. Such assembly techniques increase the weights and sizes of the loads to be handled and, consequently, the need for increasing the sizes and capacities of the cranes necessary to accomplish the tasks.[4]

Long booms and operating radii have caused conditions under which the operator can no longer see the landing locations of the loads. This places increased responsibility on the signal worker and increases error possibilities. Is it feasible for some loads to be safety-controlled by conventional tag lines being held by one or two people? The answer is "No!"

Personal Protective Equipment

Ear Protection.   Construction sites are incredibly noisy. It seems that there cannot be a quiet construction job: nails are pounded; earth is moved; rivets are exploded. With all that noise one can see the need for hearing protection. A hearing conservation program should be established. Every construction worker should be fitted with proper hearing protection devices and required to wear them. Workers should be given audiometric tests on a regular basis in accordance with 29 CFR 1910.95.

Eye and Face Protection.   Eye and face protection is often necessary in the construction industry. According to the National Safety Council, disabling injuries to the face accounted for 22% of all face injuries.[5] Almost one-third of the face injuries were caused by metal objects which weighed over 1 pound. Accidents resulted in cuts, lacerations, puncture wounds, and fractures (including broken teeth). Eye injuries occurred primarily in construction and manufacturing industries (70%). In 1980 there were

Section	Requirements	Compliance			Comments
		Yes	No	N/A	
4887a	Access to cage, or machine house. Access must be a conveniently placed stationary ladder, stairs or platform. A stepover may not be more than 12 inches. If installation was made prior to January 1, 1973, then the gap must not exceed 20 inches. The stepover must be located such that a person entering or leaving the cab will not be exposed to a shear hazard.				
4887b	Access to bridge and gantry cranes. A ladder, stairs, or other safe means shall provide a convenient access to bridge walkway. Where cage is attached to and below bridge girders, no portion of the cage or cage platform shall be in the projected area between the girders.				
4887c	Access to revolving cabs, or machine houses. An adequate means shall be provided to permit the operator to enter or				

FIGURE 23.1. Partial checklist for crane operations. To be published in Gloss, D. S., Wardle, M. G., Gloss, M. R., and Gloss, D. S. II. *Handbook of Occupational Safety and Health Standards* Volume III John Wiley and Sons, 1985.

486

Section	Requirements	Compliance			Comments
		Yes	No	N/A	
	leave the crane cab or machine house irrespective of the position of the cab and to safety reach the floor.				
4887d	Access to boom or bridges for servicing. Each boom or bridge must be equipment with at least one of the following a a substantial walkway or platform b platform hand holds c grab irons d mobile platform e permanent elevated platform attached to the building at the same height as bridge.				

FIGURE 23.1. (*Continued*)

110,999 disabling eye injuries. However, the majority of the injured were not wearing eye protection devices.[3]

Eye and face protection is the most widely used form of needed protection. See Figure 23.2 for recommendations the National Safety Council has made for eye and face equipment.[5] The National Safety Council further states that eye protective devices must be considered as optical instruments and that they should be carefully selected, fitted, and used. Furthermore, NSC asserts that contact lenses should never be considered a replacement for safe protective equipment for the eyes. The American Society for the Prevention of Blindness has issued a statement saying that contact lenses have no place in the construction site. If corrective lenses are required, it is preferable to grind the correction into a goggle lens. The cost of eye protection devices is negligible when measured against the possibility of blindness.

SELECTION OF EYE AND FACE PROTECTIVE EQUIPMENT

1. **GOGGLES**, Flexible Fitting, Regular Ventilation
2. **GOGGLES**, Flexible Fitting, Hooded Ventilation
3. **GOGGLES**, Cushioned Fitting, Rigid Body
*4. **SPECTACLES**, Metal Frame, with Sideshields
*5. **SPECTACLES**, Plastic Frame, with Sideshields
*6. **SPECTACLES**, Metal-Plastic Frame, with Sideshields

** 7. **WELDING GOGGLES**, Eyecup Type, Tinted Lenses (Illustrated)
   7A. **CHIPPING GOGGLES**, Eyecup Type, Clear Safety Lenses (Not Illustrated)
** 8. **WELDING GOGGLES**, Coverspec Type Tinted Lenses (Illustrated)
   8A. **CHIPPING GOGGLES**, Coverspec Type, Clear Safety Lenses (Not Illustrated)
** 9. **WELDING GOGGLES**, Coverspec Type, Tinted Plate Lens
   10. **FACE SHIELD** (Available with Plastic or Mesh Window)
11. **WELDING HELMETS

*Non-sideshield spectacles are available for limited hazard use requiring only frontal protection.
**See appendix chart "Selection of Shade Numbers for Welding Filters."

APPLICATIONS		
OPERATION	**HAZARDS**	**RECOMMENDED PROTECTORS:** Bold Type Numbers Signify Preferred Protection
ACETYLENE—BURNING ACETYLENE—CUTTING ACETYLENE—WELDING	SPARKS, HARMFUL RAYS, MOLTEN METAL, FLYING PARTICLES	7, 8, **9**
CHEMICAL HANDLING	SPLASH, ACID BURNS, FUMES	**2**, 10 (For severe exposure add 10 over 2)
CHIPPING	FLYING PARTICLES	**1**, 3, 4, 5, 6, 7A, 8A
ELECTRIC (ARC) WELDING	SPARKS, INTENSE RAYS, MOLTEN METAL	**9, 11** (11 in combination with 4, 5, 6, in tinted lenses, advisable)
FURNACE OPERATIONS	GLARE, HEAT, MOLTEN METAL	7, **8, 9** (For severe exposure add 10)
GRINDING—LIGHT	FLYING PARTICLES	**1**, 3, 4, **5, 6**, 10
GRINDING—HEAVY	FLYING PARTICLES	**1**, 3, 7A, 8A (For severe exposure add 10)
LABORATORY	CHEMICAL SPLASH, GLASS BREAKAGE	**2** (10 when in combination with 4, 5, 6)
MACHINING	FLYING PARTICLES	**1**, 3, 4, **5, 6**, 10
MOLTEN METALS	HEAT, GLARE, SPARKS, SPLASH	7, **8** (10 in combination with 4, 5, 6, in tinted lenses)
SPOT WELDING	FLYING PARTICLES, SPARKS	**1**, 3, 4, **5, 6**, 10

FIGURE 23.2. Eye and face equipment recommendations (*this material is reproduced with permission from American National Standard Practice for Occupational and Educational Eye and Face Protection*).

Protective Footwear. Protective footwear is divided into five types: safety-toe shoes, conductive shoes, foundry shoes, explosive-operations shoes, and electrical hazards shoes. OSHA requires that the safety-toe shoe be used for work involving the handling of heavy materials. To protect feet from the heaviest impacts, workers should wear metatarsal (or overfoot) guards in addition to the safety shoes.[2]

Head Protection. Have you ever seen the sign "This Is a Hard Hat Job"? There were over 20,000 head injuries on construction sites in 1980. In one survey of persons with head injuries, less than 16% of the workers were wearing head protection.[3] Where was safety engineering?

SITE SAFETY AND HEALTH SERVICES

The construction industry must provide adequate sanitation for employees, including potable water for drinking, washing, and food preparation; adequate disposal of sewage and garbage; adequate personal service facilities; sanitary food service; and satisfactory heating and ventilation. This is not always easy on a construction site, but trailers and other accoutrements can provide adequate facilities.

Medical Care and First Aid

First aid and medical care must also be provided. A good practice is to require medical examinations before job placement to ensure that the capabilities of the prospective employees are compatible with the demands of the specific jobs. If heavy work is required, the prospective employees' physical work capacity should be measured also. It is easier and less expensive to prevent injuries than it is to provide medical care for them. Periodic health evaluations for hazardous jobs and early treatment of any illness or injury should also be encouraged. The evaluation of lead levels for pipe fitters should be made at the start of construction when molten lead is to be used and periodically thereafter.

Emergency phone numbers must be posted near telephones. An emergency information chart like the one shown in Figure 23.3 may be helpful. Stretchers and blankets should be available for immediate transportation of ill or injured employees.

At least one and preferably more employees on each shift must be trained in first aid. This is required if the site is not in close proximity to a medical treatment facility (29 CFR 1910.151). Local health departments, the American Red Cross, the U.S. Bureau of Mines, and St. John's Ambulance have training programs that will satisfy OSHA first aid requirements. Medical personnel must be readily available by phone or on site for advice and consultation.

Emergency Information

Fire

Telephone fire department _____

Nearest alarm box at   _____

Crime

Telephone police _____

Injury/Illnesses

Avoid infection of minor injuries; always get medical attention or skilled first aid.

Doctor _____
    Office _____ Tel. _____
    Residence _____ Tel. _____

_____
_____

Hospital _____
    Address _____ Tel. _____
Ambulance _____
    Address _____ Tel. _____

(In emergencies, get medical attention and transportation elsewhere if necessary.)

In all cases fire, crime, accident, or sickness, promptly notify:

1. Name _____ Office Tel. _____
    Address _____ Res. Tel. _____

or

2. Name _____ Office Tel. _____
    Address _____ Res. Tel. _____

FIGURE 23.3. Emergency information chart (*from DHEW, 1978*).

First aid supplies approved by a consulting physician must be readily available. The supplies should be in sanitary containers with individually sealed, sterile packages for material such as gauze, bandages, and dressings. Other items often needed are adhesive tape, triangular bandages (to be used as slings), inflatable plastic splints, scissors, and a mild soap for cleansing of wounds of cuts.

Suitable facilities for quick drenching or flushing of the eyes and body must be provided within work areas where a person may be exposed to corrosive materials, such as acids or caustics.

FIRE PROTECTION

Every construction site is unique. It differs in terrain, location, type of construction, construction method, and site preparation. Structures under construction are more vulnerable to fire than are finished structures. There are many operations during construction that use combustible substances and direct flames. These operations include gasoline heating as well as welding and soldering. Construction sites are frequently littered with flammable debris, and construction materials themselves are often flammable. Fire spreads rapidly because of the absence of fire detection equipment (smoke and flame detectors), suppression equipment (standpipes, sprinklers, and fire extinguishers), and containment means (fire doors).

Fire protection in construction is critical. Figure 23.4 is a chart showing fire extinguishers by class and how to use them.

The employer is responsible for establishing an adequate fire protection program covering all phases of construction and/or demolition work. Effective fire-fighting equipment must be available to meet all fire hazards. These are trash and paper (class A), oil and flammable liquids (class B), and electrical (class C).

Fire-fighting equipment must be conspicuously located, inspected monthly, appropriately maintained, and readily accessible; it must be replaced promptly if found defective. Appropriate fire extinguishers must be available for hot work operations such as welding. The employer is responsible for having a system to warn employees of a fire and for having means available (phone number posted) to promptly obtain the services of the local fire department.

Fire drills should be conducted throughout construction, as means of egress will differ during the various construction phases, and tradespeople will be in different locations on the site. Smoking is prohibited in areas of potential fire hazards, and "No Smoking" signs must be conspicuously posted.

Only approved containers and portable tanks should be used to store flammable and combustible liquids; these liquids must not be stored in or around means of egress such as stairways and exits. At least one approved fire extinguisher must be located within 10 feet of any room or inside storage area used to store such liquids. Furthermore, an approved fire extinguisher must not be more than 75 feet from any outside storage area.

SUGGESTED LEARNING EXPERIENCES

1. Observe a construction site of your choice, assessing safety. Were your observations what you expected them to be? Discuss.

Kind of fire		Approved type of extinguisher							How to operate
Decide the class of fire you are fighting...	...Then check the columns to the right of class	Match up proper extinguisher with class of fire shown at left							
		Foam (solution of aluminum sulphate and bicarbonate of soda)	Carbon dioxide (carbon dioxide gas under pressure)	Soda acid (bicarbonate of soda solution and sulphuric acid)	Pump tank (plain water)	Gas cartridge (water expelled by carbon dioxide gas)	Multipurpose dry chemical	Ordinary dry chemical	
Class A fires — Use these extinguishers — Ordinary combustibles • Wood • Paper • Cloth, etc.		A B	X	A	A	A	A B C	X	Foam: Don't play stream into the burning liquid. Allow foam to fall lightly on fire. Carbon dioxide: Direct discharge as close to fire as possible. First at edge of flames and gradually for forward and and upward
Class B fires — Use these extinguishers — Flammable liquids, grease • Gasoline • Paints • Oils, etc.		A B	B C	X	X	X	A B C	B C	Soda – acid, gas cartridge: Direct stream at base of flame. Pump tank: Place foot on footrest and direct stream at base of flame.
Class C fires — Use these extinguishers — Electrical equipment • Motors • Switches, etc.		X	B C	X	X	X	A B C	B C	Dry chemical: Direct at the base of the flame. In the case of class A fires, follow up by directing the dry chemicals at remaining material that is burning.

FIGURE 23.4. Approved type of fire extinguishers (*from NIOSH*).

492

2. If you were a worker in a construction group, what safety procedures would you want provided for you? Which of these would you demand?

3. As a safety engineer for a small construction project, such as a 200-room motel, develop a plan for sanitation facilities for the employees.

4. Develop an alcoholism program for the construction industry. Be sure to include your plans for enlisting all needed personnel and your own philosophy about alcoholism.

5. Describe the provisions for first aid and medical care that you as a safety professional would want all construction sites to have.

6. What basic information about electricity would you as a safety professional want all construction workers to know? How would you provide it?

7. Develop an employee training program on face and eye protection for a group of construction workers. Present this program to your class.

8. Critique the above employee training presentation from the viewpoint of a new employee. Include what you would have done differently and the areas that you found beneficial. Be sure to share your observations with the presenter(s).

9. What facts regarding fire protection and control would you as a safety professional want every construction employee to know?

10. Write a term paper discussing the various hazards in the construction industry. Then select a single hazard which you think is the most serious, and provide a rationale for this. In the final section of your paper, describe how you would approach the control of this hazard.

REFERENCES

1. McCall, B. *Safty First—At Last*. New York: Vantage Press, 1975.
2. National Safety Council. *Accident Prevention Manual for Industrial Operations* (7th ed.). Chicago, 1977.
3. National Safety Council. *Accident Facts*. Chicago, 1981.
4. O'Rourke, D. J. Hazards in craining. *Professional Safety*, 1980, **25**, 13–18. (July)
5. Wang, C. L., and Kilaski, H. J. The safety and health record in the construction industry. *Monthly Labor Review*, 1978, **93**, 3–9. (Mar.)

24 TRANSPORTATION

Transportation safety includes motor vehicle, railroad, and aviation safety, and each aspect will be discussed in this chapter.

Highway safety administrators continually work to reduce the carnage on the roadways. Annually, over 50,000 persons are killed and many billion dollars worth of motor vehicles and property are damaged or destroyed. To prevent accidents, driver performance must be closely monitored, and unsafe drivers must be removed from the road. This may be accomplished either by suspending their licenses or by insurance carriers refusing to insure them. Drivers must be trained to drive defensively. Safety engineers who are in charge of fleet safety must be concerned not only with the company employees and vehicles but with the general public as well. The public and the motor vehicle manufacturers must be educated in the need for built-in safety devices. The public has been relatively indifferent to highway safety and to using current safety devices. The manufacturers thus far resisted installing new safety features in the vehicles which they produce.

Railroad safety confronts problems caused by inadequate maintenance of the million miles of roadbeds. Safety must not only be concerned with rolling stock, which may be damaged, but also concerned with passengers who may be hurt and with goods which may be destroyed. Railroads carry so many hazardous chemicals in tank cars that safety engineers must be concerned with the potential for disasters. The railroad cars may become time bombs, or they may discharge hazardous or toxic materials which can affect people living within the vicinity of the railroad. Incidents of spills may temporarily or permanently affect the land and waterways which abut the railroad.

Aviation safety is widely discussed. A major accident tends to trigger redoubled efforts toward aviation safety. Over 2.5% of general aviation aircraft have accidents each year, and over 1,000 aircraft are destroyed annually. Only through continual, extensive training and worker screening can safety engineers expect to reduce aircraft accidents.

This chapter discusses each of these modes of transportation in terms of the magnitude of the accident problem, human error as a contributing factor in the related accidents, management policies, physical examination requirements, training requirements, maintenance requirements, accident record keeping, and special investigative procedures.

SIZE OF THE PROBLEM

The problem of transportation safety is greater than any other safety problem facing the United States. Transportation accidents cause more than 4 times the number of deaths that occur in industry. It is the leading cause of death for persons 1 to 24 years of age. In 1980 there were more transportation-related disabling injuries than workplace disabling injuries.[4]

Motor Vehicles

The carnage on our roads and highways makes the motor vehicle look like Attila the Hun, who was known for his brutality. Few names, like the Black Plague and Adolf Hitler, are associated with more deaths. Although the motor vehicle is more similar to the Black Plague in that the deaths appear to be without relationship to race, age, or sex, this is not entirely true for motor vehicle homicides. More men are killed than women; more young adults are killed than other age groups; and more caucasians are killed in the United States than other races. The motor vehicle shows no mercy to its victims.

Since 1903 there have been 2,041,000 persons killed upon the highways of the United States. In the 28,500 days from 1903 to 1981 there have been 72 persons killed every single day of every year, and the slaughter goes on. In 1981 the rate was approximately 2 times that amount, or 139 persons killed in motor vehicle accidents daily. The cost of motor vehicle accidents in a single year, 1981, is shown in Table 24.1.

What do the statistics tell us? One out of every five vehicles on the road was involved in an accident in 1980. One out of every 2,552 vehicles was involved in a fatal accident. There was $40.6 billion spent,[4] which includes lost wages, medical expenses, insurance administrative expenses, and property damage from motor vehicle accidents. Not included are the costs of public agencies such as police, fire departments, and courts; indirect losses to employers, which are deprived of employees who are involved in off-the-job accidents; the value of cargo losses in commercial vehicles; and the damages awarded by the courts in excess of the actual losses.

In most accidents, the vehicle, the driver, and the road are principal

TABLE 24.1
Motor Vehicle Accidents, 1981

Facts	Figures
Deaths	50,800
Disabling injuries	1,900,000
Costs	$40,600,000,000
Total motor vehicle accidents	18,000,000
Total motor vehicles involved	30,000,000
Total motor vehicles registered	165,700,000

SOURCE: National Safety Council, *Accident Facts*. Chicago, 1982, with permission.

causes. It is the interaction of these factors which usually sets up the series of events which leads to an accident. Improper driving is the major cause of motor vehicle accidents. This includes speeding too fast for conditions, driving on the wrong side of the road, and passing stop signs. Undoubtedly, you can think of others. What leads to improper driving? The most frequently occurring answer is alcohol. Driving while intoxicated (DWI) is the cause of at least 50% of the fatal motor vehicle accidents. According to the National Highway Traffic Safety Administration, 40 to 55% of the fatally injured drivers have blood levels of alcohol high enough to indicate intoxication. Not even considered is the degree of impairment that intoxication below the stated "intoxication" levels causes.

Railroads

In comparison to motor vehicle accident rates, the railroads cause little death and disability to either the traveling public or railroad workers. However, there is substantial property damage associated with railroad accidents.

There were approximately 600 deaths (only 5 were passengers) each year associated with railroads in the years 1975–1978. In addition, there were 900 deaths from collision between motor vehicles (moving or stalled) and railroad vehicles at grade crossings in 1980. There were also 70 pedestrians killed that year in grade-crossing accidents. Therefore, railroad accidents accounted for 1,500 deaths annually.

The railroads are not particularly safe places to work. Out of every 100 persons employed in 1980, there were 11.66 recordable accidents. This is substantially greater than the overall industry figure of 7.81 and almost as high as the construction industry's 12.03. As far as lost-time accidents, the railroad experienced 6.5 accidents per 100 workers. These 6.5 accidents resulted in 126 days lost, or 19 days per accident.[4]

Aviation

Civil aviation accident and death rates have steadily declined since 1945. This period has been one of rapid growth for all types of aviation. In the years 1976–1980, as shown in Table 24.2, passenger deaths fluctuated widely. During the past 50 years, 1979 was the year with the highest number of passenger deaths (262) in domestic service.

In terms of safety, the number of passenger fatalities is often misleading. If a plane is destroyed and all aboard are killed, the number may vary widely. The number of fatal accidents in scheduled air transports from 1976–1980 was eight. In 1978, there were three fatal airline accidents with 13 persons killed.

MOTOR VEHICLES

Selection of Drivers

Selection of competent drivers requires that management have some means of evaluating the motor and judgmental skills of applicants. It must ascertain (1) driver's experience and performance on previous jobs, (2) job knowledge and technical know-how, (3) attitude toward safety, and (4) job performance through the probationary period. Specific evaluation tools are application forms, interviews, reference checks, license checks, physical examinations, written tests, and road and yard tests.[3]

Previous Driving Experience. In selecting a new driver, management's job is to determine first of all how the applicant has driven in the past. This is the best single indicator of how he or she will drive in the future. Of all the possible selection criteria, those that tell of the person's past driving performance are the most important indicators.

After the application and the interview, some reference checks should be made. Checking prior work references establishes the validity of the information obtained, and it can be done simply and inexpensively over the telephone. If the candidate driver is presently employed by the company, his or her permission to contact the present employer should be obtained. Persons have been dismissed from a job when the present employer finds out they are looking for a new job. The following should be determined:

1. Were they employed by the company as stated?
2. What type of work were they engaged in?
3. What were their absentee records?
4. What were their wage records?
5. What were their reasons for leaving?
6. Would the company rehire them?

TABLE 24.2
Passenger Fatality Rates on U.S. Airlines, 1976–1980, for All Scheduled Airlines

Year	Fatal Accidents	Fatal Accidents Per Thousand Revenue Departures	Number of Deaths	Per Million Revenue Passenger Miles	Passenger Fatalities	
					Per Million Revenue Passengers Carried	Per Thousand Revenue Aircraft Departures
Domestic Flights						
1976	1	0.0002	1	—	0.0049	0.0002
1977	1	0.0002	60	0.0004	0.2703	0.0128
1978	3	0.0006	13	0.0001	0.0512	0.0027
1979	2	0.0004	259	0.0012	0.8840	0.0504
1980	1	0.0002	11	0.0001	0.0403	0.0021
1976–1980	8	0.0003	344	0.0004	0.2756	0.0141
International Flights						
1976	1	0.0043	35	0.0010	2.0588	0.1489
1977	—	—	—	—	—	—
1978	—	—	—	—	—	—
1979	1	0.0042	59	0.0011	2.4583	0.2480
1980	—	—	—	—	—	—
1976–1980	2	0.0018	94	0.0004	0.9038	0.0827
All Flights						
1976	2	0.0004	36	0.0002	0.1263	0.0075
1977	1	0.0002	60	0.0003	0.2500	0.0122
1978	3	0.0006	13	0.0001	0.0473	0.0026
1979	3	0.0006	318	0.0012	1.0032	0.0591
1980	1	0.0002	11	—	0.0297	0.0021
1976–1980	10	0.0003	438	0.0004	0.2582	0.0142

SOURCE: National Safety Council, *Accident Facts*. Chicago, 1981, with permission.

Knowledge of Traffic Regulations.   Knowledge of traffic regulations is essential for safe driving. Some questions that an interviewer might ask are about school bus regulations, right of way, and speed zone laws.

Driving Tests.   Driving tests consist of yard tests and road tests; either of which may be utilized depending upon the job description. The yard test is used to determine a driver's skill in handling the equipment without going into traffic. Some exercises often included are the parallel park and the alley dock. The yard test simply shows whether the applicant can maneuver the equipment. The road test (in traffic) should be made over a predetermined route which approximates the kind of driving that the applicant will be required to do if hired.

Road tests permit observation of an applicant under actual driving conditions. As a preemployment test, it measures the following: (1) the driver's ability to drive in traffic, (2) observation of the driver's general attitude while driving, (3) determination of the need for further training, and (4) detection of any unsafe driving habits. Two types of road tests are used: (1) the "in-traffic" road test and (2) the "driving range" test. The latter requires the establishment of a relatively permanent driving course and must of necessity use a considerable amount of space. Ordinarily, the driving range test is prepared so as to cover all driving situations normally encountered when operating a truck. The applicant demonstrates ability according to the following: While traversing a rectangular course (usually about 800 feet long and 350 feet wide), the examinee is observed performing some of the following: (1) gear-shifting, (2) stopping within 1 foot of a marker, (3) making right, left, and "U" turns, (4) driving and backing in narrow straight lanes, (5) driving and backing in a zigzag course, and (6) backing to within 1 foot of a loading platform. During these basic maneuvers and possibly others pertinent to the operations to be performed, a close watch is kept to check whether the examinee has given the usual signal (to indicate intent to other drivers) for each of the basic operations, stopping, slowing, and changing direction.

The in-traffic road test is very practical for most firms. The test should take sufficient time (10 minutes is generally considered an absolute minimum) in order to make a fair sampling of the examinee's performance under actual driving situations. All the normally difficult maneuvers encountered in actual driving experience should be included. The same procedure should be followed when testing all drivers.

Customary items for inclusion in an in-traffic road test include the following:

1. Starting
2. Turning (at least three right- and left-hand turns each)
3. Stopping (including ordinary and emergency stops)

4. Backing
5. Turning in the opposite direction on the street
6. Parking (parallel to the curb and between two other vehicles or objects)
7. Approaching a variety of intersections
8. Starting and stopping on hills (if no hill is available, the examinee should be asked to demonstrate handling the vehicle as if it were on a hill)
9. The examinee should be observed for manner of handling the vehicle when passing and being passed, how speed is controlled under varying conditions, attentiveness and smoothness of operating performance, and safety attitude toward other drivers and pedestrians

See Figure 24.1 for an example of a road test.

Physical Examination.   Another important factor in driver selection is a complete physical examination, including the field of vision, hearing ability, reaction time, visual acuity, and a lumbar x-ray. The lumbar x-ray is taken to ascertain if any preemployment back injury exists.

A physical examination should be given to each applicant by a qualified physician. Drivers who will be employed in interstate commerce are required to have a medical examination according to federal regulations. Employers must have on file for each driver a medical certificate showing him or her to be physically qualified. Minimum physical requirements for drivers engaged in interstate commerce have been established.

Driver training is another aspect of a fleet safety program. This will include information regarding company policies, accident reporting, defensive driving, federal regulations that apply to the industry or task, state and local traffic laws, record keeping, preventive maintenance, and emergency procedures.

Company Policies

As with all other aspects of safety, company policies are important for the employee to understand thoroughly in order to know what framework he or she is operating under, what guidelines have proved to be effective for others, and, overall, what specific behaviors are expected.

Accident Reporting

The driver must make a complete report on involvement in any accident to the safety engineer. A separate record folder should be maintained for each driver, showing in chonological order the driver's entire "accident" experience since employment by the company. These reports should be analyzed

ROAD TEST SCORE SHEET

Name _____ 225 – _____ = _____
                                              Errors   Total Score

Address _____

Examined by _____   Type of Vehicle _____

Remarks _____

Instructions

Require the examinee to enter the vehicle, start the motor, leave the curb, perform basic operations of driving in traffic, and then park the vehicle. The test route should be at least five miles long and should include at least five intersections, five right turns, five left turns, one turn-around, a hill, a railroad grade crossing. The purpose and nature of the test should be explained to the examinee before being given; test results should be discussed with the examinee. The test vehicle should be in proper operating condition. The *lowest* passing grade is 160; highest grade is 225.

BEFORE OPERATION (14)

–1 Fails to check tires, oil, water
–1 Fails to enter curb side where practicable
–1 Fails to check instruments
–1 Fails to adjust seat
–1 Fails to adjust mirror
–1 Fails to test horn

–2 Fails to check stop and driving lights
–1 Fails to check hand brakes
–2 Fails to check foot brakes
–1 Fails to disengage clutch when starting engine
–1 Fails to warm up cold engine
–1 Fails to check windshield wiper

LEAVING CURB (8)

–2 Fails to look back
–2 Fails to release hand brake
–2 Fails to signal
–2 Pull out in front of traffic

SPEED CONTROL (26)

–7 Fails to observe speed laws
–7 Speeds through intersections
–7 Too fast for conditions
–5 Erratic (feeds gas by spurts)

RIGHT TURNS (15)

–3 Improper lane
–3 Fails to signal in advance
–3 Fails to look both ways
–3 Swings to wide
–3 Turns too fast

OVERTAKING AND PASSING (25)

–4 Fails to use hand and horn signals
–7 Cuts in too soon
–7 Passes in intersections
–7 Follows too closely

LEFT TURNS (15)

–3 Improper lane
–3 Fails to signal in advance
–3 Fails to look both ways
–3 Cuts corner
–3 Turns too fast

USE OF CONTROLS (12)

–3 Rides clutch
–3 Clashes gears
–3 Both hands not on wheel
–3 Races engine

SLOWING AND STOPPING (17)

–4 Fails to signal intent
–2 Does not look in mirror
–2 Fails to use engine as a brake
–2 Excessive brake application
–5 Stops abruptly
–2 Does not come to full stop at stop sign

BACKING (7)

–2 Fails to sound horn where necessary
–2 Fails to look before backing
–2 Backs too fast
–1 Does not back straight and smoothly

FIGURE 24.1 Road test form for rating of skill *(from Kramer, 1947, pp. 118–119).*

PARKING (6)

− 1 Fails to signal
− 1 Backs more than twice
− 2 Rubs tires against curb
− 1 Fails to set hand brake
− 1 Over 6 inches from curb (one check
       for each 6 inches)

DRIVING ON GRADE (15)

− 3 Rolls back when standing or starting
       on grade
− 3 Coasts down grade
− 3 Unreasonable speed on grade
− 3 Clashes gears
− 3 Selects or uses wrong gear

TURNING AROUND (4)

− 1 Fails to signal in advance
− 1 Does not look both ways
− 1 Backs more than twice
− 1 Turns wheel while vehicle is station-
       ary

ATTENTION AND ATTITUDE (22)

− 4 Fails to keep eyes on road
− 3 Excessive talking
− 6 Excessively nervous
− 6 Loses temper
− 3 Rests arm on window

GENERAL (29)

− 5 Depends on others for safety
− 5 Fails to grant right-of-way
− 5 Disobeys traffic sign or signal
− 5 Inconsiderate of pedestrians
− 1 Unnecessary use of horn
− 5 Fails to allow vehicles to pass
− 3 Fails to keep on right side of road or
       proper lane of traffic

CROSSING RAILROAD TRACKS (10)

− 2 Does not look both ways
− 2 Fails to slow down
− 3 Fails to stop where required
− 1 Shifts gears on tracks
− 2 Cross tracks to fast

FIGURE 24.1.   (*Continued*)

by the safety specialist, and any common causes or other significant infor-mation should be brought out and used as a basis for corrective action. Drivers should be held strictly accountable for following all safety regula-tions, and drivers with a record of repeated accidents should be transferred to other jobs or discharged. Accidents are regarded as preventable in terms of driver responsibility unless in the safety supervisor's opinion that driver did everything reasonably possible to prevent the accident.

Standard accident frequency rates in motor vehicle operations are com-puted on the basis of the number of accidents per 100,000 miles of opera-tion. This is calculated by multiplying the number of accidents by 100,000 and dividing that product by the number of vehicle-miles driven. This is shown in Table 24.3.

The objective of defensive driving (which is very important in fleet safety) is to prevent or avoid accidents and not to cause them.

It is necessary to keep driver and accident records on fleet safety.

Accident records start with accident reports submitted by drivers. For serious accidents, investigations are usually made by the company and the insurance carrier. Management should collect, analyze, and summarize these accident reports to detect any trends that may show. The information and/or reports are evaluated for (1) type of accident, (2) immediate causes, and (3) driver chargeability. Causes are generally classified in terms of driver, road conditions, and type of vehicle. Of primary importance is proper analysis of driver-related causes.

TABLE 24.3
Computation Methods for Determining Accident/Incident Rate

1. Annual vehicle accident frequency per million miles: Multiply the annual number of accidents for a given year by 1 million miles, and divide by the actual mileage covered in the year.

$$\text{Accident frequency} = \frac{\text{annual no. of accidents} \times 1 \text{ million miles}}{\text{annual mileage of driver}}$$

2. Annual vehicle-accident loss rate: Divide gross revenue for one year by the total dollar losses of vehicle accidents for that year.

$$\text{Vehicle-accident loss rate} = \frac{\text{annual gross revenue}}{\substack{\text{annual dollar losses} \\ \text{from vehicle dollars}}}$$

3. Annual accident rate per driver: Divide the total number of accidents for one year by the total number of drivers.

$$\text{Accident rate per driver} = \frac{\text{total number of accidents}}{\text{number of drivers}}$$

4. Annual employee-injury loss rate: Divide the annual gross revenue for one year by employee-injury dollar losses for that year.

$$\text{Employee-injury loss rate} = \frac{\text{annual gross revenue}}{\substack{\text{annual dollar losses from} \\ \text{employee injuries}}}$$

SOURCE: Petersen, D. *Techniques of Safety Management.* New York: McGraw-Hill, 1970. With permission.

Errors which lead to accidents were analyzed by the National Safety Council. They found that such factors as speed control, lane usage, and passing accounted for many accidents. An extensive list of errors that cause accidents is shown in Table 24.4.

In addition to causing accidents, driver behavior can affect public relations. Many trucks bear advertising of the organization using them. This advertising may influence the public positively or negatively depending on how the driver behaves on the road. Driving with courtesy and consideration enhances an organization's goodwill. Certain behavior, like splashing pedestrians and blocking cross-walks, negatively affect an organization's image. A list of such behavior is shown in Table 24.5.

Federal regulations must be understood to preclude employees from unwittingly violating them. Federal regulations are particularly important in interstate commerce, when state lines are crossed. It is also important, of course, to follow state and local traffic laws at all times. Accurate record keeping is essential to provide reliable documentation when it is needed or requested. Emergency procedures must be thoroughly understood, for in an emergency there is no time to look up what should be done. In vehicle maintenance, the emphasis is on preventive maintenance.

Maintenance

Approximately 15 of every 100 motor vehicle "accidents" are attributable to mechanical defects—that is over 4 million accidents per year. A program is required for keeping motor vehicle fleet equipment in good condition. A preventive maintenance program involves overhaul, repair, or replacement of major pieces of equipment or their components, based on mileage, years of use, or expected wear.

Preventive maintenance also includes the periodic inspection on either a daily, a weekly, or a monthly basis to disclose equipment defects, followed by immediate correction. Daily inspection should be made by the driver before taking the vehicle out of the garage. The usual practice is to ask the driver to complete a printed checklist and submit the slip to the dispatcher before leaving. The checklist should ordinarily cover the items shown in Table 24-6. When this daily inspection discloses any deficiencies, they should be repaired before the vehicle is allowed to leave. Upon returning to the garage, the driver should log in any defects in the vehicle noted during the day.

Equipment Requirements

The federal government Department of Transportation has developed extensive safety requirements for motor vehicles. These set the minimum standards that vehicles must meet. In addition, safety task analysis of a particular maintenance job is shown in Table 24.7.

Cycles

Cycles are popular both for pleasure and transportation. They, however, contribute substantially to motor vehicle deaths, as shown in Table 24.8. The major types of cycles are motorcycles, mopeds (motorbikes), and bicycles. Motorcycles comprise 3.5% of the motor vehicles on the road but account for 8.5% of the fatalities. The majority of vehicle fatalities are caused by head injuries. However, cyclists are attempting to use less protection by not wearing helmets.

The number of pedal cycles (mostly bicycles) has increased by about 50% in the past 25 years—from 42.3 million in 1968 to an estimated 62 to 65 million in 1981. Currently, there is more than 1 bicycle (100 million) for every 2 registered vehicles (164.9 million). Bicycles were associated with more than 500,000 hospital-treated injuries in 1980. Estimates of the aver-

TABLE 24.4
Errors That Lead to Accidents

1. Before starting
   a. Failure to first check clearances, front, rear, overhead
   b. Failure to signal when pulling out from curb
   c. Failure to check for break in traffic before moving
2. Speed control
   a. Too fast for volume of traffic
   b. Too fast for condition of road surface
   c. Too fast for condition of visibility (due to weather or road)
   d. Too fast for condition of light (dusk/darkness)
   e. Too fast for neighborhood or roadside environment
   f. Too fast for street/highway layout and traffic signals
   g. Too slow for speed of traffic stream
3. Lane usage
   a. Failure to select proper lane
   b. Failure to drive in middle of lane
   c. Abrupt lane change
   d. Failure to signal intent to change lanes
   e. Weaving
4. Passing
   a. Misjudging speed and nearness of oncoming vehicle
   b. Failure to check to rear before pulling out to pass
   c. Overtaking and passing too slowly
   d. Cutting in too quickly after passing
   e. Failure to signal intention of passing
   f. Unnecessary passing
   g. Racing other vehicle trying to pass you
   h. Passing in an intersection
5. Turning
   a. Turning from wrong lane
   b. Failure to let oncoming traffic clear before turning left
   c. Failure to block area to right of vehicle on right turns
   d. Over-running curb on right turns
   e. Abrupt turn on slippery road surface leading to skid
   f. Failure to signal intention to turn
6. Stopping
   a. Failure to make smooth gradual stop
   b. Failure to signal stop
   c. Failure to stop in time
   d. Abrupt breaking on slippery road surface leading to skid
7. Parking
   a. Parking in unsafe or illegal place
   b. Parking with front or rear of vehicle protruding into traffic
   c. Failure to secure unattended vehicle on hill
   d. Failure to properly mark disabled vehicle

TABLE 24.4. *(Continued)*

8. Specific signaling errors
    a. Failure to signal
    b. Signal too late
    c. Wrong signal
    d. Failure to use horn
    e. Excessive or improper use of horn
9. Errors in clearance judgment
    a. Following vehicle ahead too closely
    b. Failure to check clearance to rear w
    c. Failure to check right side clearance
    d. Failure to check left side clearance
    e. Failure to check top clearance
    f. Failure to yield space in any traffic
10. Errors in observation
    a. Failure to observe object or pedestrian in path of vehicle
    b. Failure to observe traffic at rear of vehicle while moving
    c. Failure to observe to left and right of vehicle at locations from which vehicles or pedestrians could enter path of vehicle
    d. Inadequate observation, failure to see vehicle or pedestrian approaching
    e. Observation too late
    f. Failure to anticipate parked vehicles pulling out
11. Lack of personal control
    a. Inattention—any cause
    b. Distraction—any cause
    c. Driving while drowsy
    d. Reacting emotionally to driving situations
    e. Driving under influence of alcohol or drugs
    f. Driving while ill
12. Lack of knowledge and awareness of equipment, load, route
    a. Failure to inspect equipment (before, during, after trip)
    b. Being unfamiliar with equipment
    c. Being unfamiliar with load
    d. Being unfamiliar with route
    e. Failure to secure doors or cargo

SOURCE: National Safety Council, *Motor Fleet Safety Manual*. Chicago, IL, 1972, with permission.

age number of cycling injuries requiring medical attention or causing one or more days of restricted activity have ranged as high as 1 million annually. Collisions with motor vehicles were responsible for more than 90% of all fatalities among cyclists in 1978, underscoring the increasing risk to the cyclist in today's heavy traffic.[2,4]

A study assessing bicycle injuries reported that the most common soft-tissue injuries involved the head and face and the most frequent skeletal

peed control

eeding speed limit and passing motorists who are abiding by the law

peeding through towns and residential areas

Driving too slow so as to back up traffic

Errors in following

  a. Following passenger cars too closely

  b. Tailgating by two or more trucks

  c. Following too close at night without dimming lights

3. Errors in passing

  a. Passing when unnecessary

  b. Cutting in too sharply after passing

4. Errors in lane use

  a. Crowding center line

  b. Straddling two lanes

  c. Drifting across lane dividers

  d. Weaving from one lane to another

  e. Running two or three abreast on limited-access highways

  f. Use of wrong lane for speed or traffic flow

5. Errors that block traffic

  a. Blocking crosswalk

  b. Double parking in heavy traffic

  c. Not pulling off road to allow buildup traffic to pass

6. Errors in noise and smoke abatement

  a. Excessive diesel smoke

  b. Excessive engine noise

  c. Unnecessary use of air horns

  d. Racing engines to hurry pedestrians across intersection

  e. Splashing pedestrians in sloppy weather

SOURCE: National Safety Council, *Motor Fleet Safety Manual*, Chicago, IL, 1972, with permission.

injuries involved the wrist and hand.[3] Other significant factors were mechanical or structural defects in the bicycles, collisions with motor vehicles, riding on a surface covered by loose sand and gravel, and errors in turning or going downhill. The report suggests that establishment by the bicycle industry of minimum standards in bicycle design, strength, and performance would prevent many accidents.

The increase in bicycle and motor vehicle traffic accentuates the necessity for developing more bikepaths and bikeways to separate cycles from motor vehicles. More and more employees are using pedal cycles to get to and from work, which increases the frequency of cycle safety issues.

TABLE 24.6
Checklist for Preventive Maintenance

Acceptable	Defective	
		Exterior
—	—	Headlights
—	—	Tail lights
—	—	Parking lights
—	—	Tires
—	—	Wheel lugs
—	—	Spare tire
—	—	Oil, grease, or water on ground
		Under Hood
—	—	Oil level
—	—	Power steering fluid level
—	—	Antifreeze level
—	—	Fan belts
—	—	Air conditioner belts
—	—	Radiator hoses
—	—	Power steering hoses
—	—	Heater hoses
—	—	Pollution control system
—	—	Battery level
—	—	Battery cable
—	—	Spark plug wires

AVIATION

General avaiation refers to operation of civil aircraft owned and operated by persons or corporations other than those engaged in air carrier operations.

Air Carriers

The second type of aviation in the United States is that of air carriers. U.S. air carriers comprise three groups:

1. Certified route air carriers (for example, United Airlines or American Airlines)
2. Supplemental (charter) air carriers
3. Commercial operators of large aircraft (overnight package delivery, like Flying Tigers)

TABLE 24.7
Example of a Safe Transportation Operating Procedure

Safe operating procedure. The employer shall establish a safe operating procedure for servicing multipiece wheels and assure that employees are instructed in and follow that procedure. The procedure shall include, but not be limited to, the following elements:

1. Tires shall be completely deflated before being demounted.
2. Tires that have been driven underinflated shall be deflated to 10 pounds per square inch gauge (psig) or less before being removed from the vehicle axle.
3. Rubber lubricant shall be applied to the head and rim mating surfaces for assembly and inflation of the wheel.
4. When a tire is partially inflated without a restraining device for the sole purpose of seating the lock rig, air pressure shall not exceed 10 psig.
5. Whenever a tire is in a restraining device during inflation, the employee shall not rest or lean any part of his or her body or equipment on or against the restraining device.
6. After tire inflation, the tire, rim, and ring(s) shall be inspected while still within the restraining device to make sure that they are properly seated and locked. If further adjustment to the tire, rim, or ring(s) is necessary, the tire shall be deflated before making the adjustments.
7. No attempt shall be made to correct the seating of side and lock ring(s) by hammering or forcing the components while the tire is pressurized.
8. Whenever high-temperature heat, for example, from a welding torch, is applied to any part of the rim base, rings, or lugs, either on or off the vehicle, the tire shall first be completely deflated.
9. Inflated wheels being moored or stored in a service area shall be positioned such that the trajectory path of a lock ring will not pass through an area normally occupied by employees.
10. Tires shall be deflated prior to being removed from a vehicle axle, and no employees are to be in the trajectory path during such deflation.

SOURCE: Federal Register, *Servicing Multipiece Rim Wheels; Proposed Rulemaking.* Department of Labor, Occupational Safety and Health Administration, Washington, D.C., 1979.

TABLE 24.8
Cycle Deaths

Year	Motorcycles and Bicycles	Motorbikes
1976	3,000	900
1977	3,870	1,100
1978	4,530	1,200
1979	4,080	1,200
1980	4,480	1,200

SOURCE: Metropolitan Life Insurance Company. Cycling accidents in the United States. *Statistical Bulletin of the Metropolitan Life Insurance Company*, 1981, **62**, 4–9. (Oct.–Dec.) With permission.

510

TABLE 24.9
Accidents and Rates by Aircraft Make and Model: U.S.
Certificated Route Air Carriers, 1970–1979

	Accidents		Accident Rate per 100,000 Aircraft Hours	
Make/Model	Total	Fatal	Total	Fatal
B-747	30	2	0.921	0.061
B-707	45	2	0.506	0.081
B-720	6	1	0.527	0.088
B-727	74	8	0.366	0.040
B-737	14	1	0.463	0.033
DC-8	34	5	0.627	0.092
DC-9	36	9	0.395	0.088
DC-10	14	4	0.600	0.172
L-1011	13	2	0.980	0.151
L-180	5	2	3.401	1.361
CV-580, 600, 640	27	3	1.349	0.150
CV-880	4	1	0.952	0.238
BAC 1-11	6	0	0.709	0.000
DHC-6	4	2	1.587	0.794
F-27, FH-227	8	4	0.940	0.470
YS-11	4	0	0.71	0.00
Total	324	53	0.543	0.084

SOURCE: National Transportation Safety Board. *Annual Review of Aircraft Accident Data*. U.S. Carrier Operations 1979. U.S. Department of Transportation, Washington, D.C., 1981.

The air carriers had 31 accidents in 1979. The airlines had 26 accidents, 5 of which were fatal accidents which resulted in 352 fatalities. Of these, 273 occurred in the American Airlines DC-10 accident in Chicago where the airplane was in the initial climb on takeoff and had an engine fall off. The supplemental air carriers had one accident which resulted in three fatal injuries to the crew. The commercial operators of large aircraft had four nonfatal accidents in 1979.

Accident Rates by Make and Model.   Accidents and accident rates by aircraft make and model used in all operations of the certificated route carriers are shown in Table 24.9. These aircraft accounted for 91% of the accidents and 96% of the total hours flown in the ten-year period.

Types of Accidents.   Types of accidents in the 10-year period 1970 through 1979 are listed in Table 24.10 in descending order of frequency.

TABLE 24.10
Total and Fatal Accidents, 1970–1979

Type of Accident	All Accidents		Fatal Accidents	
	Frequency	Percent	Frequency	Percent
Turbulance	108	27.8	1	1.8
Miscellaneous/other	39	10.8	6	10.5
Collided with wires, trees, towers	30	8.3	8	16.0
Engine failure/ malfunction	24	6.9	3	5.3
Collision with ground/ water	22	6.1	16	28.1
Gear collapse	16	4.4	1	1.8
Fire or explosion on the ground	16	4.4		
Collision between aircraft—on ground	12	3.3	2	3.5
Ground loop	10	2.8		
Overshoot	9	2.5	2	3.5
Hard landing	8	2.2	1	1.8
Fire or explosion, in flight	7	1.9	2	3.5
Airframe failure on ground	5	1.4		
Midair collision	5	1.4	4	7.0
Stall/spin/spiral	5	1.4	3	5.3
Evasive maneuver	4	1.1		
Nose over/down	4	1.1		
Propeller/rotor failure	5	1.4	3	5.3
Prop/rotor accident to person	3	0.8	2	3.5
Uncontrolled altitude deviation	3	0.8		
Wheels-up landing	3	0.8		
Gear retracted	3	0.8		

SOURCE: National Transportation Safety Board. *Annual Review of Accident Data. U.S. Carrier Operations 1979.* U.S. Department of Transportation, Washington, D.C., 1981.

The first eight accident types—turbulence; miscellaneous/other; collided with wires, trees, towers; collision with ground/water; engine failure or malfunction; gear collapsed, fire or explosion on the ground; and collision between aircraft (both on ground)—accounted for 73.8% of all the accidents.

Similarly, for fatal accidents, the accident types—collision with ground/ water; collided with wires, trees, towers; midair collision; stall/spin/spiral;

TABLE 24.11
Phase of Operation, 1970–1979: Total Accidents and Fatalities

	All Accidents		Fatal Accidents	
Operation Phase	Number	Percent of Total	Number	Percent of Total
Static	35	9.6	4	7.0
Taxi	36	9.9	4	7.0
Takeoff	41	11.3	8	14.0
In flight	152	41.9	16	28.1
Landing	98	27.0	24	42.1
Unknown/not reported	1	0.3	1	1.8
Total	363	—	57	

SOURCE: National Transportation Safety Board. *Annual Review of Aircraft Accident Data. U.S. Air Carrier Operations 1979.* U.S. Department of Transportation, Washington, D.C., 1981.

engine malfunction; propeller/rotor failure; and miscellaneous/other—accounted for 66.7% of all the fatal accidents.

Phases of Operation. The percentage distributions of total and fatal accidents by phase of operation are shown in Table 24.11. For total accidents, most of the accidents occurred during the in-flight phase of operation, while in fatal accidents most of the accidents occurred during landing.

Cause/Factors. The percentage distributions of causes or related factors for total and fatal accidents are shown in Table 24.12. For the 10-year period 1970 through 1979, "personnel" (maintenance, weather, and traffic control personnel) was the most frequently cited cause/factor in U.S. certificated route air carrier accidents, followed by "weather" and the "pilot." In fatal accidents, the pilot, personnel, and weather were the most frequently cited causes/factors.

Civil Aviation

The continuing high rate of serious accidents in civil aviation requires analysis if it is to be contained and/or reduced. Table 24.13 shows an overview of general aviation accidents in 1979.

Types of Accidents. Engine failure or malfunction is cited as the first accident type in 24% of general aviation accidents occurring in 1979. Because of the nature of this accident classification, it is usually followed by a second type of accident where actual damage or injury occurred. (See Table

TABLE 24.12
Causes/Factors in Aircraft Accidents for Certificated
Route Air Carriers, 1970 through 1979, in Percent[a]

Causes/Factors	Accidents	
	Total	Fatal
Personnel	50.0	48.1
Weather	46.5	46.2
Pilot	38.6	61.5
Landing gear	10.2	7.7
Powerplant	9.1	5.8
Airport/airways/facilities	8.5	5.8
Systems	6.7	7.7
Miscellaneous	6.4	7.7
Airframe	2.9	5.8
Instrument/equipment	2.9	3.9
Terrain	1.8	0.0
Undetermined	1.0	3.9
Rotorcraft	0.3	1.9

SOURCE: National Transportation Safety Board. *Annual Review of Aircraft Data. U.S. General Aviation Calendar Year 1979*. U.S. Department of Transportation, Washington, D.C., 1981.

[a] The percentage totals exceed 100 percent because multiple causes/factors can be cited in any accident.

24.14.) The primary second types of accidents associated with the 987 engine failure/malfunction accidents in order of frequency are nose over/down; collision with trees; collision with ground/water—uncontrolled; hard landing; and gear collapsed. Although engine failure/malfunction accidents account for 24% of total general aviation accidents, this accident type accounts for only 80, or 11.56%, of the 678 fatal general aviation accidents.

Ground loop was cited as a first type of accident 427 times (10.51%) and 361 times in conjunction with a second type of accident. The principal second types in order of frequency of occurrence are nose over/down; gear collapsed; collision with snowbank; collision with ditches; and collision with trees. Ground loop accidents accounted for two fatal accidents in 1979 and were cited in only five accidents where serious injuries but no deaths occurred.

Hard landing is the third-highest first accident type with 253 accidents in 1979. The most common second accident types following hard landing are gear collapsed and nose over/down.

TABLE 24.13
Overview of U.S. General Aviation, 1978–1979

Description	1978	1979	Percentage Change
Aircraft-hours flown	29,402,269	43,340,001	+ 9.97
Eligible aircraft	198,778	210,329	+ 5.82
Total accidents	4,494	4,023	− 19.46
Aircraft involved	4,557	4,063	− 10.84
Aircraft damage[a]			
Destroyed	1,227	1,055	− 14.02
Substantial	3,284	2,956	− 9.99
Minor	29	20	0
None	17	23	+ 35.29
Injury Index			
Fatal	793	678	− 14.50
Serious	458	395	+ 13.76
Minor	662	603	− 8.91
None	2,581	2,347	− 9.07
Injuries[b]			
Fatal	1,770[c]	1,367	− 22.77
Serious	858	700	− 18.41
Minor	1,317	1,077	− 18.22
None	5,599	4,901	− 12.47

SOURCE: National Transportation Safety Board. *Annual Review of Aircraft Data. U.S. General Aviation Calendar Year 1979.* U.S. Department of Transportation, ashington, D.C., 1981.

[a] Includes all aircraft involved in collisions.
[b] Includes persons aboard any aircraft involved in a collision with general aviation aircraft and persons injured on the ground.
[c] Includes a midair collision between a Pacific Southwest Airline 727 and a Cessna 172 over San Diego on September 25, 1978, in which 144 persons died.

Collision with ground/water controlled was cited as a first accident type 244 times, and collision with trees 203 times. The number of second type of accident citations following these accidents is insignificant.

The five leading first accident types—engine failure/malfunction; ground/water loop—swerve; hard landing; collision with ground/water—controlled; and collision with trees—account for 52.05% of all accident type citations and 41.75% of the fatal accidents. However, the five leading accident types—collision with ground/water—uncontrolled; engine failure/

TABLE 24.14
Types of Accidents: 10 Most Frequent Types of Accidents, 1979

Type of Accident	Accident Frequency			
	Total	Percent	Fatal	Percent
Engine failure or malfunction	987	24.4	80	11.6
Ground/water loop— swerve	427	10.5	2	
Hard landing	233	6.2		
Collision with ground/ water—controlled	244	6.0	122	17.6
Collided with trees	203	5.0	77	11.1
Overshoot	158	3.9		
Stall/spin and stall	154	3.8	102	14.7
Collided with wires/ poles	149	3.7	30	4.34
Nose over/down	137	3.4		
Collision with ground/ water—uncontrolled	130	3.2	112	16.2

SOURCE: National Transportation Safety Board. *Annual Review of Aircraft Data. U.S. General Aviation Calender Year 1979.* U.S. Department of Transportation, Washington, D.C., 1981.

malfunction; collision with trees; and stall/spin—account for 64.03% of the fatal accidents but less than half of the total accidents.

Three first-type accidents continue to appear in the top-five first accident types for both total and fatal accidents for a single first accident type. These types include controlled collision with ground/water involving 122 of 678 fatal accidents, or 18%; uncontrolled collision with ground/water followed closely, accounting for 16% of fatal accidents; and engine failure malfunction accounted for 11% of the citations.

Phases of Operation. The landing phase of operation continues to dominate the numbers of total accidents recorded by phase of operation, while the majority of fatal accidents continue to happen during the in-flight phase of operation. There were 1,643 total accidents that occurred during the landing phase, a decline from 1,880 accidents during this same phase in 1978. The in-flight phase of operation had the second-highest number of total accidents, 1,402, which represents a decline from 1,528 accidents in 1978 but shows a slight increase in percentages over 1978. Engine failure/ malfunction led all other accidents in this phase of operation. Accidents occurring during takeoff numbered 807, a decrease from the 1978 figure of 889.

With reference to fatal accidents by phase, the in-flight phase has the

TABLE 24.15
Total and Fatal Accidents by Phase of Operations

Phase of Operation	Total Accidents		Fatal Accidents	
	Frequency	Percent	Frequency	Percent
Landing	1,643	40.5	138	20.0
In flight	1,402	35.4	430	62.2
Takeoff	807	19.9	86	12.7
Taxi	131	3.2	1	0.1
Static	37	0.9	6	0.9
Unknown	41	1.0	29	4.2

SOURCE: National Transportation Safety Board. *Annual Review of Aircraft Data. U.S. General Aviation Calendar Year 1979.* U.S. Department of Transportation, Washington, D.C., 1981.

highest number with 430 accidents, 62.1% of all fatal accidents. Fatal landing and takeoff accidents numbered 138 and 88, respectively. Six fatal accidents were recorded in the static phase of operation, and one fatal accident was recorded in the taxi phase of operation. (See Table 24.15.)

Accident Investigation. Accident investigation of any plane crash is undertaken by the National Transportation Safety Board (NTSB) employing FAA regulations. When the NTSB studied the Continental Airlines/Air Micronesia Boeing 727 which landed short of the runway at Yap Airport in the Western Caroline Islands on November 21, 1980, it reported on all aspects of possible aircraft failure and possible pilot error.[6] It studied the survival aspects, including the restraint system and the evacuation procedures. It also studied the report of a "very severe landing." In addition, it conducted a landing gear failure analysis. It assessed the descent rate, landing procedures, emergency evacuation training, and captain's training. Then an analysis of the accident was performed. From this information, specific areas were further assessed. These included the approach method, the thrust reduction, the transition training, and the proper glide path. The conclusions from this in-depth analysis are presented in Table 24.16.

The probable cause of this accident was the captain's premature reduction of thrust in combination with flying a shallow approach slope. This led to an improper touchdown aim point. These actions resulted in a high rate of descent and a touchdown on upward sloping terrain short of the runway threshold. This exceeded the design strength and caused the right landing gear to fail. Contributing to the accident were the captain's lack of recent experience in flying B-727s and a transfer of his DC-10 landing habits and techniques to the operation of the B-727 aircraft. This is a negative example of transfer of training, which was previously discussed with

TABLE 23.16
Findings of NTSB in Crash of B-727 in November 1980

The flight crew was properly certificated and qualified to conduct the flight.

The aircraft was properly certificated and maintained in accordance with prescribed procedures.

The aircraft touched down on the right main landing gear 13 ft short of the approach end of the landing runway.

The right main landing gear separated at initial ground contact.

The area of initial touchdown of the right main landing gear tires sloped upward about 4.07°.

The combined forces of the excessive sink rate and an upsloping touchdown point exceeded the design strength of the right main landing gear.

The captain flew a flat, dragged-in final approach with about a 1.5° glide slope which required excess thrust.

The first and second officers and the mechanic in the cockpit jumpseat were concerned about the approach being low.

The captain reduced the throttles to idle at 50 ft above the runway elevation and short of the runway threshold.

The landing was the first unsupervised landing at Yap for the captain.

The captain had been flying DC-10s for about $3\frac{1}{2}$ years prior to November 1980.

The captain had not landed a B-727 for 61 days before the date of the accident.   He made one landing, at Saipan, on the day of the accident.

Fire erupted around the damaged right wing area as the aircraft came to a stop.

The crash forces in the accident were not sufficient to cause serious impact injuries to the occupants.

The evacuation was completed in about 55 seconds.

The flight attendants were not aware of how to open the aft airstair exit door using the emergency system.

Immediately following the accident investigation, the airline implemented new training techniques to include "hands-on" training on the aft airstair exit emergency opening system.

SOURCE: National Transportation Safety Board Aircraft accident report: Continental Airlines/ Air Micronesia, Inc. Boeing 727-92C, N18479 YAP Airport, YAP, Western Caroline Islands, November 21, 1980. NTSB-AAR-81-7. Washington DC: National Transportation Safety Board, 1981.

safety training. This is also an excellent example of why cockpits of commercial aircraft should be uniform.

The transportation industry is very important in contemporary United States; however, safety on our roadways, railroads, and airways must be maximized.

SUGGESTED LEARNING EXPERIENCES

1.   Discuss your rights as a user of highways, your responsibilities as a user, and your obligations as a safety professional.

2. If you were the direction of the National Transportation Safety Board, what would be your priorities? How would you go about developing them?

3. If you were the director of a national corporation, what training programs for highway safety would you institute? Discuss your reasons.

4. If you were the safety director at an airport, how would you envision your position and role? Discuss.

5. What suggestions do you have regarding human error and the resultant transportation accidents? Develop a framework utilizing your suggestions.

6. You receive a report stating that alcohol-related accidents are on the increase. As a chairperson of a task force studying this problem, develop and present a comprehensive program to reduce such accidents.

7. Debate the amount of regulation there should be for light aircraft. Be sure to include the ethical issues regarding air suicide.

8. With the computer age enmeshing us, what are your ideas for why the majority of the transportation industry remains so uncontrollable for safety.

9. As a safety professional, develop a strategy for prevention of bicycling accidents.

10. What role could a safety professional have on an airline's decision-making board? Discuss.

REFERENCES

1. Federal Register. *Servicing Multipiece Rim Wheels; Proposed Rulemaking*. Department of Labor, Occupational Safety and Health Administration, ashington, D.C., 1979.

2. Flora, J. S., and Abbott, R. D. National trends in bicycle accidents. *Journal of Safety Research*, 1979, **11**, 22–27.

3. Kramer, M. S. *Safety Supervision in Motor Vehicle Fleets*. New York: Association of Casualty and Surety Companies, 1947.

4. National Safety Council. *Accident Facts*. 1981.

5. National Transportation Safety Board. *Annual Review of Aircraft Accident Data. U.S. Air Carrier Operations 1979*. U.S. Department of Transportation, Washington, D.C., 1981.

6. National Transportation Safety Board. NTSB issues findings in crash of 727. *Aviation Week and Space Technology*, April 12, 1982, 117–124.

25 PETROCHEMICAL INDUSTRY

This chapter is concerned with petrochemicals used in the production of petroleum products which include gasoline, kerosene, distillate fuel oils, residual fuel oils, lubricating oils, asphalt, coke, and pitch. These are produced from crude oil and fractionation products of crude oil by distillation, redistribution, cracking, and other processes.

The raw material for the petroleum refining industry is crude oil, a substance which is an extremely variable mixture of thousands of hydrocarbon compounds—alkanes, cycloalkanes, aromatics—containing from 1 to 40 or more carbon atoms, along with small quantities of sulfur compounds—including hydrogen sulfide (H_2S), vanadium, arsenic, mercury, carbon dioxide (CO_2), and nitrogen—which are collectively referred to as impurities. Oil from Mexico and Texas usually has a high-sulfur content. High-sulfur oil produces sulfur dioxide (SO_2) when it burns. Sulfur dioxide contributes substantially to air pollution and when mixed with water becomes sulfuric acid (H_2SO_4), a component of "acid rain." The Pennsylvania crude oils are generally rich in long-chain alkanes with excellent lubricating properties. You may hear a famous golf professional advertising the beneficial lubricating properties of Pennsylvania oil. Crude oil composition varies according to the geological strata of its origin. The storage of crude oil is one refinery operation. Crude oils are processed into intermediates that are further refined, blended, purified, and packaged for sale to chemical plants, consumers of motor fuels and heating oils, and construction industries, among others. Refinery operations leading to the intermediates and final products include the following: fractionation, cracking, molecular rearrangement, extraction procedures, and product finishing.

Fractionation is a method of separating various components of crude oil by distillation. The crude oil is separated into fractions with specific boiling

521

temperatures. For example, heating fuel fractions distill from approximately 200–350°C and may contain a mixture of aliphatic and aromatic hydrocarbons with 10–23 carbon atoms (for example, hexadecane, heptadecane, naphthalene, and polycyclic aromatic hydrocarbons including benzopyrene). Increased boiling temperature is associated with decreasing volatility and capability of being vaporized and increasing molecular weight. Each cut contains many aromatic paraffins and cycloparaffins. Vacuum distillation of residual oils produces lubricating oils and heavy fuel oil components for blending.

Cracking (thermal decomposition of a petroleum fraction) is used to increase the yield of isooctanes [highly flammable liquids—$(CH_3)_2CH(CH_2)_4CH_3$] by the breakdown and reforming of heavier hydrocarbons in the heating fuel and residual oil cuts. The material used for cracking depends upon the demand for gasoline at the expense of other petroleum products. There are several other means of changing the petroleum cut:

1. Molecular rearrangement (polymerization, alkylation, reforming, and isomerization reactions). This is utilized to enrich gasoline for "antiknock" properties.
2. Extraction procedures are employed for separating aromatic from paraffinic hydrocarbons in a cut.
3. Solvent dewaxing removes long-chain alkanes from jet fuel and kerosene.
4. Solvent refining is employed in removing benzopyrenes and other polycyclic aromatic hydrocarbons from lubricating oils to yield white mineral oils used industrially and pharmaceutically.

The final process in the petrochemical operation is called "product finishing." It includes desulfurization, sweetening of gasolines, finishing lubricating oils, blending various intermediate stocks, and packaging. Desulfurization is the removal of objectional impurities which might be malodorous or corrosive. Sweetening of gasoline is the blending of gasoline to meet fuel octane requirements. Blending of intermediate stocks is the blending of diesel fuel cuts with cracked stocks to improve hexadecane content ("Cetane number" is a rating of diesel fuel that is expressed as the percentage of cetane that must be mixed with liquid methylnaphthalene to produce the same ignition performance as the diesel fuel being tested. Compare with octane number.) Packaging is the addition of various substances (additives) to impart color, decrease corrosiveness, or improve the function of the petroleum product.

SIZE OF THE PROBLEM

Estimates of the number of exposed persons in the petrochemical industry are uncertain. The maximum number of workers in the entire petroleum refining industry was reported as 157,000, of which 97,000 were classed as "production" workers. These figures are difficult to compare to the total of 100,179 reported to the National Safety Council.[7] According to the National Safety Council, petroleum refinery workers had a total recordable case rate of 6.20 per 100 full-time employees. This is much lower than the overall rate for all industries of 7.94 cases. It is higher than pipelines, except natural gas (SIC 46) and natural gas transmission (SIC 4922) which were 4.87 cases and 5.38 cases, respectively. The petroleum industry's total of lost workday cases was 2.72. This incidence rate was still below the overall industrial rate of 3.54.[7]

Occupational Illnesses

Most studies to date have examined the risks of developing excessive cancers, noting that the carcinogen benzopyrene is present in crude petroleum, residues, soots, and air pollution around refineries. One study explored cancer mortality, over a 29-year period, in employees of a Texas oil refinery. There were 377 cancer deaths reported among 15,437 employees. The mortality rates were compared with those of the general adult male population of the same geographic area and were no higher than expected. There was no greater incidence of lung cancer deaths in refinery employees than among employees in production, transportation, sales, or exploration for the same company. It was concluded that the air pollutants in refinery operations were not a lung cancer hazard to employees, based on cancer mortality records.[9]

A more recent study disagreed. A national survey compared cancer mortality rates of approximately 50,000 white males in 39 U.S. counties in which the petroleum industry is heavily concentrated with a demographically matched control population of white males not heavily exposed to oil refinery pollution, over the years 1950–1969. For a total of 23 cancer sites, mortality was significantly higher ($p \leq 0.01$) among petroleum refinery workers and nearby residents than among controls. The highest mortality rates were for cancers of the nasal cavity, nasal sinuses, and lung. The cause of the excessive cancer mortality was not determined, but polycyclic aromatic hydrocarbons involved in the manufacture of petroleum, and present in the air around refineries, were implicated.[6]

Cancer of the pancreas has been associated with occupations involving exposures to petroleum and petrochemicals.[8] Indications suggest that the petrochemical industry poses a significant cancer risk to workers.

OCCUPATIONAL HAZARDS

Occupational hazards in the petrochemical industry include heat, electricity, welding, confined areas, scaffolds, cables and chains, hand tools, and moving vehicles. The emphasis in this section is on some of the hazards and safety design possibilities for offshore oil drilling platforms.

The same precautions that apply in manufacturing hold true for the occupational hazards listed above, with added precautions when handling huge loads, working with and handling flammable substances, and, in some cases, working underwater.

In confined areas, examination of tanks and vessels may expose the worker to toxic vapors and fumes or cause asphyxiation from insufficient oxygen levels. When using hand tools, chains, cables, and scaffolds, the worker must be properly protected. Asbestos, which may be used to line pipes, has been found to be toxic and carcinogenic.

Fires and Explosions

There are three types of fires and explosions on offshore platforms: small fires, large fires, and blowouts. The probability is that small fires are the least dangerous, blowouts the most dangerous. All three should be reduced to a minimum. The fire should be prevented from spreading. Alarms should be sounded, people evacuated, if necessary, and the fire extinguished.

Protection.   The three types of protection are complete, partial, or delayed effect. Every protective element is designed to give partial protection, but it is the sum of these protective elements that decides whether the protection is adequate. In the case of a fire, the protection may, depending upon where the fire occurs, be designed to delay the effect of the fire. This delay gives the platform crew the necessary time to organize fire fighting, evacuate the platform, and get external help to put out the fire. This delay is a type of partial protection, being selective either in effects, in magnitude, in structure, or in activities to protect.

Blowouts.   Blowouts cause great concern. Adequate figures for the frequency of blowouts are available only for the Gulf of Mexico. There was 1 blowout per 200 to 400 drilling wells. A systematic study of factors contributing to the risk of blowouts showed that it is possible to reduce the probability of a blowout in the drilling phase to less than one-tenth its present rate.

The frequency of blowouts in the North Sea was estimated at 1 per 5,000 drilled wells during drilling and 1 blowout per 50,000 well-years during production. The present experience from the North Sea indicates that the risk for a blowout is less than that shown by the figures for the Gulf of Mexico.[2]

OFFSHORE PLATFORMS

Offshore platforms consist basically of two components: (1) the drilling and operating facilities, which constitute the topside, and (2) the supporting structure and its foundations. The topside consists of the drilling rigs, oil- and gas-processing equipment, transportation pumps, and utilities and living quarters for as many as hundreds of workers.[5] (See Figure 25.1.) All the major platforms now have helicopter landing pads. After the oil is processed, it is either pumped directly ashore through a submarine pipeline or stored until it can be transferred to tankers. In the latter case, sufficient storage is usually provided to allow for intervals between tankers so that there is no need to curtail the flow from the wells.[2]

Topside facilities for the platform are huge. They weight from 16,000 to 40,000 tons and have a total floor space of from 75,000 to 400,000 square feet. These massive facilities must be positioned above the crest of the highest wave expected to come once in 100 years. For the North Sea, this calls for positioning the bottom of the lowest deck at least 80 feet above mean water level.[5]

The second component of an offshore platform, and the one that has elicited the most ingenuity, is the supporting structure. Such structures must secure the topside facilities against the assault of winds, waves, currents, and, in some instances, seismic disturbances. The main environmental forces that need to be taken into account are, in almost every case, those that result from wave impact. (Notable exceptions are the types of structures that must be built along the West Coast of the United States, where seismic forces are the major concern.) In establishing the design, two wave conditions are critical. The first is the maximum wave that might come as a single event during the lifetime of the project. This predicted wave determines the maximum strength of the structure. The second condition is the cumulative effect of several million waves per year whose period happens to match the fundamental oscillation frequency of the platform itself. Although such resonant forces are individually small, they are amplified dynamically by the structure and, therefore, determine the fragile lifetime that must be designed into the platform.[5]

Finally, the supporting structure must have a secure foundation. In soft seabeds it may be necessary to drive piles more than 500 feet into the seafloor before adequate support is achieved. The overall length of the pile, including the pile follower between the pile and the pile driver above water, can therefore exceed 1,000 feet and call for handling hundreds of tons of material in one continuous assembly.[5]

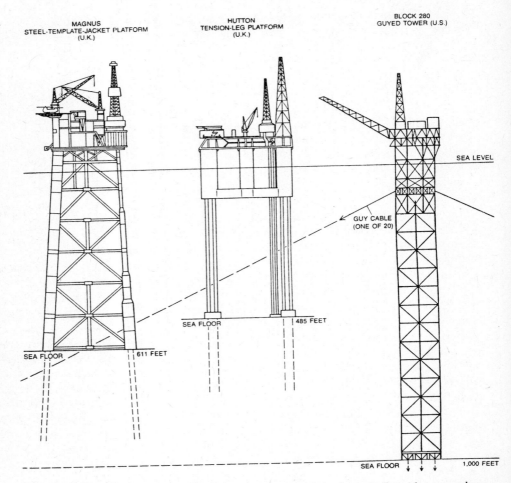

FIGURE 25.1. Four offshore oil platforms are compared schematically with two onshore structures of comparable height: the Eiffel Tower, completed in 1880, and one of the towers of the World Trade Center in New York. From the viewpoint of environmental conditions, the comparison is highly inadequate. Onshore buildings are subjected to lateral forces no greater than heavy winds. The offshore towers must take the impact of waves as much as 100 feet high. Three of the offshore platforms are designed for neighboring oil fields in the North Sea. The Stratfjord B platform rests on four massive concrete columns with storage tanks at the base. Since it depends solely on its own mass for stability, it is described as a gravity-base platform. The base of the Magnus platform, known as a steel-template jacket, is the heaviest structure of its type yet fabricated: It will weigh 41,000 tons. The Hutton structure, known as a tension-leg platform, will consist of a buoyant hull tethered to the sea by slender steel tubes at its four corners. The fourth of the platforms is the Block 280. It is designed for service in the Gulf of Mexico in water 1,000 feet deep (*from Ellers, 1982*).

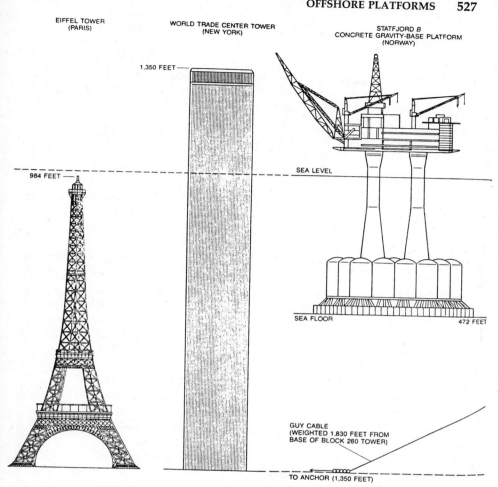

FIGURE 25.1. (*Continued*)

These offshore platforms pose innumerable safety hazards that must be controlled. This is still a relatively new technology, and much more information will be forthcoming regarding safety to keep pace with advancing technology.

Toxic and Hazardous Chemicals

Three types of toxic chemicals used in oil refining will be discussed: carcinogens, inorganic substances, and organic substances.

Antimony. Antimony (Sb) is a silvery-white, soft metal. It is often used as an alloy of copper and tin. Antimony is an irritant, especially in the eyes and mucous membranes. It has a slow excretion rate from the body; hence it accumulates, and chronic poisoning can result. Also, inhaling it can lead to myocardial degeneration and dysfunction. Respiratory tract, lung, and gastrointestinal disorders have also been reported.

Hydrogen Sulfide. Hydrogen Sulfide (H_2S) is a gas which is a by-product of many chemical reactions, including petroleum processing and decaying organic matter. Hydrogen sulfide causes respiratory paralysis and death, and in lower concentrations it can cause pneumonia, keratitis, conjunctivitis, nausea, and headache. Hydrogen sulfide has a distinguishing "rotten egg" odor.

Nickel. Nickel (Ni) is a silvery metal which is used in plating to make a highly polished corrosion resistant finish. It is used in steel to make stainless steel and causes lung or nasal cancer in humans. Nickel fumes are a respiratory irritant. Nickel compounds, such as nickel chloride ($NiCl_2$), nickel nitrate [$Ni(NO_2)_2$], and nickel sulfate ($NiSO_4$) are skin irritants which cause "nickel itch." Nickel in its metallic form is a carcinogen. Nickel carbonyl [$Ni(CO)_4$] gas causes acute pneumonitis, liver toxicity, lung cancer, and cancer of the nasal sinuses in exposed workers.[10]

Benzene. Benzene (C_6H_6) an organic substance, is an acute poison which causes narcosis. Chronic exposure leads to bone marrow toxicity and aplastic anemia, which is frequently fatal. Leukemia caused by benzene exposure has also been reported. Exposure of 3,600 workers to benzene at two Monsanto facilities in the St. Louis area has been suspected of causing chromosome damage and leukemia. Other Monsanto employees in Alabama and Texas have also been exposed to benzene. Dow Chemical was charged with withholding research findings on benzene when it testified during OSHA hearings in 1977. At the Dow Freeport plant in Texas, chromosome damage increased with exposure to benzene by as much as 30% of those exposed to more than 25 parts per million.[6]

Cyclohexane. Cyclohexane is another organic substance found in some workplaces. Its vapors may irritate the eyes and the mucous membranes.

Gasoline. Gasoline is an acute irritant to the conjunctiva. It can also cause asphyxiation in high concentrations of vapor. It is a recognized carcinogen, possibly due to its benzene content.

Aromatic Hydrocarbons. Aromatic hydrocarbons include many substances found in the petrochemical industry. Some of these are benzene, alkyl-substituted benzenes, naphthalene, indanes, Tetralins, alkyl-substituted naphthalenes, acenaphthenes, fluorines, and tricyclic compounds such as phenanthracenes and anthracenes. Polycyclic aromatic hydrocarbons including benzopyrenes are among the isolated and identified compounds of the class which occur in crude petroleum and in most petroleum products from gasolie to asphalt and pitch.

Naphthalene. Naphthalene ($C_{10}H_8$) is a hydrocarbon. It is also called naphthalin. It is used as a moth repellant and is also used in chemical manufacturing. Napthalene is a hemolytic agent which destroys the red blood cells.[4] Exposure to it also causes skin and eye irritations, headaches, confusion, nausea, and corneal injury. Ingestion of naphthalene can cause severe hemolytic anemia.[9] Exposure to a high concentration of fumes may cause cataracts.

Tetraethyl Lead. Tetraethyl lead is a gasoline additive used to prevent knocking in internal combustion engines. Tetraethyl lead (Pb $(C_2H_5)_{11}$) is absorbed through the skin and may be ingested. Tetraethyl lead affects the nervous system and may lead to psychological disturbances like psychosis and manistates. The symptoms this form of lead produces are disimilar from organic lead.

Trichloroethylene. Trichloroethylene (Cl_2C—$CHCl$), another organic compound, is an industrial solvent. Trichloroethylene fumes are frequently inhaled. The liquid may penetrate the skin. It is a central nervous system depressant and a respiratory tract irritant. It may cause dizziness, tremors, loss of coordination, confusion, abdominal cramps, and coma.[4]

As can be seen from this brief description of substances used in the petrochemical industry, many workers are exposed to injury or illness. Before control measures can be instituted, a thorough health hazard evaluation is necessary.[3] These techniques will only be defined here. They are descirbed in detail in Chapters 7 to 14.

As new technology has a tendency to produce larger tankers, petrochemical plants, and offshore platforms, the prevention of possible accidents is becoming more and more important. This calls for a systematic method of representing certain events that are or can cause accidents. System safety engineering is becoming critical in the design of these units. Chapter 27 is devoted to a discussion of systems safety and its importance in the elimination and control of hazards.

A hazard analysis is most appropriate when there is some question about the presence of a particular substance in the workplace. Before an extensive study, a preliminary hazard analysis should be conducted to attempt to gather and organize logically the facts that are known; its goal is to decide what variables should be included in the study and what methods should be used to determine whether hazards exist. This approach is valuable in the investigating of new or modified operations to determine if potential hazards are being introduced into the system.

SUGGESTED LEARNING EXPERIENCES

1. Make a poster depicting the process of oil refinery and inherent hazards.

2. Make a poster taking any one chemical used in industry and depicting its uses, hazards, and control measures.

3. If you were a safety engineer on an offshore oil platform, how would you envision your role?

4. If you were a safety engineer in a large petrochemical industry, discuss how you would go about setting priorities.

5. Considering the documented evidence of the petrochemical industry having many possibilities of occupational disease, how would you as a safety engineer present this information to involved workers?

6. Prepare a debate discussing what role and leverage management should have in deciding when childbearing-age workers cannot work in areas that are exposed to known reproductive hazards (even though these specific jobs may pay a higher wage than other comparable positions in this petrochemical industry).

7. As a safety engineer in the petrochemical industry, what specific safety precautions in relation to heat stress would you want to inform the workers about?

8. What do you think are the greatest workplace risks to the petrochemical worker? Document.

9. How would you as a safety engineer conduct a health hazard assessment after an oil spill? Choose the material involved, and detail each step you would do and why.

10. As a safety engineer, what do you see as your responsibility to protecting the environment? Where do your role and that of a consumer differ? merge? What philosophy about living and/or life do you bring into this discussion?

REFERENCES

1. American Conference of Government Industrial Hygienists. *Documentation of the Threshold Limit Values of Substances in Workroom Air.* Cincinnati, Ohio, 1977.

2. Borse, E. Design basis accidents and accident analysis with particular reference to offshore platforms. *Journal of Occupational Accidents*, 1979, **2**, 227–243.

3. Brown, S. L. Chemical pollution dossiers for environmental decisions. In R. Conley (Ed.), *Environmental Risk Analysis for Chemicals*. New York: Van Nostrand-Reinhold, 1982.

4. Darby, G. H., Dukich, A., Margena, C. W., Hill, H. G., Hsiac, S. H., Liss-Suter, D., Mason, R., and Miller, R. *Information Profiles on Potential Occupational Hazards*. NIOSH Report, Washington, D.C., 1977.

5. Ellers, F. B. Advanced offshore oil platforms. *Scientific American*, 1982, **246**(4), 39–49.

6. Hanis, N. M., Holmes, T. M., Shallenberger, L. B., and Jones, K. E. Epidemiologic study of refinery and chemical plant workers. *JOM, Journal of Occupational Medicine*, 1982, **44,** 203–212.

7. National Safety Council. *Accident Facts,* Chicago: 1981.

8. Ruttenberg, R., and Hudgins, R. *Occupational Safety and Health in the Chemical Industry*. New York: Council on Economic Priorities, 1981.

9. Thomas, T. L., Decoufle, P., and Mours-Eraso, R. Mortality among workers employed in petroleum refining and petrochemical plants. *JOM, Journal of Occupational Medicine*, 1960, **22,** 97–103.

10. U.S. Department of Health, Education, and Welfare, NIOSH. *Criteria for a Recommended Standard. Occupational Exposure to Nickel.* Washington, D.C.: GPO, 1977, pp. 77–164.

26 SHIPBUILDING, SHIP REPAIRING, AND SHIP BREAKING

Injury rates in ship construction, repair, and demolition are among the highest of all industries. Shipbuilding is labor-intensive, and large groups of relatively unskilled workers who know little about safety are hired.

The work is arduous, and physical hazards, such as closed spaces, unguarded holes, and extensive debris, are common. Besides the inherent dangers of the job—walking–working surfaces, scaffolding, welding, cutting, and confined spaces—there are also substantial amounts of hazardous or toxic materials like cadmium and lead.

SIZE OF THE PROBLEM

The shipbuilding industry employs hundreds of thousands of persons in the United States in the building and repair of ships. The total recordable occupational injuries (any injury such as a cut, fracture, sprain, or amputation), occupational illness (any abnormal condition or disorder caused by exposure to environmental factors associated with the workplace), and occupational deaths for 1980 were 30.41 per 100 workers. That is, every 10 workers suffered more than 3 recordable conditions that year. Trucking had even more recordable conditions. This is not that much different from the 1976–1980 average of 30.01 recordable cases per 100 full-time employees. Only SIC Codes 4221 (Standard Industrial Classification) and 1791 had higher rates. These were farm product warehousing and storage and structural steel erection. The telephone industry, by comparison, had an aver-

TABLE 26.1
Incidence of Shipyard Injuries, 1967

Part of Body	Rate per 1000 Employees
Eye	2,426
Hand	1,241
Leg—foot	893
Muscles—strains, sprains	599
Skin—burns, scalds	299
Scalp	225
Back	224
Lungs—welding fumes	222
Face	149
Trunk	149
Other	132
Total	6,559

SOURCE: International Labor Organization (ILO). Rates are per annum, in a British shipyard when protective equipment, such as helmets for welders, was used only if needed for work.

age incidence rate of 2.15 cases for 1976–1980. Even construction had only an average incidence rate of 12.09 cases. Shipbuilding is probably one of the most hazardous occupations in the United States.[9]

In occupational deaths, shipbuilding is not nearly as deadly as many other industries. The chances of being killed at work were 5/100,000. The chances of being killed at work in mining and construction were over 2/10,000. The industry with the most fatalities is still coal mining with a rate of over 6/10,000. See Chapter 2 for a discussion of coal mine accidents.

Only seven other industries, including lumber and wood products, railroad transportation, air transportation, general government, railroad equipment, meat products, and trucking, had higher lost-time incidence rates.

In addition to the industry as a whole, specific shipyard work also involved many of the same risks as the construction industry. Table 26.1 shows the record of injuries in a shipyard during a year in which limited personal protective equipment was used. There were more than six injuries per worker. This tabulation highlights the need for adequate protective equipment, by means of which at least some of these accidents could have been prevented.

The safety engineer may find that, in shipbuilding, workers, supervisors, and management are uninterested and even hostile toward safety. Production is their goal, and safety is often not their concern. The safety engineer's work is sometimes limited to removing the worst situations but necessarily allowing less dangerous hazards to persist because of lack of cooperation. Only after OSHA performs a wall-to-wall inspection and many $10,000 penalties have been assessed shipyards may become safety conscious.

A critical review of shipbuilding in the United States reveals that the Occupational Safety and Health Act of 1970 has been instrumental in changing the attitude toward safety and health of both management and employees. As a result, some insurance companies now offer lower insurance rates to companies that demonstrate attention to safety through organized health and safety programs.

OCCUPATIONAL SAFETY AND HEALTH

This chapter will describe some of the hazards associated with shipbuilding and the regulations developed to control them. Some physical hazards cannot be eliminated because of the confined space or the nature of the work. They include scaffolding, surface residues, and structural hazards.

Scaffolds and Ladders

The use of scaffolding has contributed to accidents because of workers or materials falling, improper use of guardrails, and lack of proper maintenance. The choice of material used for scaffolding depends upon the type of work to be performed, the calculated weight to be supported, the surface on which the scaffolding is placed, and the substance previously stored in the confined space.

Scaffolds, ladders, and other working surfaces must conform to the requirements of OSHA. These requirements include the following: All scaffolds and their supports, whether of lumber, steel, or other material, shall be capable of supporting the load they are designed to carry with a safety factor of not less than 4. Scaffold use should comply with ANSI Standard A10-8.[3] Other requirements are spelled out in Figure 26.1.

Defective ladders have been causing accidents as long as safety records have been maintained. Thus the use of ladders with broken or missing rungs or steps, broken or split side rails, or other faulty or defective construction is prohibited. When ladders with such defects are discovered, they shall be immediately withdrawn from service. Portable metal ladders shall not be used near electric conductors or for electric arc welding. Ladders used in shipbuilding or in any other industry should comply with ANSI Standards.[1,4,5]

Walking–Working Surfaces

Another major safety hazard in ship construction is the walking–working surface. There are many openings on decks before hatches are put in place

Section	Requirement	Compliance			Comments
		Yes	No	N/A	
1910.28	SAFETY REQUIREMENTS FOR SCAFFOLDING				
1910.28a	GENERAL REQUIREMENTS FOR ALL SCAFFOLDS				
1910.28a(1)	Scaffolds shall be furnished and erected for persons engaged in work that cannot be done safely from the ground or from solid construction.				
1910.28a(2)	The footing or anchorage for scaffolds shall be sound, rigid, and capable of carrying the maximum intended load without settling or displacement.				
1910.28a(2)	Unstable objects such as barrels, boxes, loose brick, or concrete blocks shall not be used to support scaffolds or planks.				
1910.28a(3)	Guardrails and toeboards shall be installed on all open sides and ends of platforms more than 10 feet above the ground or floor except: (i) Scaffolding completely within the interior of a building and covering the entire floor area of any room therein and not having any side exposed to a hoistway, elevator shaft, stairwell, or other floor openings, and (ii) Needle-beam scaffolds and floats in use by structural iron workers.				
1910.28a(3)	Scaffold guardrails should all be of 2- × 4-inch lumber or equivalent, installed no less than 36 inches and not more than 42 inches above the work platform.				
1910.28a(3)	Scaffold guardrail supports shall be at intervals not to exceed 10 feet.				
1910.28a(3)	Scaffold toeboards shall be a minimum of 4 inches in height.				

FIGURE 26.1. Safety Checklist: Derived from General Industry Occupational Safety and Health Standards (29 CFR 1910). Subpart D: Walking, Working Surfaces. To be published in Gloss, D. S., Wardle, M. G., Gloss, M. R., and Gloss, D. S. II *Handbook of Occupational Safety and Health Standards* John Wiley & Sons, Volume I 1984.

Section	Requirement	Compliance			Comments
		Yes	No	N/A	
1910.28a(4)	Scaffolds and their components shall be capable of supporting without failure at least four times the maximum intended load.				
1910.28a(5)	Scaffolds shall not be altered or moved horizontally while they are in use or occupied.				
1910.28a(6)	Any scaffold damaged or weakened from any cause shall be immediately repaired and shall not be used until repairs have been completed.				
1910.28a(7)	Scaffolds shall not be loaded in excess of the working load for which they are intended.				
1910.28a(8)	All load-carrying timber members of scaffold framing shall be a minimum of 1,500 f. (Stress Grade) construction grade lumber.				
1910.28a(9)	All scaffold planking shall be Scaffold Grade as recognized by grading rules for the species of wood used.				
1910.28a(9)	The maximum permissible spans for 2- × 9-inch or wider planks used on scaffolds are shown below:				

	Working Load (psf)	Permissible Span (ft)	Compliance		
			Yes	No	N/A
Full	25	10			
Thickness	50	8			
Lumber	75	6			
Nominal	25	8			
Thickness	50	9			
Lumber					

Section	Requirement	Compliance			
1910.28a(9)	The maximum permissible span for 1¼ × 9-inch or wider plank which is used for scaffolds and is of full thickness is 4 feet with medium loading of 50 psf.				

FIGURE 26.1.   (*Continued*)

Section	Requirement	Compliance			Comments
		Yes	No	N/A	
1910.28a(10)	Nails or bolts used in the construction of scaffolds shall be of adequate size and in sufficient numbers at each connection to develop the designed strength of the scaffold. Nails shall not be subjected to a straight pull and shall be driven full length.				
1910.28a(12)	All planking or platforms on scaffolds shall be overlapped (minimum 12 inches) or secured from movement.				
1910.28a(12)	An access ladder or equivalent safe access shall be provided for scaffolds.				
1910.28a(13)	Scaffold planks shall extend over their end supports not less than 6 inches or more than 18 inches.				
1910.28a(14)	The poles, legs, or uprights of scaffolds shall be plumb, and securely and rigidly braced to prevent swaying and displacement.				
1910.28a(15)	Materials being hoisted onto a scaffold shall have a tag line.				
1910.28a(16)	Overhead protection shall be provided for men on a scaffold exposed to overhead hazards.				
1910.28a(17)	Scaffolds shall be provided with a screen between the toeboard and the guardrail, extending along the entire opening, consisting of No. 18 gauge U.S. Standard Wire $\frac{1}{2}$-inch mesh or the equivalent, where persons are required to work or pass under the scaffolds.				
1910.28a(18)	Employees shall not work on scaffolds during storms or high winds.				
1910.28a(19)	Employees shall not work on scaffolds which are covered with ice or snow, unless all ice or snow is removed and planking sanded to prevent slipping.				
1910.28a(20)	Tools, materials and debris shall not be allowed to accumulate on scaffolds in quantities enough to cause a hazard.				

FIGURE 26.1. (*Continued*)

Section	Requirement	Compliance			Comments
		Yes	No	N/A	
1910.28a(21)	Only treated or protected fiber rope shall be used for or near scaffolds involving the use of corrosive substances or chemicals.				
1910.28a(22)	Wire or fiber rope used for scaffold suspension shall be capable of supporting at least six times the intended load.				

Rope Strength in Pounds

Rope (inches)	6 × 19 Wire Plow	Manila Rope
$\frac{1}{4}$	5000	600
$\frac{5}{16}$	7800	1000
$\frac{3}{8}$	11000	1350
$\frac{7}{16}$	14600	1750
$\frac{1}{2}$	18800	2650
$\frac{5}{8}$	28800	4400
$\frac{3}{4}$	41200	5400
$\frac{7}{8}$	56000	7700
1	73000	9000

Section	Requirement	Compliance			Comments
		Yes	No	N/A	
1910.28a(23)	When acid solutions are used for cleaning buildings over 50 feet in height, wire rope supported scaffolds shall be used.				
1910.28a(24)	The use of shore scaffolds or lean-to scaffolds is prohibited.				
1910.28a(26)	Scaffolds shall be secured to permanent structures, through use of anchor bolts, reveal bolts, or other equivalent means. Window cleaners' anchor bolts shall not be used.				
1910.28a(27)	Special precautions shall be taken to protect scaffold members, including any wire or fiber ropes, when using a heat producing process.				
1910.28b	GENERAL REQUIREMENTS FOR WOOD POLE SCAFFOLDS				
1910.28(1)	Scaffold poles shall bear on a foundation or sufficient size and strength to spread the load to prevent settlement.				
1910.28b(1)	Scaffold poles shall be set plumb.				

FIGURE 26.1. (*Continued*)

Section	Requirement	Compliance			Comments
		Yes	No	N/A	
1910.28b(2)	Where wood poles are spliced, the ends shall be squared.				
1910.28b(2)	Where wood poles are spliced the end shall rest squarely on the lower section.				
1910.28b(2)	Where wood poles are spliced, wood splice plates shall be placed on at least two adjacent sides.				
1910.28b(2)	Wood splice plates shall overlap the abutted ends equally.				
1910.28b(2)	Wood splice plates shall be the same width as the poles.				
1910.28b(2)	Wood splice plates shall be at least 4 feet in length.				
1910.28b(2)	Wood splice plates shall be the same cross section as the wood poles or larger.				
1910.28b(3)	Independent pole scaffolds shall be set as near to the wall of the building as practicable.				
1910.28b(4)	All pole scaffolds shall be securely guyed or tied to the building or structure.				
1910.28b(4)	Where the height or length of a pole scaffold exceeds 25 feet, the scaffold shall be secured at intervals not greater than 25 feet vertically and horizontally.				
1910.28b(5)	Putlogs or bearers used on pole scaffolds shall be set with their greater dimensions vertical, long enough to project over the ledgers of the inner and outer rows of poles at least 3 inches for proper support.				
1910.28b(6)	Every wooden putlog on single pole scaffolds shall be reinforced with a $\frac{3}{16}$- $\times$ 2-inch steel strip or equivalent secured to its lower edge throughout its entire length.				
1910.28b(7)	Ledgers used on pole scaffolds shall be long enough to extend over two pole spaces.				
1910.28b(7)	Ledgers used on pole scaffolds shall not be spliced between the poles.				

FIGURE 26.1. (*Continued*)

Section	Requirement	Compliance			Comments
		Yes	No	N/A	
1910.28b(7)	Ledgers used on pole scaffolds shall be reinforced by bearing blocks securely nailed to the side of the pole to form a support for the ledger.				
1910.28b(8)	Diagonal bracing shall be provided to prevent the poles from moving in a direction parallel with the wall of the building or from buckling.				
1910.28b(9)	Cross bracing shall be provided between the inner and outer sets of poles in independent pole scaffolds.				
1910.28b(9)	The free ends of pole scaffolds shall be cross braced.				
1910.28b(10)	Full diagonal face bracing shall be erected across the entire face of pole scaffolds in both directions.				
1910.28b(10)	The braces shall be spliced at the poles.				
1910.28b(11)	Platform planks used on pole scaffolds shall be laid with their edges close together so the platform will be tight with no spaces through which tools or fragments of material can fall.				
1910.28b(12)	When planking on pole scaffolds is lapped, each plank shall lap its end supports at least 12 inches.				
1910.28b(12)	When the ends of planks used on pole scaffolds abut each other to form a flush floor, the butt joint shall be at the centerline of a pole.				
1910.28b(12)	The abutted ends of planks used on pole scaffolds shall rest on separate bearers.				
1910.28b(12)	Intermediate beams shall be provided on pole scaffolds where necessary to prevent dislodgement of planks due to deflection, and the ends shall be nailed or cleated to prevent their dislodgement.				
1910.28b(13)	When a pole scaffold turns a corner, the platform planks that meet the corner putlog at an angle shall be laid first, extending over the				

FIGURE 26.1. (*Continued*)

Section	Requirement	Compliance			Comments
		Yes	No	N/A	
	diagonally placed putlog far enough to have a good safe bearing, but not far enough to involve any danger from tipping.				
1910.28b(13)	When a pole scaffold turns a corner, the planking which runs at right angles shall be laid so as to extend over and rest on that layer of planking.				
1910.28b(14)	When moving pole scaffold platforms to the next level, the old platform shall be left undisturbed until the new putlogs or bearers have been set in place, ready to receive the platform planks.				
1910.28b(15)	Guard rails on pole scaffolds shall meet the following requirements: i. Shall be installed on all scaffolds more than 10 feet above the floor or ground. ii. Made from 2 × 4 material or equivalent. iii. Not less than 36 inches or more than 42 inches high. iv. Have a midrail made from 1- × 4-inch lumber, when required. v. Have a toe board 4 inches in height. vi. Have a wire mesh screen between the toeboard and guardrail and extending along the entire opening where persons are required to work or pass under scaffolds. vii. Have a wire mesh screen consisting of No. 18 Gauge U.S. Standard wire and of one-half inch mesh.				
1910.28b(16)	All wood pole scaffolds 60 feet or less in height shall be constructed and erected in accordance with Tables listed.				

FIGURE 26.1. (*Continued*)

Minimum Nominal Size and Maximum Spacing of Members of Single Pole Scaffolds—Light Duty

		Compliance		
		Yes	No	N/A
	20 feet			
Uniformly distributed load	Not to exceed 25 lbs. per sq. ft.			
Poles or	2 by 4 in.			
Pole spacing (longitudinal)	6 ft. 0 in.			
Maximum width of scaffold	5 ft. 0 in.			
Bearers or putlogs to 3 ft. 0 in. width	2 by 4 in.			
Bearers or putlogs to 5 ft. 0 in. width	2 by 6 in. or 3 by 4 in.			
Ledgers	1 by 4 in.			
Planking	1¼ by 9 in. (rough)			
Vertical spacing of horizontal members	7 ft. 0 in.			
Bracing, horizontal and diagonal	1 by 4 in.			
Tie-ins	1 by 4			
Toeboards	4 in. high (minimum)			
Guardrail	2 by 4 in.			
	60 feet			
Uniformly distributed load	Not to exceed 25 lbs. per sq. ft.			
Poles or uprights	4 by 4 in.			
Pole spacing (longitudinal)	10 ft. 0 in.			
Maximum width of scaffold	5 ft. 0 in.			
Bearers or putlogs to 3 ft. 0 in. width	2 by 4 in.			
Bearers or putlogs to 5 ft. 0 in. width	2 by 6 in. or 3 by 4 in. (rough)			
Ledgers	1¼ by 9 in.			
Planking	2 by 9 in.			
Vertical spacing of horizontal members	7 ft. 0 in.			
Bracing, horizontal and diagonal	1 by 4 in.			
Tie-ins	1 by 4 in.			
Toeboards	2 by 4 in.			
Guardrails	2 by 4 in.			

All members except planking are used on edge.

FIGURE 26.1.   (*Continued*)

Minimum Nominal Size and Maximum Spacing of Members of Single Pole Scaffolds—Medium Duty

		Compliance		
		Yes	No	N/A
Uniformly distributed load	Not to exceed 50 lbs. per sq. ft.			
Maximum height of scaffold	60 ft.			
Poles or uprights	4 by 4 in.			
Pole spacing (longitudinal)	8 ft. 0 in.			
Maximum width of scaffold	5 ft. 0 in.			
Bearers or putlogs	2 by 9 in. or 3 by 4 in.			
Spacing of bearers or putlogs	8 ft. 0 in.			
Ledgers	2 by 9 in.			
Vertical spacing of horizontal members	9 ft. 0 in.			
Bracing, horizontal	1 by 6 in. or 1¼ by 4 in.			
Bracing, diagonal	1 by 4 in.			
Tie-ins	1 by 4 in.			
Planking	2 by 9 in.			
Toeboards	4 in. high (min.)			
Guardrail	2 by 4 in.			
All members except planking are used on edge.				

Minimum Nominal Size and Maximum Spacing of Members of Single Pole Scaffolds—Heavy Duty

		Compliance		
		Yes	No	N/A
Uniformly distributed load	Not to exceed 75 lbs. per sq. ft.			
Maximum height of scaffold	60 ft.			
Poles or uprights	4 by 4 in.			
Pole spacing (longitudinal)	6 ft. 0 in.			
Maximum width of scaffold	5 ft. 0 in.			
Bearers or putlogs	2 by 9 in. or 3 by 5 in. (rough)			
Spacing of bearers or putlogs	6 ft. 0 in.			
Ledgers	2 by 9 in.			
Vertical spacing of horizontal members	6 ft. 6 in.			
Bracing, horizontal and diagonal	2 by 4 in.			
Tie-ins	1 by 4 in.			
Planking	2 by 9 in.			
Toeboards	4 in. high (min.)			
Guardrail	2 by 4 in.			
All members except planking are used on edge.				

FIGURE 26.1. (*Continued*)

Minimum Nominal Size and Maximum Spacing of Members
of Independent Pole Scaffolds—Light Duty

		Compliance		
		Yes	No	N/A
	20 Feet			
Uniformly distributed load	Not to exceed 25 lbs. per sq. ft.			
Poles or uprights	2 by 4 in.			
Pole spacing (longitudinal)	6 ft. 0 in.			
Pole spacing (transverse)	6 ft. 0 in.			
Ledges	2 by 4 in.			
Bearers to 3 ft. 0 in. span	2 by 6 in. or 3 by 4 in.			
Bearers to 10 ft. 0 in. span	$1\frac{1}{4}$ by 9 in.			
Vertical spacing or horizontal members	7 ft. 0 in.			
Bracing, horizontal and diagonal	1 by 4 in.			
Tie-ins	1 by 4 in.			
Toeboards	4 in. high			
Guardrail	2 by 4 in.			
	60 Feet			
Uniformly distributed load	Not to exceed 25 lbs. per sq. ft.			
Poles or uprights	4 by 4 in.			
Pole spacing (longitudinal)	10 ft. 0 in.			
Pole spacing (transverse)	10 ft. 0 in.			
Ledgers	$1\frac{1}{4}$ by 9 in.			
Bearers to 3 ft. 0 in. span	2 by 4 in.			
Bearers to 10 ft. 0 in. span	2 by 9 in. (rough) or 3 by 8 in.			
Planking	2 by 9 in.			
Vertical spacing of horizontal members	7 ft. 0 in.			
Bracing, horizontal and diagonal	1 by 4 in.			
Tie-ins	4 in. high (min.)			
Guardrail	2 by 4 in.			
All members except planking are used on edge.				

FIGURE 26.1. (*Continued*)

**Minimum Nominal Size and Maximum Spacing of Members
of Independent Pole Scaffolds—Medium Duty**

		Compliance		
		Yes	No	N/A
Uniformly distributed load	Not to exceed 50 lbs. per sq. ft.			
Maximum height of scaffold	60 ft.			
Poles or uprights	4 by 4 in.			
Pole spacing (longitudinal)	8 ft. 0 in.			
Pole spacing (transverse)	8 ft. 0 in.			
Ledgers	2 by 9 in.			
Vertical spacing of horizontal members	6 ft. 0 in.			
Spacing of bearers	8 ft. 0 in.			
Bearers	2 by 9 in. (rough) or 2 by 10 in.			
Bracing, horizontal	1 by 6 in. or 1¼ by 4 in.			
Bracing, diagonal	1 by 4 in.			
Tie-ins	1 by 4 in.			
Planking	2 by 9 in.			
Toeboards	4 in. high (min.)			
Guardrail	2 by 4 in.			
All members except planking are used on edge.				

FIGURE 26.1.   (*Continued*)

or holes are covered. These are very dangerous. A fall from the deck to the bottom of a hole may be as much as 40 feet (12.2 m). In such a fall severe injury or death would result. Some of the related safety requirements are as follows:

1. When employees are working in the vicinity of flush manholes and other small openings of comparable size in the deck and other working surfaces, such openings shall be suitably covered or guarded to a height of not less than 30 inches (76 cm).

2. When employees are working around open, unprotected hatches and other large openings, the edge of the opening shall be guarded in the working area to a height of 36 to 42 inches (92 to 107 cm).

3. When the employees are exposed to elevated decks, platforms, scaffolds, and similar work areas more than 5 feet (1.5 m) above a solid working surface, the edges shall be guarded by a guardrail.

4. When employees are working near the unguarded edges of decks of vessels afloat, they shall wear personal flotation devices.

Another major walking–working surface on vessels is the means of access or egress. An OSHA requirement for gangways is that they have at least 20 inches (51 cm) or walking surface.

Surface Residues

In ship construction surface residues are of particular importance. The steel may rust before it is protected. The outside of the hull is protected with antifouling materials. The primer put on the metal may be red lead or zinc chromate. The paint itself may contain titanium and lead. In ship breaking the fumes are intensified as these substances are burned during cutting.

Welding and Cutting

Ship hulls and bulkheads are generally made of plate steel which is very heavy. It is cut to the correct size in the shop and brought to the work site where it is then welded. Often it is tacked or spot-welded at first to secure it in place; then the entire perimeter is welded. The superstructures of many vessels are now made of aluminum to save weight.

Most welding and cutting operations in shipbuilding, ship repairing, and ship breaking use manually operated equipment. The safety engineer should be familiar with the ANSI standard.[2]

Gas Welding and Cutting. Gas welding is a process of combining two or more pieces of metal by heating the metals to a very high temperature and fusing them together. A combination of acetylene and oxygen is usually used to achieve these temperatures.

Cutting is the process of severing a piece of metal into a number of parts. It may be accomplished by heating the metal to a very high temperature with the acetylene-oxygen combination and then using the oxygen under pressure to blow the molten metal out of the piece, thus separating it. The accuracy of the operation is attained by altering the nozzle size.

The oxygen and actylene are usually supplied in large cylinders, but in many shipbuiding and ship breaking operations, oxygen and acetylene are stored in large tanks with hose lines running to the various operations.

Oxygen itself will not burn, but it supports combustion. In the presence of oxygen combustible materials will burn more rapidly than in air. Oxygen combines with acetylene to produce a very high temperature.

Acetylene is a hydrocarbon. That is, it consists of carbon and hydrogen. Its formula is C_2H_5. In the presence of oxygen it can produce a temperature as high as 3,300°C (5,400°F). Acetylene is a flammable gas and must be treated with caution. The flammable limits for acetylene are 2.5% for the lower explosive limit (LEL) and 89% for the upper explosive limit (UEL). These terms were discussed in Chapter 11.

Arc Welding and Cutting.   Arc welding is a process for combining metals by heating with an electric arc. It is used for joining many ferrous (steel) and nonferrous (aluminum) metals. In arc welding, two leads are used to handle the current. One lead is connected to the electrode, the other to the work.

Safety Issues in Welding

There are a number of safety issues in welding. The safe handling of the compressed cylinder is one of them. Molten metal is produced in welding, and droplets may splatter against the welder. To protect against these, he or she should wear a welder's apron, welder's gloves, and a face shield.

Metal fumes are produced in welding. The metals are oxidized and converted to particles which are deposited in different parts of the respiratory system. Large variations occur depending on ventilation rates and whether inhalation is by nose or mouth. Metal exposures in shipyards are often high. Two of the most hazardous are lead and cadmium, which have been discussed in Chapter 14.

Fire Protection.   Welding uses high temperatures that are hot enough to melt steel and hot enough to ignite other substances in the environment. The welder, the supervisor, and the safety engineer should determine what fire protection equipment is necessary. In addition, a fire watch should be present whenever welding is performed in the vicinity of flammable material.

Confined Areas.   Ships have many confined spaces which make welding hazardous. Either general mechanical or local exhaust ventilation must be used. When sufficient ventilation cannot be provided without blocking the means of access, workers in the confined space must use respirators and have a co-worker outside to maintain communication and provide emergency assistance.

GENERAL WORKING CONDITIONS

General working conditions in shipbuilding, ship repairing, and ship breaking are typically hazardous. This section discusses inadequate ventilation in confined spaces, housekeeping, illumination, heat stress, cold stress (including windchill), and noise.

Confined Space

A confined space is a space which by design has limited openings for entry and exit; has unfavorable natural ventilation that could contain or produce dangerous air contaminants; and is not intended for continuous occupancy. Many confined spaces include but are not limited to storage tanks, compartments of ships, process vessels, pits, silos, vats, degreasers, reaction vessels, boilers, ventilation and exhaust ducts, sewers, tunnels, underground utility valves, and pipelines. Inadequate ventilation means that the worker is exposed to unacceptable levels of contaminants especially in confined spaces.

The major environmental control for a confined space is accomplished by ventilating which will be used to determine the potential hazards that arise due to the product stores or produce, suspected contaminants, the work to be performed, and the design of the confined space. It will include the following:

1. The blower controls are to be located at a safe distance from the confined space.
2. Equipment must have an audible warning signal when there is a ventilation failure.
3. Airflow measurements are to be made before each work shift to ensure that a safe environmental level is maintained.

Any contaminated air that is exhausted from a confined space should not become a hazard for workers in the surrounding area. If flammable concentrations are present, all electric equipment must comply with the requirements of The National Electric Code (NFPA, no. 70) for hazardous locations. The electric equipment must be explosion-proof. Where continuous ventilation is not a part of the operating procedure, the air in the confined area must be tested until continuous acceptable levels of oxygen and contaminants are maintained for three tests at 5-minute intervals.

Continuous general ventilation must be maintained where toxic atmospheres are produced as part of a work procedure, such as welding or painting, or where a toxic atmosphere may develop due to the nature of the confined space, as in the case of desorption from walls or evaporation of residual chemicals. General ventilation is an effective procedure for distributing contaminants from a local generation point throughout the work space to obtain maximum dilution. However, special safety precautions shall be taken if the ventilating system partially blocks the exit opening. These precautions include a method for providing respirable air to each worker for the time necessary to exit and a method of maintaining communications.

Local exhaust ventilation shall be provided when general ventilation is not effective due to restrictions in the confined space or when high concentrations of contaminants occur in the breathing zone of the worker. Local high concentrations may occur during work activities such as welding, painting, and chemical cleaning. The use of respiratory protection will be determined by the safety engineer. When freely moving exhaust hoods are used to provide control of fumes generated during welding, such hoods shall maintain a velocity of 100 cubic feet per minute in the zone of the welding. The effective force of freely moving exhaust hoods is decreased by approximately 90% at a distance of one duct diameter from the exhaust opening. Therefore, to obtain maximum effectiveness, the welder must reposition the exhaust hood when changing welding locations to keep the hood in close proximity to the fume source.

Special precautions must be taken when outgassing or vaporization of toxic and/or flammable substances is likely. If the vapor-generating rate can be determined, the exhaust rate required can be calculated to dilute the atmosphere below 10% of the lower explosive level. This shall be the lowest ventilation rate. If the area of concern is relatively small, diffusion of the contaminants may be controlled by enclosure with a relatively low volume exhaust for control or by exhaust hoods located as close as possible to the area of vaporization or outgassing. If the area to be ventilated is too extensive to be controlled by local exhaust, then general ventilation procedures must be used to control the contaminant level. When the problem of outgassing is due to the application of protective coatings of paint, ventilation shall be continued until the buildup of a flammable and/or toxic atmosphere is no longer possible.

Confined spaces such as tanks, vaults, and compartments of ships usually have limited access. The limited access increases the risk of injury. Gases which are heavier than air, such as chlorine and propane, may lie in a tank for hours or even days undisturbed unless they are flushed out. Because some gases are odorless, the hazard may be overlooked—with fatal results. Gases that are lighter than air, such as hydrogen and helium, may also be trapped at the top of an enclosed type of confined area, especially those with access from the bottom.

The extent of precautions taken and the standby equipment needed to maintain a safe work area will be determined by the means of access and rescue. The following should be considered: type of confined space to be entered, access to the entrance, number and size of openings, barriers within the space, the occupancy load, and the time requirement for exiting in event of fire, or vapor incursion, and the time required to rescue injured workers. Planning for an emergency in a confined space must be done before the first worker enters.

Housekeeping

In shipbuilding, as in most other industries, housekeeping plays a vital role in safety. Working and walking surfaces may already have obstructions, such as cables and hoses, which should be covered by planks. Additional clutter, such as material, equipment and trash, accumulates. Just imagine the amount of trash that might accumulate after five days if 200 workers each brought a daily paper on board. Those 1,000 newspapers might also be a major fire hazard.

Illumination

Illumination is a basic requirement in shipbuilding. Many areas are without any natural light. If work is performed around the clock, there is much time when natural light is not available. Appropriate lighting standards have been set by ergonomists for various types of activities and are shown in Chapter 9. In addition, these standards provide much information for all types of ergonomic problems encountered by safety engineers.

Walkways leading to working areas, as well as the working areas themselves, must be adequately illuminated. Temporary lights must meet the following requirements:

1. They must be equipped with heavy-duty electric cords. The lights must not be suspended by their electric cords unless cords and lights are designed for this means of suspension. Splices which have insulation equal to that of the cable are permitted.
2. They must be equipped with guards to prevent accidental contact with the bulb, except that the guards are not required when the construction of the reflector is such that the bulb is deeply recessed.
3. Cords shall be kept clear of working surfaces and walkways or other locations in which they are readily exposed to damage.

Exposed non–current-carrying metal parts of temporary lights furnished by the employer shall be grounded either through a third wire in the cable containing a circuit conductor or through a separate wire which is grounded at the source of the current. Where temporary lighting from sources outside the vessel is the only means of illumination, portable emergency lighting equipment shall be available to provide illumination for safety of employees. Employees shall not be permitted to enter dark spaces without a suitable portable light. The use of matches and open flame is prohibited. Temporary lighting of stringers or streamers shall be so arranged as to avoid overloading of branch circuits. Each branch circuit must be equipped with overcurrent protection of capacity not exceeding the rated current-carrying capacity of the cord used.

The results of a survey concerned with visual fatigue among 8,032 shipbuilders in southern Spain have been reported. Visual fatigue (asthenopia) was found in 865 persons (10.7% of the population considered); the average age of the subjects affected was 39 years.

Thermal Effects

Thermal effects in the confined areas of shipbuilding must be monitored. Four factors influence the interchange of heat between worker and environment: (1) air temperature, (2) air velocity, (3) moisture contained in the air, and (4) radiant heat. Because of the nature and design of most confined spaces, moisture content and radiant heat are difficult to control. As the body temperature rises progressively, workers will continue to function until the body temperature reaches 38.3–39.5°C (100.9–103.1°F). When this body temperature is exceeded, the workers are less efficient and are prone to heat exhaustion, heat cramps, or heatstroke. In a cold environment certain physiologic mechanisms come into play, which tend to limit heat loss and increase heat production. The most severe strain in cold conditions is chilling of the extremities so that activity is restricted. Special precautions must be taken to prevent frostbite, trench foot, and general hypothermia.

Protective insulated clothing for both hot and cold environments will add additional bulk to the worker; this must be considered in allowing for movement in confined space and exit time. Therefore, air temperature of the environment becomes an important consideration when evaluating working conditions in confined spaces.

Noise

Noise is intensified in confined spaces because the interior causes sound to reverberate; thus the worker is exposed to higher noise levels than in an open environment. This noise increases the risk to hearing. Noise not loud enough to cause damage may still disrupt communication with the emergency standby person outside the confined space. If workers inside are not able to hear commands or warnings, the probability of severe accidents can increase. Noise, however, is a major problem for the whole shipbuilding industry because of the lack of stress on engineering control of noise, poor enforcement of hearing protector use, and the uncertainty of getting annual monitoring audiograms on all employees exposed to hazardous noise.

If noise exposure exceeds 95 decibels, workers should only be allowed to be exposed for short durations; this was discussed further in Chapter 14. The audiograms of 1,392 noise-exposed shipyard workers were analyzed. Hearing impairment was greatest at frequencies about 4,000 hertz (4,000 cps). A hearing conservation program for the shipbuilding industry has been proposed consisting of (1) noise measurements and analyses, (2) audiometry, (3) personal protective measures, (4) education, and (5) engineering control methods.[10]

The noise conservation program proposed includes posting of noise hazard signs, education of employees and medical personnel, audiometric testing, appropriate interpretation of each audiogram, proper disposition of affected employees, and adequate physician involvement in the overall program. Noise is measured regularly using a variety of equipment. Audiometric testing of all personnel is done annually, and ear protective devices are required where the noise level warrants. An active education program is followed to familiarize personnel with hearing facts. Engineering noise-control techniques are used. OSHA will be instituting much of what is suggested. (See 29 CFR 1910.95C for a draft of the standard for hearing conservation.)

Where the daily noise exposure is composed of two or more periods of noise exposure of different levels, their combined exposure should be considered, rather than the individual effect of each. The following formula should be used to determine a time-weighted average (TWA):

$$\frac{C_1}{T_1} + \frac{C_2}{T_2} + \frac{C_n}{T_n} \leq 1$$

where C_{12} is the total time in hours of exposure at a specified noise level and T indicates the total time of exposure at that level.

Suppose a shipfitter is exposed to 4 hours per day of 90 decibel noise, 2 hours per day of 95 decibel noise, and 2 hours per day of 97 decibel noise. Would that exceed the allowable limit?

$$\tfrac{4}{3} + \tfrac{2}{4} + \tfrac{2}{3} = \tfrac{18}{12} = 2.5$$

Therefore, the shipfitter is exposed to excessive noise. The safety engineer should ensure that this shipfitter is reassigned for part of the day to a less noisy environment so that his TWA does not exceed one.

Each of the above sections covered an area of major safety and health concern for the shipbuilding industry.

Ship Machinery

The ship's machinery also provides many areas of possible hazards for the worker. These include the ship's boilers, piping systems, propulsion machinery, and deck machines (such as cranes or winches). Only the boilers and piping will be discussed here.

Boilers.   The boilers are the shipboard equipment which heats the water under pressure. The steam, in turn, is used to drive the propulsion machinery that makes the ship move. Live steam can cause serious burns. A number of safety considerations must be taken in repairing the ship's boilers. Before work is performed in the fire, steam, or water spaces of a boiler where employees may be subject to injury from the direct escape of steam or boiling water, the worker, the supervisor, and the safety engineer shall ensure that the following precautions are taken:

Lock out the system. The valves connecting the dead boiler with the live system or systems must be secured, blanked (a barrier is put in so that the flow is prevented), and tagged, to indicate that employees are working in the boiler. This tag must not be removed or the valves unblanked until the work in the boiler is completed. Where valves are welded instead of bolted, at least two isolation and shutoff valves connecting the dead boiler with the live system or systems shall be secured, locked, and tagged.

Drain connections to atmosphere on all the dead interconnecting systems must be opened for visual observation of drainage. A warning sign calling attention to the fact that employees are working in the boilers will be hung in a conspicuous location in the engine room. This sign is not to be removed until it is determined that the work is completed and all employees are out of the boilers.

Before work is performed on a valve, fitting, or section of piping in a piping system where employees may be subject to injury from steam, water, or oil, the worker, supervisor, and safety engineer must ensure that the same precautions are taken as those for working with boilers except that the warning sign may be in some other location than in the engine room.

Tools

Tools are used in every phase of shipbuilding, ship repairing, and ship breaking. They range from a screwdriver to a wrecking ball. Tools may cause accidents. Tools are the instruments involved in many hand, face, and eye injuries. Using ungrounded tools or using them in wet areas may result in electric shock.

Abrasive tools, such as wheel or disk grinders, are used in a variety of shipboard operations, such as rounding sharp edges and corners.

The maximum angular exposure of the grinding wheel periphery and sides shall be no more than 90 degrees; however, when work requires contact with the wheel below the horizontal plane of the spindle, the angular exposure shall not exceed 125 degrees. Safety guards shall be strong enough to withstand the effect of a bursting wheel. Floor and bench-mounted grinders shall be provided with work rests that are rigidly supported and readily adjustable. Such work rests shall be kept a distance not to exceed ⅛ inch from the surface of the wheel. Cup-type wheels used for external grinding shall be protected by either a revolving cup guard or a band-type guard. All other portable abrasive wheels used for external grinding shall be provided with safety guards.

Toxic and Hazardous Materials

Toxic and hazardous materials in the shipbuilding industry could be a whole chapter in and of itself.[8]

Lead. Lead (Pb) is used extensively in shipyards in the form of red lead. Lead and alloys of lead are also used in solder where it is mixed with tin, in batteries where it is used with sulfuric acid, and in bronze such as naval bronze.

Lead fumes and dust are readily absorbed in the body by inhalation and by ingestion. The symptoms of lead poisoning in gastrointestinal disorders are constipation followed by weakness, then anemia, and, finally, paralysis. Lead poisoning causes mental retardation typically in children but only on rare occasions in adults (except from organic lead compounds).

Cadmium.   Cadmium (Cd) is used extensively on ships as a metal coating to protect the steel. Cadmium fumes cause irritation of the nose and throat which is followed by cough, chest pain, sweating, chills, dyspnea (shortness of breath), and weakness. Death may ensue.

Cadmium fumes are a severe pulmonary irritant that has caused fatal pulmonary edema. A small concentration is enough to cause much harm. Most acute intoxications have been caused by inhalation of cadmium fume concentrations which did not provide warning symptoms. Respiratory equipment needed to protect the worker is shown in Chapter 10.

Asbestos.   Ships require large amounts of thermal insulating materials for hot water pipes, refrigeration units, and steam and heating pipes. Insulation is also placed under open decks to minimize solar heat transfer below decks and is especially necessary under flight decks of aircraft carriers. Fire-resistant materials are generally selected because of the dangers resulting from fires on board ship. Asbestos has been one of the most generally used materials because of its favorable characteristics. Its use, however, is diminishing as other materials have been developed that can replace it. These newer materials are magnesia bricks, fibrous glass products, and rock wool. To date, there is little evidence that these latter substances are hazardous. Asbestos, on the other hand, is hazardous because of its capacity to produce chronic lung disease and cancer. Its use may still be required in some areas where it is the only known adequate material. It is still present in many of the ships afloat, and it does cause a problem when repair work is performed and the old insulation is removed. It is also a problem when ships are scrapped.

Perhaps the most dramatic example of this problem occurred when the British navy performed a major overhaul. This involved removing 3 inches of asbestos from beneath the flight deck of the *Ark Royal,* Britain's largest aircraft carrier. The men who were removing the material wore ventilated plastic suits. The material was thoroughly soaked with water before removal in order to minimize exposure. It was immediately gathered in plastic bags for safe disposal. The area was then vacuumed to remove the residual contaminating material. This job was accomplished with minimal risk to the workers and others on board ship.[9]

Asbestos has been used extensively. Another study shows that the workers who installed asbestos were exposed. Pipe coverers in new ship construction in a New England shipyard, occupationally exposed to low levels of asbestos dust for 20 years, were studied.[7] Dust exposure had been near the then recommended TLV. Asbestosis was 11 times more prevalent among pipe coverers than among controls. This study and others showed the need for a lower TLV in order to make the workplace safe, which was done by OSHA. Now the permissible eight-hour time-weighted average (TWA) for airborne asbestos fibers cannot exceed 2 fibers longer than 5 micrometers per cc of air. The ceiling concentration is an amount above which no worker may be exposed at any time. The ceiling concentration is 10 fibers longer than 5 micrometers per cc.[3]

Carbon monoxide.   Carbon monoxide (CO) is an odorless, colorless gas that has approximately the same density as air and is formed from incomplete combustion of organic materials such as wood, coal, gas, oil, and gasoline; it can be formed from microbial decomposition of organic matter in sewers, silos, and fermentation tanks. Carbon monoxide is an insidious toxic gas because of its lack of warning signs. Early signs of carbon monoxide intoxication are nausea and headache, which are easily mistaken. Carbon monoxide may be fatal at 1,000 ppm in air and is considered dangerous at 200 ppm because it forms carboxyhemoglobin in the blood which prevents the distribution of oxygen in the body.

Carbon monoxide is a relatively abundant, colorless, odorless gas; therefore, any untested atmospheres must be suspect. Carbon monoxide must be tested for specifically.

The normal atmosphere is composed approximately of 20.9% oxygen, 76.1% nitrogen, and 1% argon with small amounts of various other gases. Reduction of oxygen (O_2) in a confined space may be the result of either consumption or displacement.

Solvents.   Solvents such as benzene are a workplace hazard. Benzene (C_6H_6) affects the bone marrow, causing leukemia after an incubation period of 6 to 14 years. Solvents are used for cleaning and degreasing. Adequate ventilation must be provided.

The initial step in chemical cleaning usually is the conversion of the scale or sludge into a liquid state which may cause poisonous gases to be liberated. In 1974, several employees were cleaning a boiler tank prior to repairing a leak using a cleaning fluid, Vestan 675. The cleaning action caused the release of ammonia (NH_3) fumes that were not properly exhausted. The men were hospitalized with severe chest pain but recovered.[6]

When toxic solvents are used, the following measures must be employed:

1. The cleaning operation shall be completely enclosed to prevent the escape of vapor into the working space.
2. Either natural ventilation or mechanical exhaust ventilation shall be used to remove the vapor at the source and to dilute the concentra-

tion of vapors in the working space to a concentration which is safe for the entire work period.

3. Employees must wear respirators when working with hazardous solvents that require respirators.[6]

Removal of Paints. When ships go into dry dock for overhaul, one of the first activities is to scrape the bottom and to remove the rust. In this operation, the worker is exposed to lead and other substances. Employees engaged in the removal of paints, preservatives, rusts, or other coatings must use the appropriate personal protective equipment, such as respirators to prevent inhaling toxic dusts. When using paint and rust removers containing strong acids or alkalies, employees must wear face shields.

Flammable Liquids. Flammable liquids are a hazard of the shipbuilding, ship repairing, and ship breaking industry. In all cases when liquid solvents, paint and preservative removers, paints, or vehicles which are flammable are used, the following precautions shall be taken:

1. Smoking, open flames, arcs, and spark-producing equipment shall be prohibited in the area.
2. Ventilation shall be provided in sufficient quantities to keep the concentration of vapors below 10% of their lower explosive limit. Frequent tests shall be made by a competent person to ascertain the concentration.
3. Scrapings and rags soaked with these materials shall be kept in a metal container.
4. Only explosion-proof lights are permissible.
5. All power and lighting cables must be inspected to ensure that the insulation is in excellent condition (free of all cracks and worn spots), that there are no connections within 50 feet of the operation, that lines are not overloaded, and that they are suspended with sufficient slack to prevent undue stress or chafing.
6. Suitable fire-extinguishing equipment shall be immediately available in the work area and shall be maintained in a state of readiness for instant use.

Radiation. Radiation is another shipbuilding hazard. Exposures to x-rays or radioisotopes may become more of a problem in the future. Many ship specifications require x-ray inspections of part or even all welds on the ship. This means extensive use of radiation. Usually, only a small number of skilled workers are involved, but the potential hazard exists, and a sound radiation protection program is indicated. The further development of nuclear power for use in ships will also mean an increase in this potential hazard. At present, this is limited almost entirely to military vessels, but it may not be so in the future.[2]

Irritants.   Examples of primary irritants are chlorine (Cl_2), ozone (O_3), hydrochloric acid (HCL), hydrofluoric acid (HF), sulfuric acid (H_2SO_4), nitrogen dioxide (NO_2), ammonia (NH_3), and sulfur dioxide (SO_2). A secondary irritant is one that may produce systemic toxic effects in addition to surface irritation. Examples of secondary irritants include benzene (C_6H_6), carbon tetrachloride (CCl_4), ethyl chloride (CH_3Ch_2Cl), trichloroethane (CH_3CCl_3), trichloroethylene (Cl_2CCHCl), and chloroprene (allyl chloride, or CH_2CHCH_2Cl).

Prolonged exposure at irritant or corrosive concentrations in a confined space may produce little or no evidence of irritation. This has been interpreted to mean that the worker has become adapted to the harmful agent involved. In reality, it means that there has been a general weakening of the defense reflexes from changes in sensitivity, due to damage of the nerve endings in the mucous membranes of the conjunctiva and upper respiratory tract. The danger of this situation is that the worker is usually not aware of any increase in the exposure to toxic substances.

The most important factor in the materials discussed above is for the worker to be aware of what he or she is being exposed to so the necessary controls can be taken.

Because of the nature of the work, the worker in the shipbuilding industry is at high risk for workplace hazards. This is not an easy industry to control, but it is one which needs firm and thorough controls if its workers are going to receive the health and safety to which they are entitled.

SUGGESTED LEARNING EXPERIENCES

1. For you as a safety engineer, what would be the most overwhelming safety problem in the shipbuilding, ship repairing, and ship breaking industry? Why?

2. Develop a safety training program for this industry, and include how you would proceed to implement it.

3. Do you believe the statement that life or limb is of little consequence in the shipbuilding industry? Discuss.

4. What incentives to control costs on workplace injuries in the shipbuilding industry would you as a safety engineer develop?

5. Detail what tasks you envision for a safety engineer for one month in the shipbuilding, ship repairing, and ship breaking industry.

6. Discuss workers' rights in relation to confined spaces.

7. What kind of monitoring or responsibility would you delegate and to whom in relation to drinking alcohol during the workday? ustify and explain.

8. What kind of training would you as a safety engineer advocate for the prospective shipbuilding employee before he or she begins actual work on a ship?

9. What would you have done to avoid the following situation from occurring? Trichloroethane, also known as methyl chloroform, was substituted for trichloroethylene because of the high toxicity of the latter. The work crew was involved in tank cleaning and an assembly operation. The technique of cleaning the interior of the tanks varied among workers. Some would moisten a pad with solvent and would wipe by hand the metal surfaces by reaching through an opening on the end of the tank; some would use pads on the end of a shaft; and others would climb inside to clean. One particular worker would saturate a pad with solvent and lower himself head first into the down-tilted tip of the tank and clean as fast as possible. This worker was found with his legs protruding from the upper end of the 450-gallon tank and was unresponsive. He was removed immediately and was given artificial respiration until a physician arrived and pronounced him dead.

10. In most shipyards, the wearing of personal protective equipment is frowned on by management as too costly, by supervisors as slowing production, and by workers as cowardly. How would you go about changing these positions?

REFERENCES

1. American National Standards Institute. *Code for Fixed Ladders.* (ANSI 14.3-1956). New York, 1956.

2. American National Standards Institute. *Safety in Welding and Cutting.* (ANSI Z49.1-1959). New York, 1959.

3. American National Standards Institute. *Safety Requirements for Scaffolds* (ANSI A10.8-1960). New York, 1960.

4. American National Standards Institute. *Safety Code for Portable Wood Ladders* (ANSI 14.1a-1977). New York, 1977.

5. American National Standards Institute. *Safety Code for Portable Metal Ladders* (ANSI 14.2a-1977). New York, 1977.

6. Ferris, B. G., and Heimann, H. Shipyard health problems. *Environmental Research,* 1976, **1,** 140–150.

7. Murphy, R. L., Ferris, B. G., Burgess, W. A., orcester, J., and Gaensler, E. A. Effects of low concentrations of asbestos: Clinical, environmental, radiologic and epidemiologic observations in shipyard pipe covers and controls. *New England Journal of Medicine,* 1971, **285,** 1271–1278.

8. NIOSH. *Occupational Health Guidelines for Chemical Hazards.* Cincinnati, Ohio, 1981.

9. National Safety Council. *Accident Facts,* Chicago, Ill., 1981.

10. Nowack, R., and Dahl, D. The course of noise-induced hearing loss in shipbuilding workers. *Zeitschrift fuer die Gesamte Hygiene und Ihre Granzgebiete,* 1971, **17,** 483–488.

27 SYSTEM SAFETY ENGINEERING

System engineering encompasses consideration of all the various methods of accomplishing a desired result. System engineering views each system as an integrated whole even though it is composed of diverse, specialized structures and functions. It further recognizes that a system may have several objectives, and the balance between them may differ widely. It seeks to optimize the system functions according to stated objectives and to achieve maximum compatibility of its parts.

The system is in no way limited in terms of the size, the nature of the physical problem, or the manner in which the whole is subdivided. The system might be economic, social, or political, or it might be an engineering system for the design, manufacture, and operation of a particular product. The success of the system, as opposed to the parts, is the primary objective. Although a system may be composed of several subsystems, the objective is to integrate these subsystems into the most effective configuration possible.

The objectives of system design may be any or all of the following:

1. Total safety
2. Ease of operation
3. Ease of maintenance
4. Low initial cost
5. Low life-cycle cost
6. High reliability (very long meantime between failures)

Thus a new automobile may need to be crushworthy up to 40 miles per hour. This means that the occupants should be able to survive a crash at 40 miles per hour. The attainment of this safety goal may negatively affect ease of operation, ease of maintenance, and low initial cost.

Every effort must be made to provide the safest possible product. The control or elimination of hazards or potential hazards in the design of this product is the job of the system safety engineer.

System safety engineering is concerned with control of and elimination of hazards or potential hazards in any system. Thus the first concern of system safety is that the system is safe. The design should be as foolproof as possible; however, as stated in Chapter 1, nothing is absolutely safe under all conditions.

NEEDS ANALYSIS

The starting point of a design project is a perceived need, which may have been observed currently on the socioeconomic scene. It may be phrased in the form of a statement derived from untested observation, or it may have been developed into a statement based on market and consumer studies. The need may not yet be identified, but there may be evidence that it will be needed and that it will be labeled when economic means become available. The need may be suggested by a technical accomplishment that makes the means of its satisfaction possible. In whatever way the need has been perceived, its economic existence must be established with sufficient confidence to justify the commitment of funds necessary to explore the feasibility of developing the appropriate means. "Economic existence of a need" means that individuals and institutions will recognize the need and pay the means to satisfy it.

FEASIBILITY STUDY

A design project begins with a feasibility study to develop a set of workable solutions to the design problem. Sometimes a design concept has already been fixed. This implies one of three possibilities: a feasibility study has been made; the technical management has had so much experience with the particular design problem that further study would be superfluous; or the management, by omitting the feasibility study, is proceeding on intuition.

The steps in the study are (1) to demonstrate whether the presumed validity of the original need exists; (2) to explore the design problem engendered by the need and to identify the principal constraints and design criteria; (3) to suggest plausible solutions to the problem; and (4) to select potential solutions on the basis of physical, economic, and financial feasibility.

SAFETY DECISION

In terms of feasibility studies and making safety decisions early, the designer can use what is known about behavior. An excellent list of typical human behavior that may lead to injury from an unsafe act has been developed. (See Table 27.1.) These principles are important to keep in mind in terms of safety and performance.

DISTURBING INCIDENTS

Disturbing incidents are dramatic enough to make the news and are so impressive as to be forever etched in memory. Examples of such incidents abound. A young woman's long hair is caught in the rollers of a revolving machine that tears her scalp off her head. A huge engine falls off a DC-10; the pilot could not prevent the plane from crashing. What was the design defect that caused it in the first place? A fire escape falls when there is a fire; it does not hold the weight of the people trying to flee a burning building. A hotel bridge collapses—under the weight of people—onto the activities occurring on the floor below. Each of these is a failure in technology. Safety was ignored. The safety factor in each case was not sufficient to protect the people. A small car explodes on impact in a rear-end collision. The hazard was known by the manufacturer which chose to let people die rather than fix the defect.

Although there is a probability of design failure, precision and care must still be taken to produce the safest product no matter what the cost. Apart from human considerations, as laws become more and more stringent, soon it simply will not be cost-effective to make unsafe or unhealthy products, because legal penalties in terms of fines, costs, and punitive judgments will be exorbitant.

Changing Hazards

Hazards do change. A system that once functioned at a certain level of performance may no longer be functioning at that level; factors in the environment may have changed which, in turn, change the function of the operators or consumers. All of these influence how hazardous a system may be.

Minimizing the Risks

Background risks are those to which everyone is subjected, impartially and universally. The risks are so alarmingly high for accidents that one wonders how anyone lives to be 40. Our aim as safety engineers is to recognize hazards and to try to eliminate or control them.

TABLE 27.1
List of Typical Human Behavior That May Lead to Injury from an Unsafe Act

Behavior Description	Design Consideration
1. Many people do not consider the effects of surface friction on their ability to grasp and hold an article.	Design surface texture to provide friction characteristics commensurate with functional requirements of task or device.
2. Most people cannot estimate distances, clearances, or velocities very well; people tend to overestimate short distances and underestimate large distances.	Design products so that users need not make estimates of critical distances, clearances, or speeds. Provide indicators of these quantities where there is a functional requirement.
3. Most people do not look where they put their hands and feet, especially in familiar surroundings.	If hand or foot placement is a critical aspect of the user–product interface, design so that careless, inadvertent placement of hands or feet will not result in injury. Provide guards, restraints, warning labels.
5. Many people do not think about such things as high temperatures; becoming an electrical ground when they grasp an article which is "hot"; picking up an object that is wet; or walking carefully on a slick floor.	This lack of awareness may result from ignorance, inattention, carelessness, or sheer disregard for safety.
6. People seldom anticipate the possibility of contact with sharp edges or corners.	Except to meet functional requirements, eliminate sharp edges on surfaces or units where inadvertent human contact is remotely possible.
7. People rarely think about the possibility of fire or explosion from overheated objects or cooking oil suddenly exposed to the air.	Unless it is a functional requirement, eliminate configurations which will permit such possibilities.
8. People rarely think about the possibility of catching their clothing on a handle or other protruding object.	Same as item 6; also provide proper warnings and labels. Conduct user-task analysis, and always mount handles in optimal functional advantageous places.
9. Many people do not take the time to read labels or instructions or to observe safety precautions.	Make labels brief, bold, simple, clear. Repeat some labels on various parts of a product. Make use of color coding, fail-safe innovations, and other attention-demanding devices.

564

TABLE 27.1. (*Continued*)

10.	Many designers do not recognize the existence of response stereotypes (the average user "expects" something to operate in a certain fashion).	Become aware of the more common stereotype behaviors. Do not depart from common design on objects and items which have demonstrated utility and user acceptance just for the sake of change; for example, since standard position for ON in electric wall switch is UP, do not change to DOWN.
11.	Many people perform most tasks while thinking about something else.	Accept this as a way of life; design so that they can continue this practice. Keep it simple.
12.	Most people perform in a perfunctory manner utilizing previous patterns. Under stress they always revert to these habit patterns.	Don't change a satisfactory established design just for the sake of change.
13.	Most people will use their hands to test, examine, or explore.	Design in such a way that hands and feet used to explore won't get hurt.
14.	Many people will continue to use a faulty article even though they suspect it may be dangerous.	Do not build in "graceful degradation" characteristics. Provide clearly marked, evident fault or failure warning indicators.
15.	Very few people recognize the fact that they cannot see well enough either because of poor eyesight or because of lack of illumination.	Make all critical labels such that they can be clearly read by individuals with no better than 20-40 vision in one eye.
16.	Most people do not recheck their operational or maintenance procedures for errors or omissions.	Keep it simple; provide concise, brief instructions; design for step-by-step procedures with a minimum of procedural interaction.
17.	In emergency situations, people very often respond irrationally and with seemingly random behavior patterns.	Keep operation simple; provide for fail-safe operation; follow stereotypes and standard configuration precepts when possible.
18.	People often are unwilling to admit errors or mistakes of judgment or perception, and thus will continue a behavior or action originally initiated in error.	Keep the design simple; provide for fail-safe operations. Where functionally justifiable, design for sequence checking. Design for automatic product shutoff in the event appropriate sequence is not followed.

TABLE 27.1. (*Continued*)

19. Foolish attitudes and emotional biases often force people into apparently irrational behaviors and improper use of products.	Be aware of this.
20. A physically handicapped person will often undertake tasks and operations beyond his or her means.	Study the limitations of the physically handicapped, and design accordingly.
21. People often misread or fail to see labels, instructions, and scale markers on various items.	Follow detailed human engineering practice and techniques for placement of labels, design of scales, displays, and markers.
22. Many people become complacent after long-term successful use of or exposure to hazardous products.	Design in such a way that the fail-safe characteristics of a device cannot be avoided by simple means. Provide attention-demanding but simple, brief warning displays. Consider using dramatic warning devices to indicate potential failure: flashing lights, loud sounds. In some cases provide for fairly complicated operating procedures associated with automatic system shutdown where sequence of steps is violated. Function analysis and operator-task analysis are especially important ot preliminary design planning.
23. The span of visual comprehension is limited by certain innate human characteristics and possibly further constrained by training and experience. The average adult tends to "fill in" according to some previously experienced pattern or relationship.	Although the related behavior tendency seems to be very complex, a fairly simple design practice will provide the solution to many of the safety aspects. Keep it simple. Make labels large. Keep instructions brief, concise, and clear. Avoid long sentences. Do not make changes in configuration.

SOURCE: Adapted from R. J. Nertney and M. G. Bullock. *Human Factors in Design* (ERDA-76-45-2). U. S. Department of Energy, Energy Research and Development Administration, Washington, D.C., 1976.

The safety engineer is responsible for knowing the possible risks to all substances used in the workplace or in a product and for taking necessary precautions. It is not enough to know where to find the information; the safety engineer must know the reactions and interactions of all substances to make sure persons are not exposed unwittingly to risks.

Loss Modes.   Several loss modes have been identified: chemical and physical events, secondary chemical occurrences, autoreactions, mechanical accidents, electrical accidents, thermal accidents, and exotic physical energy events.

Chemical events are fires and explosions. Primary fires are the most familiar chemical event. Flammable material in the presence of oxygen is a candidate for a primary fire. This includes building materials, newsprint, packaging and crating materials, displays and exhibits, temporary structures, and non–benefit-producing accumulations of trash and debris which result from poor housekeeping.

Physical events include an explosion in which the shock wave is produced by purely physical phenomena, such as the sudden release of gas under pressure. An implosion is similar, except that external pressure produces an inward shock wave when a container fails. Collapse may also result from external pressure but is not related to shock waves.

Secondary chemical occurrences are the fires and explosions that occur in systems in which the fuel is normally contained, with oxygen excluded, so a trigger event is required to set up the two other sides of the fire triangle. The prior event is usually the failure of the container or containing system, so fuel escapes and combines with the oxygen outside or oxygen enters the system and combines inside. Ignition is still required to initiate the loss process, so that two events are required as opposed to one for a primary fire.

Autoreactions are common in petrochemical processing, where compounds are reacted to form the molecules of another compound. Such exposures are not limited to the petrochemical industry but may occur wherever chemicals are processed, handled, or stored. Not all the potential autoreactions, or their initiating conditions, are known. Hitherto unsuspected reactants and conditions continue to come to light. In these autoreactions there is several times more energy in steam than there is in an equivalent volume of gas at the same pressure. This energy is augmented by the heat energy in the inventory of superheated water, which flashes into steam upon failure of the containing vessel. This combination makes for a particularly long duration, heavily destructive wave front.

Mechanical accidents occur when critical components fail. Such components are connecting rods, crankshafts, bearings, and fasteners, all of which cause transient energy to be expended in destructive work upon the machine itself and/or upon objects in the vicinity. Such failures may result from normal wear and tear; may be induced by losing control of the energy inputs; or may be due to component failure from overloading or over-speeding.

Electrical accidents, or burnouts, occur in current-carrying components. When the electrical capacity of these components is exceeded, by an excessive power demand or a fault condition, the electrical energy is converted to heat and components are destroyed. All current-carrying components are vulnerable to burnout, requiring only that the power source be able to overload the components.

Thermal accidents occur when heated equipment, particularly fired process heaters and vessels, is damaged because the energy of the temperature of the heat source is greater than the strength of the transition temperature of the construction materials.

Exotic events result from the physical forces and energies of nuclear reaction, radioactivity, microwaves, and laser beams.

Exposure to Risk

It is the right of workers to know what risks they might be exposed to on the job. Twenty years ago, a pregnant nurse could work on an oncology unit where radioactive radium was being used to treat cancer patients. She had only the vaguest knowledge of the risks to herself, never even considering possible risks of her unborn child. Today, she might be told that in 15 years her child would be at risk for Hodgkin's disease. Likewise, coal miners risk silicosis, cotton workers risk byssinosis; jackhammer operators risk Reynaud's syndrome; and asbestos workers risk mesothelioma and lung cancer.

More is known every day about risk factors, and it is the safety engineer's responsibility to make sure this information is distributed to workers who are at risk. It is the system safety engineer's responsibility to keep risk factors out of the products people use.

Interference in System Safety

System safety engineering is going to have to be increasingly concerned with the entire spectrum of analysis, including not only identifying who is at risk for what but also minimizing risks as much as possible. The safety engineer has to be continually alert for new information about risks and will have to be more assertive in insisting that hazards be eliminated or controlled.

Probability of Human Error

Human error rates in simple operations are less than in complex operations because there are fewer opportunities for breakdown and possible malfunctions. It is also easier to learn a simple task than a complex one.

The important point is to master the procedure in order for the skill to become so automatic that it continues even through interference. However, vigilance is still required to minimize errors as much as possible.

SYSTEM SAFETY IN MILITARY SYSTEMS

MIL-STD-882B is paraphrased here to show military system safety requirements that must be used with any work for the Department of Defense.[2] Any company with Department of Defense contracts must conform to these requirements.

The principal objective of a system safety program within the Department of Defense is to ensure that safety, consistent with mission requirements, is designed into systems, subsystems, equipment, and facilities, hereinafter referred to as systems.

System Safety Program

A total program shall be developed in which design analyses, studies, and testing will identify system performance limitations, failure modes, safety margins, and critical operator tasks. All known facets of safety optimization, including design, engineering, education, management policy, and supervisory control, shall be considered in identifying, eliminating, or controlling hazards. System safety management and engineering shall be integrated with other management and engineering disciplines in the interest of an optimum system design. Procedures for development and integration of the system safety effort shall be applied across the management–contractor interface to assure a system safety program consistent with overall system requirements.[7]

Program Initiation Phase

Following the feasibility study, the system design begins in a preliminary development of program initiation phase. System safety tasks applicable to the program initiation phase are those required to evaluate the alternative system concepts under consideration for developing and establishing the system safety program consistent with the identified mission need and life-cycle requirements. System safety tasks include the following:

1. Prepare a system safety program plan (SSPP).
2. Evaluate all materials, design features, procedures, operational concepts, and environments under consideration which will affect safety throughout the life cycle.
3. Perform a preliminary hazard analysis (PHA) to identify hazards associated with each alternative concept.

4.   Identify possible safety interface problems.
5.   Highlight special areas of safety consideration, such as system limitations, risks, and person-rating requirements.
6.   Review safe and successful designs of similar systems for consideration in alternative concepts.
7.   Define the system safety requirements based on past experience with similar systems.
8.   Identify safety requirements that may change during the system life cycle. The life cycle of a system is the life of the system from conception in a needs analysis to final disposal.
9.   Identify any safety design analysis, test, demonstration, and validation requirements.
10. Document the system safety analyses, results, and recommendations for each promising alternative system concept.
11. Prepare a summary report of the results of the system safety tasks conducted during the program initiation phase to support the decision-making process.
12. Tailor the system safety program for the subsequent phases of the life cycle, and include detailed requirements in the appropriate demonstration and validation phase contractual documents.

Demonstration and Validation Phase

System safety tasks during the demonstration and validation phase will be tailored to programs ranging from extensive study and analyses through hardware development to prototype testing, demonstration, and validation. System safety tasks will include the following:

1.   Prepare or update the SSPP to describe the proposed integrated system safety effort planned for this phase.
2.   Perform PHA, or update the PHA performed during the program initiation phase. Prepare a PHA report of the proposed system concept in its intended use and operational environment.
3.   Identify technology, design, production, operational, and support risks affecting safety.
4.   Establish system safety criteria for verifying that requirements have been met.
5.   Participate in trade-off studies to reflect the impact on system safety requirements and risk. Recommend system design changes based on these studies to ensure that the optimum safety is achieved consistent with performance and system requirements.
6.   Identify for inclusion in the appropriate specifications any qualitative and quantitative system safety requirements. Include con-

tractor-furnished equipment, government-furnished equipment, ground-support equipment, and all interfacing and ancillary equipment.

7. Perform subsystem, system, and operating and support hazard analyses.

8. Review all test plans to assure that tests are conducted safely.

9. Ensure that identified hazards are eliminated or controlled.

10. Review training plans and programs for adequacy of safety measures.

11. Evaluate results of failure analyses and investigations recorded during the demonstration and validation phase. Recommend redesign or other corrective action.

12. Ensure that system safety requirements are incorporated into the system specification based on updated system safety studies, analyses, and tests.

13. Prepare a summary report of the results of the system safety tasks conducted during the demonstration and validation phase to support the decision-making process.

14. Continue to refine the system safety program. Prepare an SSPP for the full-scale engineering development and initial production phase.

Full-Scale Engineering Development

The system safety tasks during full-scale engineering development will include the following:

1. Ensure effective and timely implementation of the SSPP for the full-scale engineering development phase.

2. Review preliminary engineering designs to ensure that safety design requirements are incorporated and that hazards identified during demonstration and validation are eliminated or controlled.

3. Update systems safety requirements in system specifications.

4. Perform or update subsystem and operating and support hazard analyses and safety studies concurrent with the design/test effort to identify design and operating and support hazards. Recommend any required design changes and control procedures.

5. Identify testing facilities, test requirements, specifications, and criteria to ensure that design safety is verified. Review the test plans and programs to ensure safe conduct of the tests.

6. Participate in technical design and program reviews, and present results of subsystem, system, and operating and support hazard analyses.

7. Identify and evaluate the effects of storage, shelf life, packaging, transportation, handling, test, operation, and maintenance on the safety of the system and its subsystems.

8. Evaluate results of failure analyses and mishap investigations recorded during full-scale engineering development. Recommend redesign or other corrective action.

9. Identify, evaluate, and provide safety considerations for trade-off studies.

10. Review appropriate engineering documentation (drawings, specifications) to verify that safety considerations have been incorporated.

11. Review and provide safety inputs to preliminary system operation and maintenance publications.

12. Verify the adequacy of safety and warning devices, life-support equipment, and personal protective equipment.

13. Provide safety inputs to training courses.

14. Review the preliminary production engineering effort including purchase specifications, process quality control, inspection and acceptance, and test procedures to confirm that safety in the process and end product is established and maintained during production.

15. Ensure that requirements are developed for demilitarization and for safe dispersal of hazardous materials and equipment.

16. Prepare a summary report of the results of the system safety tasks conducted during the full-scale engineering development phase to support the decision-making process.[7]

SYSTEM SAFETY PROGRAM PLAN

The system safety program plan is the management document which tells what the system safety effort objectives are and the methods by which these objectives will be pursued. This plan describes how the system safety program will be established and carried out. It gives the program manager a baseline document from which to evaluate the progress of the system safety effort during the life cycle of the program. The SSPP should describe:

1. The safety management organization and how it relates to other program functions

2. The types of analyses required to identify and evaluate all hazards associated with the system

3. The specific hazards to be minimized and controlled at an acceptable level

4. The types of records to be established and maintained

The requirements for developing a system safety program plan are shown in the appendix to this chapter.

SYSTEM SAFETY MILESTONES

Each system safety program shall be planned to provide for periodic status reviews, presentations of hazard analyses and risk assessments, and evaluation of the overall effectiveness of the system safety effort. These reviews and assessments, conducted jointly by the customer and contractor, shall be performed concurrently with the appropriate program milestones. System safety shall be an agenda item of the appropriate scheduled program or design review held for the system to assess the status of compliance with the system safety requirements. These reviews shall identify any deficiencies of the system with respect to safety and provide guidance for further development. At the discretion of the managing activity, a system safety group may be established for selected systems or additional ad hoc safety reviews may be scheduled as required.

SYSTEM SAFETY CRITERIA

System safety requirements establish design and operational safety criteria for hazard elimination or control and may establish a quantitative value designating the level of system safety.

System designs and operational procedures should consider the following:

1. Review pertinent standards, specifications, regulations, design handbooks, and other sources of design guidance for applicability to the design of the system.

2. Eliminate or control hazards identified by analyses or related engineering efforts through design solution, material selection, or substitution. Potentially hazardous materials (for example, propellants, explosives, hydraulic fluids, solvents, lubricants, or fuels) should be selected to provide optimum safety characteristics.

3. Isolate hazardous substances, components, and operations from other activities, areas, personnel, and incompatible materials.

4.   Locate equipment so that access during operations, maintenance, repair, or adjustment minimizes personnel exposure to hazards (for example, hazardous chemicals, high voltage, electromagnetic radiation, cutting edges, or sharp points).

5.   Minimize hazards resulting from excessive environmental conditions (for example, temperature, pressure, noise, toxicity, acceleration, and vibration).

6.   Design to minimize human error in the operation and support of the system.

7.   Consider alternate approaches to minimize hazards that cannot be eliminated. Such approaches include interlocks, redundancy, fail-safe design, system protection, fire suppression, and protective clothing, equipment, and devices.

8.   Protect the power sources, controls, and critical components for redundant subsystems by physical separation or shielding.

9.   Provide suitable warning or caution notes in assembly, operations, maintenance, and repair instructions and distinctive markings on hazardous components, equipment, or facilities to ensure personnel and equipment protection. These should be standardized.

10. Minimize the severity of personnel injury or damage to equipment in the event of a mishap (for example, by incorporating crashworthy design features in all person-rated systems).

11. Review design criteria for inadequate or overly restrictive requirements regarding safety. Recommendations should be made for new design criteria supported by study, analyses, or test data.

PRELIMINARY HAZARD ANALYSIS

A preliminary hazard analysis (PHA) provides an initial risk assessment of a system or a piece of equipment. The PHA's purpose is to identify safety critical areas, evaluate hazards, and identify the safety design criteria to be used. The PHA effort shall be initiated during the preliminary design phase or the earliest life-cycle phase so that safety considerations are included in trade-off studies. (Trade-off studies are engineering studies of various approaches to accomplishing a task so that the best approach or alternative will be taken.) Based upon the best data available, hazardous conditions associated with the proposed design or function should be evaluated for hazard severity, hazard probability, risks, and operational constraints. Safety provisions and alternatives needed to eliminate or control hazardous conditions should be considered. The information shall be used in the development of system safety requirements and in the preparation of performance and design specifications. Also, the PHA is the basic hazard analysis which establishes the framework for other hazard analyses and safety engineering evaluation of the design. The PHA should consider the following for identification of hazards:

1. Hazardous components (energy sources, fuels, propellants, explosives, and high-pressure conditions)
2. Safety-related interface conditions among the various elements of the system (material compatibilities—for example, dissimilar metals which may cause electrogalvanic action; static electricity, which may cause inadvertent activation or even an explosion; and electromagnetic interference)
3. Environmental constraints including normal operating environment (temperature extremes, hazardous noise, illumination, and humidity)
4. Operating test, maintenance, and emergency procedures (human error analysis of operator and maintainer functions, life-support requirements, means of egress, and ergonomic factors—which the military call "human engineering")

Other analyses made in systems safety consider high-energy potentials, catastrophic accidents, and maintenance considerations of system skills required for operation of a system.

All major system hazards are identified. Probability of occurrence is not considered at this stage, but causes which can produce the hazard are identified; assuming the hazard or accident does occur, its effect on the system is evaluated, and the effects are categorized.

Some basic hazards that have been identified are:

Fire
Explosion
Uncontrolled toxic vapor or fluid release
Uncontrolled corrosive fluid release
Electric shock to personnel
Inadvertent release of kinetic energy
Inadvertent release of potential energy
Personnel exposure to excessive heat
Personnel exposure to radiation in excess of minimum allowable dosage
Personnel exposure to excessive noise levels
Personnel eye exposure to unprotected cutting or sawing operation
Uncontrolled release of cryogenic fluids
Inadvertent fall of personnel from work platforms, ladders, towers
Uncontrolled release of vapors from metals such as beryllium
Uncontrolled personnel exposure to extreme weather conditions

Uncontrolled disposal of poisonous materials

Fragmentation of a rapidly spinning flywheel, rotating stock

In a project hazard assessment, the safety engineer should start with the most dangerous hazard identified in the PHA and determine the events which would happen before an accident can occur. When identifying these precursors, be as specific as possible in naming the subsystems and components involved. Continue this process until having identified all events that must take place prior to the most dangerous hazards. Repeat the above for each hazard listed in the PHA. An example of a PHA is shown in Figure 27.1. Figure 27.1 identifies hazards by equipment grouping. The columns in this figure give the following information for each hazard:

1. *System/Subsystem/Unit.* This column identifies the equipment grouping where the potential hazard exists, for example, receiver.

2. *System Event(s) Phase.* This column identifies the phase of system operation where the potential hazard might cause an accident. The various system phases are symbolized by a number from 1 to 5. The relationships between system phase and numeral are shown at the bottom of each page of the matrix.

3. *Hazard Description.* A brief description of the potential hazard is given under this heading, for example, electric shock.

4. *Effect on System.* This column explains what would be the effect on the system if the hazard did result in an accident, for example, personnel injury—cuts and bruises.

5. *Assigned Risk Assessment.* This heading has the two subheadings of probability and severity. Under probability, a capital letter from A to E is given. This letter is associated with a probability level. The exact association is given at the bottom of each page. This expresses the likelihood that a hazard would result in the occurrence of an accident given no corrective actions are taken to resolve the hazard.

Under severity, a Roman numeral from I to IV is given. This numeral expresses how serious the occurrence of an accident would be. At the bottom of each page of the matrix, the association between Roman numerals and severity level is given. For example, Roman numeral I would represent a catastrophic accident, such as one that would result in death.

6. *Recommended Action.* This column gives the method that will be used to alleviate the hazard. Some of these resolutions represent good engineering practices that are typically incorporated into GTE designs, such as covering exposed voltages.

7. *Effect of Recommended Action on Assigned Risk Assessment.* Under this heading are given the probability and severity levels for any residual hazard that would be left in the equipment after the recommended actions under column 6 have been taken. In other words, this column gives

the probability of the hazard still causing an accident after corrective measures have been implemented, and the severity level of the accident. The same probability and severity codes as were used under column 6 are used here.

8/9. *Remarks/Status.*   Under this column is given the status of the hazard resolution process. All identified hazards are in various stages of being resolved. Hazards will be considered closed when it has been verified that recommended actions stated under column 6 have been implemented. This verification may be through review of release drawings, testing, and/or inspection of equipment.

SUBSYSTEM HAZARD ANALYSIS

A subsystem hazard analysis (SSHA) is a hazard analysis applied to some elements of the total system. The SSHA shall be performed to identify hazards associated with component failure and functional relationship between components on equipment whose performance degradation, functional failure, or inadvertent functioning could result in a hazard. The analysis should contain a determination of the modes of failure, including all single-point failures and the effects on safety when failures occur in subsystems components. The SSHA should normally be performed during the demonstration or validation phase of a system. The SSHA should be started as soon as the detail design of the subsystem is available. The format of this hazard analysis must be carefully planned to minimize problems of integrating the SSHA into the system hazard analysis, which will be discussed later. Techniques used to accomplish a SSHA include the following.

Fault Hazard Analysis

Fault hazard analysis is an inductive method of analysis that can be used exclusively as a qualitative safety analysis. The fault hazard analysis requires a detailed investigation of the subsystem to determine component hazard modes, causes of those hazards, and the resultant effects to the subsystem and its operation.

Fault Tree Analysis

Fault tree analysis is a deductive analytical tool used to analyze all events, faults, and occurrences and all their combinations and permutations that could cause or contribute to the occurrence of a defined undesired event. The analysis is either quantitative or qualitative.

Preliminary Hazard Analysis Report

Program _____
System _____
Subsystem _____
Revision _____ Rev. Date _____

Completed by _____ D. Gloss
Organization _____ HF/SS
Date _____ 1/18/83
Page __1__ of __5__ Pages

System/ Subsystem/ Unit (1)	System Event(s) Phase[a] (2)	Hazard Description (3)	Effect on System (4)	Assigned Risk Assessment		Recommended Action (6)	Effect of Recommended Action on Assigned Risk Assessment		Remarks/Status (8/9)
				Probability[b]	Severity[c] (5)		Probability[b]	Severity[c] (7)	
Disconnect and grounding switch unit	2	Removal of ac circuit breaker	Maintainer can receive shock when replacing circuit breaker if main breaker is not turned off.	B	I	Put warning label on ac power control panel and in maintenance manual warning that main breaker must be turned off before removing ac power panel breaker.	E	I	Engineering is implementing. Safety will provide inputs to manuals as they are developed.
	1, 2, 3, 4	Sharp edges or corners	Injury of personnel	B	III	Round corners and edges	E	IV	Part of standard assembly process
	2, 3, 4	Voltages above 120 Vac.	Electric shock to personnel	C	I	Put interlock on doors, and access plates. Rack will be key interlocked.	E	I	Engineering is implementing.

Phase[a]	Hazard	Effect	Probability[b]		Severity[c]	Remarks
2, 3, 4	Low Voltage at or below 120 Vac.	Electric shock to personnel	C	The rack will be equipped with grounding rod and will have clear cover over cable.	I	Engineering is implementing.
				Barriers over all exposed voltages at or above 30 Vac.		Engineering is implementing.
				Label on barrier indicating voltage underneath cover.		Engineering is implementing.
				Power off switch in rack so ac power can be removed for maintenance.		Engineering is implementing.
				Engineering is implementing warning labels. Safety will provide inputs to manuals as they are developed.	E I	Because of maintenance requirements, high voltage interlock in rack will not remove low voltage ac power. Warnings will be placed on units where ac power will remain on after.

[a]Enter one or more system event phase

1. Operating   4. Installation and checkout
2. Maintenance   5. Other—see remarks
3. Test

[b]Enter probability of occurrence

A. Frequent   D. Remote
B. Reasonably probable   E. Extremely improbable
C. Occasional   F. Impossible

[c]Enter severity classification category

I. Catastrophic   III. Marginal
II. Critical   IV. Negligible

FIGURE 27.1. An example of a preliminary hazard analysis for an item of electromechanical hardware.

Preliminary Hazard Analysis Report

Program _____
System _____
Subsystem _____
Revision _____ Rev. Date _____

Completed by _____ D. Gloss _____
Organization _____ HF/SS _____
Date _____ 1/18/83 _____
Page _____ 2 _____ of _____ 5 _____ Pages

System/ Subsystem/ Unit (1)	System Event(s) Phase[a] (2)	Hazard Description (3)	Effect on System (4)	Assigned Risk Assessment		Recommended Action (6)	Effect of Recommended Action on Assigned Risk Assessment		Remarks/Status (8/9)
				Probability[b] (5)	Severity[c]		Probability[b] (7)	Severity[c]	
Disconnect and grounding switch unit		Low voltage at or below 120 Vac. (continued)							Interlock is engaged. Warning will also be placed in maintenance manuals.
						A guard shall be placed over the power on/off switch to prevent inadvertent turning off/on.			Engineering is implementing.
	2, 3, 4	Equipment may be dropped when being removed for maintenance.	Personnel injury or damage to drawer components	B	II	Put stops on drawer slides.	E	II	Engineering is implementing.

System event phase[a]	Hazard	Effect	[b]		Recommended action/control	[c]		Remarks
1	Meters shorting	Personnel injury	C	II	Meters shall be protected from voltage overload.	E	II	Engineering is implementing.
2, 3, 4	Weight	Personnel injury	B	II	If any drawer or component of the antenna disconnect weighs in excess of 25 pounds, a label shall be provided giving weight and lifting instructions. Warning will also be placed in manuals.	D	II	Engineering is implementing. Safety will provide inputs to manuals as they are developed.
1, 2, 3, 4	Rack falling	Personnel injury Equipment damage	C	I	Place heavy unit low in rack.	F	IV	Engineering is implementing.
2, 3, 4	Fanblades	Personnel injury	B	II	Use barrier or grill.	D	IV	Engineering is implementing.
2, 3, 4	Hot surfaces Resistor at 200°C	Personnel injury	B	II	Put label on cover stating how long it takes resistors to cool to safe.	E	II	Engineering is implementing.

[a]Enter one or more system event phase
1. Operating   4. Installation and checkout
2. Maintenance   5. Other—see remarks
3. Test

[b]Enter probability of occurrence
A. Frequent   D. Remote
B. Reasonably probable   E. Extremely improbable
C. Occasional   F. Impossible

[c]Enter severity classification category
I. Catastrophic   III. Marginal
II. Critical   IV. Negligible

FIGURE 27.1.   (*Continued*)

The fault tree analysis is unique in that it reasons backward from a postulated accident. Clearly, the selection of events to be analyzed will be conditioned by the analyst's knowledge and imagination. Also, since the analysis is not easy, it is unlikely that many will be done unless the accident is apt to have intolerable consequences. Clearly, fault tree analysis is not an infallible means for resolving all hazard problems; however, it is a useful tool and, under the right circumstances, can be a powerful aid when hazard control decisions must be made.

Beginning with a predicted accident, the fault tree analysis tracks the sequence of possible events which could lead to the unwanted happening. A model of the system is used to trace contributory factors. When the network is simulated, the developing diagram becomes a free-form figure constructed of increasingly significant failure modes, leading to the accident selected for analysis. The analysis proceeds through AND-OR logic gates using conventional computer logic symbols.

The fault tree method can routinize tracing the more vulnerable paths in the analyzed system. Decisions must then be made concerning how to strengthen the system. This may involve replacing with more durable substitutes those components which will probably fail most frequently. Redundant devices, or preventive maintenance, scheduled to interrupt the expected failure frequency, may also be used.

The completion of a fault tree analysis leads to other considerations relative to controlling the system. Human errors remain to be dealt with.

Management Oversight and Risk Tree

The management oversight and risk tree (MORT) is a system safety program originally developed for the Nuclear Regulatory Commission (NRC). MORT is designed to accomplish the following objectives:

1. Prevent safety-related oversights, errors, and omissions.
2. Express risks in quantitative form (to the maximum possible degree), and refer these risks to proper management levels for appropriate actions.
3. Make effective allocation of resources through the safety program to individual hazard control efforts.

MORT is based on the concept that all losses of the type that are called "accidents" or "incidents" arise from two sources: oversights and assumed risks. MORT results in changes which are judged necessary to prevent accidents or incidents. If existing program elements are judged adequate, no change occurs.

MORT and the derived systems have been validated in thousands of applications to accident investigations, program evaluations, and improvement plans in both industrial and academic environments.

Sneak Circuit Analysis

Sneak circuit analysis is conducted on hardware and software to identify latent (sneak) circuits and conditions that may cause some unsafe function to happen or to inhibit some desired functions without a component failure. This analysis employs use of the electrical flow patterns of all circuits.

SYSTEM HAZARD ANALYSIS

A system hazard analysis (SHA) is performed on the various subsystem interfaces to determine safety critical areas of the whole system. Techniques similar to those of the SSHA are generally used. Such analysis should include a review of the following:

1. Compliance with safety criteria (Does the system meet the safety standards set by the Occupational Safety and Health Administration, the Armed Forces, the National Fire Protection Association, and others?)
2. Possible combinations of failures that could result in a hazardous condition, including the failure of a safety device
3. Degradation in the safety of a subsystem or the total system through normal operation of another system
4. Assessment of all safety critical areas for hazard severity, hazard probability, safety precedence, and risk

Hazard Severity

Hazard severity categories are defined to provide a qualitative measure of the worst potential consequences resulting from human error, environmental conditions, design inadequacies, procedural deficiencies, and system, subsystem, or component failure as follows:

1. *Catastrophic.* May cause death or system loss
2. *Critical.* May cause severe injury, severe occupational illness, or major system damage
3. *Marginal.* May cause minor injury, minor occupational illness, or minor system damage
4. *Negligible.* Will not result in injury, occupational illness, or system damage

These hazard severity categories provide guidance to a wide variety of programs. However, adaptation to a program may be required. This adaptation may include definite transition points between categories and further definition of the degree of injury, illness, or system damage.

Hazard Probability

The probability that a hazard will occur during the planned life of a system can be described in terms of potential occurrence per unit of time, events, or activities. A quantitative hazard probability may be derived from research, analysis, and evaluation of historical safety data from similar systems. These were previously described in Chapter 19.

Precedence

Generally, hazard prevention and control actions should be performed in the following order:

1. Incorporate safety criteria and requirements into the design to reduce the probability of an event occurring that would contribute to creating a hazard.
2. If necessary, design special safety devices into the system to prevent a hazard.
3. Include safety procedures in test procedures and operating and maintenance instructions.
4. Provide safety warning signs and restrictive barriers where needed to protect personnel.

Risk Assessment

A risk assessment establishes priorities for corrective actions and resolution of conflicting safety issues. It is based on hazard severity, hazard probability, and safety precedence.

OPERATING AND SUPPORT HAZARD ANALYSIS

Operating and support hazard analysis is performed to identify and control hazards and determine safety requirements for personnel, procedures, and equipment used in installation, maintenance, testing, overhauling, transportation, storage, and operation. The Operating and Support Hazard Analysis is usually begun during the demonstration and validation phase of development.

SYSTEM SAFETY CHECKLIST

The system safety checklist is a list of known hazards or high-risk accident situations that have been elaborated in a tabular format for ease of evaluation. See Figure 27.2 for an example. The checklist offers the safety en-

Section	Requirement	Compliance			Comments
		Yes	No	N/A	
1910.28q	LADDER-JACK SCAFFOLDS				
1910.28q(1)	All ladder-jack scaffolds shall be limited to light duty and shall not exceed a height of 20 feet above the floor or ground.				
1910.28q(2)	All ladders used in connection with ladder-jack scaffolds shall be heavy-duty ladders and shall be designed and constructed in accordance with sections 1910.25 (portable wood ladders) and 1910.26 (portable metal ladders) of CFR 29.				
1910.28q(3)	The ladder jack of a ladder-jack scaffold shall be so designed and constructed that it will bear on the side rails in addition to the ladder rungs, or if bearing on rungs only, the bearing area shall be at least 10 inches on each rung.				
1910.28q(4)	Ladders used in conjunction with ladder jacks shall be so placed, fastened, held, or equipped with devices so as to prevent slipping.				
1910.28q(5)	The wood platform planks of ladder-jack scaffolds shall be not less than 2 inches nominal in thickness.				
1910.28q(5)	Both metal and wood platform planks used on ladder-jack scaffolds shall overlap the bearing surface not less than 12 inches.				
1910.28q(5)	The span between supports for platform planks used on ladder-jack scaffolds shall not exceed 8 feet.				
1910.28q(5)	The width of ladder-jack scaffold platforms shall not be less than 18 inches.				

FIGURE 27.2 Safety Checklist: Derived from General Industry Occupational Safety and Health Standards. (29 CFR 1910). To be published in Gloss, D. S., Wardle, M. G., Gloss, M. R. and Gloss, D. S. II. *Handbook of Occupational Safety and Health Standards.* Vol. I John Wiley & Sons, 1984.

Section	Requirement	Compliance			Comments
		Yes	No	N/A	
1910.28q(5)	The platform width on a ladder-jack scaffold shall not be less than 18 inches.				
1910.28q(6)	Not more than two persons shall occupy any given 8 feet of any ladder-jack scaffold at any one time.				
1910.28r	WINDOW-JACK SCAFFOLDS				
1910.28r(1)	Window-jack scaffolds shall be used only for the purpose of working at the window opening through which the jack is placed.				
1910.28r(2)	Window jacks shall not be used to support planks placed between one window jack and another or for other elements of scaffolding.				
1910.28r(3)	Window-jack scaffolds shall be provided with suitable guardrails unless safety belts with lifelines are attached and provided for the workman.				
1910.28r(3)	Window-jack scaffolds shall be used by one man only.				
1910.28s	ROOFING BRACKETS				
1910.28s(1)	Roofing brackets shall be constructed to fit the pitch of the roof.				
1910.28s(2)	Roofing brackets shall be secured in place by nailing in addition to the pointed metal projections. The nails shall be driven full length into the roof.				
1910.28s(2)	When rope supports are used with roofing brackets, they shall consist of first-grade manila of at least three-quarter-inch diameter, or equivalent.				
1910.28s(3)	When roofing brackets are used, a substantial catch platform shall be installed below the working area of roofs more than 20 feet from the ground to				

FIGURE 27.2. (*Continued*)

Section	Requirement	Compliance			Comments
		Yes	No	N/A	
	eaves with a slope greater than 3 inches in 12 inches without a parapet.				
1910.28s(3)	The catch platform used in conjunction with roofing brackets shall extend 2 feet in width beyond the projection of the eaves and shall be provided with a safety rail, midrail, and toe-board. Unless employees working upon the roof are protected by a safety belt attached to a life-line.				
1910.28t	CRAWLING BOARD				
1910.28t(1)	Crawling boards shall have 1 × 1½ inch cleats. The cleats on crawling boards shall be equal in length to the width of the board and spaced at equal intervals not to exceed 24 inches.				
1910.28t(1)	The cleats on crawling boards shall be fastened with nails driven through and clinched on the underside.				
1910.28t(2)	A crawling board shall extend from the ridge pole to the eaves when used in connection with roof construction, repair or maintenance.				
1910.28t(2)	A firmly fastened lifeline of at least three-quarter-inch rope shall be strung beside each crawling board for a handhold.				
1910.28t(3)	Crawling boards shall be secured to the roof by means of adequate ridge hooks or equivalent effective means.				
1910.28u	FLOAT OR SHIP SCAFFOLDS				
1910.28u(1)	Float or ship scaffolds shall support not more than three men and a few light tools, such as those needed for riveting, bolting, and welding.				

FIGURE 27.2. (*Continued*)

Section	Requirement	Compliance			Comments
		Yes	No	N/A	
1910.28u(2)	The platform of float or ship scaffolds shall be not less than 3 feet wide and 6 feet long, made of three-quarter-inch plywood, equivalent to American Plywood Association Grade B-B, Group I, Exterior.				
1910.28u(3)	Under the platform of a float or ship scaffold, there shall be two supporting bearers made from 2- × 4-inch, or 1- × 10-inch rough, selected lumber, or better.				
1910.28u(3)	The supporting bearers of a float or ship scaffold shall be free of knots or other flaws and project 6 inches beyond the platform on both sides.				
1910.28u(3)	The ends of the platform on a float or ship scaffold shall extend about 6 inches beyond the outer edges of the bearers.				
1910.28u(3)	Each supporting bearer of a float or ship scaffold shall be securely fastened to the platform.				
1910.28u(4)	An edging of wood not less than $\frac{3}{4} \times 1\frac{1}{2}$ inches, or equivalent, shall be placed around all sides of the platform on a float or ship scaffold to prevent tools from rolling off.				
1910.28u(5)	Supporting ropes for a float or ship scaffold shall be 1-inch diameter manila rope or equivalent, free from deterioration, chemical damage, flaws, or other imperfections.				
1910.28u(5)	Rope connections for a float or ship scaffold shall be such that the platform cannot shift or slip.				
1910.28u(5)	If two ropes are used with each float or ship scaffold, they should be arranged so as to				

FIGURE 27.2. *(Continued)*

Section	Requirement	Compliance			Comments
		Yes	No	N/A	
	provide four ends which are to be securely fastened to an overhead support.				
1910.28u(5)	If two ropes are used with each float or ship scaffold, each of the two supporting ropes shall be hitched around one end of a bearer and pass under the platforms to the other end of the bearer where it is hitched again, leaving sufficient rope at each end for the supporting ties.				
1910.28u(6)	Each workman who uses a float or ship scaffold shall be protected by a safety lifebelt attached to a lifeline.				
1910.28u(6)	The lifeline shall be securely attached to substantial members of the structure (not scaffold), or to securely rigged lines, which will safely suspend the workmen in case of a fall.				

FIGURE 27.2.   (*Continued*)

gineer a method of assessing hazards using a format which makes it hard to overlook a potential hazard. The other reasons for using a checklist are the following:

1. Clarity
2. Ease of implementation
3. Consistency of evaluation

SUMMARY

System safety is a valuable method of ensuring that safety goes in before the product goes out. It is applicable to all products from toys for tots to battleships for the U.S Navy and to all the intermediate items. It also provides a systematic methodology for examining the workplace and ensuring that no hazard is overlooked.

SUGGESTED LEARNING EXPERIENCES

1. Collect stated views of system safety engineers on the role of the professional organization in ensuring sound safety practice.

2. Write a system safety program plan for ensuring the a superwidget will be safe. A superwidget is an advanced model of a widget (a make-believe product).

3. Pretend that you are going to develop a new restraint system for your automobile. Examine the one presently in your car. Review the literature on restraint systems. Perform a preliminary hazard analysis.

4. Examine any lawn mower or snowblower. How safe is it? Pretend that you are the system safety engineer on a design team to develop a safe one. Perform appropriate hazard analyses for the various subsystems.

5. Familiarize yourself with the symbols used in a fault tree analysis or MORT. Perform one of these methods on one of the subsystems in the lawn mower or snowblower for which you performed a hazard analysis.

6. How would you go about implementing a system safety precedence on the lawn mower or snowblower? If you were going to increase the precedence level by one step for each subsystem, what effects would it make in the design of these products?

7. A company making children's shoes already has a safety engineer on its staff. Convince the manager that it would be appropriate to hire you as a system safety engineer.

8. Compare the analysis process as used by a philosopher, a mathematician, a nurse, and a system safety engineer.

9. Conduct a safety audit of your classroom and of your dormitory?

10. Construct a flow diagram illustrating the process of system safety engineering for a new system.

REFERENCES

1. Johnson, W. S. *MORT Safety Assurance Systems*. New York: Marcel Dekker, 1980.

1. U.S. Department of Defense. *System Safety Program*, Military Standard 882B. Washington, D.C., 1977.

APPENDIX 27.1

<table>
<tr><td colspan="2">DATA ITEM DESCRIPTION</td><td colspan="2">2   IDENTIFICATION NO(S)</td></tr>
<tr><td colspan="2">TITLE</td><td>AGENCY</td><td>NUMBER</td></tr>
<tr><td colspan="2">SYSTEM SAFETY PROGRAM PLAN</td><td>DOD</td><td>DI-H-7047</td></tr>
</table>

DESCRIPTION/PURPOSE

System Safety Program Plan (SSPP) is a detailed description of the tasks and activities of system safety management and system safety engineering required to identify, evaluate, and eliminate or control hazards throughout the system life cycle. The purpose of the SSPP is to provide a basis of understanding between the contractor and the managing activity as to how the system safety effort will be accomplished to implement MIL-STD-882 A.

APPLICATION/INTERRELATIONSHIP

1. The System Safety Program Plan provides detailed instructions for compliance with MIL-STD-882A to ensure that adequate consideration is given to safety during all life-cycle phases of the program and to establish a formal, disciplined program to achieve the system safety objectives.

2. Each SSPP must be tailored to the specific acquisition. Therefore Preparation Instructions may be modified by the DRL (DD Form 1423) to require only certain paragraphs of this DID.

3. Data Items which relate to this data item description are DI-H-7048 System Safety Hazard Analysis Report, DI-H-7049 Safety Assessment Report, and DI-H-7050 System Safety Engineering Report.

4. This data item description replaces DI-R-3531, UDI-R-8435, DI-H-1320A, UDI-H-20180, UDI-H-20417B and UDI-H- 23390.

4 APPROVAL DATE
29 November 1978

5 OFFICE OF PRIMARY RESPONSIBILITY
Air Force - 10

6 DDC REQUIRED

8 APPROVAL LIMITATION

9 REFERENCES (Mandatory as cited in block 10)
MIL-STD-882A

MCSL NUMBER(S)

PREPARATION INSTRUCTIONS

1. <u>The System Safety Program Plan (SSPP)</u>. The SSPP shall incorporate the following essential features, expanded upon as necessary, to indicate the extent of the system safety program.

2. <u>General Program Requirements</u>. The SSPP shall define a program to satisfy the requirements of MIL-STD-882A, paragraphs 4.1 and 5.1, and shall:

   a. Describe the scope of the overall program and the related system safety program.

   b. Describe the tasks and activities of system safety management and engineering and the interrelationship between system safety and other functional elements of the program. System safety program requirements and tasks included in other contractual documents shall be cross-referenced in the SSPP to avoid duplication of effort.

   c. List the contractor and Government documents which will be applied either as directives or guidance in the conduct of the system safety program.

3. <u>System Safety Organization</u>. The SSPP shall describe the implementation of MIL-STD-882A, paragraph 5.1.2 and 5.2, and shall describe:

   a. The system safety organization or function within the organization of the total program using charts to show the organizational and functional relationships, and lines of communication.

   b. The responsibility and authority of system safety personnel, other contractor organizational elements involved in the system safety effort, sub-contractors, and system safety groups. Identify the organizational unit responsible for executing each task. Identify the authority in regard to resolution of all identified hazards.

   c. The staffing of the system safety organization for the duration of the contract to include manpower loading and the qualifications of assigned personnel.

DD FORM 1664   PAGE 1 OF 2 PAGES

10. PREPARATION INSTRUCTIONS (continued)

    d. The procedures by which the contractor will integrate and coordinate the system safety efforts including dissemination of the system safety requirements to action organizations and subcontractors, coordination of subcontractor's system safety programs, integration of hazard analyses, program and design reviews, program status reporting, and system safety groups.

    e. The process through which contractor management decisions will be made to include notification of critical and catastrophic hazards, corrective action taken, mishaps or malfunctions, waivers to safety requirements, program deviations.

10.4. System Safety Program Milestones. The SSPP shall describe the implementation of MIL-STD-882A, paragraph 5.3 and shall:

    a. Identify safety milestones to permit evaluation of the effectiveness of the system safety effort at critical safety check points such as preliminary design reviews, critical design reviews, etc.

    b. Provide a program schedule of safety tasks showing start and completion dates, reports, reviews, and manloading which shall be kept current with other program milestones.

    c. Identify integrated system activities (i.e., design analyses, tests, and demonstrations) applicable to the system safety program but specified in other engineering studies to preclude duplication. Included as a part of this section shall be the estimated manpower loading required to do these tasks.

10.5. System Safety Requirements. The SSPP shall describe the implementation of MIL-STD-882A, paragraph 5.4 and shall:

    a. Describe or reference the engineering requirements and design criteria for safety. Describe safety requirements for supporting equipment, and procedures for all appropriate phases of acquisition up to, and including, disposal. List the safety standards and system specifications containing safety requirements through which the contractor intends to comply.

    b. Describe the risk assessment procedures. The hazard severity categories, hazard probability levels, and the system safety precedence to be followed in satisfying safety requirements shall be in accordance with MIL-STD-882A, paragraph 5.4, unless specified otherwise. State any qualitative or quantitative measures of system safety which the contractor is required to meet. Include system safety definitions which deviate or are in addition to MIL-STD-882A.

    c. Describe the controls that shall be used to ensure compliance or justify waivers/deviations with general design and operational safety criteria, and the closed-loop procedures to ensure hazard resolution.

10.6. Hazard Analyses. The SSPP shall describe the implementation of MIL-STD-882A, paragraph 5.5 and shall describe:

    a. The analysis technique and format that will be used in qualitative or quantitative analysis to identify hazards, their causes and effects, and recommended corrective action.

Page 2 of 3 pages

592

10. **PREPARATION INSTRUCTIONS** (continued)

    b. The depth within the system to which each technique will be used including hazard identification associated with the system, subsystem, components, personnel, ground support equipment, government furnished equipment, facilities, and their interrelationship in the logistic support, training, maintenance, and operational environments.
    c. The integration of subcontractor hazard analyses and techniques with overall system hazard analyses.

10.7. <u>System Safety Data</u>. The SSPP shall describe the implementation of MIL-STD-882A, paragraph 5.6 and shall:

    a. Describe the approach for searching, disseminating, and analyzing pertinent historical hazard or mishap data.
    b. Identify deliverable data by attaching a copy of the appropriate sheets of the DD Form 1423.
    c. Identify non-deliverable data and describe the procedures for accessibility by the managing activity and retention of data of historical value.

10.8. <u>Safety Testing</u>. The SSPP shall describe the implementation of MIL-STD-882A, paragraph 5.7, and shall describe:

    a. The test requirements for ensuring that safety is adequately demonstrated.
    b. Procedures for ensuring feedback of test information for review and analysis for use in design modifications.
    c. The review procedures established by contractor's system safety organization to ensure safe conduct of all tests.

10.9. <u>Training</u>. The SSPP shall describe the implementation of MIL-STD-882A, paragraph 5.8 and shall describe the safety training for engineering, technician, operating, and maintenance personnel.
10.10. <u>Audit Program</u>. The SSPP shall describe the implementation of MIL-STD-882A, paragraph 5.9 and shall describe the techniques and procedures to be employed by the contractor to ensure that the objectives and requirements of the system safety program are being accomplished.
10.11. <u>Mishap Reporting and Investigation</u>. The SSPP shall describe the mishap reporting and investigation procedures established by the contractor to alert the Government of mishaps that occur during the life of the contract.
10.12. <u>System Safety Interfaces</u>. The SSPP shall identify, in detail, the interface between system safety and all other applicable safety disciplines such as: Nuclear Safety, Range Safety, Explosive and Ordnance Safety, Chemical and Biological Safety, Laser Safety, etc. These interfaces shall be attached as addendums to the basic SSPP.

Page 3 of 3 pages

INDEX